International Series in Operations Research & Management Science

Volume 314

More information about this series at http://www.springer.com/series/6161

Chandrasekar Vuppalapati

Machine Learning and Artificial Intelligence for Agricultural Economics

Prognostic Data Analytics to Serve Small Scale Farmers Worldwide

Springer

Chandrasekar Vuppalapati
Computer Engineering
San Jose State University
San Jose, CA, USA

ISSN 0884-8289 ISSN 2214-7934 (electronic)
International Series in Operations Research & Management Science
ISBN 978-3-030-77484-4 ISBN 978-3-030-77485-1 (eBook)
https://doi.org/10.1007/978-3-030-77485-1

This Springer imprint is published by the registered company Springer Nature Switzerland AG
The registered company address is: Gewerbestrasse 11, 6330 Cham, Switzerland

The author dedicates his works, book, and efforts of this book to his late father Murahari Rao Vuppalapati, who had inspired him to write the book and guided the author's pursuits of life, and to his loving mother Hanumayamma Vuppalapati, who was the inspiration to develop company and dairy products for the betterment of humanity and future generations.

Preface

Agriculture is one of the largest worldwide employers and one of the biggest industries, contributing to 4% in 2018, of global gross domestic product (GDP), and in some developing countries, it can account for more than 25% of the GDP. Globally, food and agriculture businesses form a $5-trillion global industry that is getting bigger with the largest agricultural output, in 2018, from China ($978.5 billion), India ($397 billion), the Unites States ($178.6 billion), Indonesia ($133.5 billion), and Nigeria ($84.2 billion). Agricultural development is one of the most powerful tools to end extreme poverty, boost shared prosperity, and feed a projected 9.7 billion people by 2050. Unlike other sectors, the growth in the agriculture sector is two to four times more effective in raising incomes among the poorest. Analyses in 2016 found that 65% of poor working adults made a living through agriculture.

Nevertheless, agriculture has its own challenges. First, farming is a very tough and back-breaking industry. Farming is sensitive to local resource changes and global macrolevel economic factors. Second, farming has several uncertain threat vectors. Threat vectors include change in people consumption patterns, changes in people's caloric intake, changes in weather and climate, changes in commodity markets, tariff battles, changes in trade policies, changes in oil and crude prices, changes in currency valuations, new government regulations and ordinances, changes in stock market, and changes in family and life events, especially for small-scale farmers. Third, availability of human resources is a great challenge and poses huge constraint to sustainability of farming, especially small-scale farming. Finally, given the non-availability of macrolevel financial data and volatility of markets, small-scale farmers are bankrupted more often in recent times and have committed self-destruction and suicides in many cases. These life-taking issues could be avoided if small-scale farmers are equipped with the same information and analytical capabilities as currently available to mid- and large-scale farms; we can save our farmers. Farming is for ardent and hardworking individuals and not for chicken souls.

To harvest the benefits of advanced analytics and prognostics capabilities to the worldwide small farmers, the traditional AI software development paradigms must be fully upgraded to facilitate and empower small-scale farmer with informational, statistical, and analytical capabilities to function well in the dynamic, albeit complex, agricultural and economic markets. The goal of the book, *Machine Learning and Artificial Intelligence for Agricultural Economics: Prognostic Data Analytics to Serve Small-Scale Farmers Worldwide*, is to prepare current and future data science, software engineering, and agriculture scientists' teams with the skills and tools to fully utilize economic models to develop software capabilities that help to create lifesaving applications for small farmers. The book introduces essential agricultural economic concepts from the perspectives of full-scale software development with the emphasis on creating niche blue ocean products. It also:

- Outlines agricultural economic models for small farmers
- Outlines artificial intelligence software architectures and patterns to develop AI applications in agricultural economics
- Provides comprehensive AI algorithms and techniques to fully utilize agricultural datasets and World Development Indicators
- Provides exclusive AI deployments to serve small-scale farmers across the globe
- Explains AI solution development from agricultural and constrained device paradigms

Finally, it is our fiduciary responsibility to help farmers who are feeding us despite sufferings. The purpose of the book is to enable software engineers, data scientists, and agriculture economists to come together and develop applications to serve small-scale farmers and provide prognostic data analytics for the well-being and sustainability of small farmers.

Fremont, CA, USA Chandrasekar Vuppalapati

Acknowledgment

The author is deeply indebted to the love, support, and encouragement of his wife Anitha Ilapakurti and the active support of his daughters, Sriya Vuppalapati and Shruti Vuppalapati, who helped draft the first version of the manuscript.

The author is deeply thankful to the support of his elder brother, Rajasekar Vuppalapati, and the joyful support of his younger brother, Jaya Shankar Vuppalapati. Additionally, the author is also very thankful to his elder sisters Padmavati Vuppalapati and Sridevi Vuppalapati.

The author sincerely thanks Santosh Kedari and his team in India for conducting field studies in India for Dairy and Sanjeevani Healthcare analytics products and bringing the market feedback for the betterment of services.

The author is deeply indebted to Sharat Kedari, who helped in setting cloud and machine learning algorithms. In addition, the author is thankful for the support, hard work, and dedication of Vanaja Satish, who helped to develop ML models. Finally, the author is sincerely and deeply thankful to Sandhya Vissapragada for her contributions in data science and in creating machine learning models on Microsoft Azure and Google Cloud.

Abbreviations

AI	Artificial Intelligence
ARIMA	Autoregressive integrated moving average
BEA	The U.S. Bureau of Economic Analysis
BLS	Bureau of Labor Statistics
CDF	Cumulative Distribution Function
CVT	Computer Vision
FAO	The Food and Agriculture Organization
FRED	Federal Reserve Economic Data
GDP	The Gross Domestic Product
ICT	Information Communication Technology
IFAD	The International Fund for Agricultural Development
IMF	The International Monetary Fund
KL Divergence	Kullback-Leibler divergence
LAC	The Latin America and Caribbean
MGP	Minimum Guarantee Prices
ML	Machine Learning
MLP	Multi-layer perceptron
MSP	Minimum Support Price
NLP	Natural Language Processing Techniques
OECD	Organization for Economic Co-operation and Development
OLS	Ordinary Least Squares
PCA	Principal Component Analysis
PCE	The personal consumption expenditures
PDF	Probability Distribution Function
PPP	Purchasing Power Parity
SVM	Support Vector Machines
S&P 500 Index	The Standard & Poor's Financial Index
The HUNT Study	The Trøndelag Health Study (The HUNT Study)
TS	Time Series
UN	The United Nations
UNICEF	The United Nations International Children's Emergency Fund

USDA	The U.S. Department of Agriculture
VAR	Vector Autoregression
WFP	The World Food Programme
WHO	The World Health Organization
WPS	The world Bank Policy Research Working Paper

Contents

Section I
Artificial Intelligence

Chapter 1
Introduction

This chapter covers:

- What is artificial intelligence (AI)?
- Machine learning (ML).
- Type of analytics.
- Supervised, unsupervised, and deep learning.
- Metrics to evaluate classification and regression models.
- Agricultural machine learning use cases:
  - ML-enabled adaptive edge device coefficients.
  - Wine multiclass classification.
- The role of AI and ML in agriculture.
- AI Economics.
- The call for policy makers!

This chapter introduces artificial intelligence and covers techniques that constitute AI and maps AI to classical machine learning (ML) techniques. Next, it introduces techniques to process agricultural datasets and walks through field-collected sensor data processing. The chapter also introduces processing of simple classification and regression techniques. Finally, it introduces AI readiness of the countries and calls for a framework to net present value of AI project.

Artificial intelligence (AI) stands out as a transformational technology of our digital age—and its practical application throughout the economy is growing apace [1]. One of the chief reasons why AI applications are getting prominence and industry acceptance is in its software ability to learn, albeit continuously, from real-world use and experience and its capability to improve its performance. It is no wonder that the applications of AI span from complex high-technology equipment manufacturing to personalized exclusive recommendations.

Artificial intelligence applications enable agriculture, one of the oldest and most important professions in the world, to be more productive and enable to achieve sustainability, especially for small farmers. Worldwide, agriculture is a $5 trillion industry (table contains top 15 countries with agriculture value added to their respective economies) and is fully ready for embracing a digital, data fueled, smart intelligent transformation to overcome increasing demand and several disruptive

C. Vuppalapati, *Machine Learning and Artificial Intelligence for Agricultural Economics*, International Series in Operations Research & Management Science 314,
https://doi.org/10.1007/978-3-030-77485-1_1

forces[1] and challenges [1]. The demand for food is growing at the same time the supply side faces constraints in land [2], decrease in harvested land,[2] and farming inputs, constrained or limited water, fertilizers, and availability of human resources. AI will become even more important as the world's growing population creates an increasing demand for agricultural production and increasing challenges that include:

- Feeding the world's population that is on track to reach 9.7 billion by 2050, requiring a corresponding 70% increase in calories available for consumption, even as the cost of the inputs needed to generate those calories is rising.
- By 2030, the water supply will fall, due to population, drought, and deforestation, 40% short of meeting global water needs; and rising energy, labor, and nutrient costs are already pressuring profit margins and sustainability of small farmers. About one-quarter of arable land is degraded and needs significant restoration before it can again sustain crops at scale.
- And then there are increasing environmental pressures, such as climate change and the economic impact of catastrophic weather events, and social pressures, including the push for more ethical and sustainable farm practices, such as higher standards for farm-animal welfare and reduced use of chemicals and water.

To address these demand, challenges, and forces poised to further roil the industry, agriculture must embrace a digital transformation that is infused with augmented intelligence driven by artificial intelligence and machine learning. Needless to mention, agricultural is the next frontier[3,4] of the AI and machine [3, 4]. AI for agricultural macroeconomics is at the core. AI can augment human capabilities to improve agriculture productivity including crop monitoring, macroeconomic predictions, robotics, and prediction/prognostics analytics. In the future, we may eat food that has been planted, watered, monitored, picked, sorted, and delivered without human intervention [5].

Agriculture is one of the most fertile industries there are for artificial intelligence (AI) and machine learning.

[1] Agriculture's connected future: How technology can yield new growth – https://www.mckinsey.com/industries/agriculture/our-insights/agricultures-connected-future-how-technology-can-yield-new-growth

[2] How AI Is Transforming Agriculture – https://www.forbes.com/sites/cognitiveworld/2019/07/05/how-ai-is-transforming-agriculture/?sh=7ce84c004ad1

[3] Agricultural Intelligence: What AI Can Do for Smallholder Farmers – https://foodtank.com/news/2018/10/agricultural-intelligence-what-ai-can-do-for-smallholder-farmers/

[4] 10 Ways AI Has The Potential To Improve Agriculture In 2021 – https://www.forbes.com/sites/louiscolumbus/2021/02/17/10-ways-ai-has-the-potential-to-improve-agriculture-in-2021/?sh=7f611e297f3b

Top 15 Countries: Agriculture, Forestry, and Fishing Value Added (US$)

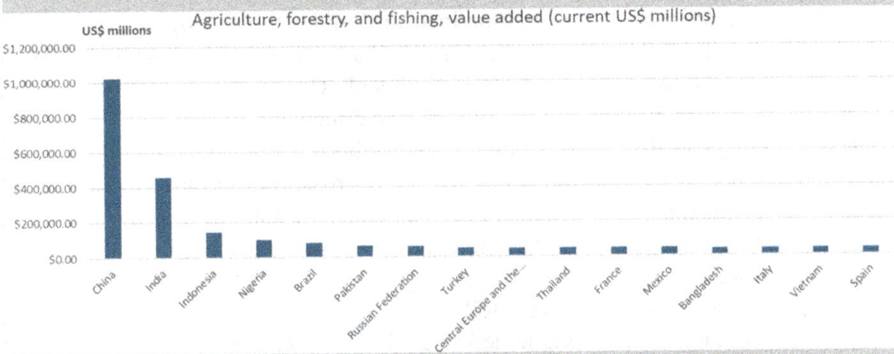

Data Source[5]: Agriculture corresponds to International Standard Industrial Classification (ISIC) divisions 1–5 and includes forestry, hunting, and fishing, as well as cultivation of crops and livestock production. Value added is the net output of a sector after adding up all outputs and subtracting intermediate inputs. It is calculated without making deductions for depreciation of fabricated assets or depletion and degradation of natural resources. The origin of value added is determined by the ISIC, revision 3. Data are in current US dollars.

Country name	Agriculture, forestry, and fishing, value added (current US$ millions)
China	$1,020,014.78
India	$459,461.28
Indonesia	$142,329.21
Nigeria	$98,166.59
Brazil	$81,621.69
Pakistan	$61,324.06
Russian Federation	$58,616.71
Turkey	$48,907.95
Central Europe and the Baltics	$46,064.20
Thailand	$43,398.56
France	$43,370.68
Mexico	$42,993.88
Bangladesh	$38,366.86
Italy	$38,339.09
Vietnam	$36,554.97
Spain	$36,438.90

[5] The World Bank – Agriculture, forestry, and fishing, value added (current US$) – World https://data.worldbank.org/indicator/NV.AGR.TOTL.CD?locations=1W

[6] What is Artificial Intelligence? – https://www.brookings.edu/research/what-is-artificial-intelligence/

What Is AI?

Artificial intelligence (AI) has been defined and understood in many ways, as it encompasses a vast field within its realm [6]. Perspectives of AI also vary based upon many factors, use cases, and actors involved. Therefore, no definition of AI can be so comprehensive that it will cover all the dimensions of AI.

Alan Turing generally is credited with the origin of the concept when he speculated in 1950 about *thinking machines* that could reason at the level of a human being[6] [7]. Turing was followed up a few years later by John McCarthy, who first used the term "artificial intelligence" to denote machines that could think autonomously. He described the threshold as "getting a computer to do things which, when done by people, are said to involve intelligence."

Stuart Russel and Peter Norvig in defining AI[7] have used the following taxonomy about systems that [8] [9]):

- Think like humans (e.g., cognitive architectures or neural network).
- Think rationally like humans (e.g., inferences optimization).
- Act like humans (e.g., language processing, knowledge representation, automated reasoning and learning).
- Act rationally like humans (e.g., intelligent software agents, embodied robots that achieve goals via perception, planning, reasoning, learning, etc.).

In computer science, artificial intelligence (AI), sometimes called machine intelligence, is intelligence demonstrated by machines, in contrast to the natural intelligence displayed by the humans. According to Gartner Market Research,[8] artificial intelligence applies advanced analysis and logic-based techniques, including machine learning, to interpret events, support and automate decisions, and take actions. The typical definition of AI looks something like the composite design of AI, machine learning, and deep learning (see Fig. 1.1).

AI embodies techniques that enable computers to mimic and predict human behavior. Simply put, AI is the ability of a machine to perform cognitive functions that associate with human minds, such as perceiving, reasoning, learning, and problem-solving. A full end-to-end AI solution—what we sometimes call a system of intelligence or a smart system—can ingest human-level knowledge (e.g., via machine reading and computer vision) and use this information to automate and accelerate tasks that were previously performed only by humans [10].

[7]What is Artificial Intelligence? An Informed Definition – https://emerj.com/ai-glossary-terms/what-is-artificial-intelligence-an-informed-definition/#:~:text=In%20%E2%80%9CArtificial%20Intelligence%3A%20A%20Modern,many%20different%20splintered%20fields%20%E2%80%93%20speech

[8]Artificial Intelligence (AI) – https://www.gartner.com/en/information-technology/glossary/artificial-intelligence

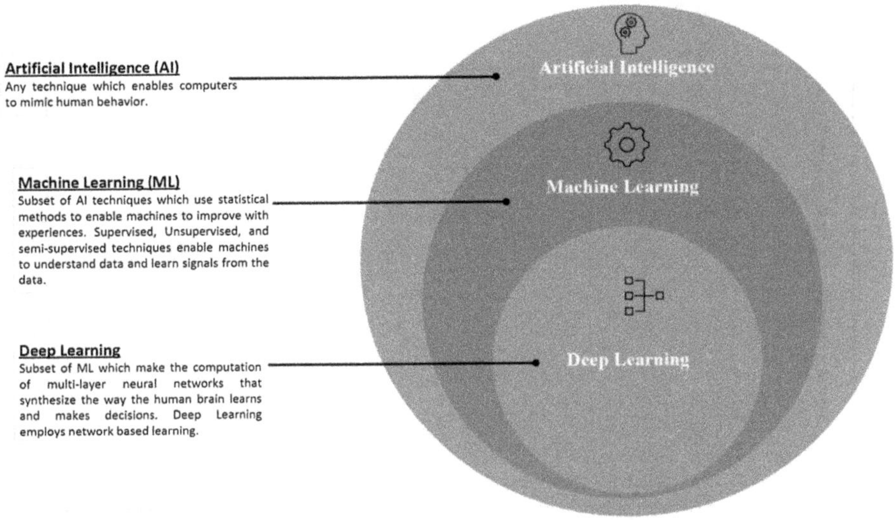

Fig. 1.1 Artificial Intelligence

Machine Learning

Machine learning (ML) is the subset of AI that use statistical techniques to enable machines to improve with experience. Machine learning algorithms detect patterns and learn how to make predictions and recommendations by processing data and experiences, rather than by receiving explicit programming instruction [11]. The algorithms also adapt in response to new data and experiences to improve efficiency and efficacy over time. ML, importantly, enables to autocalibrate, correct, or re-build itself if it sees the data that is being drifted, of course, from original trained model. The availability of such features dawns a new era of commuting that adapts itself to course correct or to hyper-scale efficiently as it sees new.

There are many vendors' enabled platforms which are available to develop machine learning-based applications, and here are the top players—as per Gartner's 2020 Magic Quadrant[9] [12].

As you can see, the Gartner Magic Quadrant (see Fig. 1.2) provides the major players, niche, visionaries, and leaders. One area many of these vendors are

[9] Gartner's 2020 Magic Quadrant For Data Science And Machine Learning Platforms Has Many Surprises – https://www.forbes.com/sites/janakirammsv/2020/02/20/gartners-2020-magic-quadrant-for-data-science-and-machine-learning-platforms-has-many-surprises/?sh=18b2268c3f55

Fig. 1.2 Gartner Magic Quadrant

focusing is automatic machine learning (AutoML). The AutoML[10,11,12] enables machines to create new models based on automated methods for model selection and/or hyperparameter optimization [13]. Unlike traditional ML development paradigm where the data scientist understands the data, performs exploratory analysis, and fits with the best statistical / ML model, the AutoML, on the other hand, does all the abovementioned tasks and lists the best models by fusing or ensemble of the individual models. AutoML is very powerful and provide hyper-ML productivity.

Food aid as a macrolevel ML model variable

Context: the interrelationships between production, consumption, and stocks are studied in an interdependent framework for two groups of agricultural commodities, namely, food grains and nonfood grains, with a view to obtain policy guidelines for agricultural growth. The model is formulated with aggregate information over the period 1951–1975, and both static and dynamic simulation along with multiplier analysis are carried out to assess the performance of the model. The study reveals that if there had been no PL 480 aid[a] from the US imports, food grain prices would have been higher than observed prices. Ex ante forecasts to 1985 suggest some need for adequate planning about stock building.

Some economists[13] have argued, notably, that food aid not only provides an additional surplus for agricultural/industrial development for an aid-receiving country but also results in continuous dependence on a foreign supply on account of depressed prices in the domestic market and unfavorable terms of trade for the agricultural sector vis-à-vis other sectors of economy [14].

[a]USAID and PL–480, 1961–1969 – https://history.state.gov/milestones/1961-1968/pl-480

[10] Microsoft Auto ML – https://www.microsoft.com/en-us/research/project/automl/

[11] Google AutoML – https://cloud.google.com/automl

[12] The different flavors of AutoML – https://www.h2o.ai/blog/the-different-flavors-of-automl/

[13] An econometric model of the agricultural sector of the Indian economy – https://www.sciencedirect.com/science/article/abs/pii/0161893882900278

Types of Analytics

Machine learning typically outputs four types of analytics (please see Fig. 1.3):

- Descriptive.
- Predictive.
- Prescriptive.
- Prognostics.

Descriptive Analytics

Descriptive analytics help answer the question "What happened? (or What is happening?)" in all its form: What were our sales last quarter or last month or yesterday? What were World Agricultural Supply and Demand[14] last year? Which customers required the most customer service help? Which product had the most

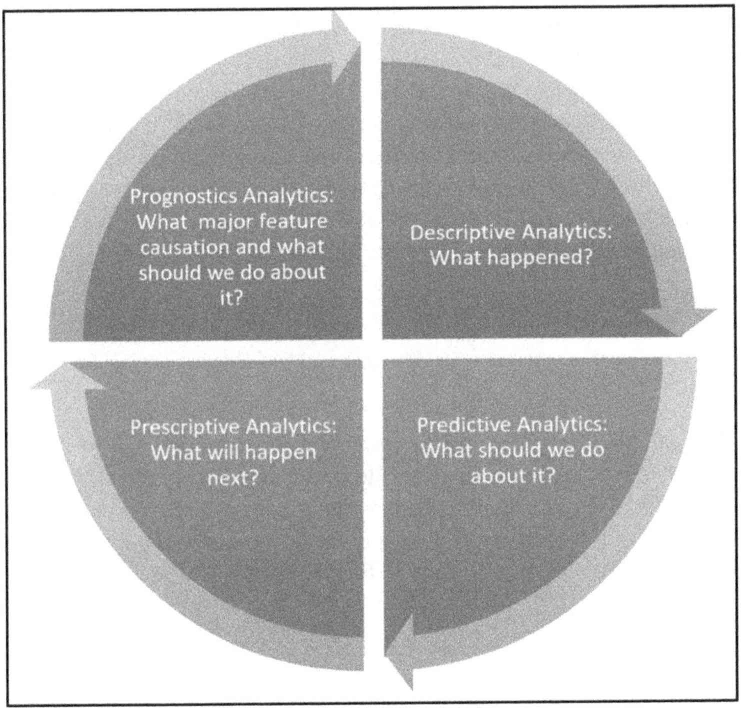

Fig. 1.3 Type of analytics

[14] USDA Data Products – https://www.ers.usda.gov/data-products/

defects? What were new accounts opened in the last quarter? Questions like these form the foundation for your entire analytics strategy. Descriptive analytics [15] characterized by traditional business intelligence (BI) and visualizations such as pie charts, bar charts, dendograms, line graphs, tables, or generated narratives. As an example of descriptive analytics (please see Fig. 1.5), consider the following dashboard that displays the rice production on milled basis (million metric tons) (updated 3/2021).

Predictive Analytics

Predictive analytics answers, "What will happen?" or "What is likely to happen next?"

Predictive analytics inherently probabilistic (regression and forecasting), and it is the domain of data scientists. Predictive analytics [15] is the process of extracting information from the data and predicting future outcomes and trends. Being able to effectively apply predictive analytics creates an opportunity for targeted resource allocation that is not driven simply by experience or a "gut feeling." Time series forecasting, one of the important examples of predictive analytics, is one of the most important topics in data science. Almost every business needs to predict the future to make better decisions and allocate resources more effectively [16]. Examples of time series forecasting use cases include crop yield prediction, commodity pricing, commodity futures, export of commodities, and many more. In the following table (please see Table 1.1) are the top 10 export destination countries of Brazil coffee roaster (quantity on average greater than 100 million tons).

Sankey chart version of the above table (please see Fig. 1.4).

Prescriptive Analytics

Prescriptive analytics take the inputs from prediction and—combined with rules and constraint-based optimization—enable better decisions about what to do. The decision might be to send an automated task to a human decision maker along with a set of next action recommendations or to send a precise next action command to another system. Prescriptive analytics is therefore best suited for situations where constraints are precise [17].

Table 1.1 Brazil coffee

Reporter countries	Partner countries	Item	Unit	Value
Brazil	United States of America	Coffee, roasted	Tons	1075.73
Brazil	Colombia	Coffee, roasted	Tons	700.4
Brazil	Italy	Coffee, roasted	Tons	438.77
Brazil	Japan	Coffee, roasted	Tons	161.83
Brazil	Argentina	Coffee, roasted	Tons	139.62
Brazil	Unspecified area	Coffee, roasted	Tons	118.71
Brazil	Russian Federation	Coffee, roasted	Tons	111.11
Brazil	Venezuela (Bolivarian Republic of)	Coffee, roasted	Tons	101.8
Brazil	Bulgaria	Coffee, roasted	Tons	100

Fig. 1.4 Brazil coffee exports—Sankey

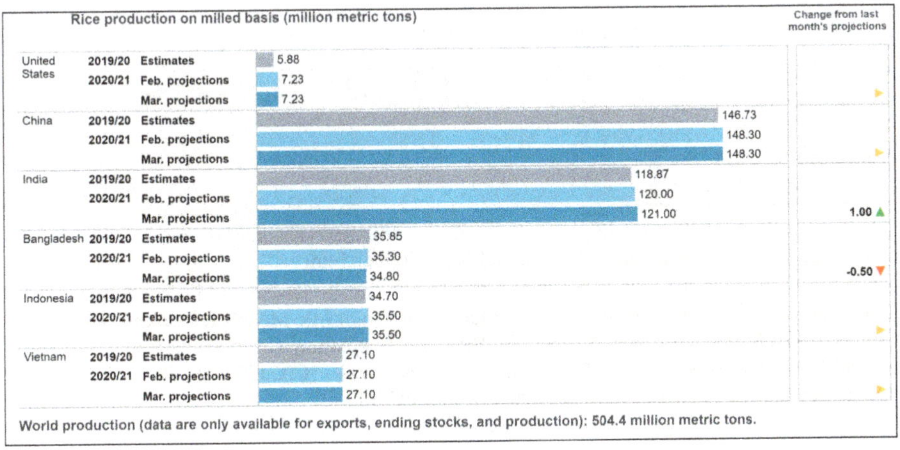

Fig. 1.5 Rice production worldwide

Prognostics Analytics

Oxford English Dictionary defines prognostics as "an advance indication of a future event, a sign or forecast of some future occurrence." Generally, it is confined to fortune or future foretellers, more have subjective or intuition driven. Data science, on the other hand, embryonically enables to model and predict the health condition of a system and/or its components, based upon current and historical system generated data or status. The chief goal of prognostics is precise estimation of Remaining Useful Life (RUL) of equipment or device.

Causation or causal analysis is the vital constituent of prognostic analytics. Prognostic analytics require predictive analytics as part of the core engine. In addition, causal relationship analysis enables the generation of recommendation as part of prognostic analytics. For example, in industrial environment, prognostic analytics predict the future degradation or damage and the Remaining Useful Life (RUL) of an asset or part, based on the measured data. The goal of prognostics is to predict the cycles remaining before the damage grows beyond the eligibility threshold. To predict the RUL, prognostics utilize the measured damage levels or an estimation of the damage up to the current cycle.[15]

In agricultural environment, prognostic analytics[16] analyze signals from historical data, macro environment factors, climatic conditions, crop phenological data to derive insights to recommend farmers on price variation, demand for a crop, and any unforeseen risks through causation [18]. The goal of prognostic analytics is to train machine learning algorithms on the large datasets to estimate prognosis on farm outcomes and potentially guide farmers to reduce overall risk in pricing and yield outcomes.[17] Prognostic analytics require predictive analytics as part of the core engine. Many algorithms have been used in prognostic systems that include [19]:

- Linear regression.
- Time series analysis.
- Vector autoregression (VAR).
- Bayesian dynamic linear models. Nonlinear regression.

[15] Prognostic analytics – https://www.oreilly.com/library/view/hands-on-industrial-internet/9781789537222/c8a23a5a-32e8-411f-b5b3-c901f57469be.xhtml

[16] J. S. Vuppalapati, S. Kedari, A. Ilapakurti, C. Vuppalapati, S. Kedari and R. Vuppalapati, "The Development of Machine Learning Infused Outpatient Prognostic Models for tackling Impacts of Climate Change and ensuring Delivery of Effective Population Health Services," 2019 IEEE International Conference on Big Data (Big Data), Los Angeles, CA, USA, 2019, pp. 2790–2799, doi: https://doi.org/10.1109/BigData47090.2019.9006570

[17] Machine learning algorithms estimating prognosis and guiding therapy in adult congenital heart disease: data from a single tertiary centre including 10,019 patients – https://academic.oup.com/eurheartj/article/40/13/1069/5301309

The purpose of prognostics analytics is to provide advance insights to farmers and reduce risk related to agriculture!

In the 2020/2021 global rice market, world production was raised 0.4 million tons to a record 504.4 million tons (milled basis), up more than 1% from a year earlier.[18]

Types of Learning

Machine learning (ML) algorithms can be built with different styles of learning to model a problem or address a business need or a pain-point [20]. The learning style is dictated by the interaction with the data environment expressed as the input to the model (e.g., contiguous streams of data with time behavior, extrapolate historical patterns to predict re-occurrence or ad hoc system with fuzziness to be modeled, or sensor integration as part of animal husbandry and providing near real-time or real-time recommendations). Internet of Things (IoT)-enabled edge analytics[19] that provide Just In Time (JIT) prognostics to small farmers not only provide yield improvement in the dairy but also provide a sustained revenue streams to improve standards of living to farmers and address global poverty issues.

Monitoring livestock's health, including vital signs, daily activity levels, and food intake, ensures their health is one of the fastest-growing aspects of AI and machine learning in agriculture [4]. Cow necklace[20] [21], a Class 10 Wearable veterinary sensor for use in capturing a cow's vital signs, providing data to the farmer to monitor the cow's milk productivity, and improving its overall health.

Supervised Learning

In supervised learning, an algorithm uses training data and feedback from humans to learn the relationship of given inputs (features) to a given output (labels) [9]. That is, the humans label desired class or output so that the machine identifies pattern between given input and labeled output. For example, in the following regression

[18] Global Rice Production – https://public.tableau.com/views/WorldAgriculturalSupplyandDemandEstimatesataGlance/MAIN_STORY?:language=en&:display_count=y&publish=yes&:origin=viz_share_link&:showVizHome=no

[19] Chandrasekar Vuppalapati, Building Enterprise IoT Applications first Edition, CRC Press; first edition (December 17, 2019), ISBN-10: 0367173859

[20] Class 10 – Hanumayamma Innovations and Technologies, Inc. – Trade Mark – https://www.trademark247.com/cow%2Bnecklace-87655622-1.html

model, a supervised learning was employed[21] to show that **rural population**, life expectancy, **foreign direct investment**, the rate of inflation, level of exports of all goods and services, and the ratio of agricultural exports to agricultural imports significantly determined the percentage contribution of agriculture to GDP in the selected Latin America and the Caribbean countries. Before use case could be introduced, here is the synopsis of regression model.

> If agriculture fails to develop at a suitable pace, this could prove to be a critical constraint to the growth of the industrial sector as well as other sectors of the economy.

The learning process continues until the model achieves a desired level of accuracy on the training data. Future incoming data (without known labels) are tested on the constructed model.

The following table summarizes supervised learning algorithms:

Algorithm	Definition
Linear Regression	Regression is concerned with modeling the relationship between variables that is iteratively refined using a measure of error in the predictions made by the model[a]. The most popular regression algorithms are [22]: • Ordinary Least Squares Regression (OLSR) • Linear Regression • Logistic Regression • Stepwise Regression • Multivariate Adaptive Regression Splines (MARS) • Locally Estimated Scatterplot Smoothing (LOESS) Highly interpretable [b], Linear Regression models between input independent variables and output variable (class or dependent variable) to generate model or classifier equation that helps to predict future values.
Logistics Regression	Logistic Regression[c] is a Machine Learning classification algorithm that is used to predict the probability of a categorical dependent variable. In logistic regression[23], the dependent variable is a binary variable that contains data coded as 1 (yes, success, etc.) or 0 (no, failure, etc.). In other words, the logistic regression model predicts P(Y=1) as a function of X. Simply put, Logistics regression is extension of Linear Regression that is used for classification tasks, meaning the output variable binary rather than contiguous[22]. Examples of Logistics regression[d] includes identification of Digits.

[21] Factors Influencing Agriculture's Contribution to GDP – https://ageconsearch.umn.edu/record/187331/files/Factors%20Influencing%20Agriculture_s%20Contribution%20to%20GDP%20Latin%20America%20and%20the%20Caribbean.pdf

Decision Tree	Decision Tree is a simple ML algorithm and does not require knowing business domain to apply [11, 22]. Highly interpretable classification or regression model that splits data-feature values into branches at decision nodes (eg, if a feature is a color, each possible color becomes a new branch) until a final decision output is made[e]. The most popular Decision Tree algorithms are: • Classification and Regression Tree (CART) • Iterative Dichotomiser 3(ID3) • C4.5 and C5.0 (different versions of a powerful approach) • Chi-squared Automatic Interaction Detection (CHAID) • Decision Stump • M5 • Conditional Decision Trees
Naïve Bayesian	Bayesian applies conditional probabilities to predict the outcome. Classification technique that applies Bayes theorem, which allows the probability of an event to be calculated based on knowledge of factors that might affect that event. The most popular Bayesian algorithms are [11, 22]: • Naive Bayes • Gaussian Naive Bayes • Multinomial Naive Bayes • Averaged One-Dependence Estimators (AODE) • Bayesian Belief Network (BBN) • Bayesian Network (BN)
Random Forest	The fundamental concept behind random forest is a simple but powerful one — the wisdom of crowds[f]. A large number of relatively uncorrelated models (trees) operating as a committee will outperform any of the individual constituent models [22]. In other words, Random Forest is a classification or regression model that improves the accuracy of a simple decision tree by generating multiple decision trees and taking a majority vote of

	them to predict the output, which is a continuous variable (e.g., age) for a regression problem and a discrete variable (e.g., either black, white, or red) for classification.
Gradient-Boosting Trees	Classification or regression technique that generates decision trees sequentially, where each tree focuses on correcting the errors coming from the previous tree model. The final output is a combination of the results from all trees [22].
XGBoost	XGBoost[g] is a decision-tree-based ensemble Machine Learning algorithm that uses a gradient boosting framework. In prediction problems involving unstructured data (images, text, etc.) artificial neural networks tend to outperform all other algorithms or frameworks. However, when it comes to small-to-medium structured/tabular data, decision tree-based algorithms are considered best-in-class right now [24].

[a]A Tour of Machine Learning Algorithms - https://machinelearningmastery.com/a-tour-of-machine-learning-algorithms/

[b]Model interpretability in Azure Machine Learning service - https://docs.microsoft.com/en-us/azure/machine-learning/service/how-to-machine-learning-interpretability

[c]Building A Logistic Regression in Python, Step by Step - https://towardsdatascience.com/building-a-logistic-regression-in-python-step-by-step-becd4d56c9c8

[d]Logistic Regression using Python (scikit-learn) - https://towardsdatascience.com/logistic-regression-using-python-sklearn-numpy-mnist-handwriting-recognition-matplotlib-a6b31e2b166a

[e]Tour of ML Algorithms - https://machinelearningmastery.com/a-tour-of-machine-learning-algorithms/

[f]Random Forest Classifier - https://towardsdatascience.com/understanding-random-forest-58381e0602d2

[g]XGBoost - https://towardsdatascience.com/https-medium-com-vishalmorde-xgboost-algorithm-long-she-may-rein-edd9f99be63d

In agricultural use cases, the most useful algorithms include decision tree and regression (linear, logistic).

Climatic parameters as a macrolevel ML model variable
Rainfall is the one of the most important input on unirrigated land [14]. Adding a macrolevel variable to the economic model has a greater explainability in the aggregate production of food grains or nonfood grains!

Decision Tree Classification Models

During the late 1970s and early 1980s, J. Ross Quinlan, a researcher in machine learning, developed a decision tree algorithm known as ID3 (Iterative Dichotomiser).

This work expanded on earlier work on concept learning systems, described by E. B. Hunt, J. Marin, and P. T. Stone. Quinlan later presented C4.5 (a successor of ID3), which became a benchmark to which newer supervised learning algorithms are often compared. In 1984, a group of statisticians (L. Breiman, J. Friedman, R. Olshen, and C. Stone) published the book *Classification and Regression Trees* (CART), which described the generation of binary decision trees. ID3 and CART were invented independently of one another at around the same time yet follow a similar approach for learning decision trees from training tuples [11]. These two cornerstone algorithms spawned a flurry of work on decision tree induction [11].

ID3, C4.5, and CART adopt a greedy (i.e., nonbacktracking) approach in which decision trees are constructed in a top-down recursive divide and conquer manner. Most algorithms for decision tree induction also follow such a top-down approach, which starts with a training set of tuples and their associated class labels. The training set is recursively partitioned into smaller subsets as the tree is being built [11].

Decision trees are a simple but powerful form of multiple variable analysis. They provide unique capabilities to supplement, complement, and substitute for [25]:

- Traditional statistical forms of analysis (such as multiple linear regression).
- A variety of data mining tools and techniques (such as neural networks).
- Recently developed multidimensional forms of reporting and analysis found in the field of business intelligence.

Decision trees are produced by algorithms that identify various ways of splitting a dataset into branch-like segments. These segments form an inverted decision tree that originates with a root node at the top of the tree [25]. A decision tree algorithm is used to classify the attributes and select the outcome of the class attribute. To construct a decision tree, both class attribute and item attributes are required. A decision tree is a tree-like structure wherein the intermediate nodes represent attributes of the data, the leaf nodes represent the outcome of the data, and the branches hold the attribute value. Decision trees are widely used in the classification process because no domain knowledge is needed to construct the decision tree. Figure 1.6 shows simple decision trees [11].

An attribute selection measure is a heuristic for selecting the splitting criterion that "best" separates a given data partition, D, of class labeled training tuples into individual classes. If we were to split D into smaller partitions according to the outcomes of the splitting criterion, ideally each partition would be pure (i.e., all of the tuples that fall into a given partition would belong to the same class). Attribute selection measures are also known as splitting rules because they determine how the tuples at a given node are to be split [11]. The following two are applied to derive attribute selection:

- Information gain.
- Gain ration.

The primary first step in the decision tree algorithm is to identify the root node for the given set of data. Multiple methods exist to decide the root node of the decision tree. Information gain and Gini impurity are the primary methods used to

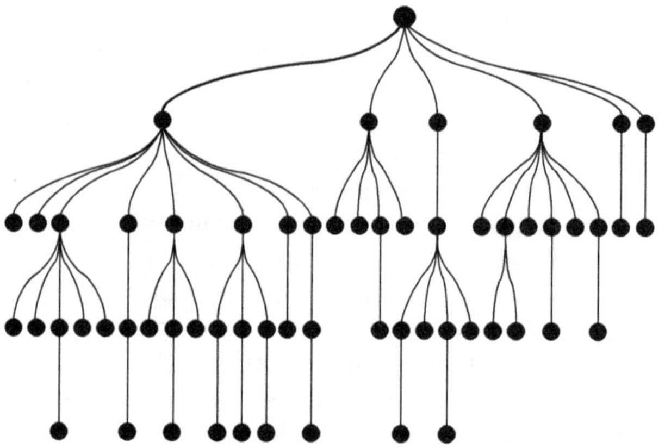

Fig. 1.6 Decision tree

identify the root node. The root node plays an important role in deciding which side of decision tree the data falls into. Like every classification method, decision trees are also constructed using the training data and tested with the test data.

Let node N represent or hold the tuples of partition D. The attribute with the highest information gain is chosen as the splitting attribute for node N. This attribute minimizes the information needed to classify the tuples in the resulting partitions and reflects the least randomness or "impurity" in these partitions. Such an approach minimizes the expected number of tests needed to classify a given tuple and guarantees that a simple (but not necessarily the simplest) tree is the expected information needed to classify a tuple in D is given by information gain.

Information Gain: Information gain is used to identify the root node and the branch nodes in the decision tree. Information gain is calculated using entropy and information. The purpose of the information gain calculation is to identify the node that has the least randomness or impurity. Entropy is calculated using the following formula:

$$\text{Info}(\text{D}) = -\sum_{i=1}^{m} pi \log_2(pi) \tag{1.1}$$

Info(D) is also known as entropy, a measure of disorder or impurity.

$$\text{Info}_{\text{A}}(D) = \sum_{j=1}^{v} \frac{|Dj|}{|D|} x \,\text{info}(Dj) \tag{1.2}$$

The information gain of an attribute is the difference between entropy and the information of that attribute. The attribute with the highest information gain is the root node, and the next level nodes are identified using the next highest information gain attributes.

$$\text{Gain}(A) = \text{Info}(D) - \text{Info}_A(D) \tag{1.3}$$

Decision Tree Classification Models Metrics

Decision Tree Classification Models Metrics are compared on the following metrics.[22]

One of the important classification metrics is confusion matrix.[23] It has following details:

	Predicted label		
True label		**No**	**Yes**
	No	**True negative** (a true negative is an outcome where the model correctly predicts the negative class).	**False negative** (a false negative is an outcome where the model incorrectly predicts the negative class).
	Yes	**False positive** (a false positive is an outcome where the model incorrectly predicts the positive class).	**True positive** (a true positive is an outcome where the model correctly predicts the positive class).

- A *true positive* is an outcome where the model correctly predicts the positive class. Similarly, a true negative is an outcome where the model correctly predicts the negative class.
- A *false positive* is an outcome where the model incorrectly predicts the positive class. And a *false negative* is an outcome where the model incorrectly predicts the negative class.

Precision: A metric for classification models. Precision[24] identifies the frequency with which a model was correct when predicting the positive class. That is:

$$\text{Precision} = \frac{\text{True Positives}}{(\text{True Positives} + \text{False Positives})} \tag{1.4}$$

Recall—A metric for classification models that answers the following question: Out of all the possible positive labels, how many did the model correctly identify[25]?

$$\text{Recall} = \frac{\text{True Positives}}{(\text{True Positives} + \text{False Negatives})} \tag{1.5}$$

---

[22] Evaluate Model – https://docs.microsoft.com/en-us/azure/machine-learning/studio-module-reference/evaluate-model

[23] Confusion Matrix – https://developers.google.com/machine-learning/glossary#confusion_matrix

[24] Precision – https://developers.google.com/machine-learning/glossary#precision

[25] Recall – https://developers.google.com/machine-learning/glossary#recall

- **Accuracy** measures the goodness of a classification model as the proportion of true results to total cases.
- **Precision** is the proportion of true results over all positive results (please see Eq. 1.4).
- **Recall** is the fraction of all correct results returned by the model (please see Eq. 1.5).
- **F-Score** computed as a weighted average of precision and recall between 0 and 1, where the ideal F-score value is 1.
- **AUC measures** the area under the curve plotted with true positives on the y-axis and false positives on the x-axis. This metrical useful as it provides one single metric to compare models.
- Average log loss is single score used to express the penalty for wrong results. It is calculated as the difference between two distributions—the true one and one in the model.
- Training log loss is a single score that represents the advantage of the classifier over a random prediction.

In addition to the training features and labels, for supervised model, we need to set a *regularization* parameter. Regularization parameter is used to counteract any *bias in the sample* and help the model generalize well by *avoiding overfitting* the model to the training data.

In the following section, decision tree technique is used to calculate temperature sensitivity coefficient for adaptive measuring purposes to detect influence of temperatures on electronic devices, a specific use case from diagnostic measuring sensor.

Use Case: Binary Decision Trees—Managing Temperature Effects in Edge IoT Deployments with ML-Enabled Adaptive Coefficients

Temperature has been known to cause changes in materials as far back as the early Stone Age some 800, 000 years ago. One of the most important temperature-related material discoveries was in 1833, when Michael Faraday discovered that silver sulfide (Ag_2S) was conductive at high temperatures and nearly insulating at low temperatures.[26] This was in direct contrast to the temperature dependence observed in materials, which become less conductive as temperature increased [26].

The findings of Faraday and other scientists have laid foundation for the impact of temperature in and on electronic systems. The main factor of the temperature on the electronic systems on the power dissipation and power density makes it among the most important of factors constraining electronic and nanoscale design [26].

[26] David Wolpert and Paul Ampadu, "Managing Temperature Effects in Nanoscale Adaptive Systems", Springer; 2012 edition (August 31, 2011), ISBN-10: 1461407478

Temperature Variations and Sensor Data Errors

There are two types of temperature variations that affect system performance: global temperature variations and local temperature variations [26]. Global temperature variations are caused by changes in ambient temperature or changes in cooling capacity. Increasing global chip temperatures will cause higher CPU clock periods, resulting in functional errors [26].

The local temperature (also called intra-die temperature variations [26]) can result in communication errors between with a large temperature differential. Intra-die temperature exceeding 50 °C have been reported [26] to causes microchip performance behavior due to temperature differential between a microprocessor core and an on-chip cache [26].

The following data captured by Class 10 Medical Diagnostics Wearable Sensor—"cow necklace." The cow necklace (please see Fig. 1.7) is a veterinary dairy sensor for use in capturing a cow's vital signs, providing data to the farmer to monitor the cow's milk productivity, and improving its overall health, and it is designed to work on extremely constrained environments and use the Internet of Things technologies. The sensors are deployed in fields of agriculture worldwide, and the data captured is specifically from India.

The above figure (please see Fig. 1.8) depicts dairy sensor data capture in the agriculture fields in Punjab and Andhra Pradesh, India. Dairy sensor captures temperature, humidity, and activity level details from dairy farms by connecting to the neck of cattle. For some dates, the data humidity values are reported more than 140%. One reason could have been (a) corruption of value reading at the sensor due to variations in temperature and humidity due to geolocation, in this case Patiala, Punjab (please note we did not observe the same behavior in Vizag—a costal of Andhra Pradesh), or (b) it could be due to voltage corruption. To weed out temperature and humidity variations, we have developed adaptive engineering data design coefficient that factors in geolocation thermal and extreme temperature changes to balance machine learning algorithm computations. Secondly, the data collection of the sensor records have shown the dates are incorrect; a factor reveals the microcontroller battery running low.

Fig. 1.7 Cow necklace

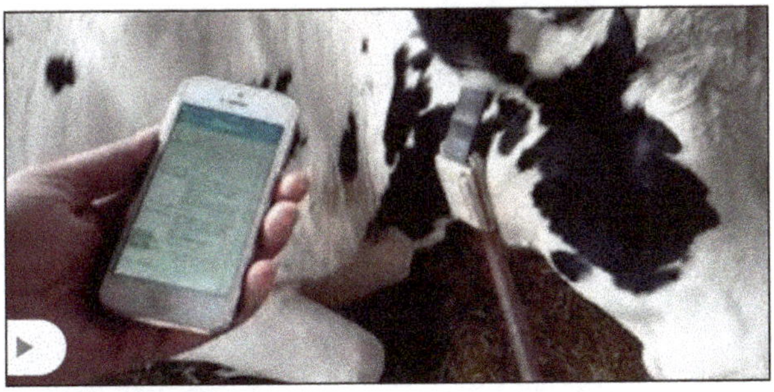

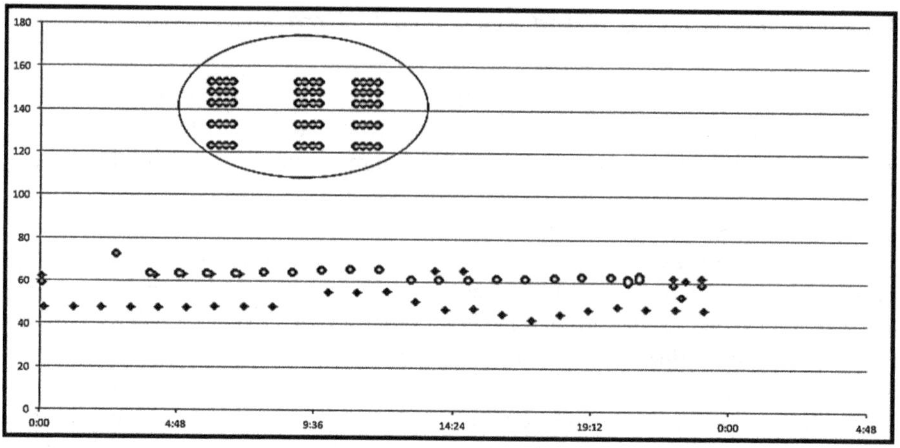

Fig. 1.8 Sensor data

Machine Learning Model Design of Adaptive System to Autocorrect

Understanding and managing temperature effects in edge IoT devices deployment is crucial for the successful working of the devices that are distributed across different geolocations with varying temperature and humidity conditions.

To improve the device performance against thermal and environmental perturbations, combination of supervised learning algorithms, such as decision trees, is used to make the device an adaptive edge to factor-in the environmental fluctuations. For instance, we have collected the following temperature data and applied machine learning to devise edge algorithms that fine-tune in real time by adjusting machine learning calculations and algorithm coefficients to improve the algorithms accuracies (please see Table 1.2).

Software code for this model can be found on GitHub Link: AdaptiveMLModelSensorDT.py (Python Code).

Binary Base Decision Tree: Adaptive Edge Thermal Calibration Coefficient

Binary classification techniques work well when the data observations belong to one of two classes or categories, such as "True" or "False." In edge device thermal calibration coefficient, "low value" and "high value" class tuple values are used.

Table 1.2 Adaptive coefficients sensor table data

Outlook	Temp (°F)	Humidity (%)	Windy	Adaptive Coefficient
Sunny	75	70	True	Low Value
Sunny	80	90	True	High Value
Sunny	85	85	False	High Value
Sunny	72	95	False	High Value
Sunny	69	70	False	Low Value
overcast	72	90	true	Low Value
overcast	83	78	False	Low Value
overcast	64	65	True	Low Value
overcast	81	75	False	Low Value
Rain	71	80	True	High Value
Rain	65	70	True	High Value
Rain	75	80	False	Low Value
Rain	68	80	False	Low Value
Rain	70	96	False	Low Value

<u>Step 1: load model specific libraries</u>:

The following code loads all important data processing, data frame, and machine learning libraries. Please note: I have used export graphviz library to display the binary decision tree.

```
from sklearn.tree import DecisionTreeClassifier, export_graphviz
from subprocess import call
import numpy as np
from sklearn import tree
from dtreeviz.trees import *
from sklearn import tree
from sklearn.model_selection import train_test_split
 from sklearn.model_selection import train_test_split
from sklearn.preprocessing import StandardScaler, MinMaxScaler
from sklearn.decomposition import PCA
mport numpy
from matplotlib import pyplot
from numpy import where
from collections import Counter
```

Step 2: load Sensor Data:

The following code loads adaptive edge coefficients dataset. It also sets feature name and label names to create the model.

```
# reading csv file
adaptivesensordf = pd.read_csv("AdaptiveMLModelSensorDT.csv")
feature_name=['Outlook','Temp(oF)','Humidity(%)','Windy']
label = 'Adaptive Coefficient'
adaptivesensordf.head()
```

	Outlook	Temp(oF)	Humidity(%)	Windy	Adaptive Coefficient	target_names	feature_names
0	1	75	70	1	0	Low Value	Outlook
1	1	80	90	1	1	High Value	Temp(oF)
2	1	85	85	0	1	High Value	Humidity(%)
3	1	72	95	0	1	High Value	Windy
4	1	69	70	0	0	Low Value	NaN

Step 3: Split the data:

Our dataset includes known values for the label, so we can use this to train a classifier so that it finds a statistical relationship between the features and the label value; but how will we know if our model is any good? How do we know it will predict correctly when we use it with new data that it wasn't trained with? Well, we can take advantage of the fact we have a large dataset with known label values, use only some of it to train the model, and hold back some to test the trained model—enabling us to compare the predicted labels with the already known labels in the test set.

In Python, the scikit-learn package contains a large number of functions we can use to build a machine learning model—including a train_test_split function that ensures we get a statistically random split of training and test data. We'll use that to split the data into 70% for training and hold back 30% for testing.

```
X, y = adaptivesensordf[feature_name].values, adaptivesensordf[label].values

X_train, X_test, y_train, y_test = train_test_split(X, y, test_size=0.30, random_state=0)
```

Step 4: Train and evaluate a binary classification model:

We're ready to train our model by fitting the training features (X_train) to the training labels (y_train). There are various algorithms we can use to train the model. In this example, we'll use decision tree classifier. In addition to the training features and labels, we'll need to set a *regularization* parameter. This is used to counteract any bias in the sample and help the model generalize well by avoiding overfitting the model to the training data.

Note: Parameters for machine learning algorithms are generally referred to as hyperparameters (to a data scientist, parameters are values in the data itself—hyperparameters are defined externally from the data!).

```
from sklearn.tree import DecisionTreeClassifier
clf = tree.DecisionTreeClassifier()

clf.fit(X_train,y_train)
```

Now we've trained the model using the training data, we can use the test data we held back to evaluate how well it predicts. Again, scikit-learn can help us do this. Let's start by using the model to predict labels for our test set and compare the predicted labels to the known labels.

Step 5: Test the model:

```
predictions = clf.predict(X_test)


from sklearn.metrics import precision_score, recall_score

print("Overall Precision:",precision_score(y_test, predictions))
print("Overall Recall:",recall_score(y_test, predictions))

from sklearn.metrics import accuracy_score
```

Output:

```
Overall Precision: 0.25
Overall Recall: 1.0
```

The most obvious thing you might want to do is to check the accuracy of the predictions—in simple terms, what proportion of the labels did the model predict correctly?

```
print('Accuracy: ', accuracy_score(y_test, predictions))
```

Output:

Accuracy: 0.4.

The accuracy is returned as a decimal value—a value of 1.0 would mean that the model got 100% of the predictions right, while an accuracy of 0.0 is, well, pretty useless!

Step 6: Test the model:

Accuracy seems like a sensible metric to evaluate (and to a certain extent it is), but you need to be careful about drawing too many conclusions from the accuracy of a classifier. Remember that it's simply a measure of how many cases were predicted correctly. Fortunately, there are some other metrics that reveal a little more about how our model is performing. Scikit-Learn includes the ability to create a classification report that provides more insight than raw accuracy alone.

```
from sklearn. metrics import classification_report

print(classification_report(y_test, predictions))
```

Output:

```
              precision    recall f1-score   support

           0       1.00      0.25      0.40         4
           1       0.25      1.00      0.40         1

    accuracy                           0.40         5
   macro avg       0.62      0.62      0.40         5
weighted avg       0.85      0.40      0.40         5
```

The classification report includes the following metrics for each class (0 and 1) note that the header row may not line up with the values!

- Precision: Of the predictions the model made for this class, what proportion were correct?
- Recall: Out of all the instances of this class in the test dataset, how many did the model identify?
- F1-Score: An average metric that takes both precision and recall into account.
- Support: How many instances of this class are there in the test dataset?

The classification report also includes averages for these metrics, including a weighted average that allows for the imbalance in the number of cases of each class.

```
from sklearn.metrics import precision_score, recall_score

print("Overall Precision:",precision_score(y_test, predictions))
print("Overall Recall:",recall_score(y_test, predictions))
```

Output:

```
Overall Precision: 0.25
Overall Recall: 1.0
```

The precision and recall metrics are derived from four possible prediction outcomes:

- True Positives: The predicted label and the actual label are both 1.
- False Positives: The predicted label is 1, but the actual label is 0.
- False Negatives: The predicted label is 0, but the actual label is 1.
- True Negatives: The predicted label and the actual label are both 0.

```
from sklearn.metrics import precision_score, recall_score
from sklearn.metrics import confusion_matrix
# Get evaluation metrics
mcm = confusion_matrix(y_test, predictions)

print('Confusion Matrix')
print(mcm)
```

Output:

```
Confusion Matrix
[[1 3]
 [0 1]]
```

The following code visually describes the output of confusion matrix:

```
from sklearn.metrics import roc_auc_score

auc = roc_auc_score(y_test,y_scores[:,1])
print('AUC: ' + str(auc))

adaptive_coefficient_cls=['Low Value','High Value']
import numpy as np
import matplotlib.pyplot as plt
plt.imshow(mcm, interpolation="nearest", cmap=plt.cm.Blues)
plt.colorbar()

tick_marks = np.arange(2)
plt.xticks(tick_marks, adaptive_coefficient_cls, rotation=45)
plt.yticks(tick_marks, adaptive_coefficient_cls)
plt.xlabel("Actual Adaptive Coefficient")
plt.ylabel("Predicted Adaptive Coefficient")
plt.show()
```

Output:

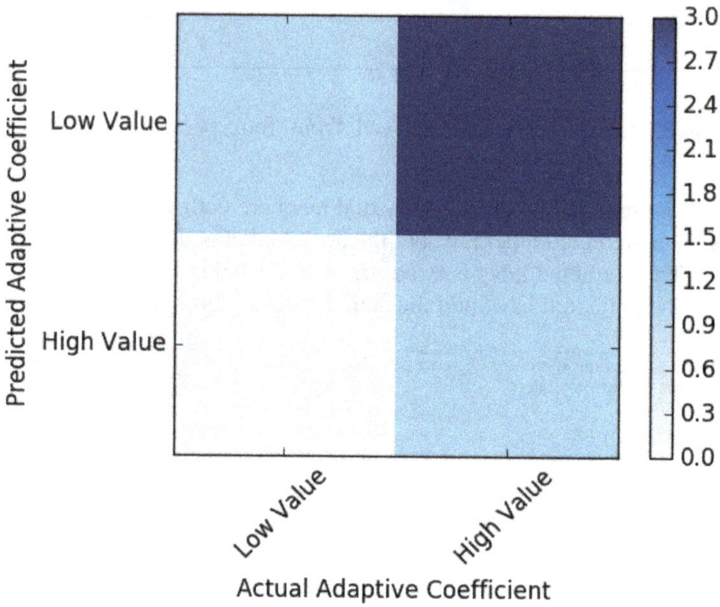

Step 7: Received operator characteristic (ROC) chart:

The decision to score a prediction as a 1 or a 0 depends on the threshold to which the predicted probabilities are compared. If we were to change the threshold, it would affect the predictions and, therefore, change the metrics in the confusion matrix. A common way to evaluate a classifier is to examine the true-positive rate (which is another name for recall) and the false-positive rate for a range of possible thresholds. These rates are then plotted against all possible thresholds to form a chart known as a *received operator characteristic (ROC) chart*, like this:

```
y_scores = clf.predict_proba(X_test)

from sklearn.metrics import roc_curve
from sklearn.metrics import confusion_matrix
import matplotlib
import matplotlib.pyplot as plt

# calculate ROC curve
fpr, tpr, thresholds = roc_curve(y_test, y_scores[:,1])

# plot ROC curve
fig = plt.figure(figsize=(6, 6))
# Plot the diagonal 50% line
plt.plot([0, 1], [0, 1], 'k--')
# Plot the FPR and TPR achieved by our model
plt.plot(fpr, tpr)
plt.xlabel('False Positive Rate')
plt.ylabel('True Positive Rate')
plt.title('ROC Curve')
plt.show()
```

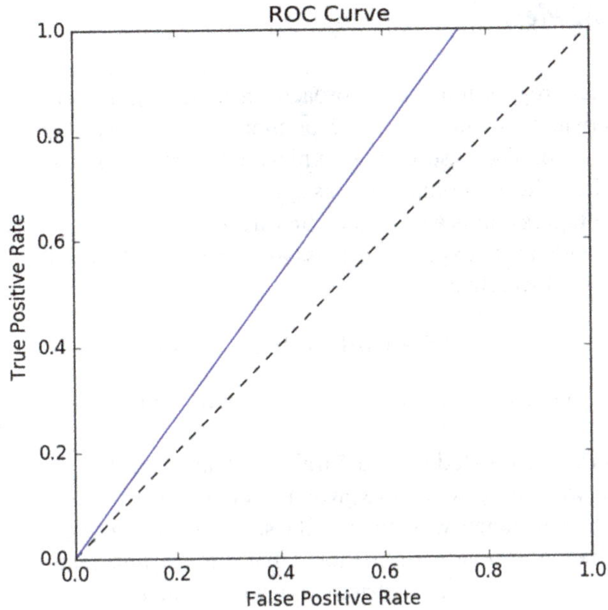

Fig. 1.9 ROC

Output:

The ROC (please see Fig. 1.9) chart shows the curve of the true- and false-positive rates for different threshold values between 0 and 1. A perfect classifier would have a curve that goes straight up the left side and straight across the top. The diagonal line across the chart represents the probability of predicting correctly with a 50/50 random prediction; so, you obviously want the curve to be higher than that (or your model is no better than simply guessing!).

The area under the curve (AUC) is a value between 0 and 1 that quantifies the overall performance of the model. The closer to 1 this value is, the better the model. Once again, scikit-Learn includes a function to calculate this metric.

```
from sklearn.metrics import roc_auc_score

auc = roc_auc_score(y_test,y_scores[:,1])
print('AUC: ' + str(auc))
```

Output:

```
AUC: 0.625
```

Regression Models

In statistics, linear regression is an approach for modeling the relationship between a scalar-dependent variable y and one or more explanatory variables denoted X (please see eq. 1.6). The case of one explanatory variable is called simple linear regression. When the outcome, or class, is numeric, and all the attributes are numeric; linear regression is a natural technique to consider. This is a staple method in statistics. The idea is to express the class as a linear combination of the attributes, with predetermined weights:

$$x = w0 + w1a1 + w2a2 + \ldots + wkak \tag{1.6}$$

where x is the class; a_1, a_2..., a_k are the attribute values; and w_0, w_1..., w_k are weights.

The weights are calculated from the training data. Here, the notation gets a little heavy, because we need a way of expressing the attribute values for each training instance. The first instance will have a class, say x(1), and attribute values, $a_1^{(1)}$, $a_2^{(1)}$..., $a_k^{(1)}$, where the superscript denotes that it is the first example. Moreover, it is convenient to assume an extra attribute a0, whose value is always 1.

Metrics for Regression Models

The metrics returned for regression models[27] are generally designed to estimate the amount of error. A model is considered to fit the data well if the difference between observed and predicted values is small. However, looking at the pattern of the residuals (the difference between any one predicted point and its corresponding actual value) can tell you a lot about potential bias in the model.

Following metrics are generally used to report regression models:

- **Negative log likelihood** measures the loss function, **a lower score is better**. Note that this metric is only calculated for **Bayesian Linear Regression and Decision Forest Regression**; for other algorithms, the value is **Infinity which means for nothing**.
- **Mean absolute error (MAE)** measures how close the predictions are to the actual outcomes; thus, a lower score is better.

[27] Scikit Regressors – https://scikit-learn.org/stable/modules/classes.html

- **Root mean squared error (RMSE)** creates a single value that summarizes the error in the model. By squaring the difference, the metric disregards the difference between over-prediction and under-prediction.
- **Relative absolute error (RAE)** is the relative absolute difference between expected and actual values, relative because the mean difference is divided by the arithmetic mean.
- **Relative squared error (RSE)** similarly normalizes the total squared error of the predicted values by dividing by the total squared error of the actual values.
- **Mean Zero One Error (MZOE)** indicates whether the prediction was correct or not. In other words: ZeroOneLoss(x,y) = 1 when x! = y; otherwise, 0.
- **Coefficient of determination, often referred to as R^2**, represents the predictive power of the model as a value between 0 and 1. Zero means the model is random (explains nothing); 1 means there is a perfect fit. However, caution should be used in interpreting R2 values, as low values can be entirely normal and high values can be suspect.

Unsupervised Learning

An algorithm explores input data without being given an explicit output variable. A model is generated by exploring the structures presented in the input data. This may be achieved by extracting general rules, going through a mathematical process to reduce redundancy, or organizing data by similarity testing. Examples of unsupervised algorithm include explore customer demographic data to identify buying patterns, arrangement of similarity of customers based on similar movie choices, and segmentation of retail chain stores based on sales categories [9].

Algorithm	Description
K-means Clustering	The k means algorithm searches for a pre-determined number of clusters within an unlabeled multidimensional dataset. It accomplishes this using a simple conception of what the optimal clustering looks like: • The "cluster center" is the arithmetic mean of all the points belonging to the cluster. • Each point is closer to its own cluster center than to other cluster centers. Those two assumptions are the basis of the k means model [11, 22].
Gaussian mixture model	A generalization of k-means clustering that provides more flexibility in the size and shape of groups (clusters).
Hierarchical clustering	Hierarchical clustering is an agglomerative (top down) clustering method. As its name suggests, the idea of this method is to build a hierarchy of clusters, showing relations between the individual members and merging clusters of data based on similarity. Splits or aggregates clusters along a hierarchical tree to form a classification system [11, 22].
Recommender System	Often uses cluster behavior prediction to identify the important data necessary for making a recommendation (Item based and Collaborative filter based)

Dimensionality Reduction	Seeks and exploits similarities in the structure of data in a manner similar to clustering algorithms but using unsupervised methods[22]. The purpose is to summarize or describe data using less information so that the dataset becomes smaller and easier to manage In some cases, people use these algorithms for classification or regression problems Here is a list of common dimensionality reduction algorithms[a]: • Principal Component Analysis (PCA) • Principal Component Regression (PCR) • Partial Least Squares Regression (PLSR) • Sammon Mapping • Multidimensional Scaling (MDS) • Projection Pursuit • Linear Discriminant Analysis (LDA) • Mixture Discriminant Analysis (MDA) • Quadratic Discriminant Analysis (QDA) • Flexible Discriminant Analysis (FDA)
Association Rules	Association rule learning methods extract rules that best explain observed relationships between variables in data. The most popular association rule learning algorithms are [11]: • Apriori algorithm • Eclat algorithm

[a]A Tour of Machine Learning Algorithms - https://machinelearningmastery.com/a-tour-of-machine-learning-algorithms/

Metrics for Clustering Models

Clustering models are different from classification and regression models in that clustering models evaluated based on distance metric and the results indicate per cluster cohesion and proximity of the points within the cluster (please see Table 1.3).

Table 1.3 Cluster modes

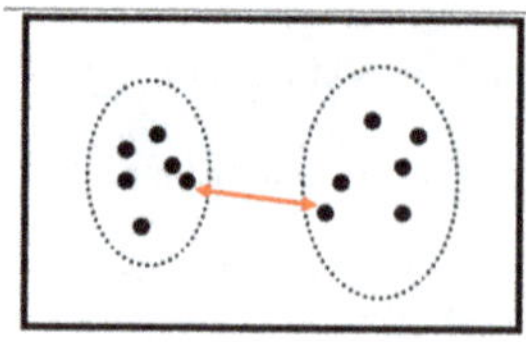	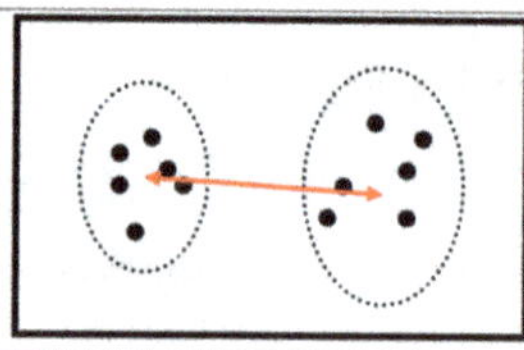	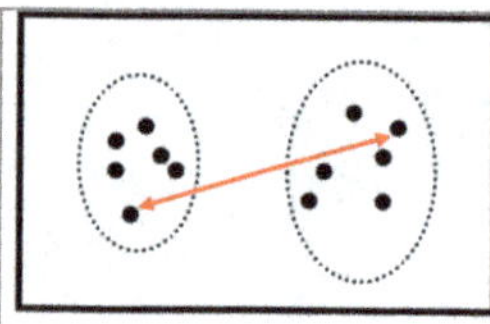
Single Link	Average Link	Complete Link

- Average distance to cluster center represents the closeness of all points in a cluster to the centroid of that cluster.
- Average distance to other centers represents how close, on average, each point in the cluster is to the centroids of all other clusters (average link[28]).
- Number of points shows how many data points were assigned to each cluster, along with the total overall number of data points in any cluster.
- Maximal distance to cluster center represent the sum of the distances between each point and the centroid of that point's cluster.

Reinforcement Learning

An algorithm learns to perform a task simply by trying to maximize rewards it receives for its actions.[29] Examples of reinforcement learning include maximizes points it receives for increasing returns of an investment portfolio [27].

Deep Learning

Deep learning is a type of machine learning that can process a wider range of data resources, requires less data preprocessing by humans, and can often produce more accurate results than traditional machine learning approaches. This group of algorithms works well with semi supervised learning problems in which the amount of labeled data is minimal. In deep learning, interconnected layers of software-based calculators known as "neurons" form a neural network. The network can ingest vast amounts of input data and process them through multiple layers that learn increasingly complex features of the data at each layer. The network can then decide about the data, learn if its determination is correct, and use what it has learned to make

[28] Data Mining: Concepts and Techniques – https://www.amazon.com/Data-Mining-Concepts-Techniques-Management/dp/0123814790

[29] Fuzzy Logic Infused Intelligent Scent Dispenser For Creating Memorable Customer Experience of Long-Tail Connected Venues – https://ieeexplore.ieee.org/document/8527046

determinations about new data. For example, once it learns what an object looks like, it can recognize the object in a new image.

Data requirements for deep learning are substantially greater [1] than for other analytics.[30] Here are the most popular deep learning algorithms[31] [22]:

- Convolutional neural network (CNN).
- Recurrent neural network (RNN).
- Long short-term memory networks (LSTMs).
- Stacked auto-encoders.
- Deep Boltzmann machine (DBM).
- Deep belief networks (DBN).

Algorithm	Description
Convolutional Neural Network (CNN)	A multilayered neural network with a special architecture designed to extract increasingly complex features of the data at each layer to determine the output. Generally used when you have an unstructured data set (e.g., images) and you need to infer information from it [22].
Recurrent Neural Network (RNN)	A multilayered neural network that can store information in context nodes, allowing it to learn data sequences and output a number or another sequence. General use cases involve when you are working with time-series data or sequences (e.g., audio recordings or text)

Neural networks are a subset of machine learning techniques. Essentially, they are AI systems based on simulating connected "neural units," loosely modeling the way that neurons interact in the brain. Computational models inspired by neural connections have been studied since the 1940s and have returned to prominence as computer processing power has increased and large training datasets have been used to successfully analyze input data such as images, video, and speech. AI

[30] Notes from AI Frontier: Applications and value of deep learning – https://www.mckinsey.com/featured-insights/artificial-intelligence/notes-from-the-ai-frontier-applications-and-value-of-deep-learning

[31] A Tour of Machine Learning Algorithms – https://machinelearningmastery.com/a-tour-of-machine-learning-algorithms/

practitioners refer to these techniques as "deep learning," since neural networks have many ("deep") layers of simulated interconnected neurons[32] [1].

Agricultural Datasets and Ensemble Learning

Ensemble models are very useful in agricultural datasets to detect the impact of climatic conditions and macroeconomic indicators impact on the yield prediction and prices of crops, respectively. Ensemble learning typically refers to methods that generate several models which are combined to make a prediction, either in classification or regression. The goal of ensemble methods is to combine the predictions of several base estimators built with a given learning algorithm to improve generalizability/robustness over a single estimator.[33] The advantage of ensembles with respect to single models has been reported in terms of increased robustness and accuracy[34] [28].

The ensemble process can be divided into three steps (please see Fig. 1.10). The first step is ensemble generation, which consists of generating a set of models. It often happens that, during the first step, several redundant models are generated. In the ensemble pruning step, the ensemble is pruned by eliminating some of the models generated earlier. Ensemble pruning has been reported, at least in some cases, to reduce the size of the ensembles obtained without degrading the accuracy. Pruning has also been added to direct methods successfully increasing the accuracy. Ensemble integration methods are combination (also called fusion) or as selection. The former approach combines the predictions of the models from the ensemble to

Fig. 1.10 Ensemble

[32] Notes from AI Frontier: Applications and value of deep learning – https://www.mckinsey.com/featured-insights/artificial-intelligence/notes-from-the-ai-frontier-applications-and-value-of-deep-learning

[33] Ensemble Methods – https://scikit-learn.org/stable/modules/ensemble.html#ensemble

[34] Ensemble approaches for regression: A survey – https://dl.acm.org/doi/abs/10.1145/2379776.2379786

obtain the final ensemble prediction. The latter approach selects from the ensemble the most promising model(s), and the prediction of the ensemble is based on the selected model(s) only.

Two families of ensemble generation methods are usually distinguished:

- Averaging methods.
- Boosting methods.

In ensemble algorithms, bagging methods form a class of algorithms which build several instances of a black box estimator on random subsets of the original training set and then aggregate their individual predictions to form a final prediction. These methods are used to reduce the variance of a base estimator (e.g., a decision tree), by introducing randomization into its construction procedure and then making an ensemble out of it. As they provide a way to reduce overfitting, bagging methods work best with strong and complex models (e.g., fully developed decision trees), in contrast with boosting methods which usually work best with weak models (e.g., shallow decision trees).

Bagging methods come in many flavors but mostly differ from each other by the way they draw random subsets of the training set[35]:

- When random subsets of the dataset are drawn as random subsets of the samples, then this algorithm is known as Pasting.
- When samples are drawn with replacement, then the method is known as Bagging.
- When random subsets of the dataset are drawn as random subsets of the features, then the method is known as Random Subspaces.
- Finally, when base estimators are built on subsets of both samples and features, then the method is known as Random Patches.

The following list contains sklearn.ensemble ensemble-based methods for classification, regression, and anomaly detection.[36]

[35] Ensemble methods – https://scikit-learn.org/stable/modules/ensemble.html#ensemble

[36] SKLearn – https://scikit-learn.org/stable/modules/classes.html

Ensemble Classifier/Regressor	Description
ensemble.AdaBoostClassifier([…])	An AdaBoost classifier.
ensemble.AdaBoostRegressor([base_estimator, …])	An AdaBoost regressor.
ensemble.BaggingClassifier([base_estimator, …])	A Bagging classifier.
ensemble.BaggingRegressor([base_estimator, …])	A Bagging regressor.
ensemble.ExtraTreesClassifier([…])	An extra-trees classifier.
ensemble.ExtraTreesRegressor([n_estimators, …])	An extra-trees regressor.
ensemble.GradientBoostingClassifier(*[, …])	Gradient Boosting for classification.
ensemble.GradientBoostingRegressor(*[, …])	Gradient Boosting for regression.
ensemble.IsolationForest(*[, n_estimators, …])	Isolation Forest Algorithm.
ensemble.RandomForestClassifier([…])	A random forest classifier.
ensemble.RandomForestRegressor([…])	A random forest regressor.
ensemble.RandomTreesEmbedding([…])	An ensemble of totally random trees.
ensemble.StackingClassifier(estimators[, …])	Stack of estimators with a final classifier.
ensemble.StackingRegressor(estimators[, …])	Stack of estimators with a final regressor.
ensemble.VotingClassifier(estimators, *[, …])	Soft Voting/Majority Rule classifier for unfitted estimators.
ensemble.VotingRegressor(estimators, *[, …])	Prediction voting regressor for unfitted estimators.
ensemble.HistGradientBoostingRegressor([…])	Histogram-based Gradient Boosting Regression Tree.
ensemble.HistGradientBoostingClassifier([…])	Histogram-based Gradient Boosting Classification Tree.

In the following use case, application of classification technique applied on agricultural dataset and comparison of base decision tree and ensemble decision tree are provided to see the accuracy differences.

Multiclass Classification: Classifying Wines and Ensemble Model

Let's see a simple example of a decision classifier and apply ensemble classifier to see the differences:

Data Source

Wine Recognition Dataset,[37] the data is from UCI Wine Data[38] [29].

Dataset Characteristics: the following table describes the characteristic of the dataset (please see Table 1.4).

[37] Wine Dataset – https://scikit-learn.org/stable/datasets/toy_dataset.html#wine-dataset

[38] UCI Wine DataDua, D. and Graff, C. (2019). UCI Machine Learning Repository [http://archive.ics.uci.edu/ml]. Irvine, CA: University of California, School of Information and Computer Science. – https://archive.ics.uci.edu/ml/datasets/wine

Table 1.4 Wine Dataset characteristics

Number of instances	178 (50 in each of three classes)
Number of attributes	13 numeric, predictive attributes and the class
Attribute information	Alcohol Malic acid Ash Alcalinity of ash Magnesium Total phenols Flavanoids Nonflavanoid phenols Proanthocyanins Color intensity Hue OD280/OD315 of diluted wines Proline

Please see below figure to see high-level view of the data:

alcohol	malic_acid	ash	alcalinity_of_ash	magnesium	total_phenols	flavanoids	nonflavanoid_phenols	proanthocyanins	color_intensity	hue	od280/od
14.23	1.71	2.43	15.6	127.0	2.80	3.06	0.28	2.29	5.64	1.04	
13.20	1.78	2.14	11.2	100.0	2.65	2.76	0.26	1.28	4.38	1.05	
13.16	2.36	2.67	18.6	101.0	2.80	3.24	0.30	2.81	5.68	1.03	
14.37	1.95	2.50	16.8	113.0	3.85	3.49	0.24	2.18	7.80	0.86	
13.24	2.59	2.87	21.0	118.0	2.80	2.69	0.39	1.82	4.32	1.04	
...	...	...	...	...	...	...	...	...	...	...	
13.71	5.65	2.45	20.5	95.0	1.68	0.61	0.52	1.06	7.70	0.64	
13.40	3.91	2.48	23.0	102.0	1.80	0.75	0.43	1.41	7.30	0.70	
13.27	4.28	2.26	20.0	120.0	1.59	0.69	0.43	1.35	10.20	0.59	
13.17	2.59	2.37	20.0	120.0	1.65	0.68	0.53	1.46	9.30	0.60	
14.13	4.10	2.74	24.5	96.0	2.05	0.76	0.56	1.35	9.20	0.61	

ows × 15 columns

Class attributes:

- class_0,
- class_1,
- class_2.

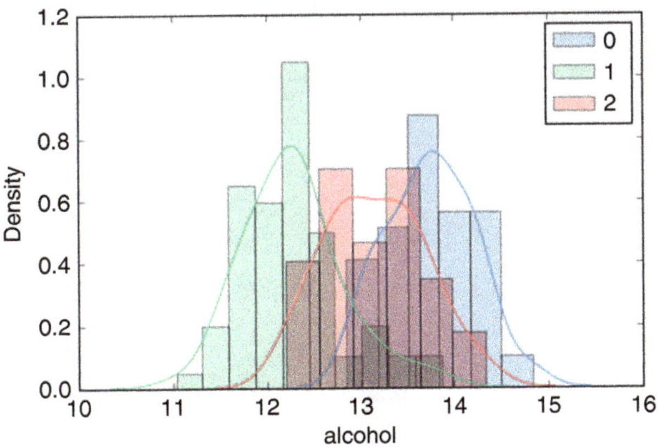

Software code for this model can be found on GitHub Link: basicdecisiontree.py (Python Code).

Model Construction: Base Decision Tree

Binary classification techniques work well when the data observations belong to one of two classes or categories, such as "True" or "False." When the data can be categorized into more than two classes, you must use a multiclass classification algorithm. Fortunately, in most machine learning frameworks, including scikit-learn, implementing a multiclass classifier is not significantly more complex than binary classification—and in many cases, the classification algorithm classes used for binary classification implicitly support multiclass classification.

Step 1: load model specific libraries:

Please note: I have used dtreeviz to display decision tree graph.

```
from sklearn.datasets import *
from sklearn import tree
from dtreeviz.trees import *
from sklearn import tree
from sklearn.model_selection import train_test_split
from sklearn.model_selection import train_test_split
from sklearn.preprocessing import StandardScaler, MinMaxScaler
from sklearn.decomposition import PCA
import numpy
from matplotlib import pyplot
from numpy import where
from collections import Counter
```

Step 2: load Wine Dataset. This is available from SKLearn library and no need to download externally.

```
winedf = load_wine()
data = winedf.data
target = winedf.target
sample_train, sample_test, label_train, label_test = \
   train_test_split(data, target, test_size=0.2, random_state=7)
print(f"Initial set observations {sample_train.shape[0]}")
print(f"Initial set target classes {sample_test.shape[0]}")
```

Output:

```
Initial set observations 142
Initial set target classes 36
```

Step 3: check class variables and count.

```
summarize class distribution
print('Counter')
counter = Counter(target)
print(counter)

 print('UNIQUE COUNTS')
(unique, counts) = numpy.unique(target, return_counts=True)
frequencies = numpy.asarray((unique, counts)).T
print(frequencies)
```

Output:

```
Counter
Counter({1: 71, 0: 59, 2: 48})
UNIQUE COUNTS
[[ 0 59]
[ 1 71]
[ 2 48]]
```

Step 4: Scatter plot the class variable data:

```
# scatter plot of examples by class label
for label, _ in counter.items():
        row_ix = where(target == label)[0]
        pyplot.scatter(data[row_ix, 0], data[row_ix, 1], label=str(label))
pyplot.legend()
pyplot.show()
```

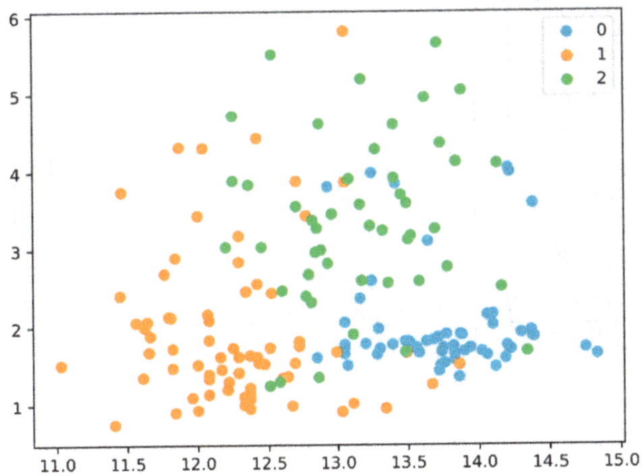

Now that we've dealt with the missing values, let's explore how the features relate to the label by creating some box charts.

```
from matplotlib import pyplot as plt
%matplotlib inline

penguin_features = ['CulmenLength','CulmenDepth','FlipperLength','BodyMass']
penguin_label = 'Species'
for col in penguin_features:
    penguins.boxplot(column=col, by=penguin_label, figsize=(6,6))
    plt.title(col)
plt.show()
```

Output:

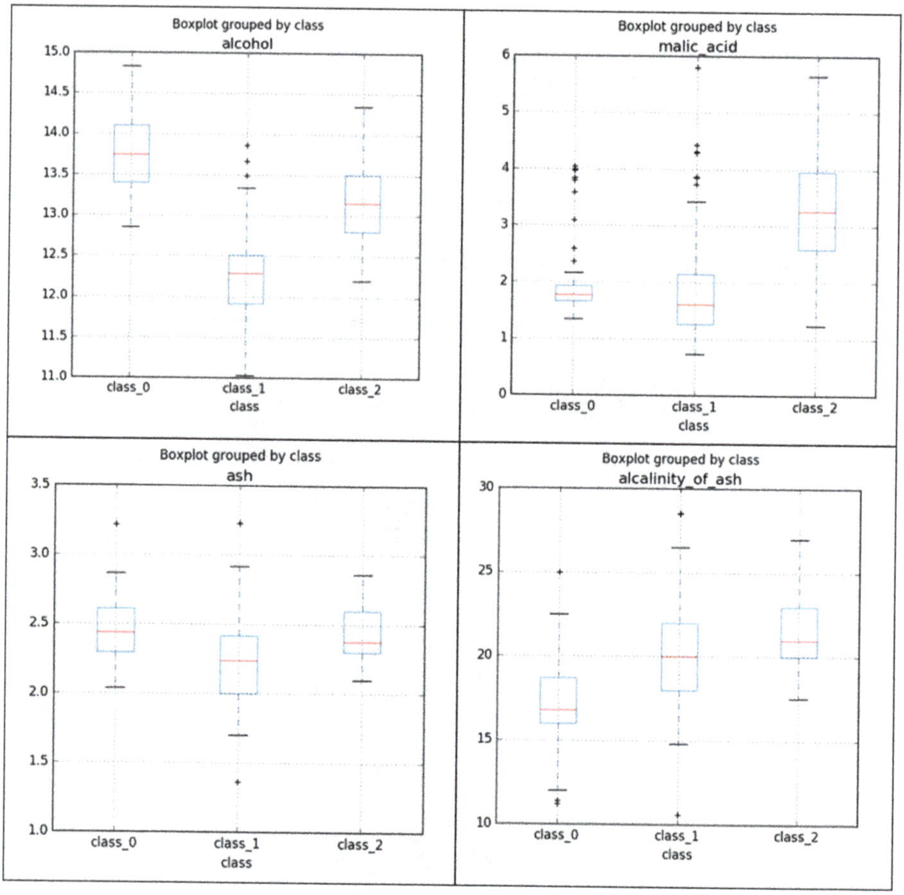

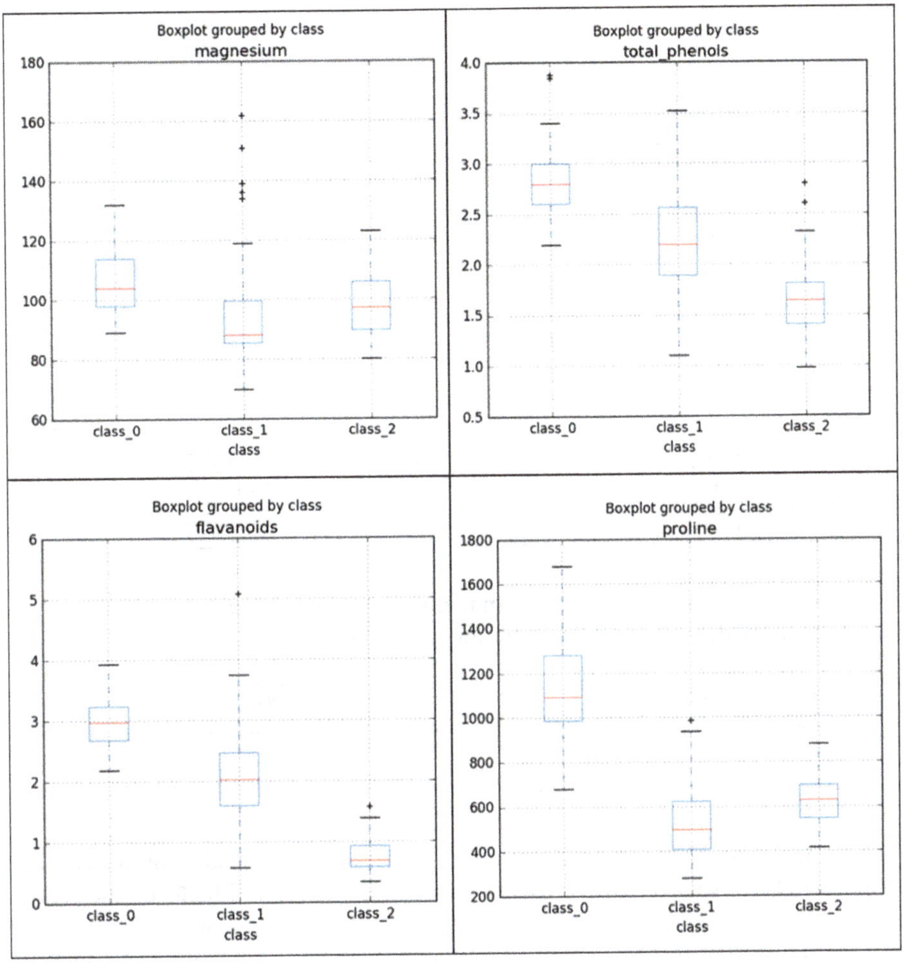

From the box plots, it looks like class 0 and 1 have similar data profiles which should help us train a good classification model.

Step 5: Prepare classifier and run the model:

Just as for binary classification, before training the model, we need to separate the features and label and then split the data into subsets for training and validation. We'll also apply a *stratification* technique when splitting the data to maintain the proportion of each label value in the training and validation datasets.

```
clf = tree.DecisionTreeClassifier(max_depth=3)

clf.fit(sample_train,label_train)

predictions = clf.predict(sample_test)
print('Predicted labels: ', predictions)
print('Actual labels:    ',label_test)


print(clf.score(sample_test, label_test))

from sklearn.metrics import accuracy_score
print('Accuracy: ', accuracy_score(label_test, predictions))
```

Output:

```
Predicted labels: [2 2 0 2 1 2 0 0 0 1 2 1 2 0 1 0 1 1 0 1 2 0 0 2 0 1 2 0 0 1 2 1 1 1 2 1]
Actual labels:    [2 2 0 2 1 1 0 0 0 1 2 1 1 0 1 0 2 1 0 1 2 0 0 1 0 1 2 0 0 2 2 1 1 1 2 1]
0.8611111111111112
Accuracy: 0.8611111111111112
```

In addition to accuracy, Scikit-Learn includes the ability to create a classification report that provides more insight than raw accuracy alone.

```
sklearn.metrics.classification_report[48](y_true, y_pred, *, labels=None, target_names=None, sample_weight=None, digits=2, output_dict=False, zero_division='warn')
```

[a]Classification report - https://scikit-learn.org/stable/modules/generated/sklearn.metrics.classification_report.html

As with binary classification, the report includes precision and recall metrics for each class. However, while with binary classification we could focus on the scores for the positive class, in this case, there are multiple classes, so we need to look at an overall metric (either the macro or weighted average) to get a sense of how well the model performs across all three classes.

You can get the overall metrics separately from the report using the scikit-learn metrics score classes, but with multiclass results you must specify which average metric you want to use for precision and recall.

```
from sklearn. metrics import classification_report
print(classification_report(label_test, predictions))
from sklearn.metrics import precision_score, recall_score
from sklearn.metrics import confusion_matrix
# Get evaluation metrics
mcm = confusion_matrix(label_test, predictions)

print ('Confusion Matrix:\n',mcm, '\n')
print('Accuracy:', accuracy_score(label_test, predictions))
```

Output:

```
            precision    recall  f1-score   support

        0       1.00      1.00      1.00        12
        1       0.85      0.79      0.81        14
        2       0.73      0.80      0.76        10

   accuracy                           0.86        36
  macro avg       0.86      0.86      0.86        36
weighted avg       0.86      0.86      0.86        36

Confusion Matrix:
 [[12  0  0]
 [ 0 11  3]
 [ 0  2  8]]
```

The confusion matrix shows the intersection of predicted and actual label values for each class—in simple terms, the diagonal intersections from top-left to bottom-right indicate the number of correct predictions.

When dealing with multiple classes, it's generally more intuitive to visualize this as a heat map, like this:

```
wine_class=list(winedf.target_names)

import numpy as np
import matplotlib.pyplot as plt

plt.imshow(mcm, interpolation="nearest", cmap=plt.cm.Blues)
plt.colorbar()
tick_marks = np.arange(len(wine_class))
plt.xticks(tick_marks, wine_class, rotation=45)
plt.yticks(tick_marks, wine_class)
plt.xlabel("Actual Species")
plt.ylabel("Predicted Species")
plt.show()
```

Output:

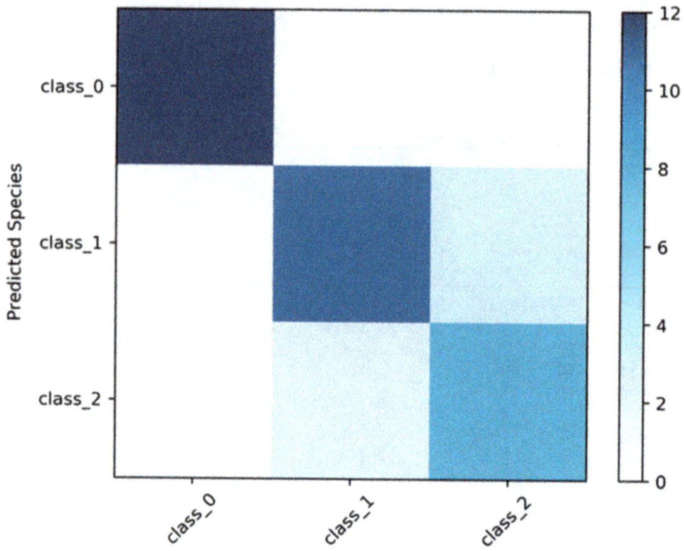

The higher the diagonal values, the better the model is.

```
viz = dtreeviz(clf, winedf.data, winedf.target, target_name='wine',
        feature_names=winedf.feature_names,
        class_names=list(winedf.target_names))

viz.save('C:\\Hanumayamma\\Book\\20200918\\ChapterCode\\Chapter1\\dtwithensembleviz.svg')
viz.view()
```

The root node for the decision tree is alcohol with flavonoids as the first level splitting node. And Hue and Proline are the second level splitting node before completion of the tree.

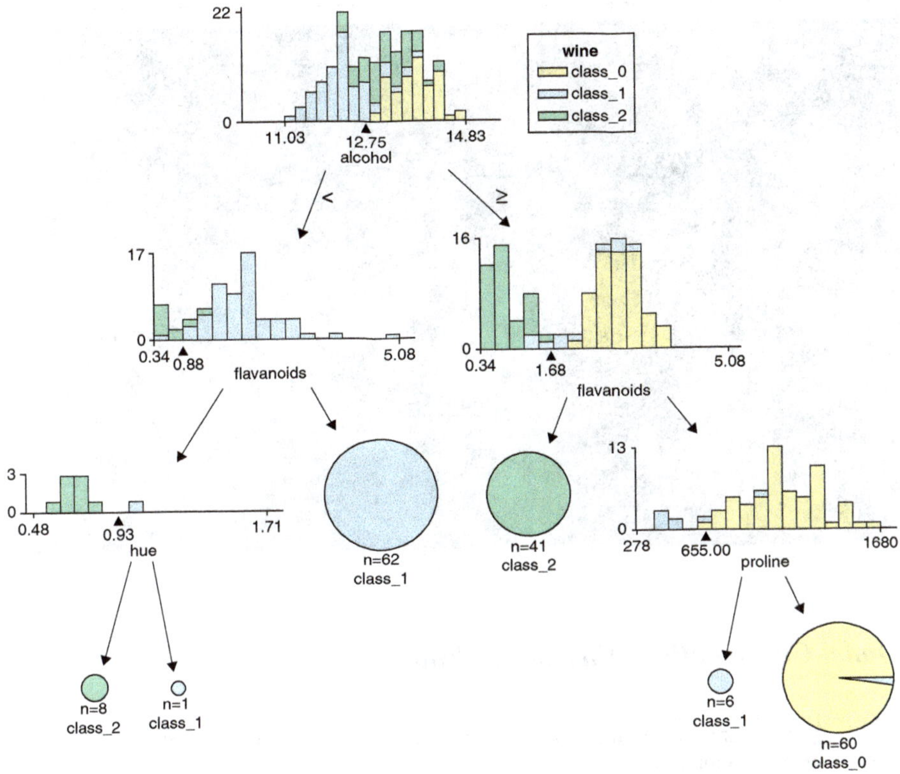

The decision tree has precision of 100% in predicting class_0, 85% class_1, and 73% class 2. The overall accuracy of the tree is 86.1%.

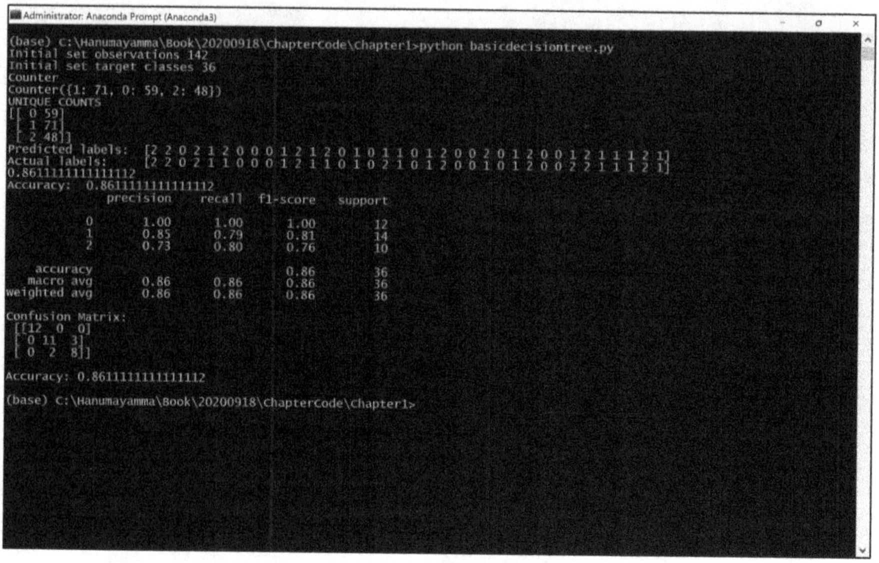

Model Construction: Ensemble Gradient Boosting Classifier

Software code for this model can be found on GitHub Link: ensembledecisiontree.py (Python Code).

For the Gradient Boost Classifier, we can re-use step1 through step 4 to load the data and inspect the data.

Dive into Model construction:

```
# -*- coding: utf-8 -*-
"""
Created on Tue Mar 16 12:52:14 2021

@author: CHVUPPAL
"""

from sklearn.datasets import *
from sklearn import tree
from dtreeviz.trees import *
from sklearn import tree
from sklearn.model_selection import train_test_split

from sklearn.model_selection import train_test_split
from sklearn.preprocessing import StandardScaler, MinMaxScaler
from sklearn.decomposition import PCA

from sklearn.ensemble import GradientBoostingClassifier
import numpy
from matplotlib import pyplot
from numpy import where


from collections import Counter

winedf = load_wine() ## load_wine(return_X_y=True)

data = winedf.data
target = winedf.target
sample_train, sample_test, label_train, label_test = \
    train_test_split(data, target, test_size=0.2)



print(f"Initial set observations {sample_train.shape[0]}")
print(f"Initial set target classes {sample_test.shape[0]}")


# summarize class distribution
```

```
print('Counter')
counter = Counter(target)
print(counter)


print('UNIQUE COUNTS')
(unique, counts) = numpy.unique(target, return_counts=True)
frequencies = numpy.asarray((unique, counts)).T
print(frequencies)


# scatter plot of examples by class label
for label, _ in counter.items():
        row_ix = where(target == label)[0]
        pyplot.scatter(data[row_ix, 0], data[row_ix, 1], label=str(label))
pyplot.legend()
pyplot.show()


clf = GradientBoostingClassifier(n_estimators=100, learning_rate=1.0,
                max_depth=1, random_state=0).fit(sample_train, label_train)


print(clf.score(sample_test, label_test))
predictions = clf.predict(sample_test)
print('Predicted labels: ', predictions)
print('Actual labels:    ',label_test)
print(clf.score(sample_test, label_test))

from sklearn.metrics import accuracy_score

print('Accuracy: ', accuracy_score(label_test, predictions))

from sklearn. metrics import classification_report

print(classification_report(label_test, predictions))

from sklearn.metrics import precision_score, recall_score

from sklearn.metrics import confusion_matrix
# Get evaluation metrics
mcm = confusion_matrix(label_test, predictions)

print ('Confusion Matrix:\n',mcm, '\n')
print('Accuracy:', accuracy_score(label_test, predictions))

wine_class=list(winedf.target_names)

import numpy as np
import matplotlib.pyplot as plt
```

```
plt.imshow(mcm, interpolation="nearest", cmap=plt.cm.Blues)
plt.colorbar()
tick_marks = np.arange(len(wine_class))
plt.xticks(tick_marks, wine_class, rotation=45)
plt.yticks(tick_marks, wine_class)
plt.xlabel("Actual Species")
plt.ylabel("Predicted Species")
plt.show()
```

Output:

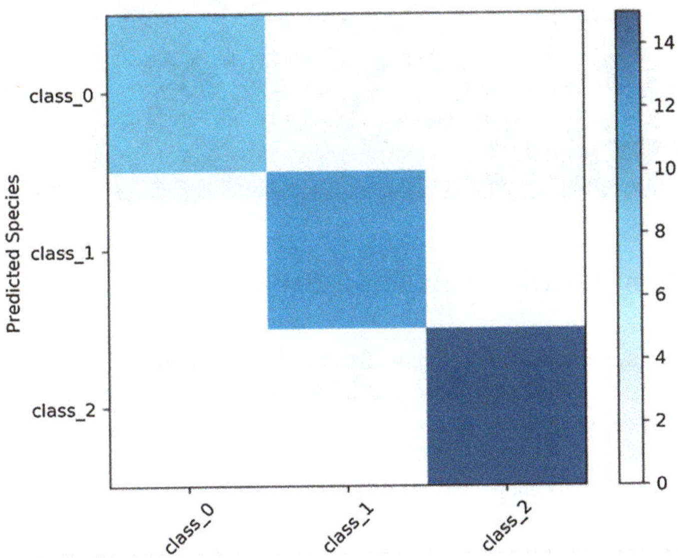

The decision tree has precision of 100% in predicting class_0, class_1, and class_2. The overall accuracy of the tree is 100%.

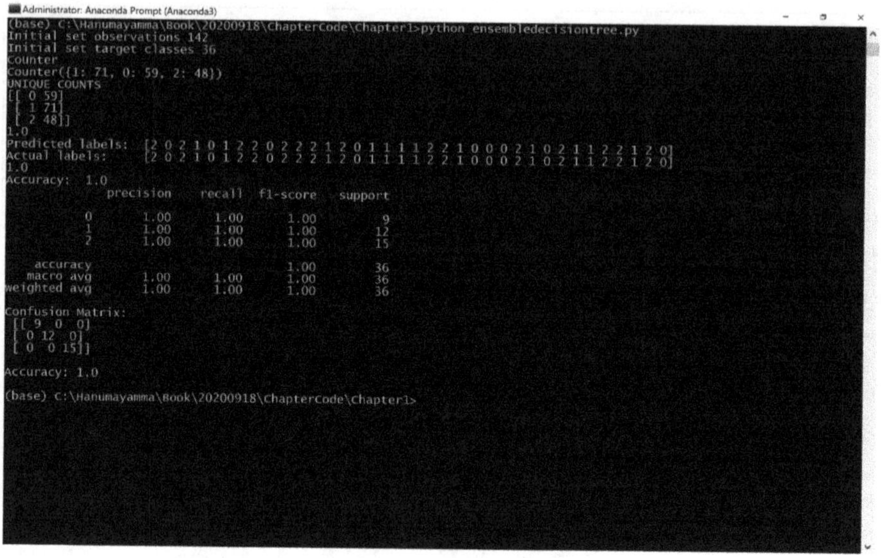

Tope 15 Destinations for Italian Exports, 2019

As an aside, here is the synopsis of wine export from Italy for the top 15 countries (units in tons): Germany leads the pack followed by France, the United States, the United Kingdom, and Switzerland.

Top 15 Destinations for Italian Wine Exports, 2019[a]	
Country	Wine Exports (tonnes), 2019
Germany	15470793
France	7436119
United States of America	7010509
United Kingdom of Great Britain and Northern Ireland	6268390
Switzerland	1951193
Canada	1532192
Austria	1192478
Netherlands	1026071
Russian Federation	906348
Sweden	894862
Japan	891047
Spain	847093
Czechia	839537
Denmark	766604
Belgium	585125

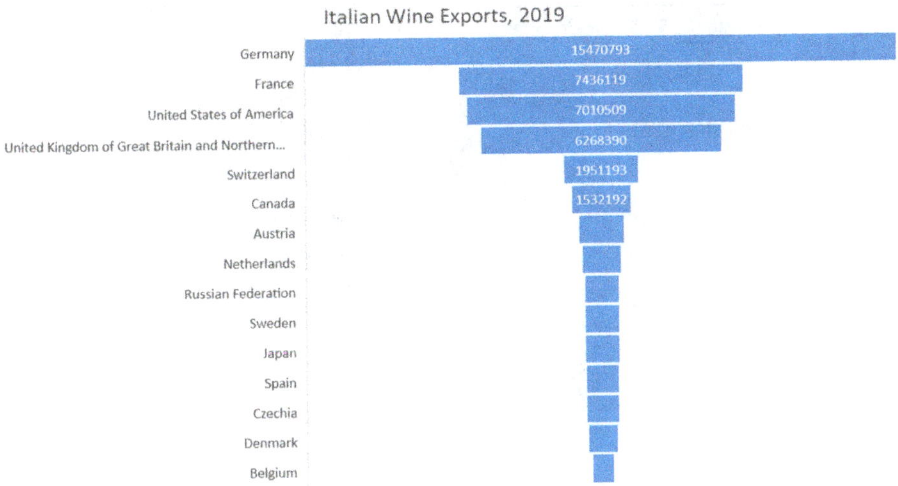

Fig. 1.11 Italian wine exports

[a]Italian Wine Exports, 2019 - http://www.fao.org/faostat/en/#data/TM

Data Source: Detailed Trade Matrix (please see Fig. 1.11) from FAOSTATS Database.[39]

Select the following field options (please see Fig. 1.12):

Reporter countries: Italy	Partner countries (region): world	Elements: export quantity
Items: wine	Years: all years	

[39] Italian Wine Exports, 2019 – http://www.fao.org/faostat/en/#data/TM

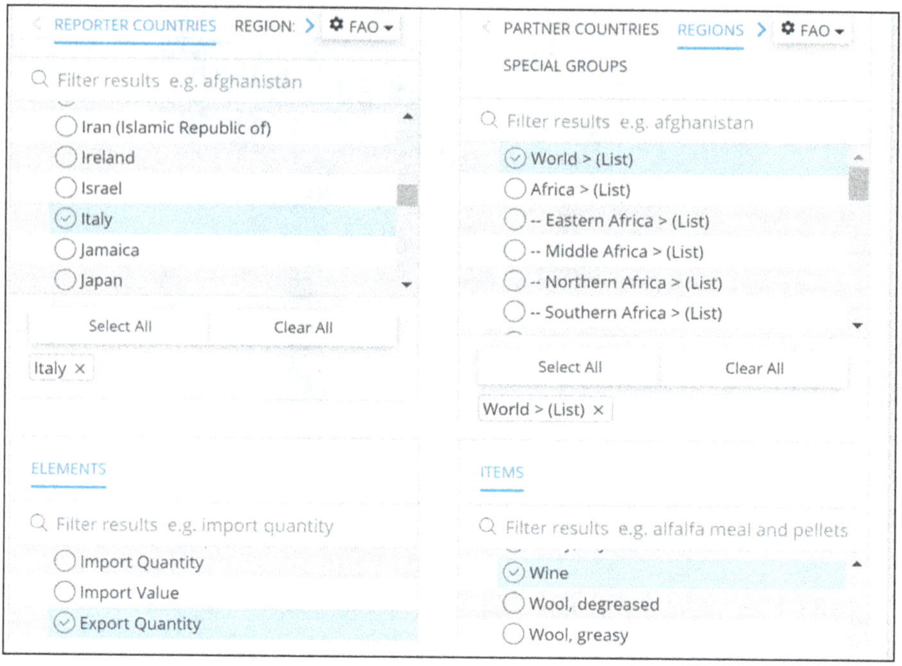

Fig. 1.12 Trade Matrix UI

Econometric Models

Econometrics deals with the measurement of economic relationships. It is an integration of economics, mathematical economics, and statistics with an objective to provide numerical values to the.

parameters of economic relationships. Economic metrics models,[40] generally, used to predict price of an agricultural commodity [14]. The interrelationships between production, consumption, and stocks of agricultural commodities are studied using econometric models. Econometric models apply pure statistical techniques to find data correlations and explain economic behavioral or pattern.

An econometric model consists of[41]:

[40] An econometric model of the agricultural sector of the Indian economy – https://www.sciencedirect.com/science/article/abs/pii/0161893882900278

[41] Introduction to Econometric Models – http://home.iitk.ac.in/~shalab/econometrics/Chapter1-Econometrics-IntroductionToEconometrics.pdf

- A set of equations describing the behavior. These equations are derived from the economic model and have two parts—observed variables and disturbances.
- A statement about the errors in the observed values of variables.
- A specification of the probability distribution of disturbances.

Endogenous and Exogenous Variables.

An endogenous is a dependent variable, and it [30] is changed or determined by its relationship with other variables within the model. Endogenous variables are important in econometrics and economic modeling because they show whether a variable causes a particular effect. Causation or causality with the other variables is the signature of an endogenous variable. As an example, rainfall to plant growth is correlated and studied by economists and ML models since the amount of rainfall is important to commodity crops such as corn and wheat.

Exogenous variables are independent variables—in an economic model, an exogenous variable is one whose value is determined outside the model and is imposed on the model. Exogenous variables do not exhibit correlation to or causation with another variable, for example, crop production yield and salary of a software engineer or personal income and color preference.

For econometric model predicting crop production, the following are endogenous variable [14]:

- O_f = Output of food grains.
- F_f = Consumption of chemical fertilizers used on food grains (thousands of tons).
- C_f = Food stock or availability of food grains.
- M_f = Import of food grains (thousand tons).
- Y = Gross domestic product.
- S_f = Changes in stocks in food grains on government.
- A_f = Area harvested (thousand hectares).
- P_f = Food grain wholesale price.
- F_a = Food aid at constant price.
- E = Foreign exchange earnings.
- G = Government expenditures.

Choosing the Right Estimator

Often the hardest part of solving a machine learning problem can be finding the right estimator for the job.[42] Different estimators are better suited for different types of data and different problems. The flowchart below is designed to give users a bit of a rough guide on how to approach problems about which estimators to try on your data. Estimator enables to learn insights from the data or able to develop a model to predict (please see Fig. 1.13).

When considering automated learning, computational resources also play a major role in determining the complexity of a task: that is, how much computation is involved in carrying out a learning task. Once a sufficient training sample is available to the learner, there is some computation to be done to extract a hypothesis or figure out the label of a given test instance.

These computational resources are crucial in any practical application of machine learning. There are two other considerations to be considered: sample complexity and computational complexity [31].

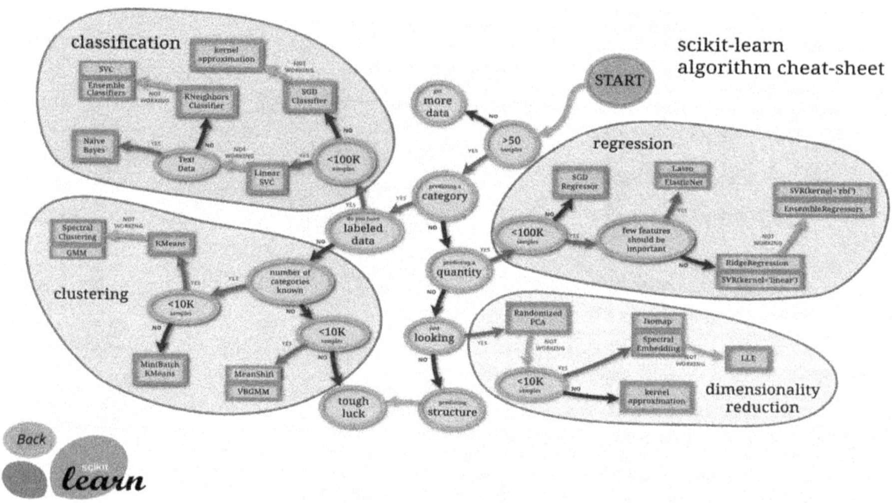

Fig. 1.13 SCIKIT learn algorithms. (Source: https://scikit-learn.org/stable/tutorial/machine_learning_map/index.html#)

[42] Choosing right estimator – https://scikit-learn.org/stable/tutorial/machine_learning_map/index.html#

Mapping AI Technique to Classical ML

As AI technologies advance, so does the definition of which techniques constitute AI (please see Fig. 1.14). The following heat-map maps the common ML techniques used in AI: classification, continuous estimation, and clustering are most used [1].

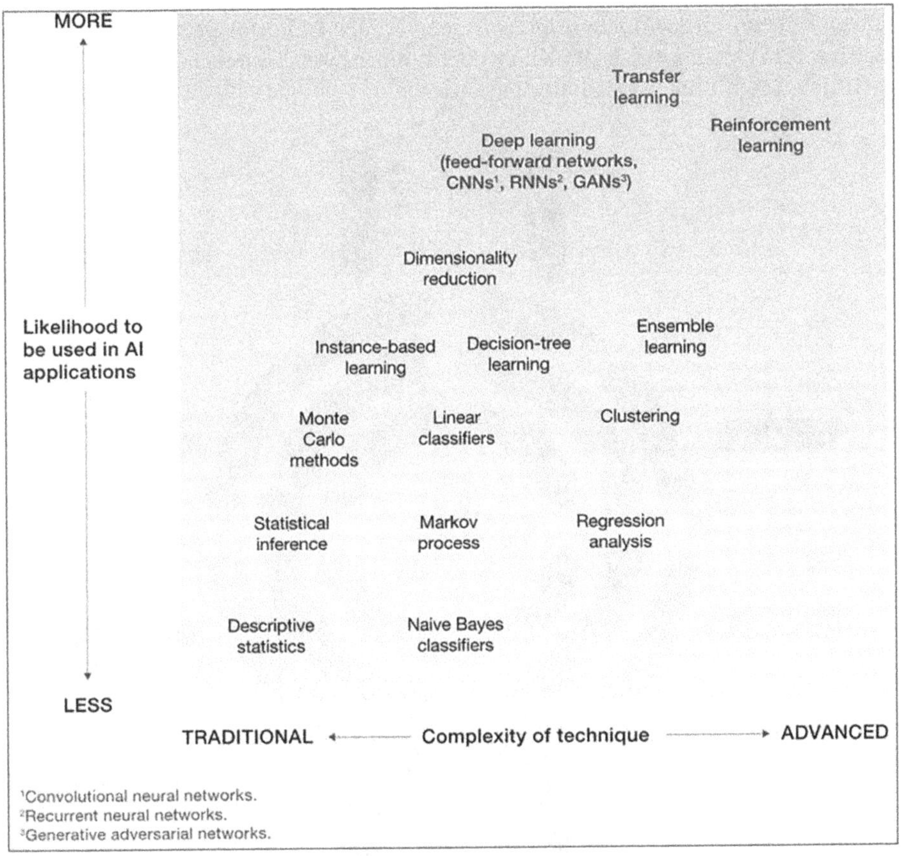

Fig. 1.14 AI mapping

AI Economics

The economics of investing in "traditional" technology initiatives are well understood. Typically, for brownfield and green field opportunities, to address market competition, Porter's five forces,[43] an organization compares the cost of opportunity or cost of conceding first movers advantage to the industry, or cost of its dominant position, or cost of a project, in most of the cases, with the net present value (NPV) of the benefits [32]. If the NPV is greater than zero and fits within the budget constraints, the project receives a green light. Generally, this is applicable in corporates as well as small office home office segments (SOHO). Porter's five forces (please see Fig. 1.15) strategy is applicable to brick-mortar industry cases as well as digitally fueled 10X (Ten X) analytics initiatives.

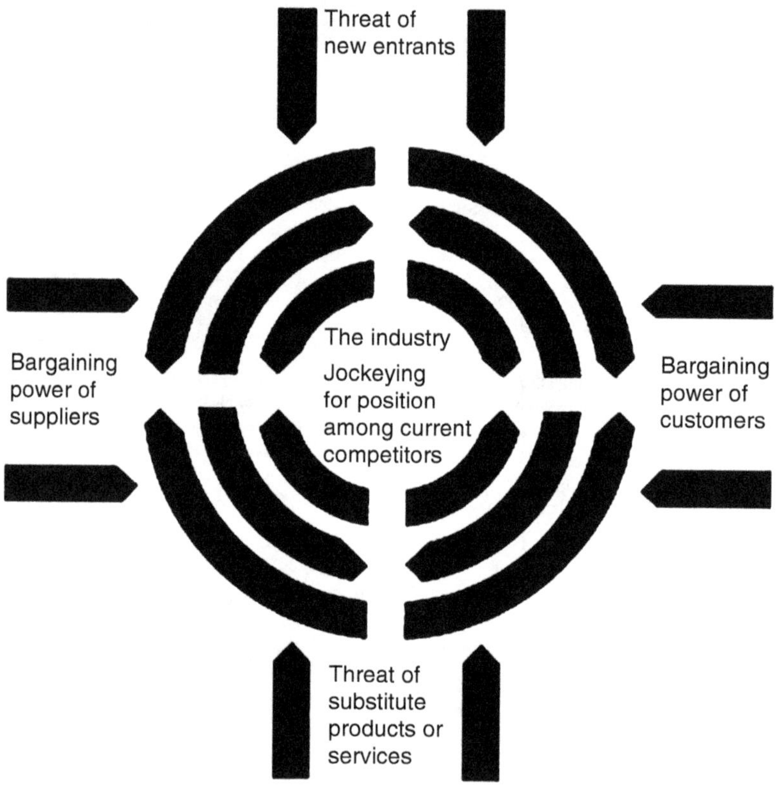

Fig. 1.15 Porter five forces model

[43] How Competitive Forces Shape Strategy – https://hbr.org/1979/03/how-competitive-forces-shape-strategy

Blue ocean[44] initiatives, on the other hand, are driven, in many cases, by the outcomes of insights derived from data-driven analytics. The risk-reward for blue ocean strategy typically is well documented as part of project or product charter and is a self-fulfilling reason to pursue, to begin with, one can attribute the fulfillment of Maslow's[45] hierarchy needs. Though Maslow's theory applies to humans intrinsically partake in behavioral motivation, the application of Maslow's theory to corporations is justified as corporations are people too[46] and they need to safety and to secure as it's the main reason for the survival [33].

The argument of NPV either through Porter's five forces or blue ocean initiatives is justified assuming that the basic human essential needs are satisfied and met. To assess value of a product that is as staple as food and that is as essential as for our humanity survival, in my opinion, the NPV will not justify the rational for investment. Additionally, given around 70% of population that is employed by agriculture and the outcomes of agriculture decide above or below poverty line of individuals and societies/countries, we need to assess the impact of product/technology to the humanity. For instance, consider the impact of ICT on humanity: information

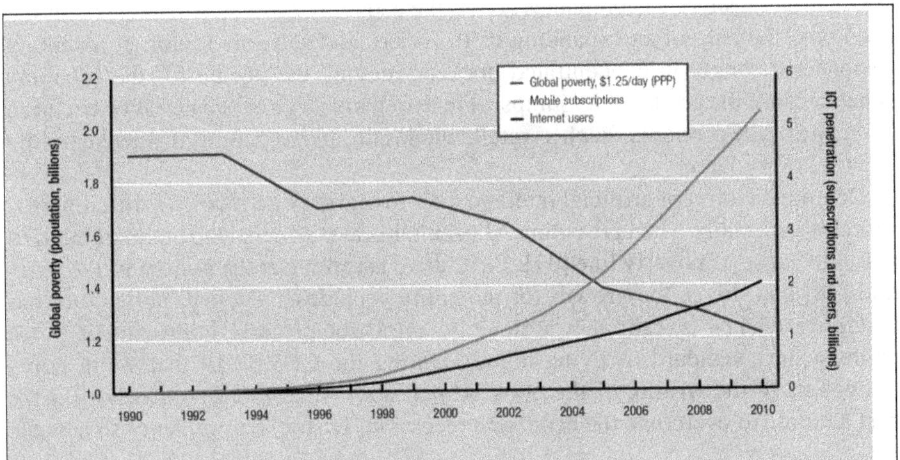

Sources: World Bank *PovCal* database (1990, 1993, 1996, 1999, 2002, 2005, 2008, 2010); authors' calculations and interpolation, ITU *World Telecommunication/ICT Indicators* database June 2013.

Fig. 1.16 ICT impact

[44] Blue-Ocean Initiatives – https://economictimes.indiatimes.com/definition/blue-ocean-strategy

[45] Maslow Triangle – https://webspace.ship.edu/cgboer/Maslow.html

[46] Corporations Are People Too: (And They Should Act Like It) Hardcover – Yale University Press (October 23, 2018)- ISBN-13: 978–0300211474

communication technologies (ICT) have shown [34] a positive impact[47] on reducing poverty and improving the life of farmers as shown in the below graph (please see Fig. 1.16)[48] [35].

Evidence from the last two decades demonstrates that ICTs, particularly broadband Internet, are an *income multiplier*. At the country level, macroeconomic data links fixed telephony, mobile telephony, Internet use, and broadband use to gross domestic product (GDP) growth in a causal relationship across developed and developing countries. And increasing the intensity of data use also drives per capita income growth. This growing body of evidence highlights the fact that we are long past the days of the "Solow paradox," when, in 1987, Nobel Prize-winning economist Robert Solow[49] noted, "you can see the computer age everywhere but in the productivity statistics" [36].

The benefits spread further as rapid declines in information, communications, and technology (ICT) equipment prices encouraged an investment boom in several other sectors, some of which, such as telecommunications and securities trading, enjoyed rapid productivity improvements. Agriculture is rightly positioned as many small farmers carry cell phones and are poised to enter advanced digital age that equips them to be served exclusive agricultural recommendations in near real time or real time. In short, as companies across the economy evolved business processes, often with the help of an expanding ICT services and software sector, productivity responded strongly in concentrated pockets, moving the needle for the economy either because the productivity jumps were extremely large or because they occurred in sectors of the economy (such as retail, wholesale, and agriculture) where employment was very large.

Democratization of artificial intelligence[50] for agriculture require a different perspective, especially when providing AI to small-scale farmers who are marginalized and at the cusp of poverty line [37]. First, there is fortune at the bottom of the pyramid [38] that drives the growth for the entire societies.[51] Second, agriculture has *multiplier effect*—investing in agriculture infrastructure and improving it thrust entire society standard of living higher. During the COVID-19 that we're going through as of the writing of the book, March 2021, investing in agriculture is the best antidote to overcome the economic recession! History has evidence to it as the

---

[47] The Paradox of ICT's Impact on Growth and Inequality – https://blogs.cisco.com/gov/gitr2015

[48] ICTs, Income In equality, and Ensuring Inclusive Growth – https://alln-extcloud-storage.cisco.com/ciscoblogs/GITR-2015-Cisco-Chapter.pdf

[49] Is the Solow Paradox back? – https://www.mckinsey.com/business-functions/mckinsey-digital/our-insights/is-the-solow-paradox-back

[50] Democratization of Artificial Intelligence for the Future of Humanity first Edition, CRC Press; first edition (January 18, 2021), ISBN-13: 978–0367524098

[51] The Fortune at the Bottom of the Pyramid: Eradicating Poverty Through Profits, Revised and Updated fifth Anniversary Edition Hardcover – Wharton School Publishing; Revised edition (October 12, 2009) – ISBN-13: 978–0137009275

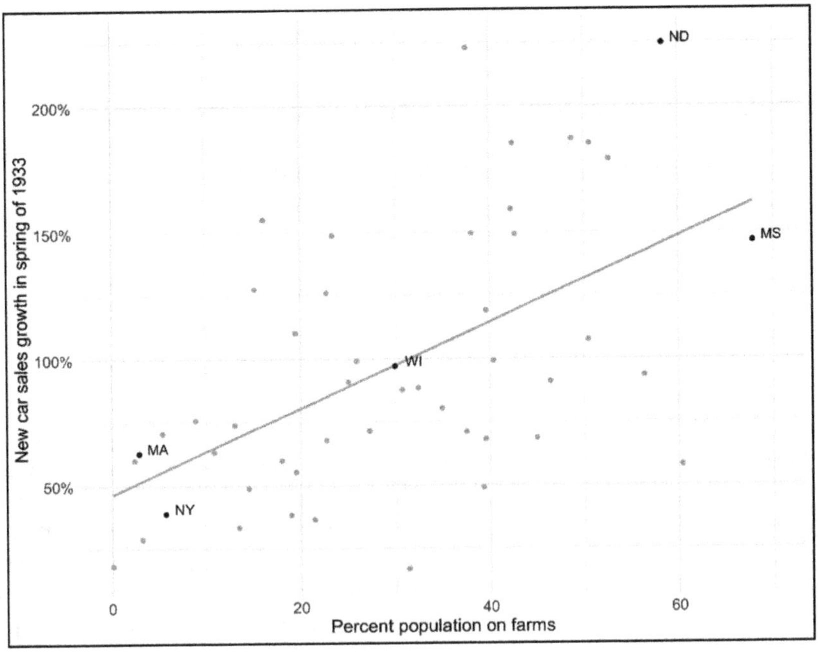

Fig. 1.17 Car sales

agriculture sector helped our economy to come out of depression during 1935.[52] Farmers clearly benefited from Roosevelt's devaluation, according to Hausman [39]. And because farmers were probably less likely to save their extra cash than most companies or other Americans, the overall economy recovered as well.

The chart above (please see Fig. 1.17) shows the connection between new-car sales and the share of farmers in each state. States with large cities, such as New York, had lower growth than states like North Dakota and Mississippi, which relied heavily on wheat and cotton, respectively. States such as Wisconsin saw relatively moderate growth, partly because its farming was concentrated in non-tradable goods like dairy[53] [39].

[52] How did agriculture help the US recover from the Great Depression in the spring of 1933? – https://www.aeaweb.org/research/farmer-recovery-great-depression

[53] How did agriculture help the US recover from the Great Depression in the spring of 1933? – https://www.aeaweb.org/research/farmer-recovery-great-depression

Recovery from the Great Depression: The Farm Channel in Spring 1933[54] **[40]**

"The depression in the manufacturing industry of the country is due chiefly to the fact that agricultural products generally have been selling below the cost of production, and thereby destroyed the purchasing power in the domestic market of nearly half of all our people. We are going to restore the purchasing power of the farmer."

Franklin D. Roosevelt,
Campaign speech[55] in Atlanta, GA, October 24, 1932.

Revenue is critical part of building Analytics software; subscription income and scale are critical KPIs. To cross the chasm[56] [41] and reach small farmers for selling analytics products, the cost of subscription to AI service should be as minimal as possible and should not result in any extra operating cost or any extra human effort as farmers are very busy and scarce commodity. Any additional investment such as buying IT equipment is expense and will deter adoption. The essential mantra of reaching farmer on analytics sale is simple and should commence with a click of a button. The opportunities are immense and unlimited. The acute need is very self-evident, especially countries with the highest agriculture percentage of economy exhibiting very low penetration of AI. This is a golden opportunity for many startups and companies in this niche segment.

The fortune at the bottom of the pyramid[57]
Low-income markets present a prodigious opportunity for the world's wealthiest companies—to seek their fortunes and bring prosperity to the aspiring poor [38].

Artificial Intelligence and Income Equality and Inclusive Growth

Countries with the highest percentage of employment in agriculture (please see Fig. 1.18) have some of the lowest AI readiness.[58] The digital gap is causing of lower productivity, rural poverty, social and political unrest, economic recessions, and increased burden to local and national governments [42].

As per the World Bank's economic classification, the following countries are classifications as low-income and lower middle-income countries (please see Tables 1.5 and 1.6):

[54] Recovery from the Great Depression – https://pubs.aeaweb.org/doi/pdfplus/10.1257/aer.20170237#page=14

[55] Campaign Speech Franklin D. Roosevelt | October 24, 1932 – https://teachingamericanhistory.org/library/document/campaign-speech/

[56] Crossing the Chasm, third Edition: Marketing and Selling Disruptive Products to Mainstream Customers – Harper Business; third edition (January 28, 2014)

[57] The Fortune at the bottom of the Pyramid – https://www.strategy-business.com/article/11518?gko=9b3b4

[58] McKinsey Global Institute (MGI) – Notes from the AI frontier Modeling the impact of AI on the world economy September-2018- https://www.mckinsey.com/~/media/McKinsey/Featured%20Insights/Artificial%20Intelligence/Notes%20from%20the%20frontier%20Modeling%20the%20impact%20of%20AI%20on%20the%20world%20economy/MGI-Notes-from-the-AI-frontier-Modeling-the-impact-of-AI-on-the-world-economy-September-2018.ashx

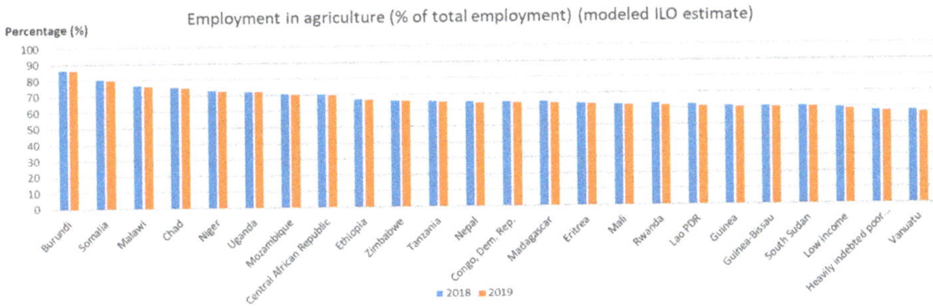

Fig. 1.18 Employment in the agriculture (Source: The World Bank)

Table 1.5 Low-income countries list

Low-income countries[a, b]		
Afghanistan	Korea, Dem. People's Rep.	Madagascar
Burundi	Rwanda	Mali
Burkina Faso	Sudan	Mozambique
Central African Republic	Sierra Leone	Malawi
Congo, Dem. Rep.	Somalia	Niger
Eritrea	South Sudan	Haiti
Ethiopia	Syrian Arab Republic	Liberia
Guinea	Chad	Tajikistan
Gambia, The	Togo	Uganda
Guinea-Bissau		Yemen, Rep.

Source: the World Bank

[a]Employment in agriculture (% of total employment) (modeled ILO estimate) – World- https://data.worldbank.org/indicator/SL.AGR.EMPL.ZS?locations=1W

[b]API_SL.AGR.EMPL.ZS_DS2_en_excel_v2_2059645.XLS – https://api.worldbank.org/v2/en/indicator/SL.AGR.EMPL.ZS?downloadformat=excel

The above two categories of countries have the highest number of small-scale farmers who are dependent on agriculture.

AI Readiness

As per McKinsey report [42], in terms of readiness for AI, the world countries are divided into four groups:

- Group 1: Active global leaders.
- Group 2: Economies with strong comparative strength.
- Group 3: Economies with moderate foundations.
- Group 4: Economies that need to strengthen foundations.

Table 1.6 Lower middle-income countries

Lower-middle income[a, b]		
Angola	Cambodia	Solomon Islands
Benin	Kiribati	El Salvador
Bangladesh	Lao PDR	São Tomé and Principe
Bolivia	Sri Lanka	Eswatini
Bhutan	Lesotho	Timor-Leste
Côte d'Ivoire	Morocco	Tunisia
Cameroon	Moldova	Tanzania
Congo, Rep.	Myanmar	Ukraine
Comoros	Mongolia	Uzbekistan
Cabo Verde	Mauritania	Vietnam
Djibouti	Nigeria	Vanuatu
Algeria	Nicaragua	Zambia
Egypt, Arab Rep.	Nepal	Zimbabwe
Micronesia, Fed. Sts.	Pakistan	Kenya
Ghana	Philippines	Kyrgyz Republic
Honduras	Papua New Guinea	Senegal
India	West Bank and Gaza	

Source: the World Bank

[a]Employment in agriculture (% of total employment) (modeled ILO estimate) – World- https://data.worldbank.org/indicator/SL.AGR.EMPL.ZS?locations=1W

[b]API_SL.AGR.EMPL.ZS_DS2_en_excel_v2_2059645.XLS – https://api.worldbank.org/v2/en/indicator/SL.AGR.EMPL.ZS?downloadformat=excel

Group 1: Active Global Leaders

Countries are currently leading the race to supply AI, and they have unique strengths that set them apart from all others. Scale effects enable more significant investment, and network effects enable these economies to attract the talent needed to make the most of AI. Together, they are responsible for most AI-related research activities. They are a long way ahead of other countries on AI-related patents, publications, and citations. They also make substantial investment in AI.

Group 1: Active global leaders
China
United States

Group 2: Economies with Strong Comparative Strength

A wide range of countries belongs to this group, including Canada, France, South Korea, and Sweden. They are relatively well positioned to capture the benefits of AI given their generally robust foundation of enablers. Many of these economies are highly motivated to embrace AI because they have been experiencing slowing productivity growth.

Group 2: Economies with strong comparative strength		
Australia	Belgium	Canada
Estonia	Finland	Germany
Iceland	Israel	Japan
Netherlands	Norway	Singapore
South Korea	Sweden	United Kingdom

Group 3 Countries: Economies with Moderate Foundations

This group, which includes India, Italy, and Malaysia, has a moderate ability to capture economic benefits from AI. While the potential for economic gains is broadly positive, these countries are in a weaker starting position than those in the first two groups, but they exhibit comparative strengths in specific areas on which they may be able to build. India, for instance, has *somewhat underdeveloped digital infrastructure and currently has a relatively low automation potential*, but it has more STEM students than the total of STEM graduates produced by all G-7 countries.

Chile	Costa Rica	Czech Republic
India	Italy	Lithuania
Malaysia	South Africa	Spain
Thailand	Turkey	

Group 4: Economies that Need to Strengthen Foundations

These countries are relatively challenged in their ability to capture the economic benefits of AI. They have somewhat limited automation potential because wages tend to be rather low, and therefore the incentive to substitute labor to boost productivity is weak. They also have relatively underdeveloped digital infrastructure, innovation and investment capacity, and digital skills and are comparatively isolated from global trade and data flows. These economies tend to prioritize stimulating economic growth, reducing poverty, and developing away from agriculture into basic, and then more advanced, manufacturing and services.

Brazil	Bulgaria	Cambodia
Colombia	Greece	Indonesia
Pakistan	Peru	Tanisha
Uruguay	Zambia	

Varying AI readiness conditions among countries imply different degrees of AI adoption and absorption and therefore economic impact [42].

It's true that if you leverage AI to improve or automate an existing process, then it's possible to measure return on investment (ROI) in the straightforward, traditional way. But there are a few characteristics of AI initiatives that make it difficult to estimate their costs and benefits.

AI-Driven Agricultural Economics

For fiscal sustainment of agricultural farms, the farm expansion and revenues are as critical as revenue streams for running a startup or managing a corporation. Put it simply, farm revenues are critical for agricultural economy. As seen above for the countries with the highest percentage of agricultural contribution to economy, the penetration of AI in those countries is the least as compared to developed countries. This causation is not accidental, but it is a systemic! Government and farms should pay attention to improve the digital infrastructure and enable the usage of AI so that they are at the same par with the rest of the world. Otherwise, we will witness unprecedented gap in terms of advancement and economies in those countries with the countries where wide AI is in place.

Two key metrics are compared to assess:

Metric 1: Agriculture, Value Added (Annual % Growth)

Definition[59]: Annual growth rate for agricultural value added based on constant local currency. Aggregates are based on constant 2010 US dollars. Agriculture corresponds to ISIC divisions 1–5 and includes forestry, hunting, and fishing, as well as cultivation of crops and livestock production. Value added is the net output of a sector after adding up all outputs and subtracting intermediate inputs. It is calculated without making deductions for depreciation of fabricated assets or depletion and degradation of natural resources. The origin of value added is determined by the International Standard Industrial Classification (ISIC), revision 3 or 4.

Statistical Concept and Methodology: Gross domestic product (GDP) represents the sum of value added by all its producers. Value added is the value of the gross output of producers less the value of intermediate goods and services consumed in production before accounting for consumption of fixed capital in production. The United Nations System of National Accounts calls for value added to be valued at either basic prices (excluding net taxes on products) or producer prices (including net taxes on products paid by producers but excluding sales or value added taxes). Both valuations exclude transport charges that are invoiced separately by producers. Total GDP is measured

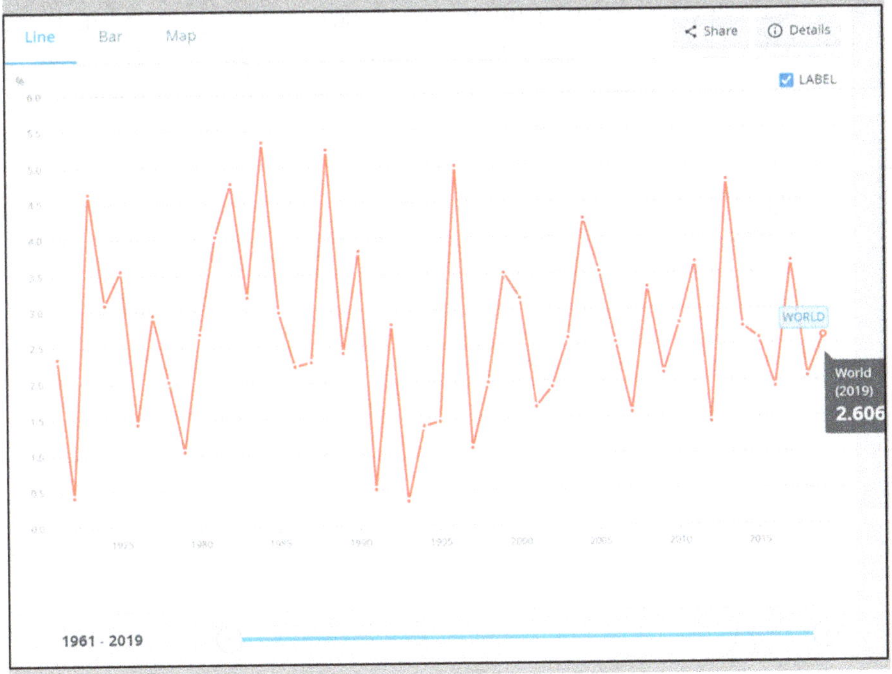

[59] Agriculture, value added (annual % growth)- https://www.indexmundi.com/facts/indicators/NV.AGR.TOTL.KD.ZG

at purchaser prices. Value added by industry is normally measured at basic prices.

Aggregation method: Weighted average.

Periodicity: Annual.

Data Source: https://data.worldbank.org/indicator/NV.AGR.TOTL.KD.ZG

Periodicity: Annual.

General Comments: Note: Data for OECD countries are based on ISIC, revision 4.

Metric 2: Employment in Agriculture (% of Total Employment) (Modeled ILO Estimate)

Definition[60]: Employment is defined as persons of working age who were engaged in any activity to produce goods or provide services for pay or profit, whether at work during the reference period or not at work due to temporary absence from a job, or to working-time arrangement. The agriculture sector consists of activities in agriculture, hunting, forestry, and fishing, in accordance with division 1 (ISIC 2) or categories A-B (ISIC 3) or category A (ISIC 4).

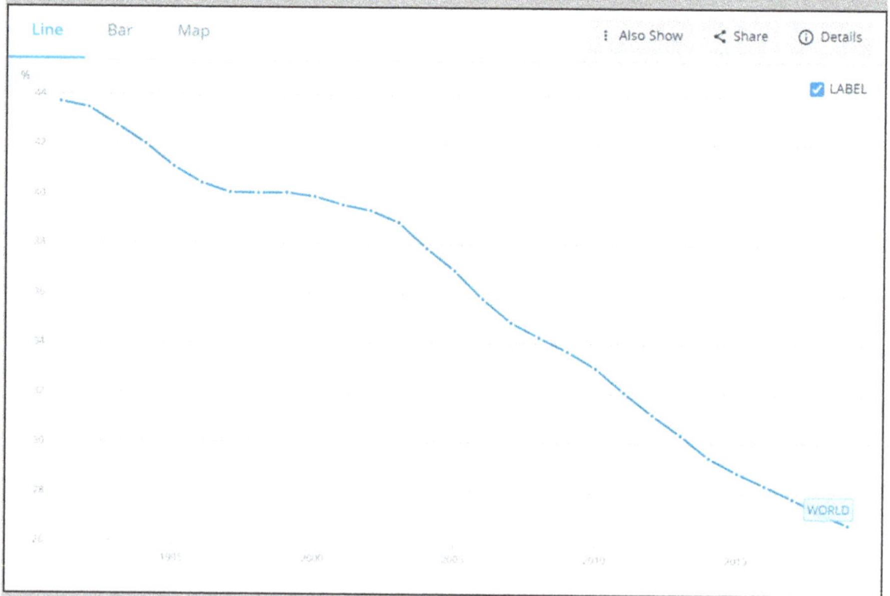

[60] Employment in agriculture (% of total employment) (modeled ILO estimate) – https://data.worldbank.org/indicator/SL.AGR.EMPL.ZS

Statistical Concept and Methodology: The International Labour Organization (ILO) classifies economic activity using the International Standard Industrial Classification (ISIC) of All Economic Activities, revision 2 (1968), revision 3 (1990), and revision 4 (2008). Because this classification is based on where work is performed (industry) rather than type of work performed (occupation), all of an enterprise's employees are classified under the same industry, regardless of their trade or occupation. The categories should sum to 100%. Where they do not, the differences are due to workers who are not classified by economic activity. The series is part of the ILO estimates and is harmonized to ensure comparability across countries and over time by accounting for differences in data source, scope of coverage, methodology, and other country-specific factors. The estimates are based mainly on nationally representative labor force surveys, with other sources (population censuses and nationally reported estimates) used only when no survey data are available.

Aggregation method: Weighted average.
Periodicity: Annual.

United States of America

AI Readiness: Group 1: Active Global Leaders	

The United States has maintained its agriculture growth (please see Fig. 1.19) and maintained as the largest exporter of paddy rice, what, and many other agricultural products, at the same time, a great marketplace and agricultural importer from the world. In the last two decades (2000 (**1.63%**)–2019(**1.36%**)), the average drop in employment in agriculture (% of total employment[61]) **is 1.42%**. The influence of AI on agriculture, in terms of data collection, data management, and insights to farmers, can be seen!

[61] Data Source – Employment in agriculture (% of total employment) (modeled ILO estimate) – https://data.worldbank.org/indicator/SL.AGR.EMPL.ZS?end=2017&locations=1W&start=1981&view=chart

Fig. 1.19 U.S. Annual growth and % of employment (Source: the World Bank)

China

AI Readiness: Group 1: Active Global Leaders	

China has achieved the robust growth[62] in agriculture commodities {Source file: API_NV.AGR.TOTL.KD.ZG_DS2_en_excel_v2_2063913.XLS[63]} such as wheat, rice, and credit to farmers (2000 (**2.03%**)–2019 (**3.01%**). In the last two decades (2000 (**50.01%**)–2019 (**25.33%**)), the average drop in employment in agriculture (% of total employment) **is 37.63**%. At the same time, China has achieved tremendous growth in AI (10X range). The influence of AI on agriculture in terms of crop productivity and creation of AI in research have contributed positively (please see Fig. 1.20).

For other countries that are mentioned in the AI readiness list, I will cover in later chapters.

---

[62] Agriculture, forestry, and fishing, value added (annual % growth) – Agriculture, forestry, and fishing, value added (annual % growth) – https://data.worldbank.org/indicator/NV.AGR.TOTL.KD.ZG?end=2017&locations=1W-AD&start=1981&view=chart

[63] API_NV.AGR.TOTL.KD.ZG_DS2_en_excel_v2_2063913.XLS – https://api.worldbank.org/v2/en/indicator/NV.AGR.TOTL.KD.ZG?downloadformat=excel

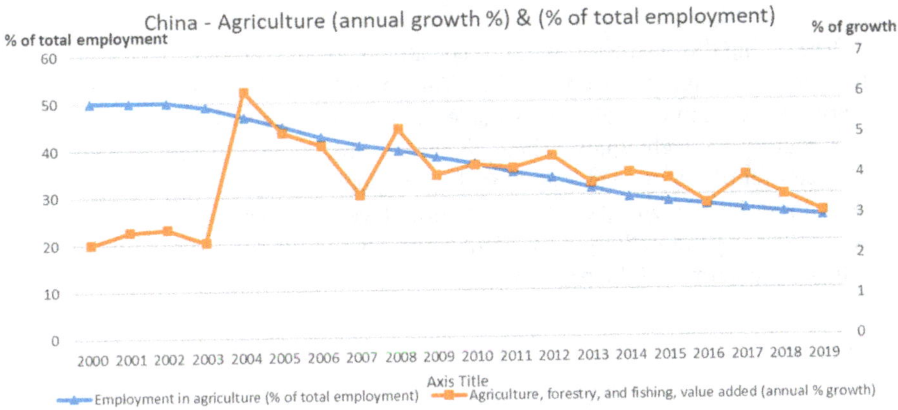

Fig. 1.20 Agriculture growth & % of total employment

Call for Policy Makers and Government Agencies!

A mix of short- and long-term factors are contributing to higher food prices. On long-term trends level, one reason increasing food prices is reduced agricultural research and development by government and international institutions. Although private sector funding of research has grown, private sector research has generally focused on innovations that private companies could sell to producers. These have often been cost reducing rather than yield enhancing technological developments. Focusing on yield enhancing agricultural techniques with the mix of information communication technologies (ICT) could lead a sustained and long-term stable agricultural development that world is desperately looking and get embraced with. Technologies such as Internet of Things (IoT), low-cost sensor development, smart agricultural technologies, and machine learning and artificial technologies are the enablers of future proof and data-centric technologies that not only address food yield requirements but also help us to fight climate change and global warming.

Farmers the world over, in dealing with costs, returns, and risks, are calculating economic agents. Within their small individual, allocative domain, they are fine-tuning entrepreneurs, tuning so subtly that many experts fail to see how efficient they are[64] [43].

[64] Relationship between milk production and price variations in the EC – http://aei.pitt.edu/36837/1/A67.pdf

After reading this chapter, you should be able to answer artificial intelligence, machine learning, and deep learning. You should be able to answer types of learning and should be able to evaluate performance metrics of classification and regression techniques. Finally, we are able to establish relationship between set of macrolevel indicators f (rural population, life expectancy, foreign direct investment, the rate of inflation, level of exports of all goods and services) and {the *ratio of agricultural exports to agricultural imports*} significantly determined the percentage contribution of agriculture to GDP.

References

1. Goedde, L., Katz, J., Ménard, A., Revellat, J.: Agriculture's connected future: How technology can yield new growth. https://www.mckinsey.com/industries/agriculture/our-insights/agricultures-connected-future-how-technology-can-yield-new-growth (2020, October 9). Accessed 18 Sept 2020
2. Walch, K.: How AI Is Transforming Agriculture. https://www.forbes.com/sites/cognitiveworld/2019/07/05/how-ai-is-transforming-agriculture/?sh=7ce84c004ad1 (2019, Jul 5). Accessed 8 Jan 2021
3. Brennan, E.: Agricultural Intelligence: What AI Can Do for Smallholder Farmers.https://foodtank.com/news/2018/10/agricultural-intelligence-what-ai-can-do-for-smallholder-farmers/ (2018, October). Accessed 20 Jan 2021
4. Columbus, L.: 10 Ways AI Has The Potential To Improve Agriculture In 2021. https://www.forbes.com/sites/louiscolumbus/2021/02/17/10-ways-ai-has-the-potential-to-improve-agriculture-in-2021/?sh=7f611e297f3b (2021, Feb 17). Accessed 10 Mar 2021
5. Medeiros, M.: Artificial intelligence and the future, https://www.nortonrosefulbright.com/en-br/knowledge/publications/6400e1ea/artificial-intelligence-and-the-future%20The%20World%20Bank%20-%20Agriculture,%20forestry (2018, April). Accessed 14 Feb 2021
6. Kumar, P.: Artificial Intelligence: Reshaping Life and Business, Chapter 2, BPB Publications; 1st edition (2019, August 17). ISBN-10: 9388511077
7. West, D.M.: What is artificial intelligence?. https://www.brookings.edu/research/what-is-artificial-intelligence/ (2018, October 4). Accessed 22 Nov 2020
8. Faggella, D.: What is Artificial Intelligence? An Informed Definition. https://emerj.com/ai-glossary-terms/what-is-artificial-intelligence-an-informed-definition/#:~:text=In%20%E2%80%9CArtificial%20Intelligence%3A%20A%20Modern,many%20different%20splintered%20fields%20%E2%80%93%20speech (2018, December 21). Accessed Oct 2020
9. Stuart, R., Norvig, P.: Artificial Intelligence: A Modern Approach (3rd edition), Pearson (2009, December 11). ISBN-10: 0136042597
10. Taddy, M.: Business Data Science: Combining Machine Learning and Economics to Optimize, Automate, and Accelerate Business Decisions 1st Edition, McGraw-Hill Education (2019, August 21). ISBN-10: 1260452778
11. Han, J., Kamber, M., Pei, J.: Data Mining: Concepts and Techniques, Morgan Kaufmann; 3rd edition (2011, June 15). ISBN-10: 9780123814791
12. Janakiram, M.S.V.: Gartner's 2020 Magic Quadrant For Data Science And Machine Learning Platforms Has Many Surprise. https://www.forbes.com/sites/janakirammsv/2020/02/20/gartners-2020-magic-quadrant-for-data-science-and-machine-learning-platforms-has-many-surprises/?sh=18b2268c3f55 (2020, Feb 20). Accessed 8 Jan 2021
13. LeDell, E.: The different flavors of AutoML. https://www.h2o.ai/blog/the-different-flavors-of-automl/ (2018, August 15). Accessed 8 Jan 2021

14. Parikh, A.: An econometric model of the agricultural sector of the Indian economy. J. Policy Model. **4**(3), 395–411 (1982). https://doi.org/10.1016/0161-8938(82)90027-8. ISSN 0161-8938
15. Vesset, D.: Descriptive analytics 101: What happened? https://www.ibm.com/blogs/business-analytics/descriptive-analytics-101-what-happened/ (2018, May 10). Accessed 30 Apr 2019
16. Chen, Y., Pechyoni, D., Taylor, A., Paunic, V.: Recurrent neural networks for time series forecasting. https://conferences.oreilly.com/artificial-intelligence/ai-eu-2018/public/schedule/detail/70226 (2018, October 9). Accessed 30 Apr 2019
17. Vesset, D.: Prescriptive analytics 101: What should be done about it? https://www.ibm.com/blogs/business-analytics/prescriptive-analytics-done/ (2018, May 15). Accessed 18 Sept 2019
18. Vuppalapati, J.S., Kedari, S., Ilapakurti, A., Vuppalapati, C., Kedari, S., Vuppalapati, R.: The Development of Machine Learning Infused Outpatient Prognostic Models for tackling Impacts of Climate Change and ensuring Delivery of Effective Population Health Services. 2019 IEEE International Conference on Big Data (Big Data), Los Angeles (2019), pp. 2790–2799. https://doi.org/10.1109/BigData47090.2019.9006570
19. Diller, G.-P., Kempny, A., Babu-Narayan, S.V., Henrichs, M., Brida, M., Uebing, A., Lammers, A.E., Baumgartner, H., Li, W., Wort, S.J., Dimopoulos, K., Gatzoulis, M.A.: Machine learning algorithms estimating prognosis and guiding therapy in adult congenital heart disease: data from a single tertiary centre including 10 019 patients. Eur. Heart J. **40**(13), 1069–1077 (2019). https://doi.org/10.1093/eurheartj/ehy915
20. Hwang, K., Chen, M.: Big-Data Analytics for Cloud, IoT and Cognitive Computing, Wiley; 1st edition (2017, August 14). ISBN-10: 9781119247029
21. Yang, X.-S., Sherratt, S., Dey, N., Joshi, A., ICICT 2019, London, Volume 2, Fourth International Congress on Information and Communication Technology (2019), eBook ISBN 978-981-329-343-4 and Softcover ISBN 978-981-329-342-7
22. Brownlee, J.: A Tour of Machine Learning Algorithms. https://machinelearningmastery.com/a-tour-of-machine-learning-algorithms/ (2019, August 12). Accessed 18 Sept 2019
23. Li, S.: Building A Logistic Regression in Python, Step by Step. https://towardsdatascience.com/building-a-logistic-regression-in-python-step-by-step-becd4d56c9c8 (2017, Sep 28). Accessed 22 Nov 2019
24. Morde, V.: XGBoost Algorithm: Long May She Reign. https://towardsdatascience.com/https-medium-com-vishalmorde-xgboost-algorithm-long-she-may-rein-edd9f99be63d (2019, April 7). Accessed 18 Sept 2019
25. de Ville, B., Neville, P.: Decision Trees for Analytics: Using SAS Enterprise Miner. SAS Inst. (2013, July 10) ISBN-10: 1612903150
26. Wolpert, D., Ampadu, P.: Managing Temperature Effects in Nanoscale Adaptive Systems, Springer; 2012 edition (2011, August 31), ISBN-10: 1461407478
27. Vuppalapati, C., Vuppalapati, R., Kedari, S., Ilapakurti, A., Vuppalapati, J.S., Kedari, S.: Fuzzy Logic Infused Intelligent Scent Dispenser For Creating Memorable Customer Experience of Long-Tail Connected Venues. 2018 International Conference on Machine Learning and Cybernetics (ICMLC), Chengdu, pp. 149–154. https://doi.org/10.1109/ICMLC.2018.8527046, http://ieeexplore.ieee.org/stamp/stamp.jsp?tp=&arnumber=8527046&isnumber=8526918 (2018). Accessed 30 Apr 2019
28. Mendes-Moreira, J., Soares, C., Jorge, A.M., De Sousa, J.F.: Ensemble approaches for regression: A survey. ACM Comput. Surv. **45**(1), Article 10 (2012, November), 40 pages. https://doi.org/10.1145/2379776.2379786
29. Dua, D., Graff, C.: UCI Machine Learning Repository. University of California, School of Information and Computer Science, Irvine, CA. http://archive.ics.uci.edu/ml (2019)
30. Kenton, W., Reviewed by Khartit, K.: Endogenous Variable, Updated Oct 30, 2020. https://www.investopedia.com/terms/e/endogenous-variable.asp#:~:text=An%20endogenous%20variable%20is%20a,within%20the%20system%20being%20studied (2020). Accessed 8 Jan 2021

31. Shalev-Shwartz, S., Ben-David, S.: Understanding Machine Learning: From Theory to Algorithms, Cambridge University Press; 1st edition (2014, May 19). ISBN-10: 1107057132
32. Porter, M.E.: How Competitive Forces Shape Strategy. https://hbr.org/1979/03/how-competitive-forces-shape-strategy (1979, March). Accessed 18 Sept 2020
33. Greenfield, K.: Corporations Are People Too: (And They Should Act Like It) Hardcover, Yale University Press (2018, October 23). ISBN-13: 978-0300211474
34. Pepper, R.: High Tech Policy – The Paradox of ICT's Impact on Growth and Inequality. https://blogs.cisco.com/gov/gitr2015 (2015, April 14). Accessed 8 Nov 2020
35. Pepper, R., Garrity, J.: ICTs, Income Inequality, and Ensuring Inclusive Growth. https://alln-extcloud-storage.cisco.com/ciscoblogs/GITR-2015-Cisco-Chapter.pdf (2015). Accessed 8 Nov 2020
36. Krishnan, M., Mischke, J., Remes, J.: Is the Solow Paradox back? https://www.mckinsey.com/business-functions/mckinsey-digital/our-insights/is-the-solow-paradox-back (2018, June 4). Accessed 8 Nov 2020
37. Vuppalapati, C.: Democratization of Artificial Intelligence for the Future of Humanity 1st Edition, CRC Press; 1st edition (2021, January 18). ISBN-13: 978-0367524098
38. Prahalad, C.K.: The Fortune at the Bottom of the Pyramid: Eradicating Poverty Through Profits, Revised and Updated 5th Anniversary Edition Hardcover, Wharton School Publishing; Revised edition (2009, October 12). ISBN-13: 978-0137009275
39. Smith, T.: As farmers go, so goes the nation How did agriculture help the US recover from the Great Depression in the spring of 1933? https://www.aeaweb.org/research/farmer-recovery-great-depression (2019, March 25). Accessed Mar 2020
40. Hausman, J.K., Rhode, P.W., Wieland, J.F.: Recovery from the great depression: The farm channel in Spring 1933. Am. Econ. Rev. **109**(2), 427–472 (2019). https://doi.org/10.1257/aer.20170237
41. Geoffrey, A.M.: Crossing the Chasm: Marketing and Selling Disruptive Products to Mainstream Customers (Collins Business Essentials), Harper Business; 3rd edition (January 28, 2014), ISBN-13: 978-0062353948
42. Bughin, J., Seong, J., Manyika, J., Chui, M., Joshi, R.: McKinsey Global Institute (MGI) – Notes from the AI frontier Modeling the impact of AI on the world economy. https://www.mckinsey.com/~/media/McKinsey/Featured%20Insights/Artificial%20Intelligence/Notes%20from%20the%20frontier%20Modeling%20the%20impact%20of%20AI%20on%20the%20world%20economy/MGI-Notes-from-the-AI-frontier-Modeling-the-impact-of-AI-on-the-world-economy-September-2018.ashx (2018, September). Accessed 8 Nov 2020
43. Schultz, R.W.: Relationship between milk production and price variations in the EC. http://aei.pitt.edu/36837/1/A67.pdf (1978). Accessed 14 Feb 2021

Chapter 2
Data Engineering and Exploratory Data Analysis Techniques

This chapter covers:

- Knowledge Discovery in Databases (KDD) framework.
- Agricultural data sources.
- Data types.
- Data mining application processing architectures.
- Exploratory data analysis (EDA).
  - Classification/regression.
  - Time series datasets.
- Feature reduction techniques.
- Imbalanced framework.
- Explainability of model.

This chapter introduces data sources and data engineering attributes for handling agricultural datasets. The chapter starts with Knowledge Discovery in Databases (KDD) framework, the reference architecture for data mining applications, feature engineering techniques, and exploratory data analysis (EDA) process. Next, the chapter introduces feature reduction techniques such as principal component analysis (PCA). In agriculture data, most of the time, data for all important features may not be available and predictor class may not be fully populated. The chapter introduces imbalanced class techniques and demonstrates the workings. Finally, the chapter concludes with explainability of ML model and the need for democratizing the artificial intelligence.

C. Vuppalapati, *Machine Learning and Artificial Intelligence for Agricultural Economics*, International Series in Operations Research & Management Science 314,
https://doi.org/10.1007/978-3-030-77485-1_2

The demand for good-quality statistical data continues to increase. Timely and reliable statistics are key inputs to the broad development strategy not only for nations of the world but also very critical for commercial enterprises. At the same token, good-quality statistical data is important for small farmers and agricultural businesses. Improvements in the quality and quantity of data on all aspects of development are essential if we are to achieve the goal of a world without poverty[1] and reaching sustainability objectives. Good data are needed to set baselines, identify effective public and private actions, set goals and targets, monitor progress, and evaluate impacts. They are also an essential tool of good government, providing means for people to assess what governments do and helping them to participate directly in the development process.

To achieve good public policy and real-time information dissemination to farmers that is meaningful and relevant, the US Department of Agriculture (USDA)-National Agricultural Statistics Service (NASS) conducts two types of agricultural data collection programs[2]—surveys and census programs. The combined survey and census data offers around 45 million data points (45,812,611) with 28,091,806 census records and 17,720,805 survey records. And the data gets updates on a weekly, monthly, and yearly basis. As of publication of book, March 30, 2021, the USDA NASS database offers 45 million data. The other important data, real-time acquisition data using sensors and SCADA systems, are in initial phases and needed to augment with survey/census data to provide a wholistic 360° view on farm operations and productivity.

Census Data Characteristics.[3]

The census of agriculture is the only source of detailed county-level agricultural data collected, tabulated, and published using a uniform set of definitions and methodology for 3200 plus counties in the United States. The census collects data on all commodities produced in the United States as well as detailed information on expenses, income, and operator characteristics. This census is conducted in all 50 states on a target population of all farms and ranches selling or intending to sell $1000 or more of agricultural products including horticulture. A census of agriculture is also conducted in Puerto Rico using a farm definition of $500 or more in sales and intentions.

Survey Data Characteristics.

The NASS conducts hundreds of surveys each year on issues including agricultural production, economics, demographics, and the environment (please see Fig. 2.1). Every 5 years NASS also conducts the Census of Agriculture, providing the only source of uniform, comprehensive agricultural data for every county in the nation.[4]

---

[1] The World Bank – https://data.worldbank.org/about

[2] USDA NASS – https://quickstats.nass.usda.gov/

[3] Census Survey Data – https://www.nass.usda.gov/Surveys/Guide_to_NASS_Surveys/Census_of_Agriculture/index.php

[4] NASS Surveys – https://www.nass.usda.gov/Surveys/index.php

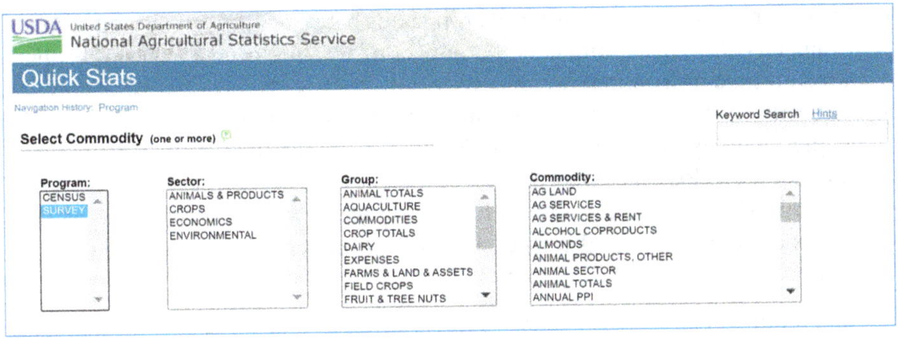

Fig. 2.1 USDA NASS Quick Stats Tool

Engineering systems that host sensors and IOT systems provide real-time sub-milliseconds or hourly data—for instance, reservoir data feeds at the state level! The data collected via survey or real time is subjected to standard data processing to extract valuable insights from it—KDD plays an important role.

Microaggregation

USDA NASS collects and publishes data about farms—To protect against data disclosure, data is only released at the county level. farms in each county are averaged together to maintain as much purity, yet still protect against disclosure.

In the following diagram:

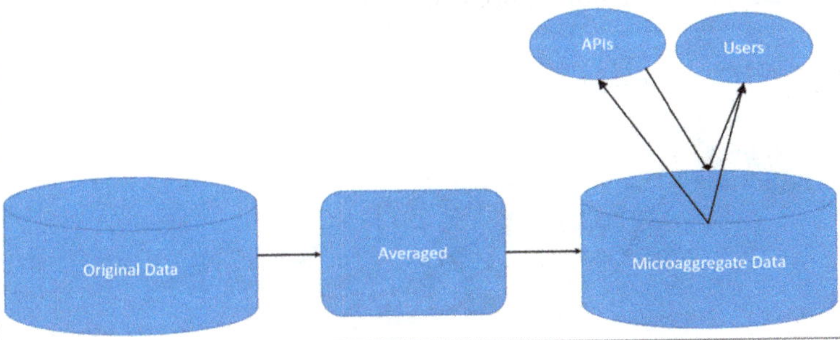

Knowledge Discovery in Databases (KDD)

Data mining consists of the use of algorithms that are directed at one, or more, of four basic areas of exploration [1]. These are:

1. Classification: the ability to classify a record or case as either belonging to a certain group or of taking on a particular value. For example, in the wine dataset, classifying a wine type based on important features.
2. Segmentation: the ability to create groups (segments or clusters) from a set of data. The groups may be created using either continuous or discrete measures, or

a mix of these measures. Cluster of crop yields based on farm inputs and climatic parameters. Another example is the impact of weather events and cluster them based on time or location!

3. Association: the ability to understand activities that may occur together (e.g., people who buy bread, also buy milk) or in sequence (e.g., people download a song, the download the entire CD).
4. Anomaly Detection: the ability to find those items, which do not fit the pattern. This is often used in fraud detection in agricultural commodity pricing or in the identification of aberrant activities.

Data mining is an interactive and iterative process.[5] Business expertise, domain experts, agricultural data scientists, and farmers must be used jointly with advanced technologies to identify underlying relationships and features in the data to uncover data signatures and to confirm the general truths of the data. For instance, with the drop in crude oil prices[6] and lower export prices for the major agricultural commodities (please see Fig. 2.2) over the same period, of course stronger/weaker dollar impacts the relationship. One can see such signature in the data [2].

A seemingly useless pattern in data discovered by data mining technology can often be transformed into a valuable piece of actionable information using business experience and expertise. The success of data mining is dependent upon teaming of relevant multi-disciplinary experts solving a particular business need.

Data mining as a step in the KDD [1] (please see Fig. 2.3):

1. Data cleansing (to remove noise and inconsistent data).

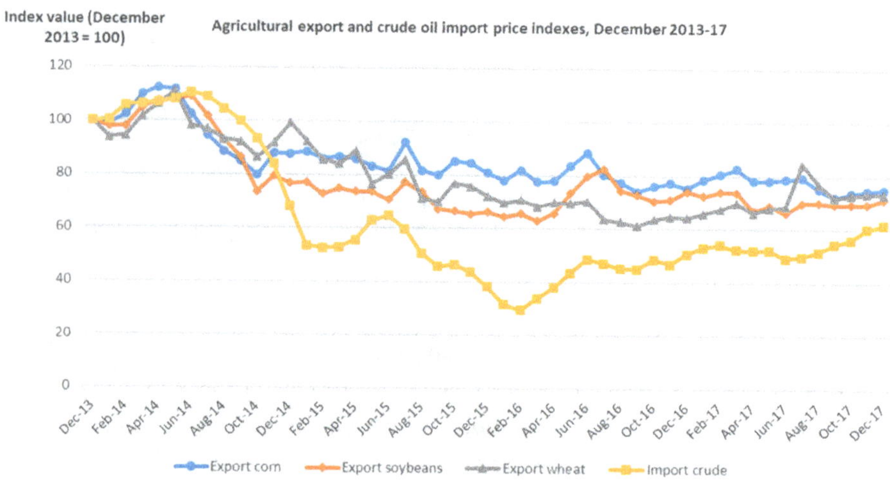

Fig. 2.2 Agricultural exports - Source: USDA and US Bureau of Labor Statistics

---

[5] Data Mining: Concepts and Techniques (The Morgan Kaufmann Series in Data Management Systems), Morgan Kaufmann; 3rd edition (July 6, 2011), ISBN-13 : 978-9380931913

[6] The relationship between crude oil prices and export prices of major agricultural commodities – https://www.bls.gov/opub/btn/volume-8/the-relationship-between-crude-oil-and-export-prices-of-major-agricultural-commodities.htm?view_full

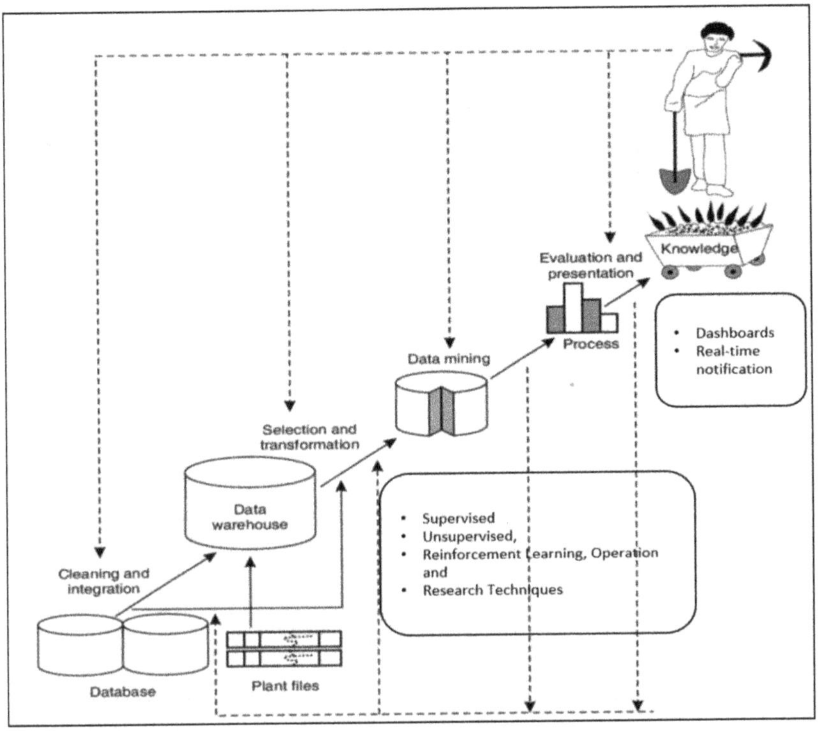

Fig. 2.3 KDD

2. Data integration (where multiple data sources are combined).
3. Data selection (where relevant features from the data are retrieved for analysis task).
4. Data transformation (where data are transformed or consolidated into forms appropriate for mining by performing summary or aggregation operations, for instance).
5. Data mining (where mathematical and statistical methods are applied to retrieve valuable information).
6. Pattern evaluation (to identify true insights and data signatures).
7. Knowledge presentation (generate recommendations and/or send just-in-time or real-time notifications).

The KDD is the most important process for setting and running machine learning experiment. The following is simplier representation of the KDD[7] (please see Fig. 2.4 and Table 2.1):

[7] Overview of the KDD Process – http://www2.cs.uregina.ca/~dbd/cs831/notes/kdd/1_kdd.html

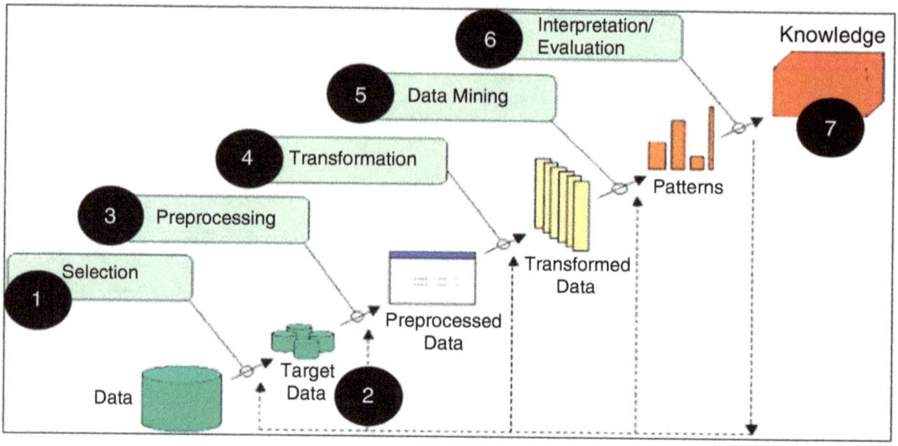

Fig. 2.4 KDD in detail

Table 2.1 KDD steps

1. Selection	Develop an understanding of The application domain The relevant prior knowledge The goals of the end user
2. Target data	Creating a target dataset: Selecting a dataset, or focusing on a subset of variables, or data samples, on which discovery is to be performed
3. Preprocessed data	Data cleaning and preprocessing Removal of noise or outliers Collecting necessary information to model or account for noise Strategies for handling missing data fields Accounting for time sequence information and known changes
4. Transformation	Data reduction and projection Finding useful features to represent the data depending on the goal of the task Using dimensionality reduction or transformation methods to reduce the effective number of variables under consideration or to find invariant representations for the data
5. Choosing data mining algorithm	Selecting method(s) to be used for searching for patterns in the data Deciding which models and parameters may be appropriate Matching a particular data mining method with the overall criteria of the KDD process
6. Data mining	Searching for patterns of interest in a particular representational form or a set of such representations as classification rules or trees, regression, clustering, and so forth
7. Interpretation/ evaluation	Interpreting mined patterns
8. Knowledge	Consolidating discovered knowledge

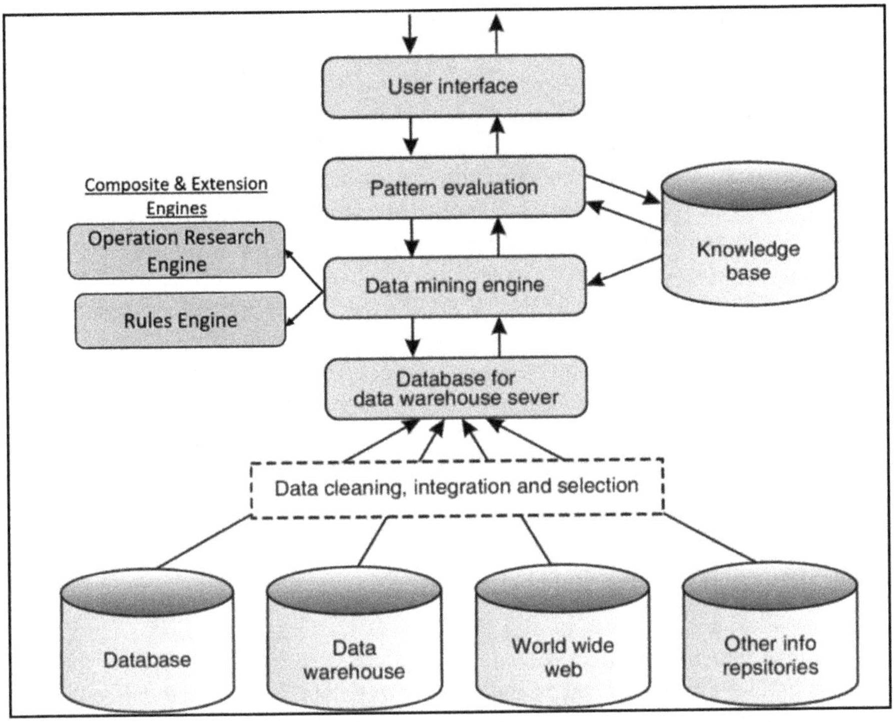

Fig. 2.5 KDD data mining

Architecture of a Data Mining System

The architecture of data mining system generally uses high-performance compute to execute the data pipeline to generate insights from the data [1].

The above is the most common architecture pattern of the data mining system. At the root of the system is data sources. Though represented as a bottom tier in the architecture stack, the data loads and frequencies are key aspects of the above or north bound components—in this case, Central Data Store, Rules Engine, and Visualization (please see Fig. 2.5).

Data sources are picked as per the business use case or analytics requirements. For instance, if we were to establish the impact of climate change on the rice yields in California, especially Sacramento Valley use case, we pick data sources that include Rice Yields in California (USDA National Agricultural Statistical Services (NASS[8])), NOAA Storm Events Database,[9] California Climatic Data (Temperature,

[8] USDA NASS – https://quickstats.nass.usda.gov/

[9] NOAA Storm Events Database – https://www.ncdc.noaa.gov/stormevents/choosedates.jsp?statefips=6%2CCALIFORNIA

Humidity, Precipitation, Drought Index),[10] and Agriculture Calendar (USDA Foreign Services[11]).

Next is application of exploratory data analysis (EDA) to filter out nulls, missing values, and cleaning of the data. Data mining engine is the next step. The EDA prepares data that is suitable for running statistical and ML algorithms. For instance, to develop prediction model for rice, given time series data, vector auto regression (VAR) models, and ensemble ML models are applied. Generally, for multivariate time series analysis of the data, identification of causation models (granger causality tests[12]) is performed! Operation Research Techniques, such as constraint programming[13] – Mixed Integer Programming (MIP),[14] KNAPSACK,[15] are applied based on the use case and optimization need! The rules engine is more of a validator and applicable if data processing is constrained by business or process rules. Pattern evaluation runs the rules on the transformed data and generates insights that can be dashboard or stored in database for further usage either by sending recommendation or generating reports! Please find the translated data mining for rice production (please see Fig. 2.6)!

Agricultural Time Series Data

Agricultural data includes macroeconomic data, field crops data, productivity, labor, governmental policies/ordinances, and commodity data. Most of the data, all most all, has the *time series signature*. That is, data is described with respect to time as a reference. Put it simply, data characteristics are presented over a timeline. For example, the World Bank[16] dataset contains collection of time series data on variety of topics to map progress and has predictability features for modeling machine learning and artificial intelligence.

Another valuable dataset is from Food and Health Organization (FAOSTAT[17]) that provides comparable performance and intrinsic indicators (time series) on crop production, prices, emissions agriculture, trade, inputs, population, macro statistics, food security, agri-environmental indicators, and others (please see Fig. 2.7). The following graph is a rice paddy exports from world's largest rice exporters[18]

[10] CA Climatic Data – https://data.noaa.gov/dataset/dataset/noaas-climate-divisional-database-nclimdiv/resource/f37ce025-7dc9-40a7-a28c-b9e420571166

[11] USDA CROP Calendar – https://ipad.fas.usda.gov/rssiws/al/crop_calendar/us.aspx

[12] granger causality tests – https://www.statsmodels.org/stable/generated/statsmodels.tsa.stattools.grangercausalitytests.html

[13] Google OR Tools – https://developers.google.com/optimization/examples

[14] Mixed Integer Programming – https://developers.google.com/optimization/mip/integer_opt

[15] KNAPSACK – https://developers.google.com/optimization/bin/knapsack

[16] The World bank Dataset – https://data.worldbank.org/

[17] FAOSTAT – http://www.fao.org/faostat/en/#data

[18] Rice exports – http://www.fao.org/faostat/en/#compare

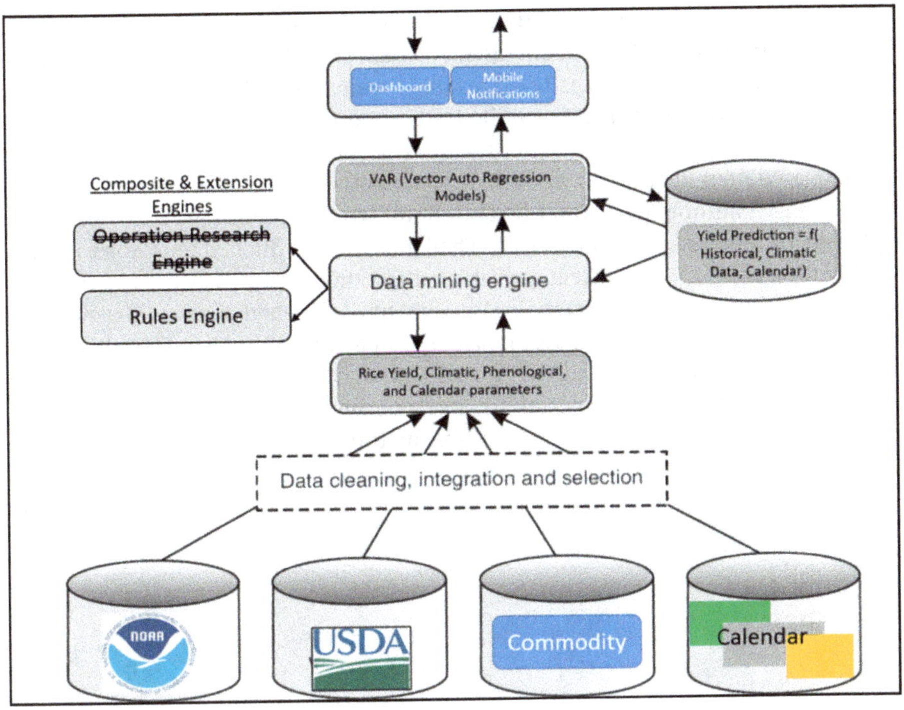

Fig. 2.6 Rice production—KDD

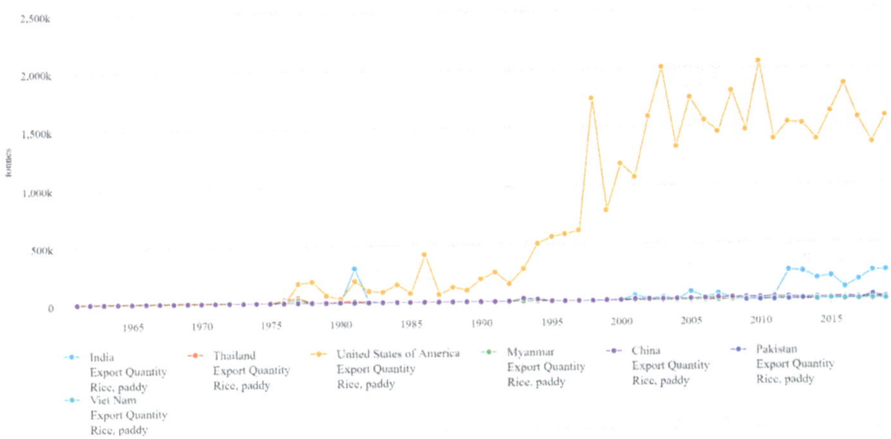

Fig. 2.7 Rice paddy exports

(1961–2019). From the trend analysis of the data, we can see the United States is consistently leading rice paddy exports since the mid-1970s and achieved over 2000 kilotons export during the early 2000s.

Another category in rice exports is milled rice. Milled rice[19] is the type most often found in stores for normal cooking. To reach milled rice category, rice kernels have been milled to remove the bran, leaving "white" or milled rice. Milled rice is also referred as polished rice. For milled rice exports, the time series has different countries and values.[20]

India exhibited huge jump in milled rice exports during 2010 and consistently maintaining 9 million tons export level (9.8 million metric tons as of 2018/2019[21]) followed by consistent exports from Thailand (7.56 million metric tons of rice worldwide), exhibited consistent growth over last three decades beginning in 1975. Rice consumption, a staple cereal across many Asian and the world countries, between 2008 and 2019, global rice consumption increased by over 50 million metric tons, from 437 million metric tons to around 490 million metric tons. In 2018/2019, the top three countries for rice consumption were China, India, and Indonesia. That year, nearly 143.8 million metric tons of rice was consumed in China (please see Fig. 2.8).

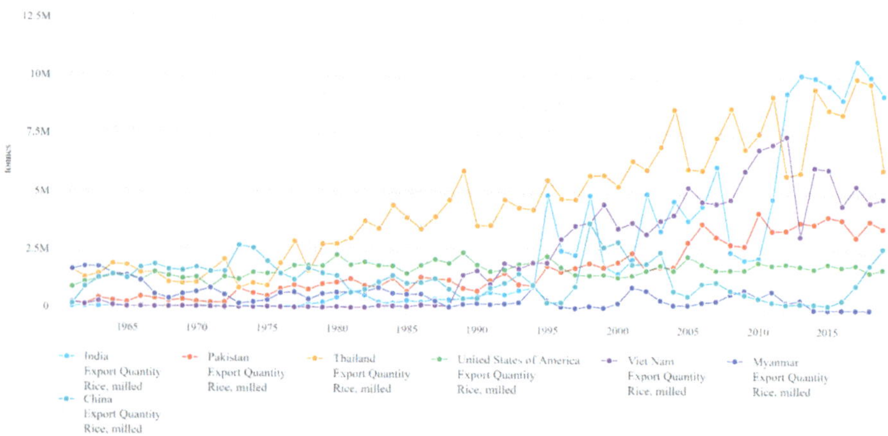

Fig. 2.8 Rice, milled exports

[19] Milled rice – https://www.ams.usda.gov/book/milled-rice

[20] Sources – http://www.fao.org/faostat/en/#compare

[21] Principal rice exporting countries worldwide in 2018/2019(in 1000 metric tons) – https://www.statista.com/statistics/255947/top-rice-exporting-countries-worldwide-2011/

Wheat Exporters[22]

Foreign exporters chipped away at the US share of the international market between 1980 and 1984; noncompetitive US prices and the strong dollar were factors in the erosion, although federal agricultural policies and aggressive marketing by competitors abroad also played a part[23] [3]. Australia's drought-reduced crop (please see Fig. 2.9) prevented it from providing much competition in markets such as China and Indonesia. It looks like it could be harvesting a similar-sized crop this year with lower carry-in from the previous harvest [4]. Time series also holds the signature of governmental subsidy and tax policies. For instance, Russia has instrumented wheat export tax policy to control high prices in domestic markets.[24]

The World Bank Data Catalog[25] includes time series, microdata, and geospatial data (please see Fig. 2.10).

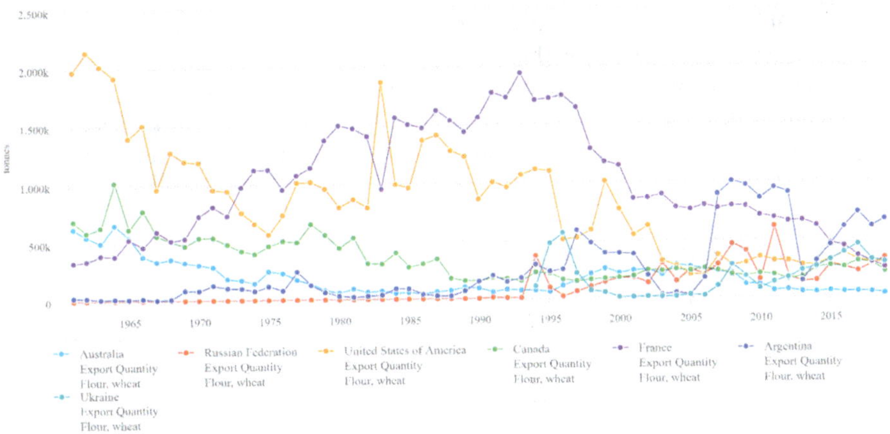

Fig. 2.9 Wheat exports

---

22 Wheat exports by country – http://www.worldstopexports.com/wheat-exports-country/

23 How US exports are faring in the world wheat market – https://www.bls.gov/opub/mlr/1985/10/art2full.pdf

24 GRAINS-WHEAT WEAKENED BY PROFIT-TAKING AS RUSSIAN EXPORT TAX ANNOUNCED–https://www.agriculture.com/markets/newswire/grains-wheat-weakened-by-profit-taking-as-russian-export-tax-announced

25 The World Bank Data Catalog – https://datacatalog.worldbank.org/

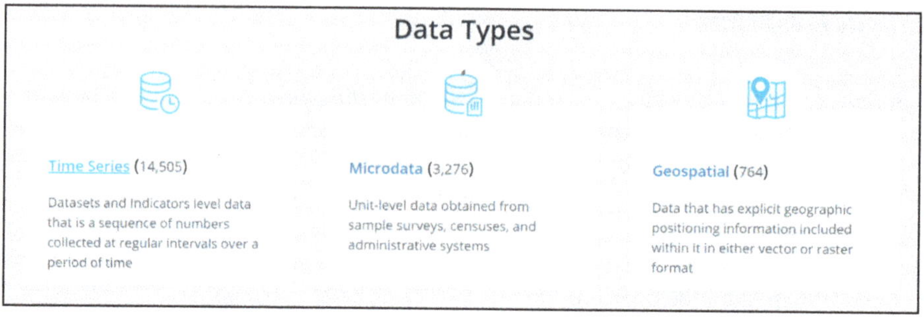

Fig. 2.10 The World Bank data types

Reduce Enteric Fermentation through New Technologies [5]
IoT sensors play an important role in identifying and providing information to Dairy Farmers on Rumination. Cow necklace sensors collect rumination threshold count that are indicators of cow health.[26] Using threshold counts, we can monitor and control agricultural greenhouse gas emissions. For instance, ruminant livestock were responsible for around half of all agricultural production emissions in 2010. Of these emissions, the largest source is "enteric methane," or cow burps. Increasing productivity of ruminants also reduces methane emissions, mainly because more milk and meat is produced per kilogram of feed. In addition, new technologies can reduce enteric fermentation.

Agricultural Data Frequencies

Agricultural data is collected at various frequencies, depending on the use case.

Hourly

Hourly data is collected for real-time supervisory control and data acquisition (SCADA) and sensor devices. Generally, these devices emit data in seconds and/or sub-milliseconds intervals. And the latency needs are faster processing and closed loop for notifications. To meet the performance requirements, the processing of the data is performed at source—edge analytics. Another agricultural sources that need real-time or near real-time collection is dairy and animal husbandry to observe cattle behavior to predict health and behavior analytics. An example of hourly data collection is from California reservoirs systems[27] (please see Fig. 2.11):

[26] Rumination – http://hanuinnotech.com/dairyanalytics.html

[27] Active Hourly Reservoirs – https://cdec.water.ca.gov/reportapp/javareports?name=HourlyRes

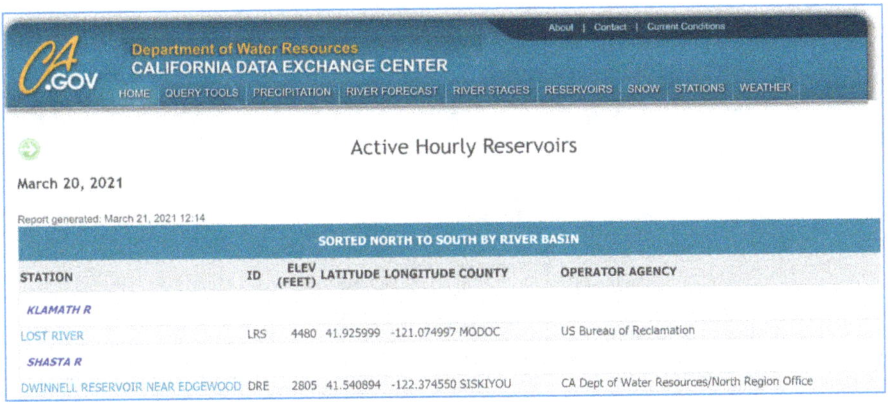

Fig. 2.11 California reservoir

The data generated by IoT sensors contain valuable information about the device, the environment, device health, diagnostics, state of firmware, state machine, and more. Generally, the data is time series with events spaced at an interval. Some of the devices might generate the data samples multiple times per second, 24 h a day, and 365 days a year. The volume of data per day might be very high. For instance, if a sensor is generating data at every microsecond level by the end of the day, the sensor would have produced 86.4 million (24 × 60 × 60 × 1000) data samples (please see Fig. 2.12). Not only is the data huge, but the velocity is also very rapid. The basic principle of IoT is that the intelligent system initiates and processes the information as early and as close to the source as possible [6].

Daily

Data frequencies are inherently dependent on the intrinsic nature of data, its report period, and its collection frequencies. Most of the macrolevel economic indicators, generally, are daily, monthly, quarterly, half yearly, and yearly.

7-Year Treasury Constant Maturity Rate,[28] 10-Year Treasury Constant Maturity Rate (DGS10),[29] gold prices, and silver prices are some of the macrolevel economic indicator prices that are set daily. The London Gold Fixing Companies set the prices for gold that are globally considered as the international standard for pricing of gold. The gold price in London is set twice a day by five LBMA Market Makers who comprise the London Gold Market Fixing Limited (LGMFL). Silver prices are set by London Silver Market Fixing Limited (LSMFL)[30] (please see Fig. 2.13).

---

[28] 7-Year Treasury Constant Maturity Rate (DGS7) – https://fred.stlouisfed.org/series/DGS7

[29] 10-Year Treasury Constant Maturity Rate (DGS10) – https://fred.stlouisfed.org/series/DGS10

[30] Silver Price: London Fixing – https://www.quandl.com/data/LBMA/SILVER-Silver-Price-London-Fixing

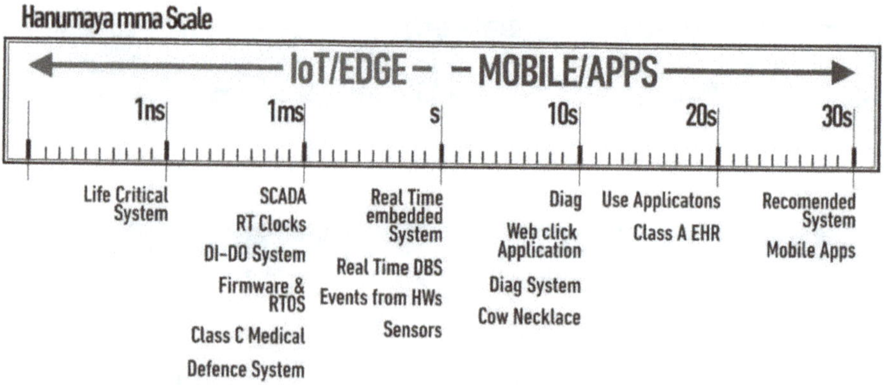

Fig. 2.12 IoT scale (edge level) [6]

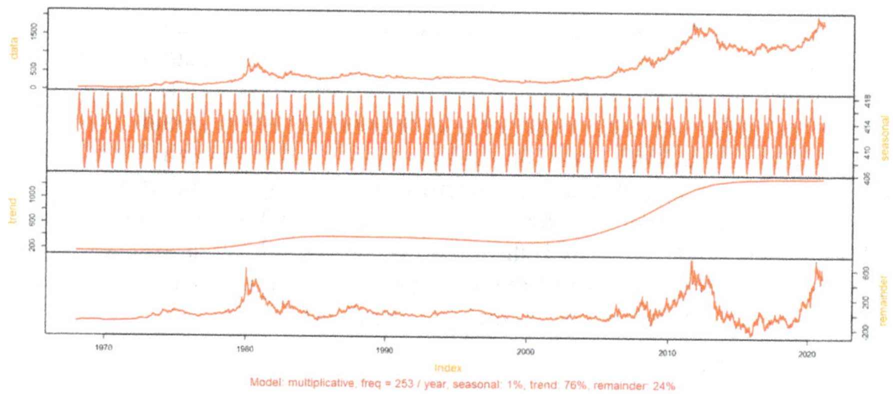

Fig. 2.13 LBMA gold prices

Weekly

Weekly frequencies are common in agricultural datasets. USDA-NASS[31] crop progress is weekly data. The USDA/NASS surveys crop progress and condition based on questionnaires returned from more than 5000 reporters and publishes percent complete (area ratio) of crop fields that have either reached or completed a specific phenological stage,[32] on ASD or state level, in a weekly report called the Crop Progress Report (NASS-CPR) [7]. Phenological stages are dictated by nature's calendar, and it answers when corn seed turns to harvest. Another example of phenology[33] includes when cherry trees bloom, when a robin builds its nest, and when leaves turn color in the fall.

[31] USDA-NASS – https://www.nass.usda.gov/

[32] MODIS-based corn grain yield estimation model incorporating crop phenology information – https://calmit.unl.edu/people/agitelson2/pdf/2013/RSE2013_MODIS_Corn_Yield.pdf

[33] Phenology – https://www.usanpn.org/about/why-phenology

For example, the following details are related to CORN CROP with weekly PCT complete (please see Fig. 2.14). The keywords selected as search criteria for the database were "SURVEY" for program, "CROPS" for sector, "FIELD CROPS" for group, "CORN" for commodity, "PROGRESS" for category, "CORN, GRAIN-PROGRESS, MEASURED IN PCT HARVESTED" for data item, "TOTAL" for domain, and "STATE" for geographic level, respectively. The selected data period was from 1981 to 2020.[34]

Except for North Dakota, where survey or progress collected from Week#13 (week ending March 29,2019), the rest of the country, especially CORN belt, the progress is surveyed from Week#29 (week ending July 21, 2010) through Week #49 (December 08, 2019). For year 2020, the progress collected from Week #29 (week ending July 19, 2020) through Week #48 (week ending November 29, 2020) (please see Fig. 2.15). The weekending dates coincides with CORN mid-season as indicated in US Crop Calendar (please see Appendix).[35]

Monthly

Consumer price index (CPI) is reported monthly (please see Fig. 2.16). The CPI for all Urban Consumers (CPIAUCSL[36]) is a measure the average monthly change in the price of good or services paid by urban/city dwellers between any two periods. The CPI can be used to recognize periods of inflation and deflation. Significant

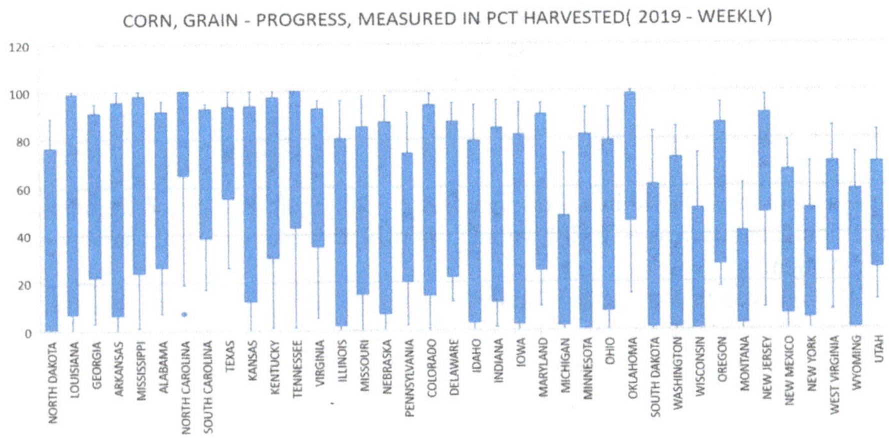

Fig. 2.14 Corn Progress—measured in PCT harvested (2019)

[34] NASS – https://quickstats.nass.usda.gov/#D1A9DE8C-2B9C-3B4F-BA88-1A4755ACE2C6

[35] US Crop Calendar – https://ipad.fas.usda.gov/rssiws/al/crop_calendar/us.aspx

[36] Consumer Price Index for All Urban Consumers: All Items in US City Average (CPIAUCSL) – https://fred.stlouisfed.org/series/CPIAUCSL

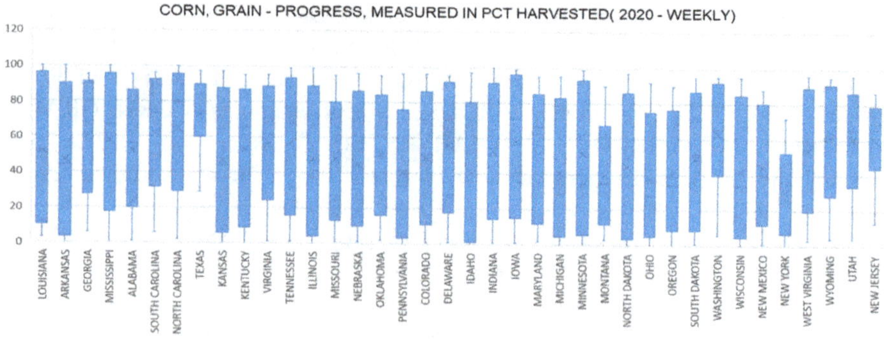

Fig. 2.15 Corn progress—measured in PCT (2020)

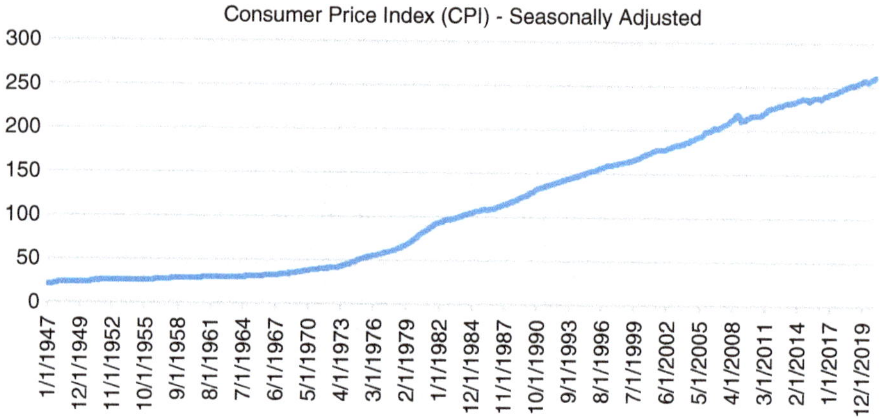

Fig. 2.16 Consumer price index

increases in the CPI within a short time frame might indicate a period of inflation, and significant decreases in CPI within a short time frame might indicate a period of deflation.

The international monetary fund[37,38] reports country indexes and weights on monthly basis (please see Fig. 2.17). The index items include consumer price index (CPI),[39] food and non-alcoholic beverage usage, alcoholic beverage and tobacco, and housing, water, electricity, gas, and other fuels.

[37] The International Monetary Fund (IMF) – https://data.imf.org/regular.aspx?key=61015892

[38] IMF Compare Countries – https://data.imf.org/?sk=4FFB52B2-3653-409A-B471-D47B46D904B5&sId=1485878802128

[39] IMF Data Access to Macroeconomic and Financial Data – https://data.imf.org/?sk=4FFB52B2-3653-409A-B471-D47B46D904B5&sId=1485878802128

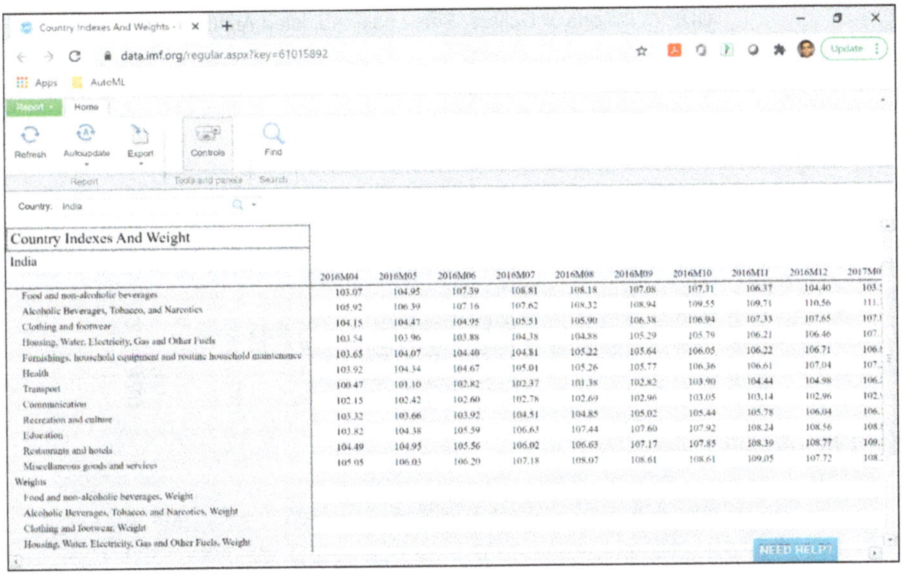

Country Indexes And Weight

India

	2016M04	2016M05	2016M06	2016M07	2016M08	2016M09	2016M10	2016M11	2016M12	2017M0
Food and non-alcoholic beverages	103.07	104.95	107.39	108.81	108.18	107.08	107.31	106.37	104.40	103.
Alcoholic Beverages, Tobacco, and Narcotics	105.92	106.39	107.16	107.62	108.32	108.94	109.55	109.71	110.56	111.
Clothing and footwear	104.15	104.47	104.95	105.51	105.90	106.38	106.94	107.33	107.65	107.
Housing, Water, Electricity, Gas and Other Fuels	103.54	103.96	103.88	104.38	104.88	105.29	105.79	106.21	106.46	107.
Furnishings, household equipment and routine household maintenance	103.65	104.07	104.40	104.81	105.22	105.64	106.05	106.22	106.71	106.
Health	103.92	104.34	104.67	105.01	105.26	105.77	106.36	106.61	107.04	107.
Transport	100.47	101.10	102.82	102.37	101.38	102.82	103.90	104.44	104.98	106.
Communication	102.15	102.42	102.60	102.78	102.69	102.96	103.05	103.14	102.96	102.
Recreation and culture	103.32	103.66	103.92	104.51	104.85	105.02	105.44	105.78	106.04	106.
Education	103.82	104.38	105.59	106.63	107.44	107.60	107.92	108.24	108.56	108.
Restaurants and hotels	104.49	104.95	105.56	106.02	106.63	107.17	107.85	108.39	108.77	109.
Miscellaneous goods and services	105.05	106.03	106.20	107.18	108.07	108.61	108.61	109.05	107.72	108.
Weights										
Food and non-alcoholic beverages, Weight										
Alcoholic Beverages, Tobacco, and Narcotics, Weight										
Clothing and footwear, Weight										
Housing, Water, Electricity, Gas and Other Fuels, Weight										

Fig. 2.17 The International Monetary Fund (IMF)—Country Indexes and Weights

Quarterly

The gross domestic product (GDP) reported quarterly. Gross domestic product (GDP), the featured measure of US output (please see Fig. 2.18), is the market value of the goods and services produced by labor and property located in the United States.[40]

Another example of quarterly reported data is employment statistics which are reported quarterly. Some of the agricultural data frequencies are monthly.

Yearly/Annual

Real Median Household Income in the United States,[41] Income and Poverty Thresholds[42] [8], and other macrolevel parameters are set annually (please see Fig. 2.19). Generally, these annual indicators provide guidance and policy inputs.

In the above graph, the weighted average poverty thresholds for individual under 65 are $13,011 and $12,261 for aged 65 and older. Poverty threshold for two people household under age 65 is $16,521, $17,622 for two people household under age 65 with one child.

[40] Gross domestic product (GDP) – https://fred.stlouisfed.org/series/GDP

[41] Real Median Household Income in the United States – https://fred.stlouisfed.org/series/MEHOINUSA672N

[42] Income and Poverty in the United States: 2019 – https://www.census.gov/data/tables/2020/demo/income-poverty/p60-270.html

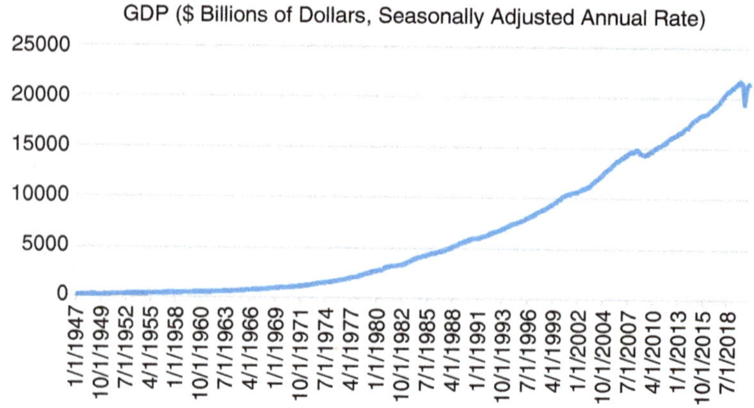

Fig. 2.18 US GDP

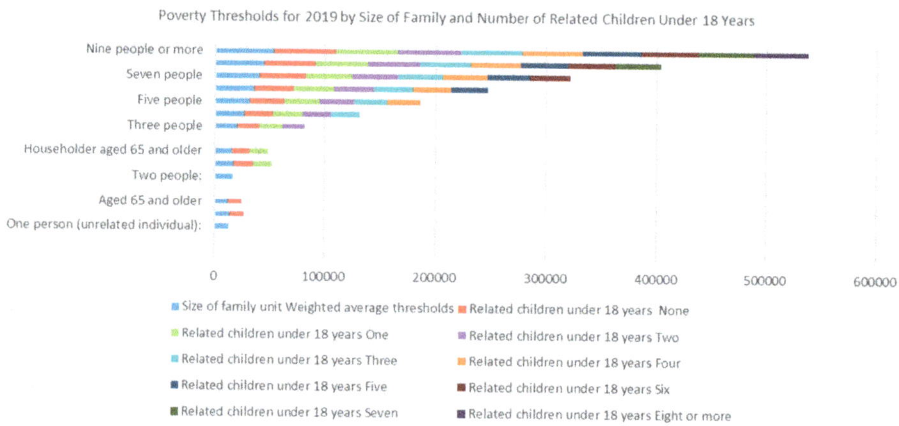

Fig. 2.19 Poverty thresholds [8]

Finally,

> Family farms comprise 96% of all US farms [9], accounting for 87% of land in farms and 82% of the value of all agricultural products sold, according to the 2017 census of agriculture farm typology report released today by the US Department of Agriculture's National Agricultural Statistics Service (NASS).[43]

Another data frequency that is very common in the agricultural datasets includes composite frequency. The *composite frequency* is a combination of one or more intrinsic time frame, for example, combination of months or combination of quarters. Consider the time frequency of phenological stages of a crop (please see Fig. 2.20). For instance, take the example of coffee bean. The phenological stages

[43] Family-owned farms account for 96% of US farms, according to the Census of Agriculture Typology Report – https://www.nass.usda.gov/Newsroom/2021/01-22-2021.php

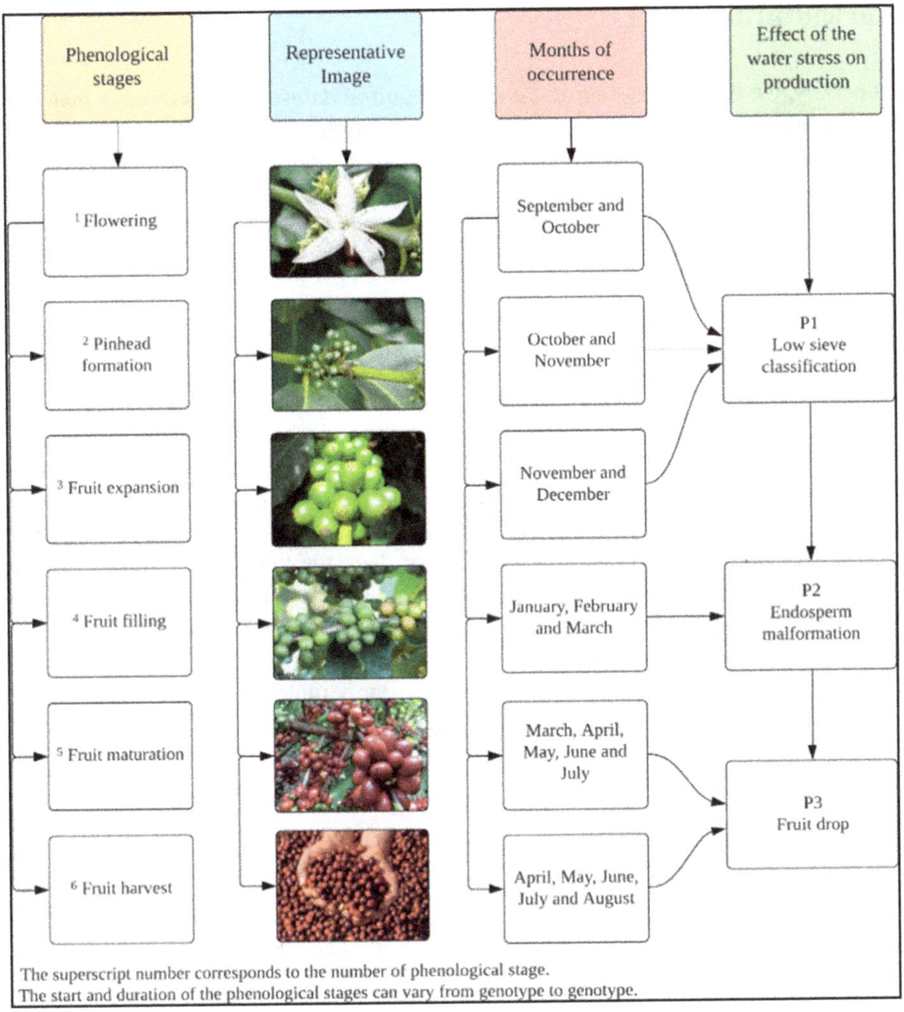

Fig. 2.20 Coffee phenological stages [10]

of coffee bean involves the stages of flowering, pinhead formation, fruit expansion, fruit filling, fruit maturation, and fruit harvest. In the below figure, phenological stages of coffee in the Brazil and effect of temperature related heat stress on the growth of coffee bean are depicted[44] [10].

[44] Impact of drought associated with high temperatures on *Coffea* canephora plantations: a case study in Espírito Santo State, Brazil – https://www.nature.com/articles/s41598-020-76713-y#Tab1

Agricultural Data Types

Generally, we find three types of data in agricultural datasets and is true for many of the machine learning products. The data types include:

- Structured data.
- Unstructured data.
- Semi-structured data.

Structured Data

Structured data is data that adheres to a schema, so all the data has the same fields or properties. Structured data can be stored in a database table with rows and columns. Structured data relies on keys to indicate how one row in a table relates to data in another row of another table. Structured data is also referred to as relational data, as the data's schema defines the table of data, the fields in the table, and the clear relationship between the two. Examples of structure data include sensor data and financial data. For example, in the following figure (please see Fig. 2.21), the sensor data is collected—as can be seen, the data has rigid structure with Record ID, Date, Time, Sensor ID, X, Y, Z, Sensor Body Temperature, Ambient Temperature, Sensor Body Humidity, and Ambient Humidity.

Unstructured Data

The organization of unstructured data is generally ambiguous. Unstructured data is often delivered in files, such as photos, images, or videos. Examples of unstructured

Record ID	Date	Time	Sensor ID	X	Y	Z	SensorBody_Temperature	Ambient_Temperature	SensorBody_Humidity	Ambient_
[103]	11/30/2019	15:00:01	HANU_NECKLACE_0001	-931	-188	-16185	11.4	11.7	69.5	66.7
[103]	11/30/2019	15:00:01	HANU_NECKLACE_0001	-931	-188	-16185	11.4	11.7	69.5	66.7
[103]	11/30/2019	15:00:01	HANU_NECKLACE_0001	-931	-188	-16185	11.4	11.7	69.5	66.7
[104]	11/30/2019	16:00:01	HANU_NECKLACE_0001	-893	-239	-16153	11.4	11.7	68.6	67
[105]	11/30/2019	17:00:01	HANU_NECKLACE_0001	-914	-198	-16153	11.4	11.7	68.4	67
[106]	11/30/2019	18:00:01	HANU_NECKLACE_0001	-961	-210	-16184	11.3	11.6	68.4	67
[107]	11/30/2019	19:00:01	HANU_NECKLACE_0001	-884	-199	-16194	11.3	11.6	68.6	67
[108]	11/30/2019	20:00:01	HANU_NECKLACE_0001	-908	-216	-16149	11.2	11.5	69	67.2
[109]	11/30/2019	21:00:01	HANU_NECKLACE_0001	-933	-210	-16169	11.3	11.5	69	67.3
[110]	11/30/2019	22:00:01	HANU_NECKLACE_0001	-935	-238	-16155	11.3	11.6	72.7	67.4
[110]	11/30/2019	22:00:01	HANU_NECKLACE_0001	-935	-238	-16155	11.3	11.6	72.7	67.4
[110]	11/30/2019	22:00:01	HANU_NECKLACE_0001	-935	-238	-16155	11.3	11.6	72.7	67.4
[110]	11/30/2019	22:00:01	HANU_NECKLACE_0001	-935	-238	-16155	11.3	11.6	72.7	67.4
[110]	11/30/2019	22:00:01	HANU_NECKLACE_0001	-935	-238	-16155	11.3	11.6	72.7	67.4
[110]	11/30/2019	22:00:01	HANU_NECKLACE_0001	-935	-238	-16155	11.3	11.6	72.7	67.4
[112]	12/1/2019	0:00:01	HANU_NECKLACE_0001	-921	-240	-16195	11.8	12	70.3	68.1
[112]	12/1/2019	0:00:01	HANU_NECKLACE_0001	-921	-240	-16195	11.8	12	70.3	68.1
[112]	12/1/2019	0:00:01	HANU_NECKLACE_0001	-921	-240	-16195	11.8	12	70.3	68.1
[112]	12/1/2019	0:00:01	HANU_NECKLACE_0001	-921	-240	-16195	11.8	12	70.3	68.1

Fig. 2.21 Real-time sensor data

data include media files such as movie files and audio files, office files such as Word documents, text files, and log files.

Semi-Structured Data

Semi-structured data is less organized than structured data and is not stored in a relational format, as the fields do not neatly fit into tables, rows, and columns. Semi-structured data contains tags that make the organization and hierarchy of the data apparent. Semi-structured data is also referred to as non-relational or NoSQL data. Examples of semi-structured data include key/value pairs, network graph data, JSON (JavaScript Object Notation), and XML (Extensible Markup Language) files. In the following diagram (please see Fig. 2.6), US Department of Agriculture-Economic Research Service (ERS) metadata JSON provided.

Agricultural Dataset Structure

Agricultural dataset structure is, broadly, represented in either tabular or relational types.

Tabular

For most people working with small amounts of data, the data table is the fundamental unit of organization. The data table, arguably the oldest data structure, is both a way of organizing data for processing by machines and of presenting data visually for consumption by humans. Common Separated Values (CSV) is the most common example of tabular data. For example, the following table (please see Table 2.2) represents commodity export and crude oil import price indexes, December 2013–2017.

Relational Data

Data stored in Relational Database Management Systems (RDBMS) generally falls into this category. The data access using SQL (Structured Query Language) is the way to access the data.

Agricultural datasets are tabular form and can be easily consumed in ML models. Nonetheless, the data noise or perturbation is very common and is highly prevalent in agricultural field and field sensor data. For instance, sensor calibration issue or

Table 2.2 Commodity prices

Month	Export corn	Export soybeans	Export wheat	Import crude
Dec-13	100	100	100	100
Jan-14	98.9	97.9	93.8	100.5
Feb-14	102.4	98.1	94.4	105.8
Mar-14	110	105	101.7	106.5
Apr-14	112.5	107.7	106.3	106.9
May-14	111.8	108.2	111	108.3
Jun-14	102.6	109.2	98.3	110.7
Jul-14	94.8	101.9	97	109.2
Aug-14	88.7	92.9	93.3	104.6
Sep-14	85.1	86.6	92.4	100.2
Oct-14	79.8	73.7	86.5	93.5
Nov-14	88	79.5	91.9	84
Dec-14	87.9	76.8	99.4	68.1
Jan-15	88.8	77.3	92.7	53.5
Feb-15	86	73	85.7	52.6
Mar-15	86.9	75	84.1	52.9

Source: US Bureau of Labor Statistics

malfunction of sensor could introduce data noise. Exploratory analysis is key to filter out noise to have good-quality statistical outputs.

Exploratory Data Analysis (EDA)

The first step in any machine learning and data mining project is to explore the data that you will use to train a model. The goal of this exploration is to try to understand the relationships between its attributes: in particular, any apparent correlation between the features and the label your model will try to predict. This may require some work to detect and fix issues in the data (such as dealing with missing values, errors, or outlier values), deriving new feature columns by transforming or combining existing features (a process known as feature engineering), normalizing numeric features (values you can measure or count) so they're on a similar scale, and encoding categorical features (values that represent discrete categories) as numeric indicators.

Exploratory employs a variety of techniques to:

- Maximize insight into a dataset (Python Dataframe.describe() statistical properties).
- Uncover underlying structure or signature (stationary or non-stationary for time series for instance and distribution is normal or others).
- Extract important variables (feature engineering with correlation or covariance steps).

- Detect outliers and anomalies (outlier detection or scatter plots).
- Test underlying assumptions (hypothesis testing T-Test and F-Test for regression).
- Develop parsimonious models (with the help of feature reduction techniques such as PCA).
- Determine optimal factor settings (hyperparameter or automated ML).

The following diagram contains EDA for classification and regression techniques. As described earlier, the following are main steps (please see Fig. 2.22):

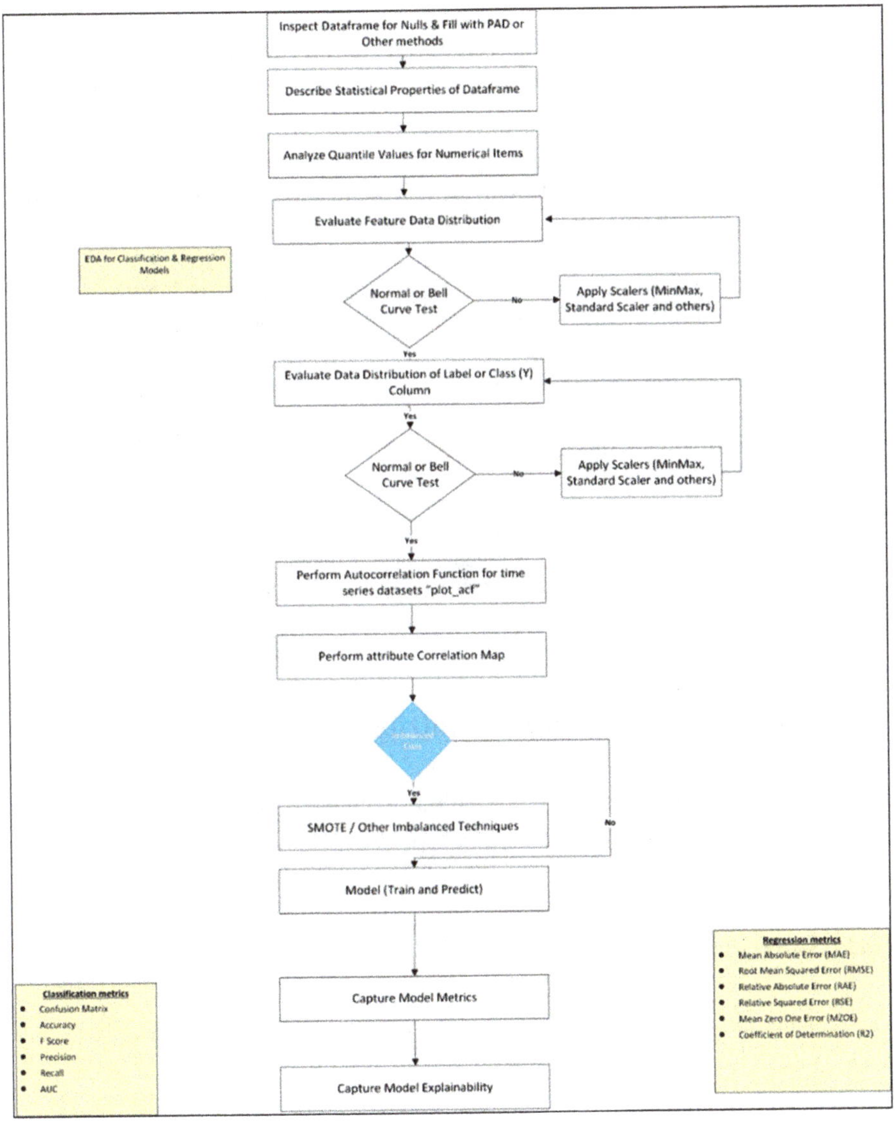

Fig. 2.22 EDA flow diagram

- Inspect Dataframe for Nulls & Fill with PAD or other methods.
- Describe statistical properties (mean, standard deviation, minimum, and maximum).
- Analyze qualitative values.
- Evaluate Feature Data Distribution.
- Normal or Bell Curve test.
- Autocorrelation function (for time series).
- Correlation map.
- SMOTE or other imbalanced techniques.

Data exploration and analysis is typically an iterative process, in which the data scientist takes a sample of data and performs the following kinds of task to analyze it and test hypotheses:

- Clean data to handle errors, missing values, and other issues.
- Apply statistical techniques to better understand the data and how the sample might be expected to represent the real-world population of data, allowing for random variation.
- Visualize data to determine relationships between variables, and in the case of a machine learning project, identify features that are potentially predictive of the label.
- Derive new features from existing ones that might better encapsulate relationships within the data.
- Revise the hypothesis and repeat the process.

Data frames are amazingly versatile and make it easy to manipulate data. Most DataFrame operations return a new copy of the DataFrame; so, if you want to modify a DataFrame but keep the existing variable, you need to assign the result of the operation to the existing variable.

Colombia Rural Agricultural Contribution to GDP!

Let's look at the Colombia's rural agriculture population and agriculture employment in ratio of overall employment influence on GDP.

Data Sources

The data source for the following model is from the World Bank Data. The main features include (please see Fig. 2.23):

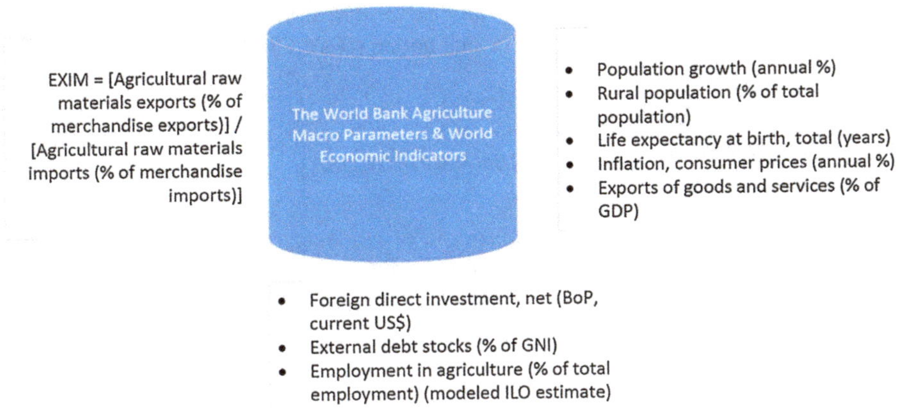

Fig. 2.23 Data sources

- The World Bank—population growth (annual %).[45]
- The World Bank—rural population (% of total population), world.[46]
- Life expectancy at birth, total (years), world.[47]
- External debt stocks (% of GNI).[48]
- Foreign direct investment, net (BoP, current US$)[49].
- Inflation, consumer prices.[50]
- Agricultural raw materials exports (% of merchandise exports).[51]
- Agricultural raw materials imports (% of merchandise imports).[52]
- Agriculture, forestry, and fishing, value added (annual % growth).[53]

---

[45] Population growth (annual %) – https://data.worldbank.org/indicator/SP.POP.GROW?end=2017&locations=1W&start=1981&view=chart

[46] The World Bank – Rural population (% of total population) – https://data.worldbank.org/indicator/SP.RUR.TOTL.ZS?end=2017&locations=1W&start=1981&view=chart

[47] Life expectancy at birth, total (years) – World – https://data.worldbank.org/indicator/SP.DYN.LE00.IN?end=2017&locations=1W&start=1981&view=chart

[48] External Debt Stocks – https://data.worldbank.org/indicator/DT.DOD.DECT.GN.ZS?end=2017&locations=1W&start=1981&view=chart

[49] FDI – https://data.worldbank.org/indicator/BN.KLT.DINV.CD?end=2017&locations=1W&start=1981&view=chart

[50] Inflation – https://data.worldbank.org/indicator/FP.CPI.TOTL.ZG?end=2017&locations=1W&start=1981&view=chart

[51] Agricultural raw materials exports (% of merchandise exports) – https://data.worldbank.org/indicator/TX.VAL.AGRI.ZS.UN?end=2017&locations=1W&start=1981&view=chart

[52] Agricultural raw materials imports (% of merchandise imports) – https://data.worldbank.org/indicator/TM.VAL.AGRI.ZS.UN?end=2017&locations=1W&start=1981&view=chart

[53] Employment in Agriculture – https://data.worldbank.org/indicator/SL.AGR.EMPL.ZS?end=2017&locations=1W&start=1981&view=chart

- Agriculture, forestry, and fishing, value added (% of GDP), world.[54]
- Agriculture, forestry, and fishing, value added (% of GDP).[55]

Descriptive Statistics and Data Distribution

When examining a variable (e.g., a sample of student grades), data scientists are particularly interested in its distribution (in other words, how are all the different grade values spread across the sample). The starting point for this exploration is often to visualize the data as a histogram and see how frequently each value for the variable occurs.

Software code for this model can be found on GitHub Link: ColombiaAgriculturalGDPModelWithFinal (Jupyter Notebook Code)

Let's load data:

```
import pandas as pd
import numpy as np
import plotly.graph_objects as go
import seaborn as sns; sns.set()
import matplotlib.pyplot as plt
from scipy import stats
from scipy.stats import pearsonr
import matplotlib.pyplot as plt
from sklearn.preprocessing import MinMaxScaler
from sklearn.model_selection import train_test_split
from sklearn.linear_model import LinearRegression
from sklearn.ensemble import RandomForestClassifier
from sklearn.metrics import r2_score
from sklearn import metrics

%matplotlib inline
mpl.style.reload_library()
mpl.style.use('classic')
mpl.rcParams['figure.facecolor'] = (1, 1, 1, 0)
mpl.rcParams['figure.figsize'] = [6.0, 4.0]
mpl.rcParams['figure.dpi'] = 100
```

```
dataColombia_df = pd.read_csv("ColombiaRegressionFinalwGDPPG.csv")
dataColombia_df.set_index("Year",inplace = True)
dataColombia_df
```

The above code loads Columbia data.

[54] Agriculture, forestry, and fishing, value added (% of GDP)– World – https://data.worldbank.org/indicator/NV.AGR.TOTL.KD.ZG?end=2017&locations=1W&start=1981&view=chart

[55] Agriculture, forestry, and fishing, value added (% of GDP) – https://data.worldbank.org/indicator/NV.AGR.TOTL.ZS

Year	Population growth (annual %)	EXIM	Rural population (% of total population)	Life expectancy at birth, total (years)	Inflation, consumer prices (annual %)	Exports of goods and services (% of GDP)	FDI	External debt stocks (% of GNI)	Employment in agriculture (% of total employment) (modeled ILO estimate)	Agriculture, forestry, and fishing, value added (% of GDP)
1991	1.961625	147.288569	30	70	30	19	-433	37	24	11
1992	1.948525	178.251091	30	70	27	17	-679	31	24	12
1993	1.924004	239.932242	29	70	22	15	-719	30	23	11
1994	1.884893	247.903267	29	70	23	15	-1298	28	23	10
1995	1.835619	211.118813	28	71	21	15	-712	28	23	10
1996	1.782274	256.324632	28	71	21	15	-2784	31	23	9
1997	1.731098	237.005634	27	72	18	15	-4753	31	22	9
1998	1.683625	276.936066	27	72	19	15	-2033	35	22	11

Measures of Central Tendency

To understand the distribution better, analyze Colombia data with respect to data distribution—central tendency, describing statistics that represent the "middle" of the data. The goal of this is to try to find a "typical" value. Common ways to define the middle of the data include:

- The mean: A simple average based on adding together all the values in the sample set and then dividing the total by the number of samples.
- The median: The value in the middle of the range of all the sample values.
- The mode: The most commonly occurring value in the sample set*.

```
# Create a function that we can re-use
def show_distribution(var_data,colname):
    from matplotlib import pyplot as plt

    # Get statistics
    min_val = var_data.min()
    max_val = var_data.max()
    mean_val = var_data.mean()
    med_val = var_data.median()
    mod_val = var_data.mode()[0]


print('Minimum:{:.2f}\nMean:{:.2f}\nMedian:{:.2f}\nMode:{:.2f}\nMaximum:{:.2f}\n'.format(min_v
al,
                                                mean_val,
                                                med_val,
                                                mod_val,
                                                max_val))

    # Create a figure for 2 subplots (2 rows, 1 column)
    fig, ax = plt.subplots(2, 1, figsize = (10,4))

    # Plot the histogram
    ax[0].hist(var_data)
    ax[0].set_ylabel('Frequency')

    # Add lines for the mean, median, and mode
    ax[0].axvline(x=min_val, color = 'gray', linestyle='dashed', linewidth = 2)
    ax[0].axvline(x=mean_val, color = 'cyan', linestyle='dashed', linewidth = 2)
    ax[0].axvline(x=med_val, color = 'red', linestyle='dashed', linewidth = 2)
    ax[0].axvline(x=mod_val, color = 'yellow', linestyle='dashed', linewidth = 2)
    ax[0].axvline(x=max_val, color = 'gray', linestyle='dashed', linewidth = 2)

    # Plot the boxplot
    ax[1].boxplot(var_data, vert=False)
    ax[1].set_xlabel('Value')

    # Add a title to the Figure
    fig.suptitle(colname + ' Data Distribution')

    # Show the figure
    fig.show()

# Get the variable to examine

for col in ['Population growth (annual %)', 'Inflation, consumer prices (annual %)',
      'Rural population (% of total population)',
      'Exports of goods and services (% of GDP)', 'EXIM', 'FDI',
```

```
        'External debt stocks (% of GNI)',
        'Life expectancy at birth, total (years)',
        'Employment in agriculture (% of total employment) (modeled ILO estimate)']:
    show_distribution(dataColombia_df[col],col)
```

Output:

The population growth (annual %) feature has following distribution properties: minimum, 0.90; mean, 1.45; median, 1.48; mode, 0.90; and maximum, 1.96. What it tells is that the population growth (annual %) is declining in Colombia with less than 1% population growth in the years 2011, 2012, and 2013.

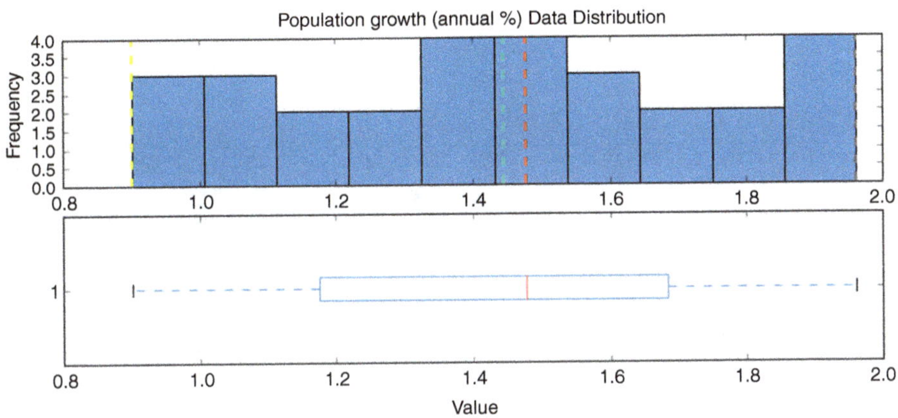

The Consumer prices (annual %) feature attribute has following properties: minimum, 2.00; mean, 10.03; median, 6.00; mode, 3.00; and maximum, 30.00. Colombia had double digit inflation during the 1990s (1990–1999), and hence you can observe skewed heavily on the left side of the graph. Nonetheless, in recent years, the inflation is in check—less than 5%.

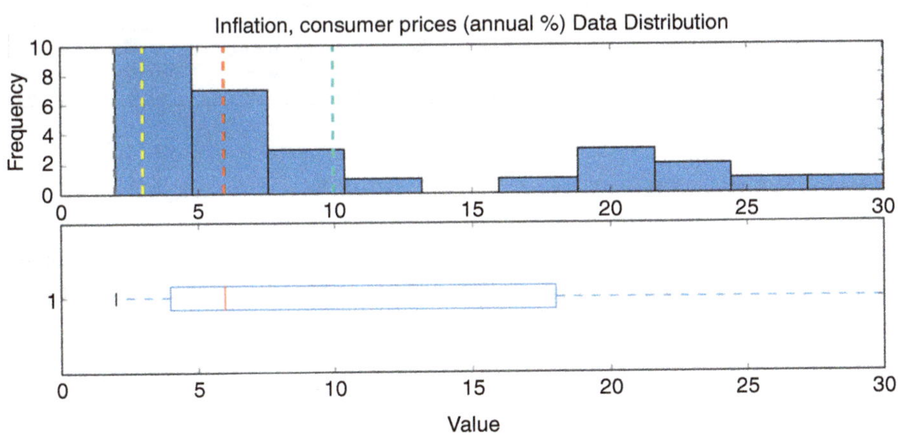

Life expectancy at birth, total (years).

Colombia witnessed increase in life expectancy—from 70 years in the 1990s to 77 years in 2019. Almost decade increase in life can be attributed to favorable economic opportunities and growth in overall GDP.

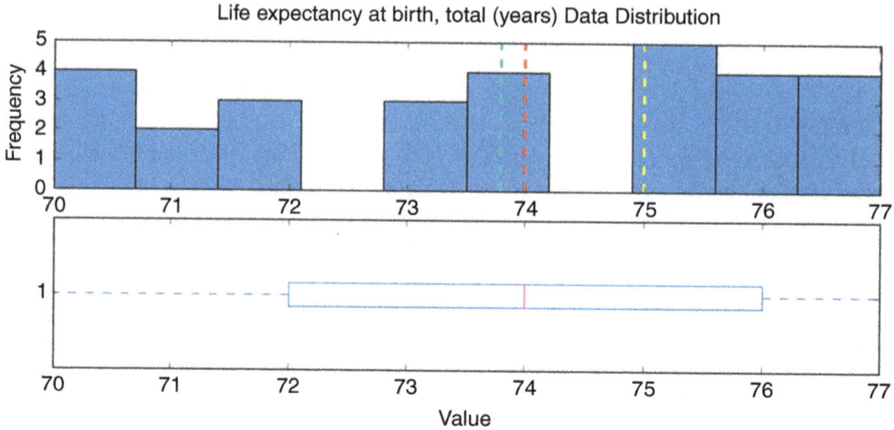

Describe Colombia Data Frame:

```
dataColombia_df.describe()
```

Output:

	Population growth (annual %)	EXIM	Rural population (% of total population)	Life expectancy at birth, total (years)	Inflation, consumer prices (annual %)	Exports of goods and services (% of GDP)	FDI	External debt stocks (% of GNI)	Employment in agriculture (% of total employment) (modeled ILO estimate)	Agriculture, forestry, and fishing, value added (% of GDP)
count	29.000000	29.000000	29.000000	29.000000	29.000000	29.000000	29.000000	29.000000	29.000000	29.000000
mean	1.445441	278.757980	24.172414	73.793103	10.034483	16.413793	-4934.379310	31.793103	20.034483	6.931034
std	0.324837	90.793514	3.402404	2.335737	8.415474	1.376259	4088.599013	7.247558	2.731859	2.477485
min	0.900335	147.288569	19.000000	70.000000	2.000000	15.000000	-15646.000000	20.000000	16.000000	4.000000
25%	1.176159	226.156779	21.000000	72.000000	4.000000	15.000000	-7506.000000	25.000000	17.000000	5.000000
50%	1.477722	256.324632	24.000000	74.000000	6.000000	16.000000	-4530.000000	31.000000	21.000000	6.000000
75%	1.683625	293.434677	27.000000	76.000000	18.000000	17.000000	-1298.000000	38.000000	22.000000	9.000000
max	1.961625	502.644864	30.000000	77.000000	30.000000	19.000000	-433.000000	44.000000	24.000000	12.000000

Box-and-Whiskers Plot

Another way to visualize the distribution of a variable is to use a box plot (sometimes called a box-and-whiskers plot). Let's create one for the 'Agriculture, forestry, and fishing, value added (% of GDP)' data.

```
# Get the variable to examine
var = dataColombia_df['Agriculture, forestry, and fishing, value added (% of GDP)']

# Create a Figure
fig = plt.figure(figsize=(10,4))

# Plot a histogram
plt.boxplot(var)

# Add titles and labels
plt.title('Data Distribution')

# Show the figure
fig.show()
```

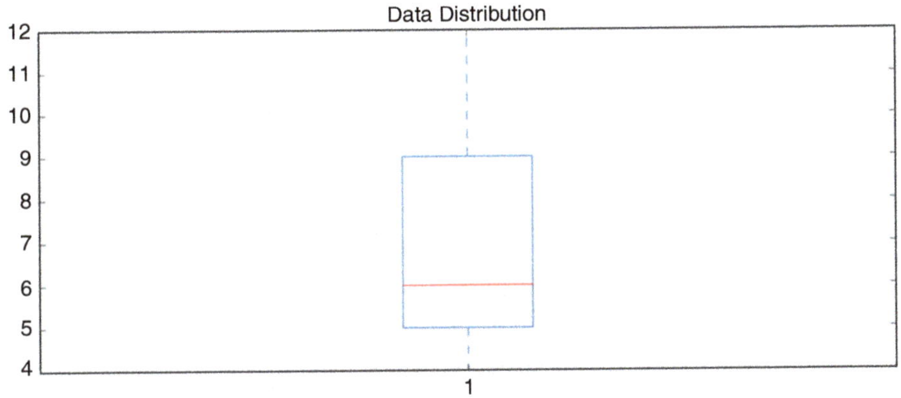

The box plot shows the distribution of the 'Agriculture, forestry, and fishing, value added (% of GDP)' values in a different format to the histogram. The box part of the plot shows where the inner two quartiles of the data reside—so in this case, half of the grades are between approximately 5 and 9. The whiskers extending from the box show the outer two quartiles; so, the other half of the grades in this case are between 4 and 5 or 9 and 12. The line in the box indicates the median value.

Density Function

To explore this distribution in more detail, probability function provides more details. For example, the Colombia data consists of 29 years data (1991–2019), and for each year there is agriculture, forestry, and fishing, value added (% of GDP). The mode represents the common population growth percentage—0.9%.

```
def show_density(var_data,colname):
    from matplotlib import pyplot as plt

    fig = plt.figure(figsize=(10,4))

    # Plot density
    var_data.plot.density()

    # Add titles and labels
    plt.title(colname + ' Density')

    # Show the mean, median, and mode
    plt.axvline(x=var_data.mean(), color = 'cyan', linestyle='dashed', linewidth = 2)
    plt.axvline(x=var_data.median(), color = 'red', linestyle='dashed', linewidth = 2)
    plt.axvline(x=var_data.mode()[0], color = 'yellow', linestyle='dashed', linewidth = 2)

    # Show the figure
    plt.show()

# Get the density of Grade

for col in ['Population growth (annual %)', 'Inflation, consumer prices (annual %)',
        'Rural population (% of total population)',
        'Exports of goods and services (% of GDP)', 'EXIM', 'FDI',
        'External debt stocks (% of GNI)',
        'Life expectancy at birth, total (years)',
        'Employment in agriculture (% of total employment) (modeled ILO estimate)']:
    show_density(dataColombia_df[col],col)
```

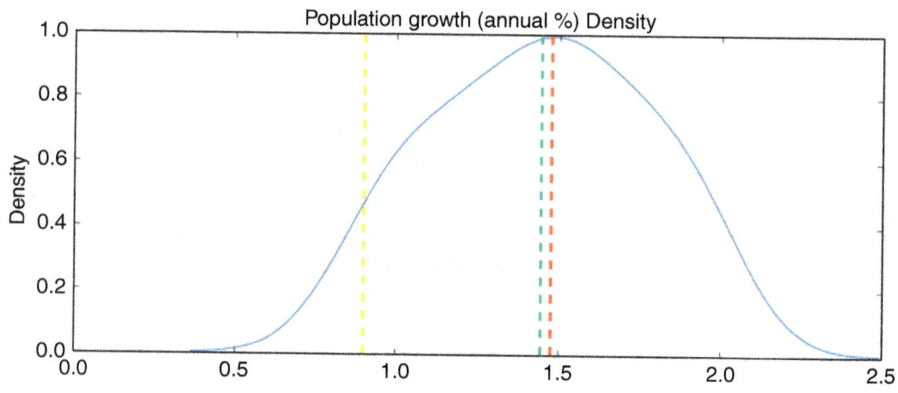

As expected from the histogram of the sample, the density shows the characteristic "bell curve" of what statisticians call a normal distribution with the mean and mode at the center and symmetric tails.

Inflation, consumer prices (annual %):

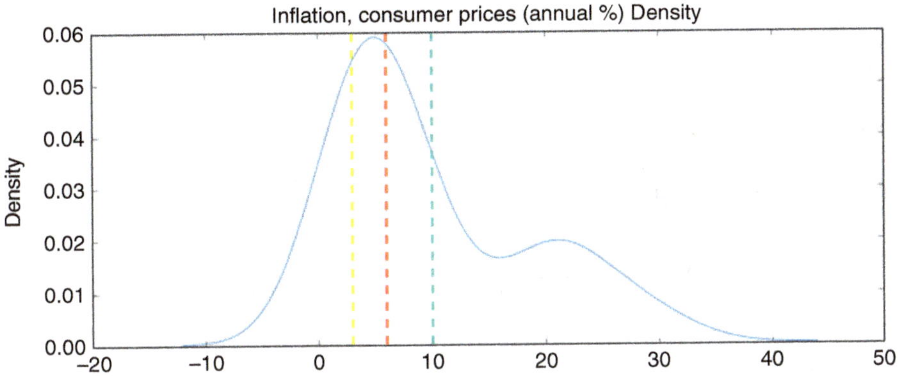

Inflation left skewed with higher economic inflation in the 1990s. The density function provides that behavior visually. The mode of 3% represents frequent inflation value.

Other variables:

Rural population skewed left with visible decline (tail of graph thinner right).	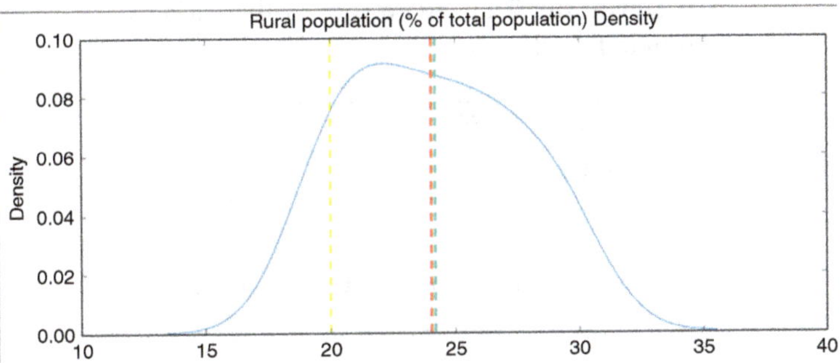
Exports of goods and services (% of GDP) with mode of 15% and graph left skewed.	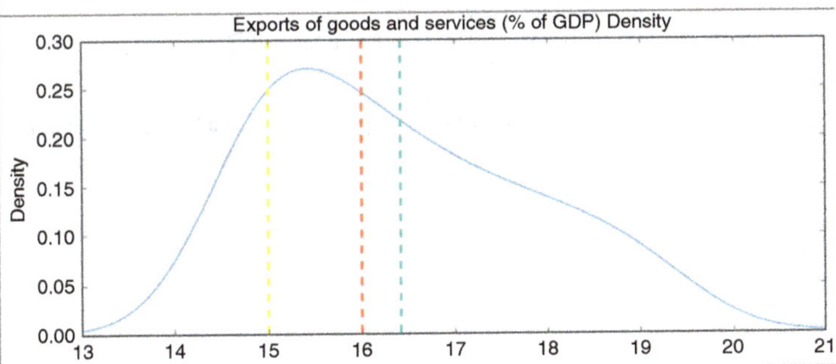

Employment in agriculture (% of total employment) (modeled ILO estimate) is with mode 22, and the frequency of 22% is right of mean/median values.	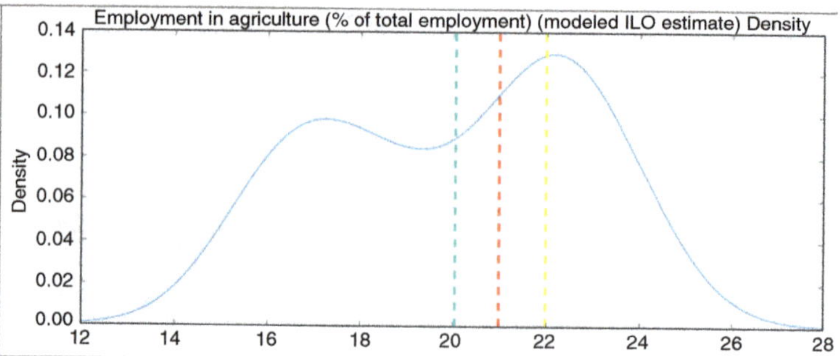

Variability

Typical statistics that measure variability in the data include:

- **Range**: The difference between the maximum and minimum. There's no built-in function for this, but it's easy to calculate using the min and max functions.
- **Variance**: The average of the squared difference from the mean. You can use the built-in var. function to find this.
- **Standard Deviation**: The square root of the variance. You can use the built-in std. function to find this.

```
def show_variability(var_data,colname):
    col = var_data
    rng = col.max() - col.min()
    var = col.var()
    mean = col.mean()
    std = col.std()
    print('\n{}:\n - Range: {:.2f}\n - Variance: {:.2f}\n - mean: {:.2f}\n - Std.Dev: {:.2f}'.format(colname,
rng, var, mean, std))


for col in ['Population growth (annual %)', 'Inflation, consumer prices (annual %)',
       'Rural population (% of total population)',
       'Exports of goods and services (% of GDP)', 'EXIM', 'FDI',
       'External debt stocks (% of GNI)',
       'Life expectancy at birth, total (years)',
       'Employment in agriculture (% of total employment) (modeled ILO estimate)']:
    show_variability(dataColombia_df[col],col)
```

Output:

Population growth (annual %): - Range: 1.06 - Variance: 0.11 - mean: 1.45 - Std.Dev: 0.32	Inflation, consumer prices (annual %): - Range: 28.00 - Variance: 70.82 - mean: 10.03 - Std.Dev: 8.42	Rural population (% of total population): - Range: 11.00 - Variance: 11.58 - mean: 24.17 - Std.Dev: 3.40	Exports of goods and services (% of GDP): - Range: 4.00 - Variance: 1.89 - mean: 16.41
EXIM: - Range: 355.36 - Variance: 8243.46 - mean: 278.76 - Std.Dev: 90.79	FDI: - Range: 15213.00 - Variance: 16716641.89 - mean: -4934.38 - Std.Dev: 4088.60	External debt stocks (% of GNI): - Range: 24.00 - Variance: 52.53 - mean: 31.79 - Std.Dev: 7.25	Life expectancy at birth, total (years): - Range: 7.00 - Variance: 5.46 - mean: 73.79 - Std.Dev: 2.34
Employment in agriculture (% of total employment) (modeled ILO estimate): - Range: 8.00 - Variance: 7.46 - mean: 20.03 - Std.Dev: 2.73			

Of these statistics, the standard deviation is generally the most useful. It provides a measure of variance in the data on the same scale as the data itself. The higher the standard deviation, the more variance there is when comparing values in the distribution to the distribution mean—in other words, *the data is more spread out*.

In any normal distribution[56]:

Approximately 68.26% of values fall within *one standard* deviation from the mean.	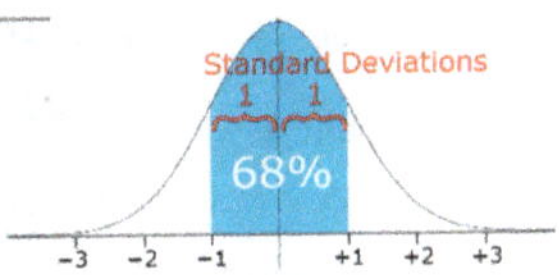
Approximately 95.45% of values fall within *two standard* deviations from the mean.	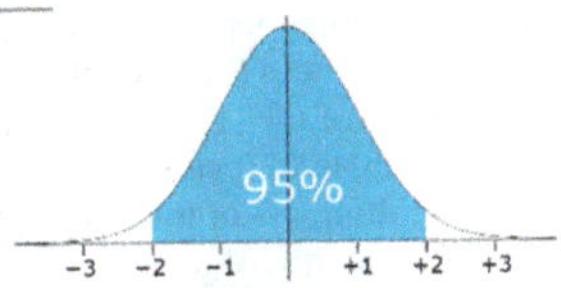
Approximately 99.73% of values fall within *three standard* deviations from the mean.	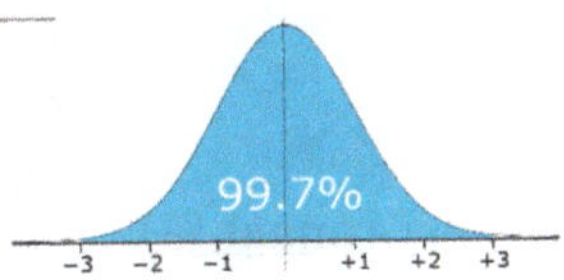

Run the cell below to see the relationship between standard deviations and the data in the normal distribution.

[56] Normal distribution – https://www.mathsisfun.com/data/standard-normal-distribution.html

```
import scipy.stats as stats

def show_pdf(var_data,colname):
    # Get the Grade column
    col = var_data
    # get the density
    density = stats.gaussian_kde(col)
    # Plot the density
    col.plot.density()
    # Get the mean and standard deviatio
    s = col.std()
    m = col.mean()
    # Annotate 1 stdev
    x1 = [m-s, m+s]
    y1 = density(x1)
    plt.plot(x1,y1, color='magenta')
    plt.annotate('1 std (68.26%)', (x1[1],y1[1]))
    # Annotate 2 stdevs
    x2 = [m-(s*2), m+(s*2)]
    y2 = density(x2)
    plt.plot(x2,y2, color='green')
    plt.annotate('2 std (95.45%)', (x2[1],y2[1]))
    # Annotate 3 stdevs
    x3 = [m-(s*3), m+(s*3)]
    y3 = density(x3)
    plt.plot(x3,y3, color='orange')
    plt.annotate('3 std (99.73%)', (x3[1],y3[1]))
    # Show the location of the mean
    plt.axvline(col.mean(), color='cyan', linestyle='dashed', linewidth=1)
    plt.axis('off')
    print('\n{}:\n'.format(colname))
    plt.show()

for col in ['Population growth (annual %)', 'Inflation, consumer prices (annual %)',
    'Rural population (% of total population)',
    'Exports of goods and services (% of GDP)', 'EXIM', 'FDI',
    'External debt stocks (% of GNI)',
    'Life expectancy at birth, total (years)',
    'Employment in agriculture (% of total employment) (modeled ILO estimate)']:
    show_pdf(dataColombia_df[col],col)
```

Output:

Population Growth (Annual %).

The horizontal lines show the percentage of data within 1, 2, and 3 standard deviations of the mean (plus or minus). In any normal distribution:

- Approximately 68.26% of values fall within one standard deviation from the mean.
- Approximately 95.45% of values fall within two standard deviations from the mean.
- Approximately 99.73% of values fall within three standard deviations from the mean.

So, since we know that the *mean* population growth (annual %) is 1.45, the standard deviation is 0.32, and distribution of population growth (annual %) is approximately normal.

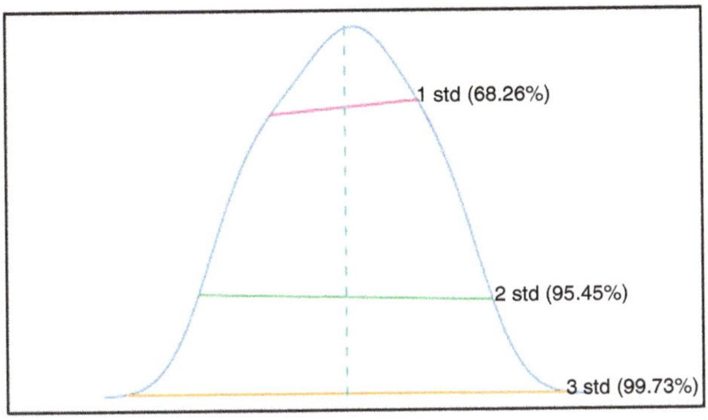

Inflation Consumer Price.

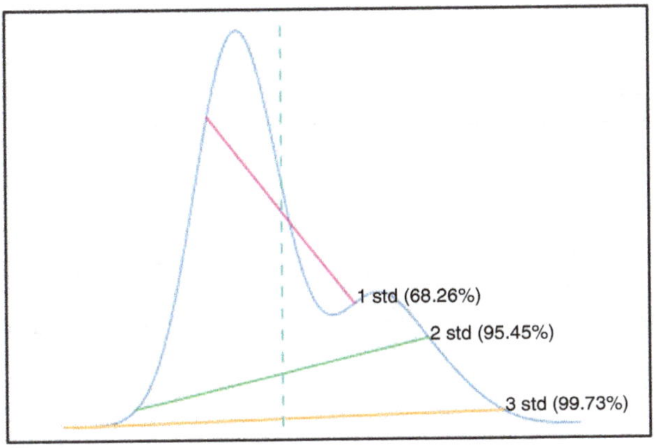

We know that the *mean* inflation is 10.45, the standard deviation is 8.42, and distribution of population growth is approximately non-normal.

Rural Population (% of Total Population).

We know that the *mean* of rural population (% of total population) is 24.17, the standard deviation is 3.40, and distribution of population growth is approximately normal (variance is close to range).

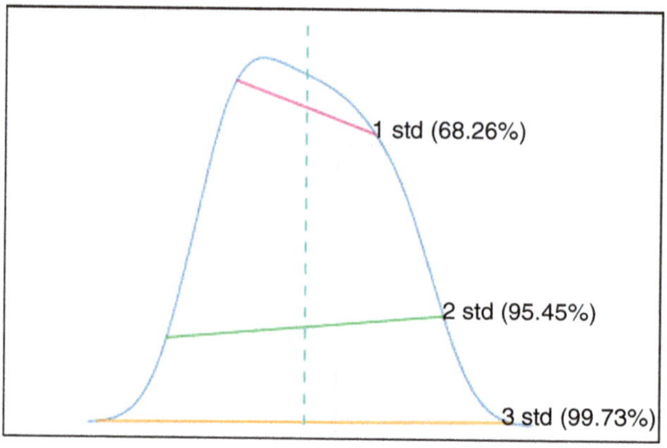

Comparing Numeric and Categorical Variables

No categorical values exist.

Correlation

```
dataColombia_df.corr()
```

Output:

	Population growth (annual %)	EXIM	Rural population (% of total population)	Life expectancy at birth, total (years)	Inflation, consumer prices (annual %)	Exports of goods and services (% of GDP)	FDI	External debt stocks (% of GNI)	Employment in agriculture (% of total employment) (modeled ILO estimate)	Agriculture, forestry, and fishing, value added (% of GDP)
Population growth (annual %)	1.000000	-0.318941	0.804144	-0.803168	0.863290	-0.434590	0.683070	0.407060	0.804469	0.891159
EXIM	-0.318941	1.000000	-0.735723	0.707933	-0.444313	-0.326141	-0.616044	0.414396	-0.758435	-0.152557
Rural population (% of total population)	0.804144	-0.735723	1.000000	-0.988523	0.881639	-0.137813	0.777446	-0.070918	0.963769	0.700544
Life expectancy at birth, total (years)	-0.803168	0.707933	-0.988523	1.000000	-0.886289	0.149795	-0.785444	0.075441	-0.944745	-0.724647
Inflation, consumer prices (annual %)	0.863290	-0.444313	0.881639	-0.886289	1.000000	-0.204796	0.594826	0.087955	0.781346	0.897720
Exports of goods and services (% of GDP)	-0.434590	-0.326141	-0.137813	0.149795	-0.204796	1.000000	-0.254168	-0.324102	-0.155917	-0.368411
FDI	0.683070	-0.616044	0.777446	-0.785444	0.594826	-0.254168	1.000000	0.106822	0.840564	0.498664
External debt stocks (% of GNI)	0.407060	0.414396	-0.070918	0.075441	0.087955	-0.324102	0.106822	1.000000	0.000373	0.359190
Employment in agriculture (% of total employment) (modeled ILO estimate)	0.804469	-0.758435	0.963769	-0.944745	0.781346	-0.155917	0.840564	0.000373	1.000000	0.628306
Agriculture, forestry, and fishing, value added (% of GDP)	0.891159	-0.152557	0.700544	-0.724647	0.897720	-0.368411	0.498664	0.359190	0.628306	1.000000

The correlation statistic is a value between −1 and 1 that indicates the strength of a relationship. Values above 0 indicate a positive correlation (high values of one variable tend to coincide with high values of the other), while values below 0 indicate a negative correlation (high values of one variable tend to coincide with low values of the other). In this case, population growth (annual %), rural population (% of total population), inflation, consumer prices (annual %), FDI, external debt stocks (% of GNI), and employment in agriculture (% of total employment) (modeled ILO estimate) showing a strongly positive correlation with the agriculture, forestry, and fishing, value added (% of GDP).

On the other hand, **life expectancy at birth, total (years), and exports of goods and services (% of GDP) are** showing a negative correlation with the agriculture, forestry, and fishing, value added (% of GDP).

Note: Data scientists often quote the maxim "correlation is not causation." In other words, as tempting as it might be, you shouldn't interpret the statistical correlation as explaining why one of the values is high. In the case of the student data, the statistics demonstrates that students with high grades tend to also have high amounts of study time; but this is not the same as proving that they achieved high grades because they studied a lot. The statistic could equally be used as evidence to support the nonsensical conclusion that the students studied a lot because their grades were going to be high.

Heatmap of correlation provides visual clue:

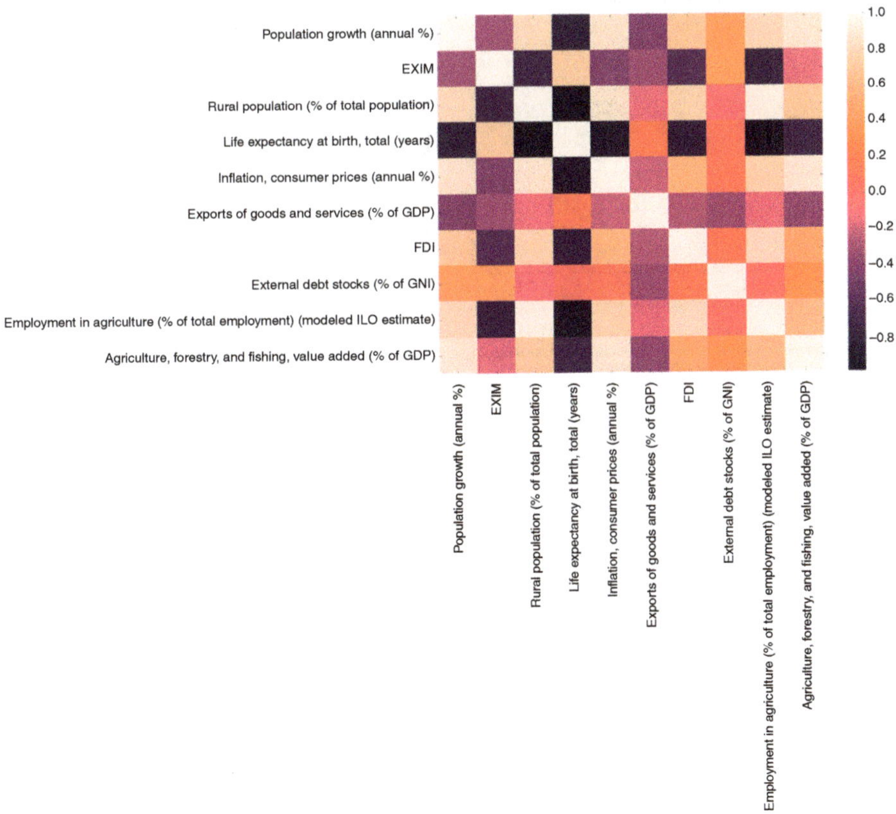

Another way to visualize the apparent correlation between two numeric columns is to use a scatter plot.

```
for col in ['Population growth (annual %)', 'Inflation, consumer prices (annual %)',
     'Rural population (% of total population)',
     'Exports of goods and services (% of GDP)', 'EXIM', 'FDI',
     'External debt stocks (% of GNI)',
     'Life expectancy at birth, total (years)',
     'Employment in agriculture (% of total employment) (modeled ILO estimate)']:
   fig = plt.figure(figsize=(9, 6))
   ax = fig.gca()
   feature = dataColombia_df[col]
   label = dataColombia_df['Agriculture, forestry, and fishing, value added (% of GDP)']
   correlation = feature.corr(label)

   plt.scatter(x=feature, y=label,color='steelblue')
   plt.xlabel(col)
   plt.ylabel('GDP per capita growth (annual %)')
   ax.set_title('GDP per capita growth (annual %)  ' + col + '- correlation: ' + str(correlation))
plt.show()
```

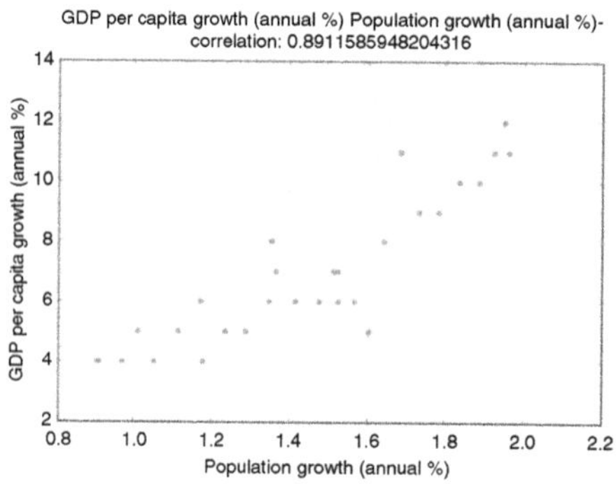

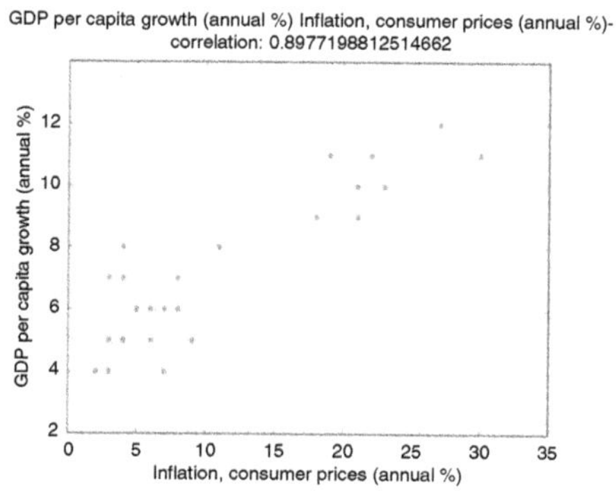

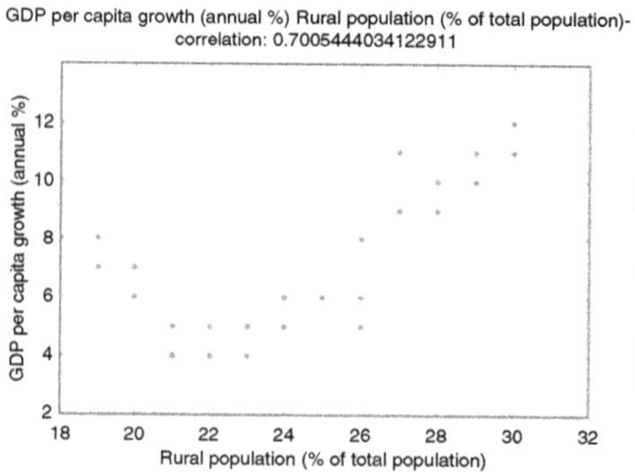

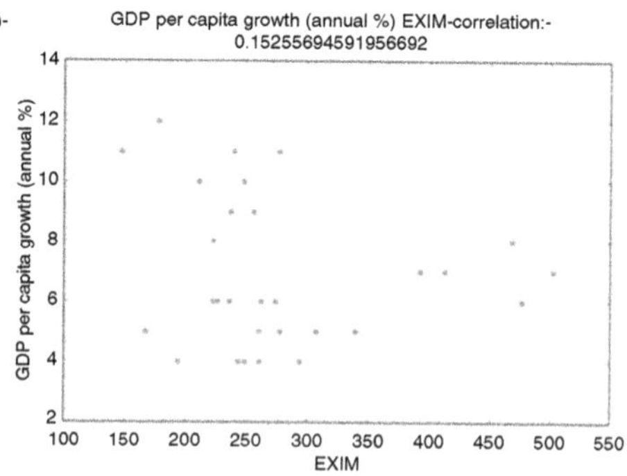

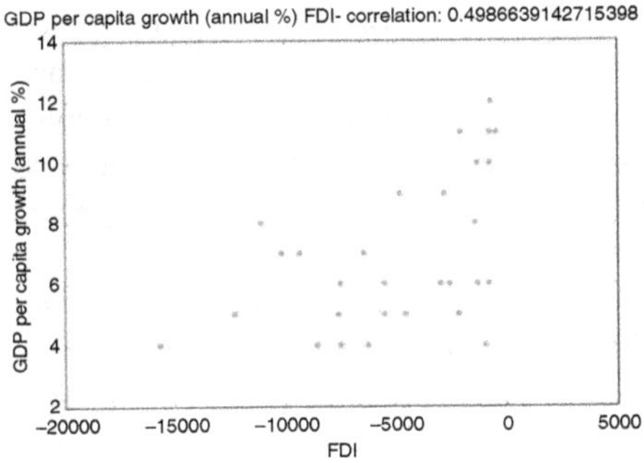

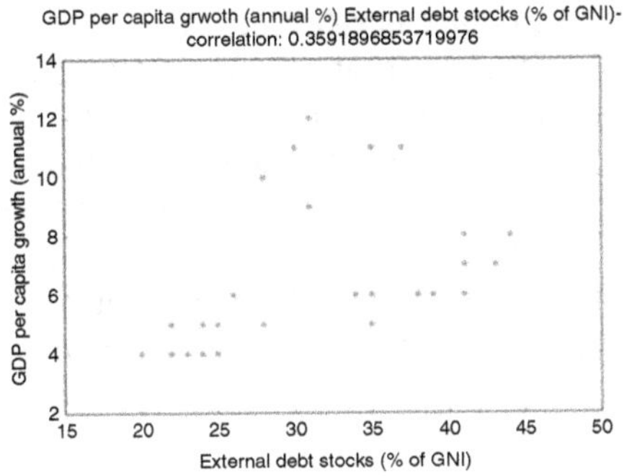

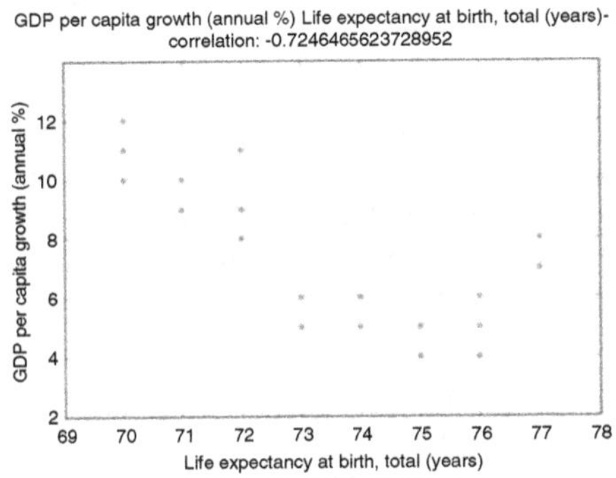

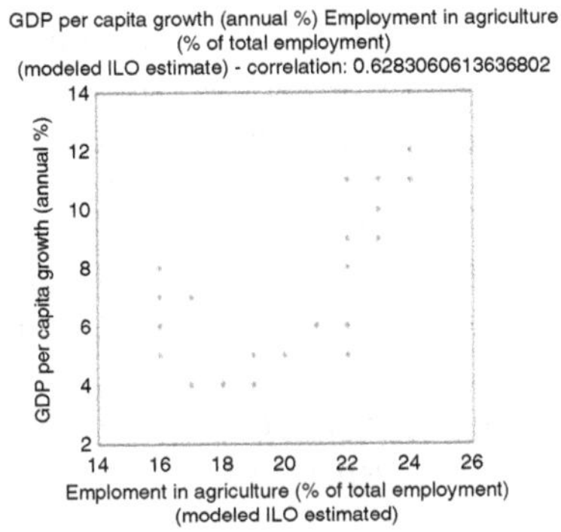

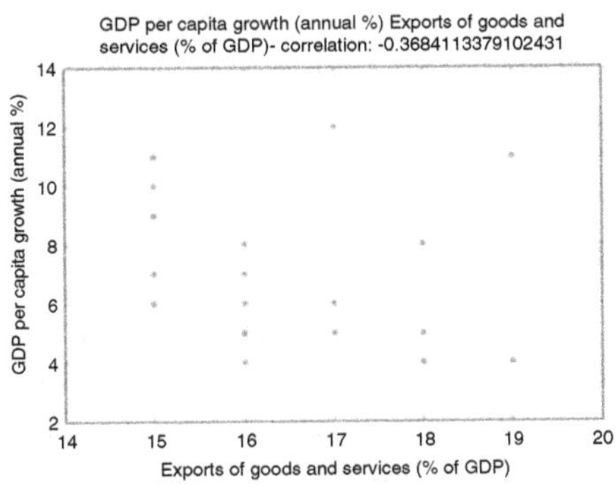

Feature Selection

Feature selection refers to the process of applying statistical tests to inputs, given a specified output or class variable, to determine which columns are more predictive of the output. The filter-based feature selection module provides multiple feature selection algorithms to choose from, including correlation methods such as Pearson's or Kendall's correlation, mutual information scores, and chi-squared values.

- **Pearson Correlation**: Pearson's correlation statistic, or Pearson's correlation coefficient, is also known in statistical models as the **r value**. For any two variables, it returns a value *that indicates the strength of the correlation*. Pearson's correlation coefficient is computed by taking the covariance of two variables and dividing by the product of their standard deviations. *The coefficient is not affected by changes of scale in the two variables.*
- **Mutual Information**: The mutual information score measures *the contribution of a variable towards reducing uncertainty about the value of another variable*. Namely, the label. Many variations of the mutual information score have been devised to suit different distributions. The mutual information score is particularly useful in feature selection because it maximizes the mutual information between the joint distribution and target variables in datasets with many dimensions.
- **Kendall Correlation**: Kendall's rank correlation is one of several statistics that measure the relationship between *rankings of different ordinal variables or different rankings of the same variable*. In other words, it measures the similarity of orderings when ranked by the quantities. Both this coefficient and Spearman's correlation coefficient are designed for use with non-parametric and non-normally distributed data.
- **Spearman Correlation**: Spearman's coefficient is a nonparametric measure of *statistical dependence* between two variables and is sometimes denoted by the Greek letter rho. The Spearman's coefficient expresses the degree to which two variables are *monotonically related*. It is also called Spearman rank correlation because it can be used with ordinal variables.
- **Chi Squared**: The two-way chi-squared test is a statistical method that measures how *close expected values are to actual results*. The method assumes that variables are random and drawn from an adequate sample of independent variables. The resulting chi-squared statistic indicates how far results are from the expected (random) result.
- **Fisher Score**: The Fisher score (also called the Fisher method, or Fisher combined probability score) is sometimes *termed the information score*, because it represents the amount of information that one variable provides about some unknown parameter on which it depends. The score is computed by measuring the variance between the expected value of the information and the observed value. *When variance is minimized, information is maximized. Since the expectation of the score is zero, the Fisher information is also the variance of the score.*
- **Count Based**: Count-based feature selection is a simple yet relatively powerful way of finding information about predictors. The basic idea underlying count-based featurization is simple: by calculating counts of individual values within a column, you can get an idea of the distribution and weight of values and, from this, understand which columns contain the most important information. Count-based feature selection is a non-supervised method of feature selection, meaning you don't need a label column. This method also reduces the dimensionality of the data without losing information.

Feature Reduction Techniques

Principal component analysis[57] is a dimension-reduction tool that can be used advantageously in such situations. Principal component analysis aims at reducing a large set of variables to a small set that still contains most of the information in the large set [11].

The technique of principal component analysis enables us to create and use a reduced set of variables, which are called principal factors.

Principal component analysis used in several agricultural use cases: ammonia (NH3) emissions from fertilized[58] soils are a costly problem that is undermining agricultural and ecological sustainability worldwide. Ammonia emissions from crop production have been reliably documented in recent years. However, insufficient efforts have been made to determine the factors most influential in facilitating NH3 emissions. The goal of this study was to identify the principal factors facilitating NH3 emissions from fertilized soils for potato production by means of principal component analysis (PCA). A dataset consisting of 14 different variables and 600 determinations of NH3 emission rates was geometrically classified into 5 zones and then analyzed with the PCA technique. The data used in this analysis originated from an incubation experiment involving four major potato production soils (two from Washington State and two from Florida, USA), five N sources, two soil water regimes, three incubation temperatures, and five measurement dates of NH3 emission rates during 28 days of incubation: Day 1, 3, 7, 14, and 28. Ammonia emission rates were classified into five distinct zones [12].

Hyperparameter Tuning

Hyperparameters are adjustable parameters that let you control the model training process. For example, with neural networks, you decide the number of hidden layers and the number of nodes in each layer. Model performance depends heavily on hyperparameters.

Hyperparameter tuning, also called hyperparameter optimization, is the process of finding the configuration of hyperparameters that results in the best performance. The process is typically computationally expensive and manual.

[57] Principal Component Analysis – https://itl.nist.gov/div898/handbook/pmc/section5/pmc55.htm

[58] Identification of factors most important for ammonia emission from fertilized soils for potato production using principal component analysis – https://www.srs.fs.usda.gov/pubs/37509

Model Interpretability

Model interpretability is very important to both model developers (data scientists, program managers) internal stakeholders (program groups, sponsors, and executive management), external stakeholders (auditors, business compliance managers), and users. The need for the model explainability stems from the fact that the models (especially deep neural networks) are black box models handling very critical production tasks, and it would be scary if model predictivity is unknown.[59] The most important aspect of model interpretability is its applicability with business company policies, industry standards, and government regulations [13, 14].

- **Explainability**: Model development team members—data scientists—need the ability to explain their models to executives and stakeholders, so they can understand the value and accuracy of their findings. They also require interpretability to debug their models and make informed decisions and cycle model refresh based on insights provided (please see Fig. 2.24).
- **During pre-model deployment** and model release, the business compliance and legal auditors require tools to validate models with respect to regulatory compliance and monitor how models' decisions are impacting humans.
- **Transparency**: Both internal and external business parties need assurance and peace of mind by having the ability to develop with explainability feature so that non-technical and business savvy users could interpret model predictions and insights—a way to ensure no bias infused in the model.

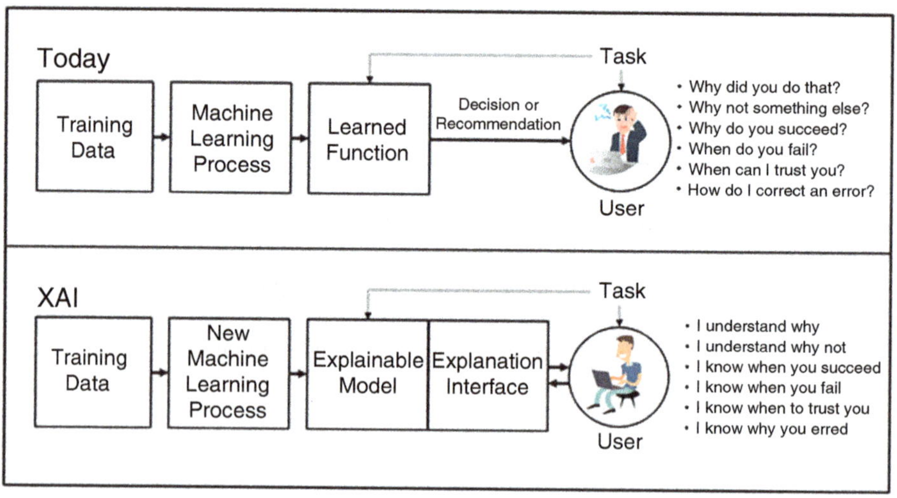

Fig. 2.24 The explainable AI [14]

[59] An overview of model explainability in modern machine learning – An overview of model explainability in modern machine learning | by Rui Aguiar | Towards Data Science

The Explainable AI (XAI)[60] [14]

Enabling the capability of explaining a machine learning model is important during two main phases of model development:

- During the training phase, as model designers and evaluators can use interpretability output of a model to verify hypotheses and build trust with stakeholders. They also use the insights into the model for debugging, validating model behavior matches their objectives, and to check for model unfairness or insignificant features.
- During the inferencing phase, as having transparency around deployed models empowers executives to understand "when deployed" how the model is working and how its decisions are treating and impacting people in real life.

Global and Local Feature Importance

Global Feature Importance: The first thing to do is try to explain the model by evaluating the overall feature importance—in other words, quantifying the extent to which each feature influences the prediction based on the whole training dataset. As an aggregate, all the features of training data are considered to develop global importance. When applied explainability from interpret community[61] API, the feature importance is ranked, with the most important feature listed first.

Local Feature Importance: with global feature importance, the data scientists develop overall view of the model. To understand and explain for individual predictions, the local explanations are used. Additionally, quantifying the extent to which each feature influenced the decision to predict each of the possible label values, local importance plays an important role.

Interpretability Techniques

- SHAP (SHapley Additive exPlanations[62]) is a game theoretic approach to explain the output of any machine learning model.[63] It connects optimal credit allocation with local explanations using the classic Shapley values from game theory and their related extensions [15] (please see Fig. 2.25).

---

[60] The Explainable AI – https://www.darpa.mil/program/explainable-artificial-intelligence

[61] Interpret Community – https://github.com/interpretml/interpret-community

[62] SHAP – https://github.com/slundberg/shap

[63] SHAP – *Shapley regression values:* Lipovetsky, Stan, and Michael Conklin. "Analysis of regression in game theory approach." Applied Stochastic Models in Business and Industry 17.4 (2001): 319–330.

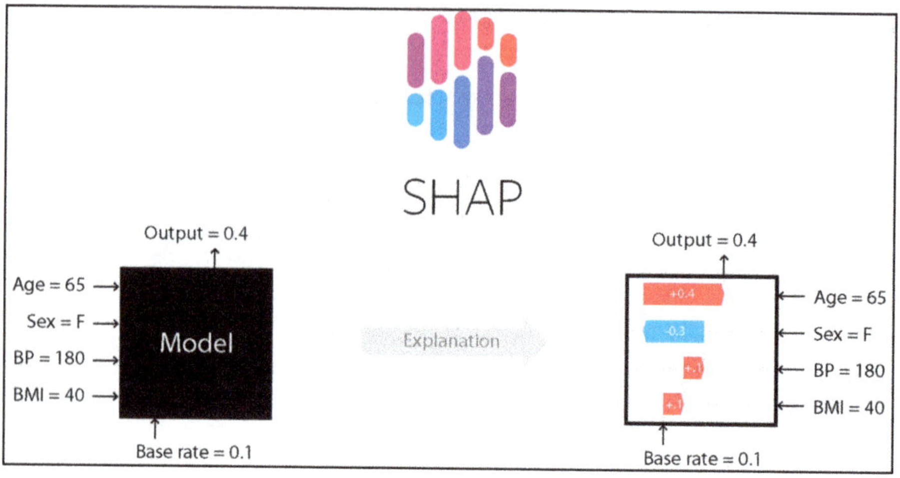

Fig. 2.25 SHAP (https://github.com/slundberg/shap)

- Global Surrogate Model[64]: A global surrogate model is an interpretable model that is trained to approximate the predictions of a black box model. We can draw conclusions about the black box model by interpreting the surrogate model. Solving machine learning interpretability by using more machine learning!
- LIME[65] (local interpretable model-agnostic explanations): Local surrogate models are interpretable models that are used to explain individual predictions of black box machine learning models. Local interpretable model-agnostic explanation (LIME) is a paper[66] in which the authors propose a concrete implementation of local surrogate models. Surrogate models are trained to approximate the predictions of the underlying black box model. Instead of training a global surrogate model, LIME focuses on training local surrogate models to explain individual predictions [16].

The idea is quite intuitive. First, forget about the training data and imagine you only have the black box model where you can input data points and get the predictions of the model. You can probe the box as often as you want. Your goal is to understand why the machine learning model made a certain prediction. LIME tests what happens to the predictions when you give variations of your data into the machine learning model. LIME generates a new dataset consisting of permuted samples and the corresponding predictions of the black box model (please see Table 2.3).

[64] Global Surrogate Model – https://christophm.github.io/interpretable-ml-book/global.html

[65] LIME – https://christophm.github.io/interpretable-ml-book/lime.html

[66] LIME – Ribeiro, Marco Tulio, Sameer Singh, and Carlos Guestrin. "Why should I trust you?: Explaining the predictions of any classifier." Proceedings of the 22nd ACM SIGKDD international conference on knowledge discovery and data mining. ACM (2016)

Table 2.3 Interpretability technique

Interpretability technique	Description	Type
SHAP tree explainer	SHAP's tree explainer, which focuses on polynomial time fast SHAP value estimation algorithm specific to trees and ensembles of trees	Model-specific
Mimic explainer (global surrogate)	Mimic explainer is based on the idea of training global surrogate models to mimic black box models. A global surrogate model is an intrinsically interpretable model that is trained to approximate the predictions of any black box model as accurately as possible. Data scientists can interpret the surrogate model to draw conclusions about the black box model. you can use one of the following interpretable models as your surrogate model: LightGBM (LGBMExplainableModel), linear regression (LinearExplainableModel), stochastic Gradient descent explainable model (SGDExplainableModel), and decision tree (DecisionTreeExplainableModel)	Model agnostic
Permutation feature importance explainer (PFI)	Permutation feature importance is a technique used to explain classification and regression. At a high level, the way it works is by randomly shuffling data one feature at a time for the entire dataset and calculating how much the performance metric of interest changes. The larger the change, the more important that feature is. PFI can explain the overall behavior of any underlying model but does not explain individual predictions	Model agnostic

Imbalanced Datasets

The challenge of working with imbalanced datasets is that most machine learning techniques will ignore, and in turn have poor performance on, the minority class, although typically it is performance on the minority class that is most important[67] [17].

Software code for this model can be found on GitHub Link: decisiontreeensemblewithadasynsmoteminority.py (Python Code)

Synthetic Minority Oversampling Technique (SMOTE)

Underrepresentation of class or target variable in agriculture datasets is a very common issue, for example, wheat predictivity datasets[68] [18], CLAND Kaggle wheat

[67] SMOTE for Imbalanced Classification with Python – https://machinelearningmastery.com/smote-oversampling-for-imbalanced-classification/

[68] Combining SMOTE sampling and Machine Learning for Forecasting Wheat Yields in France – https://hal.archives-ouvertes.fr/hal-02523637/document

dataset,[69] rice predictivity, milk, and others. To apply SCIKET machine learning techniques, first, the under sampling of class variable to be fixed and SMOTE[70] is technique to do so[71] [19].

Let's see SMOTE[72] working [20]:

Step 1: Load Libraries:

```
from sklearn.datasets import *
from sklearn import tree
from dtreeviz.trees import *
from sklearn import tree
from sklearn.model_selection import train_test_split

from sklearn.model_selection import train_test_split
from sklearn.preprocessing import StandardScaler, MinMaxScaler
from sklearn.decomposition import PCA
import numpy
import smote_variants as sv

from imblearn.under_sampling import NearMiss, RandomUnderSampler
from imblearn.over_sampling import SMOTE, ADASYN
from sklearn.ensemble import GradientBoostingClassifier
from collections import Counter
```

Step 2: Load Wine Dataset:

Load Wine Dataset and split data into train and test buckets.

```
data, target = load_wine(return_X_y=True)

sample_train, sample_test, label_train, label_test = \
   train_test_split(data, target, test_size=test_size, random_state=7)

print(f"Initial set observations {sample_train.shape[0]}")
print(f"Initial set target classes {sample_test.shape[0]}")
```

Output:

```
Initial set observations 78
Initial set target classes 100
```

[69] CLAND Kaggle Wheat Dataset – https://www.kaggle.com/ociule/crop-data-challenge-2018-cland?select=TrainingDataSet_Wheat2.csv

[70] SMOTE with Imbalance Data – https://www.kaggle.com/qianchao/smote-with-imbalance-data

[71] Oversampling SMOTE and ADASYN – https://www.kaggle.com/ruzarx/oversampling-smote-and-adasyn

[72] SMOTE: Synthetic Minority Over-sampling Technique – https://arxiv.org/pdf/1106.1813.pdf

Step 3: Apply Standard Scaler and reduce output features by applying PCA:

Apply PCA and MINMAXSCALER:

```
# Now reduce number of features to number
pca = PCA(n_components=n).fit(sample_train)
sample_train = pca.transform(sample_train)
sample_test = pca.transform(sample_test)

# Scale to the range (-1,+1)
samples = np.append(sample_train, sample_test, axis=0)
minmax_scale = MinMaxScaler((-1, 1)).fit(samples)
sample_train = minmax_scale.transform(sample_train)
sample_test = minmax_scale.transform(sample_test)
# Pick training size number of samples from each distro
training_input = {key: (sample_train[label_train == k, :])[:training_size]
            for k, key in enumerate(class_labels)}
test_input = {key: (sample_test[label_test == k, :])[:test_size]
        for k, key in enumerate(class_labels)}

# summarize class distribution
print('Counter')
counter = Counter(target)
print(counter)

print('UNIQUE COUNTS')
(unique, counts) = numpy.unique(target, return_counts=True)
frequencies = numpy.asarray((unique, counts)).T
print(frequencies)
```

Output:

```
Counter
Counter({1: 71, 0: 59, 2: 48})
UNIQUE COUNTS
[[ 0 59]
 [ 1 71]
 [ 2 48]]
```

Both PCA and MinMax scaler reduce number of features and transform data into MINMAX scales.

```
plot_data=True

if plot_data:
    try:
        import matplotlib.pyplot as plt
    except ImportError:
        raise NameError('Matplotlib not installed. Please install it before plotting')
    for k in range(0, 3):
        plt.scatter(sample_train[label_train == k, 0][:training_size],
                    sample_train[label_train == k, 1][:training_size], label=str(k))

    plt.title("PCA dim. reduced Wine dataset")
    plt.legend()
    plt.show()
```

Plot the data:

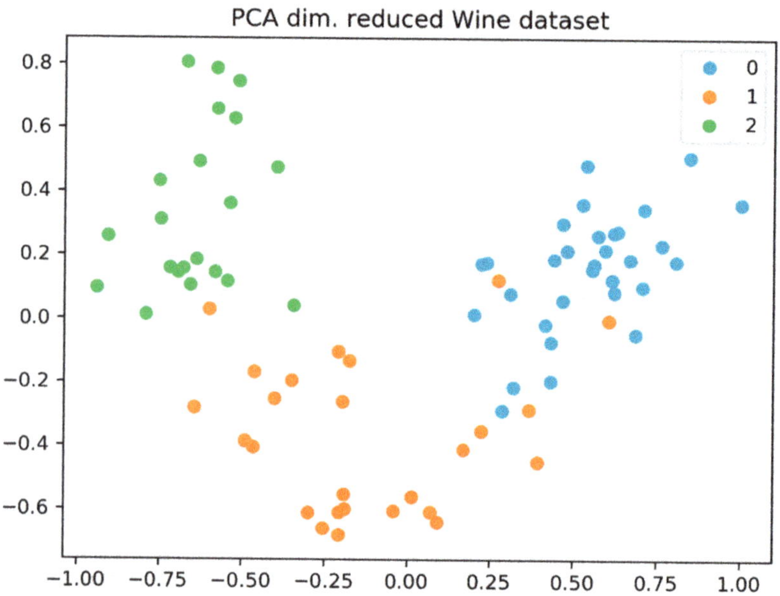

Step 4: Prepare the model and score the model:

Using ensemble Gradient Boost model, lets score:

```
clf = GradientBoostingClassifier(n_estimators=100, learning_rate=1.0,
                                 max_depth=1, random_state=0).fit(sample_train, label_train)


## clf.fit(sample_train,label_train)
print(clf.score(sample_test, label_test))
```

Output:

0.92

Step 4: Apply SMOTE and prepare the data:

Apply SMOTE minority sampling:

```
print('SMOTE')

smote_over_sample = SMOTE(sampling_strategy='minority')
smoteX, smotey = smote_over_sample.fit_resample(data, target)

sample_train, sample_test, label_train, label_test = \
  train_test_split(smoteX, smotey, test_size=test_size, random_state=7)

print(f"Initial set observations {sample_train.shape[0]}")
print(f"Initial set target classes {sample_test.shape[0]}")

# summarize class distribution
print('Counter smotey')
counter = Counter(smotey)
print(counter)

print('UNIQUE COUNTS')
(unique, counts) = numpy.unique(smotey, return_counts=True)
frequencies = numpy.asarray((unique, counts)).T
print(frequencies)
```

Output:

```
SMOTE
Initial set observations 101
Initial set target classes 100
Counter smotey
Counter({1: 71, 2: 71, 0: 59})
UNIQUE COUNTS
[[ 0 59]
 [ 1 71]
 [ 2 71]]
```

Please observe undersampled class "class 2" got class update!

Step 5: Visualize data:

Apply SMOTE minority sampling:

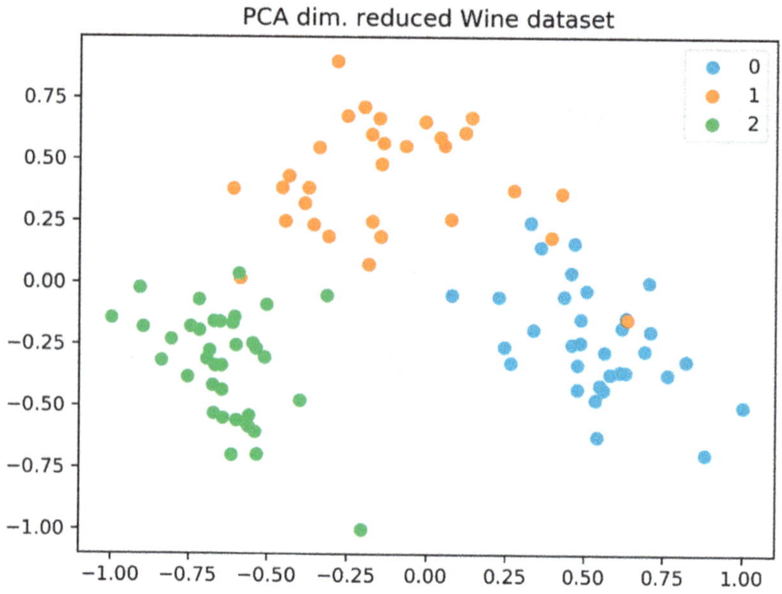

Step 6: Ensemble model and score:

On SMOTE sample data, let's apply ensemble model:

```
clf = GradientBoostingClassifier(n_estimators=100, learning_rate=1.0,
                 max_depth=1, random_state=0).fit(sample_train, label_train)


print(clf.s\core(sample_test, label_test))
```

Output:

```
0.96
```

Please note that the 4% of accuracy improved through SMOTE.

Final output:

```
(base) C:\Hanumayamma\Book\20200918\ChapterCode\Chapter1>python dtreewineimbalancesmoteensemwpca.p
Initial set observations 78
Initial set target classes 100
Counter
Counter({1: 71, 0: 59, 2: 48})
UNIQUE COUNTS
[[ 0 59]
 [ 1 71]
 [ 2 48]]
0.92
SMOTE
Initial set observations 101
Initial set target classes 100
Counter smotey
Counter({1: 71, 2: 71, 0: 59})
UNIQUE COUNTS
[[ 0 59]
 [ 1 71]
 [ 2 71]]
0.96
```

Application of SMOTE to agriculture yields immense benefits:

- Use case: Soybean productivity [21].[73]
- Combining SMOTE sampling and Machine Learning for Forecasting Wheat Yields in France[74] [18].
- Multiclass prediction.[75]
- Measuring the Impact of Climate Change on Indian Agriculture.[76]
- India Agriculture and Climate Dataset.[77]
- US: Agro-Climatic Data by County, 1981–2015.[78]
- Combining Multi-Source Data and Machine Learning Approaches to Predict Winter Wheat Yield in the Conterminous United States.

Adaptive Synthetic Sampling (ADASYN)

Another approach[79] involves generating synthetic samples inversely proportional to the density of the examples in the minority class [17]. That is, generate more synthetic examples in regions of the feature space where the density of minority examples is low and fewer or none where the density is high. This modification to SMOTE

[73] Soybean Productivity Modelling using Decision Tree Algorithms – https://citeseerx.ist.psu.edu/viewdoc/download?doi=10.1.1.259.2088&rep=rep1&type=pdf

[74] Combining SMOTE sampling and Machine Learning for Forecasting Wheat Yields in France – https://hal.archives-ouvertes.fr/hal-02523637/document

[75] A Dynamic Multi-class Neupper Classification for Multiple Crop Yield Prediction – https://www.ijrte.org/wp-content/uploads/papers/v7i6/F2713037619.pdf

[76] Impact of Weather on Agriculture – https://www.google.com/books/edition/Measuring_the_Impact_of_Climate_Change_o/xxAzR4exXi0C?hl=en&gbpv=1&printsec=frontcover

[77] INDIA AGRICULTURE AND CLIMATE DATA SET – https://ipl.econ.duke.edu/dthomas/dev_data/datafiles/india_agric_climate.htm

[78] US Agro Climatic Data – PURR – Publications: Agro-Climatic Data by County, 1981-2015: Usage (purdue.edu)

[79] SMOTE for Imbalanced Classification with Python – https://machinelearningmastery.com/smote-oversampling-for-imbalanced-classification/

is referred to as the Adaptive Synthetic Sampling Method, or ADASYN, and was proposed to Haibo He et al. in their 2008 paper named for the method titled "ADASYN: Adaptive Synthetic Sampling Approach for Imbalanced Learning."

Software code for this model can be found on GitHub link: Decisiontreeensemblewithadasynsmoteminority.Py (Python code).

Step 1: Load model specific libraries:

```
from sklearn.datasets import *
from sklearn import tree
from dtreeviz.trees import *
from sklearn import tree
from sklearn.model_selection import train_test_split

from sklearn.model_selection import train_test_split
from sklearn.preprocessing import StandardScaler, MinMaxScaler
from sklearn.decomposition import PCA

import smote_variants as sv
import numpy
from matplotlib import pyplot
from numpy import where

from imblearn.under_sampling import NearMiss, RandomUnderSampler
from imblearn.over_sampling import SMOTE, ADASYN
from imblearn.over_sampling import SVMSMOTE
from sklearn.ensemble import GradientBoostingClassifier
from collections import Counter
```

Step 2: Load Wine Dataset:

```
winedf = load_wine() ## load_wine(return_X_y=True)

data = winedf.data
target = winedf.target
```

Step 3: Split train and test data:

Split Wine Dataset into train and test data.

```
sample_train, sample_test, label_train, label_test = \
  train_test_split(data, target, test_size=test_size, random_state=7)

print(f"Initial set observations {sample_train.shape[0]}")
print(f"Initial set target classes {sample_test.shape[0]}")
```

Output:

```
Initial set observations 78
Initial set target classes 100
```

Step 4: Inspect and print dataset:

```
for key,value in winedf.items():
    print(key,'\n',value,'\n')


# summarize class distribution
print('Counter')
counter = Counter(target)
print(counter)
```

Output:

```
data
 [[1.423e+01 1.710e+00 2.430e+00 ... 1.040e+00 3.920e+00 1.065e+03]
 [1.320e+01 1.780e+00 2.140e+00 ... 1.050e+00 3.400e+00 1.050e+03]
 [1.316e+01 2.360e+00 2.670e+00 ... 1.030e+00 3.170e+00 1.185e+03]
 ...
 [1.327e+01 4.280e+00 2.260e+00 ... 5.900e-01 1.560e+00 8.350e+02]
 [1.317e+01 2.590e+00 2.370e+00 ... 6.000e-01 1.620e+00 8.400e+02]
 [1.413e+01 4.100e+00 2.740e+00 ... 6.100e-01 1.600e+00 5.600e+02]]

target
 [0 0 0 0 0 0 0 0 0 0 0 0 0 0 0 0 0 0 0 0 0 0 0 0 0 0 0 0 0 0 0 0 0 0 0 0 0
 0 0 0 0 0 0 0 0 0 0 0 0 0 0 0 0 0 0 0 0 0 0 1 1 1 1 1 1 1 1 1 1 1 1 1 1 1
 1 1 1 1 1 1 1 1 1 1 1 1 1 1 1 1 1 1 1 1 1 1 1 1 1 1 1 1 1 1 1 1 1 1 1 1 1
 1 1 1 1 1 1 1 1 1 1 1 1 1 1 1 1 1 1 1 2 2 2 2 2 2 2 2 2 2 2 2 2 2 2 2 2 2
 2 2 2 2 2 2 2 2 2 2 2 2 2 2 2 2 2 2 2 2 2 2 2 2 2 2 2 2 2 2]
```

Step 5: Visualize the data:

Let's visualize the data:

```
print('UNIQUE COUNTS')
(unique, counts) = numpy.unique(target, return_counts=True)
frequencies = numpy.asarray((unique, counts)).T
print(frequencies)
# scatter plot of examples by class label
for label, _ in counter.items():
        row_ix = where(target == label)[0]
        pyplot.scatter(data[row_ix, 0], data[row_ix, 1], label=str(label))
pyplot.legend()
pyplot.show()
```

Output:

```
Counter
Counter({1: 71, 0: 59, 2: 48})
UNIQUE COUNTS
[[ 0 59]
 [ 1 71]
 [ 2 48]]
```

The class 0 with 59 count, class 1 with 71, and class 2 with 48 counts. As you can see, the class 0 and class 2 are under-represented.

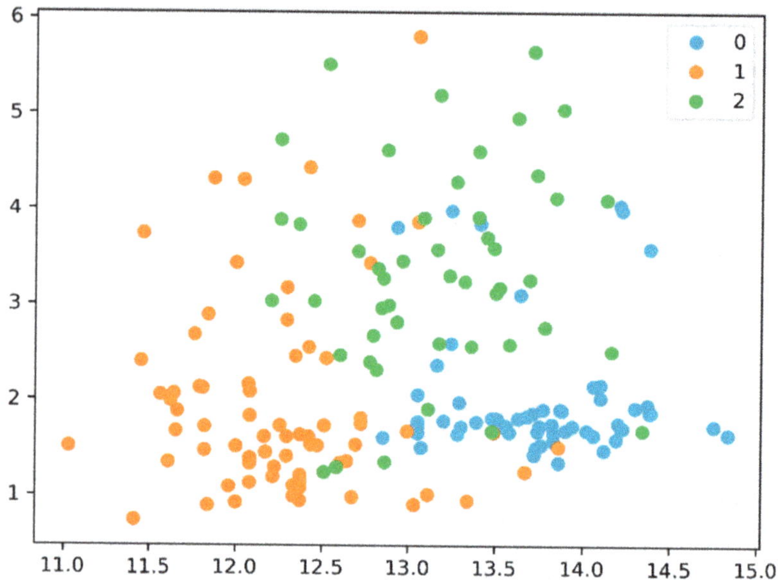

Step 6: Construct model and score:

Construct Gradient Boost model and score the model.

```
clf = GradientBoostingClassifier(n_estimators=100, learning_rate=1.0,
                    max_depth=1, random_state=0).fit(sample_train, label_train)

print(clf.score(sample_test, label_test))
```

Output: model accuracy is 94%. Please note the model score changes based on the machine and random size.

```
0.94
```

Step 7: Transform the dataset using ADASYN:

Transform the data with ADASYN and re-run the model.

```
# transform the dataset
smote_over_sample = ADASYN()
smoteX, smotey = smote_over_sample.fit_resample(data, target)

sample_train, sample_test, label_train, label_test = \
  train_test_split(smoteX, smotey, test_size=test_size, random_state=7)

print(f"Initial set observations {sample_train.shape[0]}")
print(f"Initial set target classes {sample_test.shape[0]}")

# summarize class distribution
print('Counter smotey')
counter = Counter(smotey)
print(counter)

print('UNIQUE COUNTS')
(unique, counts) = numpy.unique(smotey, return_counts=True)
frequencies = numpy.asarray((unique, counts)).T
print(frequencies)

# scatter plot of examples by class label
for label, _ in counter.items():
        row_ix = where(smotey == label)[0]
        pyplot.scatter(smoteX[row_ix, 0], smoteX[row_ix, 1], label=str(label))
pyplot.legend()
pyplot.show()
```

Output:

```
SMOTE
Initial set observations 115
Initial set target classes 100
Counter smotey
Counter({2: 73, 0: 71, 1: 71})
UNIQUE COUNTS
[[ 0 71]
 [ 1 71]
 [ 2 73]]
```

The class counts have increased: class 0, 71; class 1, 71; and class 2, 73. Visually, can be seen in the following diagram:

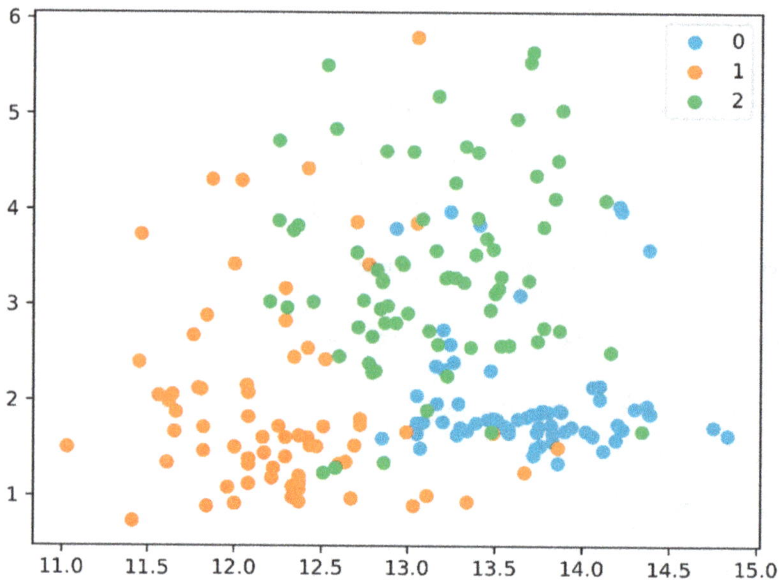

Construct the model:

```
clf = GradientBoostingClassifier(n_estimators=100, learning_rate=1.0,
                                  max_depth=1, random_state=0).fit(sample_train, label_train)

print(clf.score(sample_test, label_test))
```

Model output:

```
0.97
```

As you can see, the model accuracy increased from 94% to 97%. The most important aspect is the classes are balanced!

```
(base) C:\Hanumayamma\Book\20200918\ChapterCode\Chapter1>python decisiontreeensemblewithadasynsmoteminority.py
Initial set observations 78
Initial set target classes 100
Counter
Counter({1: 71, 0: 59, 2: 48})
UNIQUE COUNTS
[[ 0 59]
 [ 1 71]
 [ 2 48]]
0.94
SMOTE
Initial set observations 115
Initial set target classes 100
Counter smotey
Counter({2: 73, 0: 71, 1: 71})
UNIQUE COUNTS
[[ 0 71]
 [ 1 71]
 [ 2 73]]
0.97
```

Feature Engineering

Feature engineering is an essential part of machine learning model building. The success of ML models is directly dependent upon effective feature engineering process that enables translation of business or domain-specific requirements to crisper feature that are part of source data. Feature engineering is a very iterative process and takes 70% of data science team time in the data preparation phase.

Agricultural macroeconomic datasets, i.e., agricultural data and economic data, usually, are time series datasets, for example, the World Bank Commodity Pink Sheet,[80] the US Department of Agriculture (USDA) commodity data,[81] the Federal Reserve Economy Data (FRED[82]), and the price of staple commodities such as wheat, rice, soy,[83] or corn varies based on seasonality, macroeconomic conditions, or wealth of a nation [20]. Many of the data features are continuous numeric data. Application of feature engineering techniques help to unravel data insights that otherwise may not be plausible in raw data form.

Numeric Data

Numeric data typically represents data in the form of scalar values depicting observations (i.e., macroeconomic trend, GDP growth, PCE, and others), recordings (quarterly employment numbers, bank interest rates, S&P 500 Index), or measurements (sensor data from smart systems, utilization of fertilizers). Here, by numeric data, imply continuous data and not discrete data which is typically represented as categorical data.[84] Numeric data can also be represented as a vector of values where each value or entity in the vector can represent a specific feature. Integers and floats are the most common and widely used numeric data types for continuous numeric data.

Even though agricultural numeric data can be directly fed into machine learning models, the data still need to engineer features which are relevant to the macroeconomic scenario, optimization problem, and prognostic analytics domain before building a model. Let's take a commodity and apply binning techniques.

In this case, consider gold pricing.

---

[80] Commodity Markets – https://www.worldbank.org/en/research/commodity-markets

[81] Time Series Data – https://www.ers.usda.gov/data-products/supplemental-nutrition-assistance-program-snap-data-system/time-series-data/

[82] FRED – https://fred.stlouisfed.org/

[83] Soybean Productivity Modelling using Decision Tree Algorithms – https://citeseerx.ist.psu.edu/viewdoc/download?doi=10.1.1.259.2088&rep=rep1&type=pdf

[84] Continuous Numeric Data Strategies for working with continuous, numerical data – https://towardsdatascience.com/understanding-feature-engineering-part-1-continuous-numeric-data-da4e47099a7b

Data	Source
Gold[85]	London bullion market association (LBMA): The London gold fixing companies set the prices for gold that are globally considered as the international standard for pricing of gold. The gold price in London is set twice a day by five LBMA[86] market makers who comprise the London gold market fixing limited (LGMFL).

<u>Gold Price Time Series Data</u>.

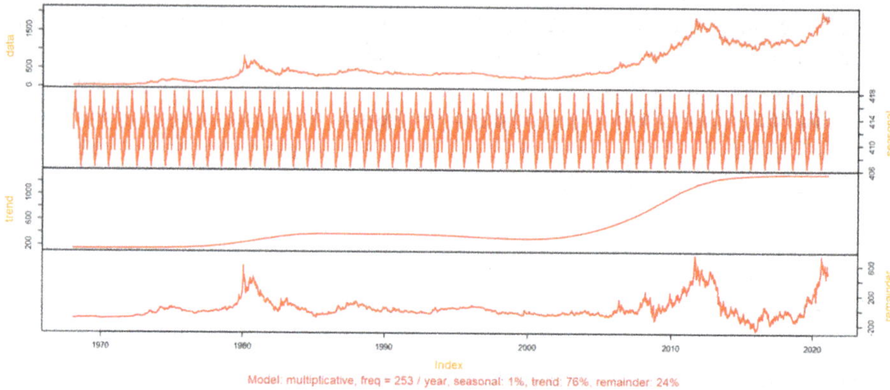

The London Bullion Market Association (LBMA) Gold Price data contains time series values from the 1970s onward. The gold price time series is multiplicative, frequency of 253 data points per year.

Software code for this model can be found on GitHub link: GoldCommodityPricingBinningTechniques.ipynb (Jupyter notebook code).

[85] Project: Gold vs Silver Price – http://rstudio-pubs-static.s3.amazonaws.com/368841_224fe698bd7c45d8a5dd9a6b639988fc.html

[86] London Bullion Market Association (LBMA) – https://www.quandl.com/data/LBMA/GOLD-Gold-Price-London-Fixing

Step 1: Load Python Libraries to process the gold data.

```
import logging
import pandas as pd
import numpy as np
from numpy import random
import nltk
from sklearn.model_selection import train_test_split
from sklearn.feature_extraction.text import CountVectorizer, TfidfVectorizer
from sklearn.metrics import accuracy_score, confusion_matrix
import matplotlib.pyplot as plt
from nltk.corpus import stopwords
import re
from bs4 import BeautifulSoup
import pandas as pd
import matplotlib.pyplot as plt
import matplotlib as mpl
import numpy as np
import scipy.stats as spstats
import seaborn as sns


%matplotlib inline
mpl.style.reload_library()
mpl.style.use('classic')
mpl.rcParams['figure.facecolor'] = (1, 1, 1, 0)
mpl.rcParams['figure.figsize'] = [6.0, 4.0]
mpl.rcParams['figure.dpi'] = 100
```

Step 2: Load gold data.

```
goldDF = pd.read_csv("LBMA-GOLD-USD.csv",parse_dates=['Date'], dayfirst=True)
goldDF = goldDF.set_index('Date')
goldDF.head(10)
```

Output:

Date	GoldPrice
2021-01-22	1853.60
2021-01-21	1867.65
2021-01-20	1854.60
2021-01-19	1843.10
2021-01-18	1833.95
2021-01-15	1853.85
2021-01-14	1840.25
2021-01-13	1852.40
2021-12-01	1861.85
2021-11-01	1847.80

Step 3: Inspect the data frame.

In this process check for NULLS and other details.

```
goldDF.isnull().sum() ## missing values
```

Output:

```
GoldPrice    1
dtype: int64
```

Step 4: Plot the gold data frame:

```
fig, ax = plt.subplots()
goldDF['GoldPrice'].hist(color='#A9C5D3', edgecolor='black',
                 grid=False)
ax.set_title('Gold Price', fontsize=12)
ax.set_xlabel('Gold Price', fontsize=12)
ax.set_ylabel('Frequency', fontsize=12)
```

Output:

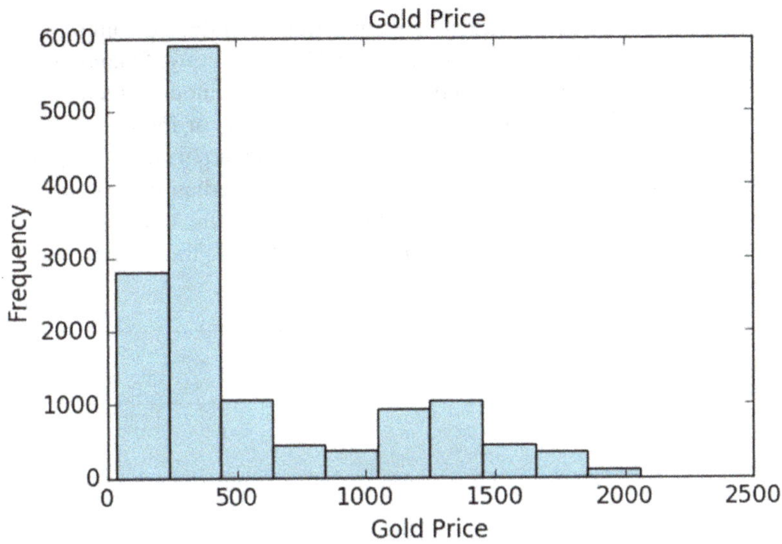

Binning

Binning or grouping data (sometimes called quantization) is an important tool in preparing numerical data for machine learning modeling and is applicable in the following scenarios:

- A column of continuous numbers has too many unique values to model effectively, so you automatically or manually assign the values to groups, to create a smaller set of discrete ranges.
- Replace a column of numbers with categorical values that represent specific ranges.
- A dataset has a few extreme values, all well outside the expected range, and these values have an outsized influence on the trained model.

The problem of working with raw, continuous numeric features is that often the distribution of values in these features will *be skewed*. In case of gold prices, the data skewed right. This signifies that some values will occur quite frequently while some will be quite rare. Binning is used for transforming continuous numeric features into discrete ones (categories). These discrete values or numbers can be thought of as categories or bins into which the raw, continuous numeric values are binned or grouped into. Each bin represents a specific degree of intensity and hence a specific range of continuous numeric values fall into it.

The above histogram depicting gold prices is slightly right skewed as expected (lesser gold price).

Adaptive Binning

In adaptive binning, as the name implies, data distribution itself decides bin ranges. Quantile-based binning is a good strategy to use for adaptive binning. Quantiles are specific values or cut points which help in partitioning the continuous valued distribution of a specific numeric field into discrete contiguous bins or intervals. Thus, q-Quantiles help in partitioning a numeric attribute into q equal partitions. Popular examples of quantiles include the 2-Quantile known as the median which divides the data distribution into two equal bins, 4-Quantiles known as the quartiles which divide the data into 4 equal bins and 10-Quantiles also known as the deciles which create 10 equal width bins.

```
quantile_list = [0, .25, .5, .75, 1.]
quantiles = goldDF['GoldPrice'].quantile(quantile_list)
quantiles
```

Output:

```
0.00      34.770
0.25     278.900
0.50     382.500
0.75     738.375
1.00    2061.500
Name: GoldPrice, dtype: float64
```

Plot bins with quantiles:

```
fig, ax = plt.subplots()
goldDF['GoldPrice'].hist(bins=30, color='#A9C5D3',
                         edgecolor='black', grid=False)
for quantile in quantiles:
   qvl = plt.axvline(quantile, color='r')
ax.legend([qvl], ['Quantiles'], fontsize=10)
ax.set_title('Gold Price Histogram with Quantiles',
         fontsize=12)
ax.set_xlabel('Gold Price', fontsize=12)
ax.set_ylabel('Frequency', fontsize=12)
```

Output:

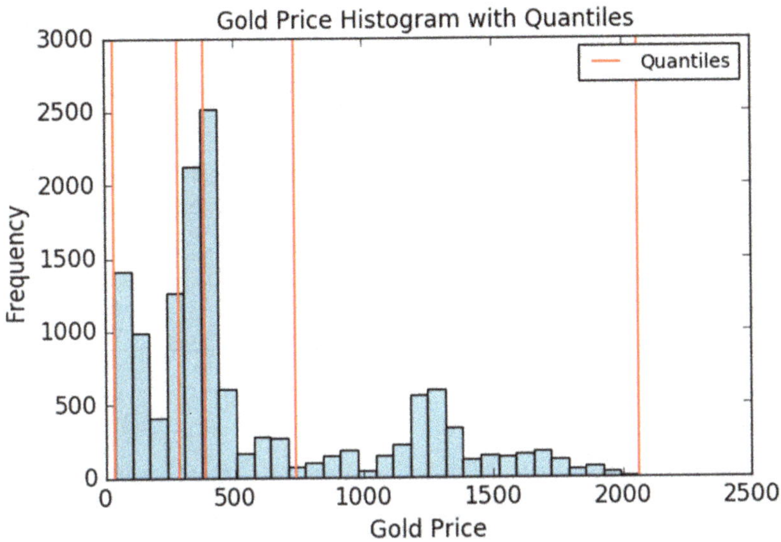

The red lines in the distribution above depict the quartile values and our potential bins.

Statistical Transformations

The log transform as well as the Box-Cox transform functions belong to the Power Transform family of functions, typically used to create monotonic data transformations. Their main significance is that they help in stabilizing variance, adhering closely to the normal distribution, and making the data independent of the mean based on its distribution as gold price is skewed toward lower numbers.

Log Transform

The log transform belongs to the power transform family of functions. This function can be mathematically represented as:

$$y = \log_b(x)$$

This can be translated to:

$$b^y = x$$

The natural logarithm uses b = e where e = 2.71828 popularly known as Euler's number. You can also use base b = 10 used popularly in the decimal system.

Log transforms are useful when applied to skewed distributions as they tend to expand the values which fall in the range of lower magnitudes and tend to compress or reduce the values which fall in the range of higher magnitudes. This tends to make the skewed distribution as normal-like as possible.

```
goldDF['GoldPrice_log'] = np.log((1+ goldDF['GoldPrice']))
goldDF[[ 'GoldPrice', 'GoldPrice_log']].iloc[4:9]
```

Output:

```
              GoldPrice       GoldPrice_log
Date
2021-01-18    1833.95         7.514773
2021-01-15    1853.85         7.525559
2021-01-14    1840.25         7.518200
2021-01-13    1852.40         7.524777
2021-12-01    1861.85         7.529863
```

Plot log transformation graph:

```
goldprice_log_mean = np.round(np.mean(goldDF['GoldPrice_log']), 2)
fig, ax = plt.subplots()
goldDF['GoldPrice_log'].hist(bins=30, color='#A9C5D3',
                   edgecolor='black', grid=False)
plt.axvline(goldprice_log_mean, color='r')
ax.set_title('Gold Price Histogram after Log Transform',
          fontsize=12)
ax.set_xlabel('Gold Price (log scale)', fontsize=12)
ax.set_ylabel('Frequency', fontsize=12)
ax.text(11.5, 450, r'$\mu$='+str(goldprice_log_mean), fontsize=10)
```

Output:

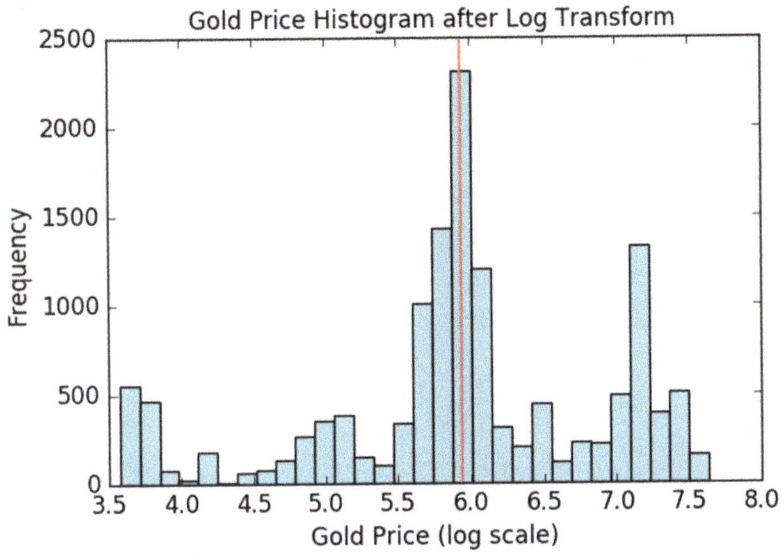

The log distribution is more normal-like or Gaussian as compared to the skewed distribution on the original data.

Box-Cox Transform

The Box-Cox transform[87] is another popular function belonging to the power transform family of functions. This function has a prerequisite that the numeric values to be transformed must be positive (similar to what log transform expects). In case they are negative, shifting using a constant value helps. Mathematically, the Box-Cox transform function can be denoted as follows [23]:

$$y = f(x, \lambda) = x^{\lambda} = \left\{ \begin{array}{ll} \frac{x^{\lambda} - 1}{\lambda} & for\ \lambda > 0 \\ \log_e(x) & for\ \lambda = 0 \end{array} \right\}$$

Such that the resulted transformed output y is a function of input x and the transformation parameter λ such that when λ = 0, the resultant transform is the natural log transform. The optimal value of λ is usually determined using a maximum likelihood or log-likelihood estimation.

Let's see generate random dataset and apply boxcox to see the normal transformation.

[87] Box, G. E. P., and D. R. Cox. "An Analysis of Transformations." Journal of the Royal Statistical Society. Series B (Methodological), vol. 26, no. 2, 1964, pp. 211–252. JSTOR, www.jstor.org/stable/2984418. Accessed 30 Jan. 2021

```
import numpy as np
from scipy import stats
import matplotlib.pyplot as plt
quantile = np.arange (0.01, 1, 0.1)

#PDF
R = stats.loggamma.pdf(5, 500, quantile)
print ("\nProbability Distribution : \n", R)


fig = plt.figure()
ax1 = fig.add_subplot(211)
x = stats.loggamma.rvs(5, size=500) + 5
prob = stats.probplot(x, dist=stats.norm, plot=ax1)
ax1.set_xlabel('')
ax1.set_title('Probplot against normal distribution')

ax2 = fig.add_subplot(212)
xt, _ = stats.boxcox(x)
prob = stats.probplot(xt, dist=stats.norm, plot=ax2)
ax2.set_title('Probplot after Box-Cox transformation')

plt.show()
```

stats.loggamma.rvs function generates some random distribution variates from a non-normal distribution and make a probability plot for it, to show it is non-normal in the tails and boxcox to transform the data so it's closest to normal.

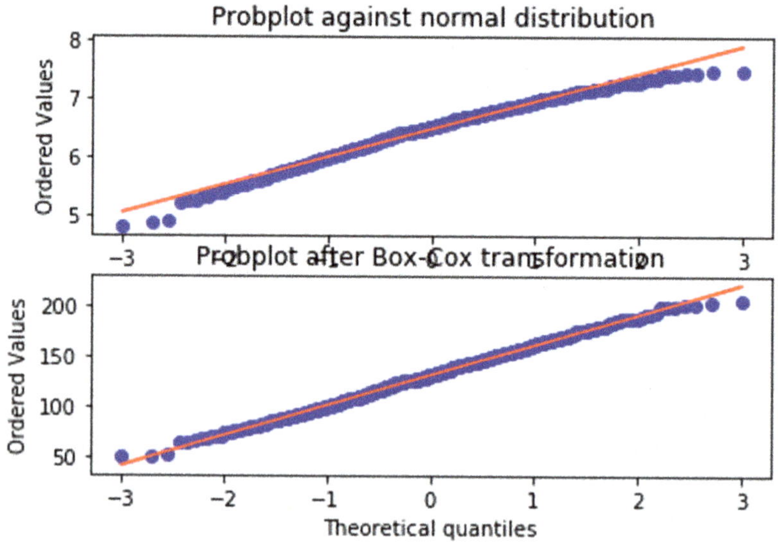

Let's now apply the Box-Cox transform to Gold Price feature in LBMA Gold[88] dataset [24]. London Bullion Market Association (LBMA) sets the prices for gold that are globally considered as the international standard for pricing of gold. The gold price in London is set twice a day by five LBMA Market Makers who comprise the London Gold Market Fixing Limited (LGMFL).

First, we get the optimal lambda value from the data distribution by removing the non-null values as follows.

```
goldprice = np.array(goldDF['GoldPrice'])
goldprice_clean = goldprice[~np.isnan(goldprice)]
l, opt_lambda = spstats.boxcox(goldprice_clean)
print('Optimal lambda value:', opt_lambda)
```

Output:

```
Optimal lambda value: 0.2369908492862791
```

Now that we have obtained the optimal λ value, let us use the Box-Cox transform for two values of λ such that λ = 0 and λ = λ (optimal) and transform the Gold Price feature[89] [22].

```
goldDF['goldprice_boxcox_lambda_0'] = spstats.boxcox(
                        (1+goldDF['GoldPrice']),
                        lmbda=0)
goldDF['goldprice_boxcox_lambda_opt'] = spstats.boxcox(
                          goldDF['GoldPrice'],
                          lmbda=opt_lambda)

goldDF[['GoldPrice', 'GoldPrice_log',
        'goldprice_boxcox_lambda_0',
        'goldprice_boxcox_lambda_opt']].iloc[4:9]
```

Output:

```
            GoldPrice    GoldPrice_log   goldprice_boxcox_lambda_0 goldprice_boxcox_lambda_opt
Date
2021-01-18   1833.95       7.514773        514773                     20.821989
2021-01-15   1853.85       7.525559        7.525559                   20.886120
2021-01-14   1840.25       7.518200        7.518200                   20.842349
2021-01-13   1852.40       7.524777        7.524777                   20.881465
```

The transformed features are depicted in the above data frame. GoldPrice_log and goldprice_boxcox_lamba_0 have the same values. Let's look at the distribution of the transformed Income feature after transforming with the optimal λ.

---

[88] Project: Gold vs Silver Price – http://rstudio-pubs-static.s3.amazonaws.com/368841_224fe698bd7c45d8a5dd9a6b639988fc.html

[89] Continuous Numeric Data Strategies for working with continuous, numerical data – https://towardsdatascience.com/understanding-feature-engineering-part-1-continuous-numeric-data-da4e47099a7b

```
goldprice_boxcox_mean = np.round(
                np.mean(
                goldDF['goldprice_boxcox_lambda_opt']),2)
fig, ax = plt.subplots()
goldDF['goldprice_boxcox_lambda_opt'].hist(bins=30,
                color='#A9C5D3', edgecolor='black', grid=False)
plt.axvline(goldprice_boxcox_mean, color='r')
ax.set_title('Gold Price after Box–Cox Transform',
         fontsize=12)
ax.set_xlabel('Gold Price (Box–Cox transform)', fontsize=12)
ax.set_ylabel('Frequency', fontsize=12)
ax.text(24, 450, r'$\mu$='+str(goldprice_boxcox_mean), fontsize=10)
```

Output:

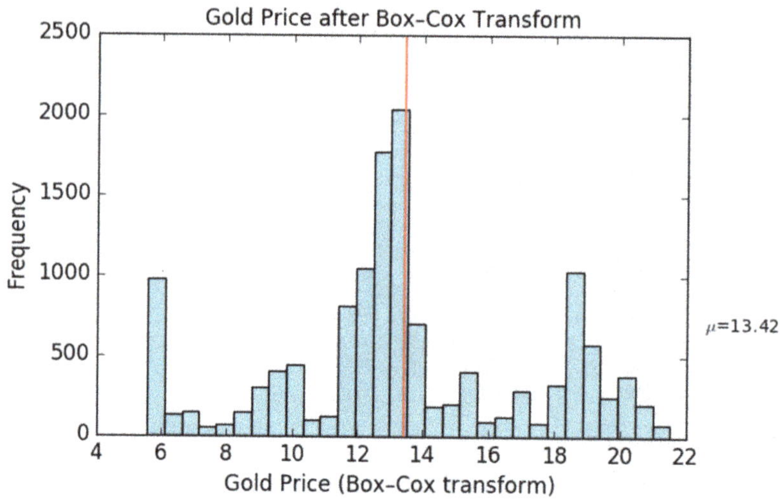

The distribution looks more normal-like like what obtained after the log transform.

Missing Data

Agricultural datasets come from various lines of sources, business, industry, academia, or field operations. Historically, the data sources are from transactional and from the reporting information systems. With the invent of Internet of Things (IoT) and smart and intelligent agriculture, the data sources, now, are more near real time or real-time, sub-milliseconds. As the sources of agriculture have expanded, so as the data issues. One could expect, therefore, the real-world agricultural datasets contain missing values, due to sensor calibration issue for IoT devices or data transformation issues when injecting into corporate or governmental data lakes, often encoded as NaN, blanks, or empty. Such datasets however are incompatible with machine learning and artificial intelligent algorithms, for instance, scikit-learn, which assume that all values in an array are numerical and that all have and hold meaning.[90]

A basic strategy to use incomplete datasets employed by many data scientists is to discard entire rows and/or columns containing missing values. Despite losing valuable information, despite forgoing underling data signal or a foreseeable prospective trend, this is a general operating practice followed by small and large data science teams. Lacking resources or even a theoretical framework, researchers, methodologists, and software developers resort to editing the data to lend an appearance of completeness. Unfortunately, ad hoc edits may do more harm than good, producing answers that are biased, inefficient (lacking in power), and unreliable[91] [25].

To maximize return on assets of information acquisition systems and to better serve customers, a novel strategy is to impute the missing values, i.e., to infer them from the known part of the data. The strategy for imputation could entail:

- Replace missing value with a placeholder value, mean, or other value[92] [26].
  - Imputation using (mean/median) values.
  - 3-Imputation Using (Most Frequent) or (Zero/Constant) Values
  - Imputation using k-NN.
  - Imputation using multivariate imputation by chained equation (MICE).
- Completely remove the rows and columns that have missing value.
- Inferring value based on statistical methods[93] (probabilistic principal component analysis (PPCA), multivariate imputation by chained equations (MICE)).

Imputation Using (Mean/Median) Values

This strategy works for column missing values and is highly effective for small numeric datasets. However, it does not factor in correlation between features and could lead to lower accuracies.

In the following code, load Bahamas agricultural dataset that relates the importance of rural population to the agriculture, forestry, and fishing, value added (annual % growth).

[90] Imputation of missing values – https://scikit-learn.org/stable/modules/impute.html

[91] Missing Data: Our View of the State of the Art – https://pubmed.ncbi.nlm.nih.gov/12090408/

[92] Imputation of missing values – https://scikit-learn.org/stable/modules/impute.html

[93] MICE vs PPCA: Missing data imputation in healthcare – https://www.sciencedirect.com/science/article/pii/S2352914819302783

Year	Inflation, consumer prices (annual %)	Rural population (% of total population)	Life expectancy at birth, female (years)	Life expectancy at birth, male (years)	GDP per capita growth (annual %)	Agriculture, forestry, and fishing, value added (annual % growth)
1990	4.66931912	20.162	73.678	66.756	−3.320656503	−1.775230039
1991	7.114681202	19.933	74.025	67.037	−5.938398218	24.19105373
1992	5.738198862	19.707	74.338	67.293	−5.640766603	−13.62491468
1993	2.72289767	19.482	74.608	67.522	−1.557221962	10.3794899
1994	1.399334237	19.26	74.831	67.723	1.362620477	1.041043772
1995	2.066994954	19.039	75	67.897	2.76268898	−2.420834325
1996	1.379137412	18.82	75.114	68.053	2.831084269	8.658968875
1997	0.544150383	18.604	75.179	68.2	0.861994441	−8.600751841
1998	1.336613366	18.389	75.204	68.343	3.548819111	−9.638554217
1999	1.25424826	18.176	75.196	68.485	5.882085305	−6.083333333
2000	1.606329417	17.993	75.16	68.623	2.755340737	9.050576752
2001	2.044989775	17.951	75.101	68.752	1.076378281	3.661513426
2002	2.173577925	17.908	75.025	68.866	1.021524148	11.93092622
2003	3.025045263	17.866	74.942	68.964	−2.989005042	9.677419355
2004	0.981914037	17.824	74.859	69.049	−0.941028833	−6.969309463
2005	1.591606181	17.782	74.782	69.124	1.496735718	−9.347079038
2006	2.389826274	17.74	74.715	69.197	0.604841969	−9.552691433
2007	2.492577949	17.698	74.659	69.275	−0.464399671	−6.370494552
2008	4.489559954	17.657	74.619	69.367	−4.111736706	3.491495076
2009	2.062734735	17.615	74.6	69.478	−5.799261534	3.287197232
2010	1.344027384	17.573	74.614	69.619	0.011342386	1.256281407
2011	3.198781417	17.524	74.67	69.799	−0.683537045	−7.775020678
2012	1.97336756	17.468	74.771	70.014	1.950585287	3.139013453
2013	1.034670351	17.404	74.914	70.26	−3.628949396	−18.60869565
2014	47.77596254	17.332	75.093	70.528	0.109572728	−5.982905983
2015	−30.24316023	17.254	75.296	70.802	−0.726544207	0.681818182
2016	−0.330038396	17.168	75.509	71.064	0.444173985	5.191873589
2017	1.499926233	17.075	75.718	71.3	2.081110777	−5.25751073
2018	2.267515101	16.975	75.911	71.499	1.989701453	−10.64552661
2019	**2.491235274**	**16.868**	..	..	**0.219421884**	**−11.91381496**

Please note that the "Life expectancy at birth, female (years)" and "Life expectancy at birth, male (years)" for 2019 are "..". That is data is missing for those columns.

Software code for this model can be found on GitHub link: Bahamas_RuralAgGrowth (Jupyter notebook code).

Step 1: Load Bahamas Dataset:

```
from sklearn.linear_model import LinearRegression
from sklearn.model_selection import StratifiedKFold
from sklearn.metrics import mean_squared_error
from math import sqrt
import random
import numpy as np
random.seed(0)

#Fetching the dataset
import pandas as pd
bahamasDF = pd.read_csv('Bahamas.csv')

bahamasDF.tail(10)
```

Output:

	Year	Inflation, consumer prices (annual %)	Rural population (% of total population)	Life expectancy at birth, female (years)	Life expectancy at birth, male (years)	GDP per capita growth (annual %)
20	2010	1.344027	17.573	74.614	69.619	0.011342
21	2011	3.198781	17.524	74.67	69.799	-0.683537
22	2012	1.973368	17.468	74.771	70.014	1.950585
23	2013	1.034670	17.404	74.914	70.26	-3.628949
24	2014	47.775963	17.332	75.093	70.528	0.109573
25	2015	-30.243160	17.254	75.296	70.802	-0.726544
26	2016	-0.330038	17.168	75.509	71.064	0.444174
27	2017	1.499926	17.075	75.718	71.3	2.081111
28	2018	2.267515	16.975	75.911	71.499	1.989701
29	2019	2.491235	16.868	NaN	NaN	0.219422

Step 2: Create train and target placeholders:

```
train = bahamasDF[['Year','Inflation, consumer prices (annual %)','Rural population (% of total
population)','Life expectancy at birth, female (years)',
        'Life expectancy at birth, male (years)','GDP per capita growth (annual %)']]
target = bahamasDF[['Agriculture, forestry, and fishing, value added (annual % growth)']]

train.tail(10)
```

Output:

	Year	Inflation, consumer prices (annual %)	Rural population (% of total population)	Life expectancy at birth, female (years)	Life expectancy at birth, male (years)	GDP per capita growth (annual %)
20	2010	1.344027	17.573	74.614	69.619	0.011342
21	2011	3.198781	17.524	74.67	69.799	-0.683537
22	2012	1.973368	17.468	74.771	70.014	1.950585
23	2013	1.034670	17.404	74.914	70.26	-3.628949
24	2014	47.775963	17.332	75.093	70.528	0.109573
25	2015	-30.243160	17.254	75.296	70.802	-0.726544
26	2016	-0.330038	17.168	75.509	71.064	0.444174
27	2017	1.499926	17.075	75.718	71.3	2.081111
28	2018	2.267515	16.975	75.911	71.499	1.989701
29	2019	2.491235	16.868	NaN	NaN	0.219422

Step 3: Simple Imputer:

In the following code, apply SCIKIT Simple Imputer to mean strategy to replace NaN with median values.

```
#Impute the values using scikit-learn SimpleImpute Class
from sklearn.impute import SimpleImputer
imp_mean = SimpleImputer( strategy='mean') #for median imputation replace 'mean' with 'median'
imp_mean.fit(train)
imputed_train_array = imp_mean.transform(train)
print(imputed_train_array)
column = imputed_train_array
print(column.size)
```

Output:

```
[[ 1.99000000e+03  4.66931912e+00  2.01620000e+01  7.36780000e+01
   6.67560000e+01 -3.32065650e+00]
 [ 1.99100000e+03  7.11468120e+00  1.99330000e+01  7.40250000e+01
   6.70370000e+01 -5.93839822e+00]
 [ 1.99200000e+03  5.73819886e+00  1.97070000e+01  7.43380000e+01
   6.72930000e+01 -5.64076660e+00]]

180
```

Convert array into Dataframe.[94]

```
imputed_traindf = pd.DataFrame(imputed_train_array, columns = ['Year','Inflation, consumer prices
(annual %)','Rural population (% of total population)','Life expectancy at birth, female (years)',
            'Life expectancy at birth, male (years)','GDP per capita growth (annual %)'])

imputed_traindf['Year'] = df['Year'].astype(int)
imputed_traindf.tail(10)
```

[94] Simple Imputer – https://scikit-learn.org/stable/modules/generated/sklearn.impute.SimpleImputer.html

Output:

	Year	Inflation, consumer prices (annual %)	Rural population (% of total population)	Life expectancy at birth, female (years)	Life expectancy at birth, male (years)	GDP per capita growth (annual %)
20	2010	1.344027	17.573	74.614000	69.619000	0.011342
21	2011	3.198781	17.524	74.670000	69.799000	-0.683537
22	2012	1.973368	17.468	74.771000	70.014000	1.950585
23	2013	1.034670	17.404	74.914000	70.260000	-3.628949
24	2014	47.775963	17.332	75.093000	70.528000	0.109573
25	2015	-30.243160	17.254	75.296000	70.802000	-0.726544
26	2016	-0.330038	17.168	75.509000	71.064000	0.444174
27	2017	1.499926	17.075	75.718000	71.300000	2.081111
28	2018	2.267515	16.975	75.911000	71.499000	1.989701
29	2019	2.491235	16.868	74.901069	69.065138	0.219422

Imputation Using (Most Frequent) or (Zero/Constant) Values

Most frequent imputation strategy[95] fits with numerical/categorical (mostly recommended) values. That is, missing categorical column values can be filled with this strategy.[96]

Software code for this model can be found on GitHub link: Bahamas_RuralAgGrowth (Jupyter notebook code).

Step 1: Create most frequent imputer:

```
#Impute the values using scikit-learn SimpleImpute Class
from sklearn.impute import SimpleImputer
imp_mean = SimpleImputer(missing_values=np.nan, strategy='most_frequent')
imp_mean.fit(train)

imputed_train_mf_array = imp_mean.transform(train)
print(imputed_train_mf_array)
column = imputed_train_mf_array
print(column.size)
```

Step 2: Imputed Most Frequent Dataframe.

[95] sklearn SimpleImputer too slow for categorical data represented as string values – https://datascience.stackexchange.com/questions/66034/sklearn-simpleimputer-too-slow-for-categorical-data-represented-as-string-values

[96] 6 Different Ways to Compensate for Missing Values In a Dataset (Data Imputation with examples) – https://towardsdatascience.com/6-different-ways-to-compensate-for-missing-values-data-imputation-with-examples-6022d9ca0779

```
imputed_trainMFdf = pd.DataFrame(imputed_train_mf_array, columns = ['Year','Inflation, consumer prices (annual %)','Rural population (% of total population)','Life expectancy at birth, female (years)',

        'Life expectancy at birth, male (years)','GDP per capita growth (annual %)'])

imputed_trainMFdf['Year'] = df['Year'].astype(int)
imputed_trainMFdf.tail(10) ['b']]
```

	Year	Inflation, consumer prices (annual %)	Rural population (% of total population)	Life expectancy at birth, female (years)	Life expectancy at birth, male (years)	GDP per capita growth (annual %)
20	2010	1.34403	17.573	74.614	69.619	0.0113424
21	2011	3.19878	17.524	74.67	69.799	-0.683537
22	2012	1.97337	17.468	74.771	70.014	1.95059
23	2013	1.03467	17.404	74.914	70.26	-3.62895
24	2014	47.776	17.332	75.093	70.528	0.109573
25	2015	-30.2432	17.254	75.296	70.802	-0.726544
26	2016	-0.330038	17.168	75.509	71.064	0.444174
27	2017	1.49993	17.075	75.718	71.3	2.08111
28	2018	2.26752	16.975	75.911	71.499	1.9897
29	2019	2.49124	16.868	73.678	66.756	0.219422

The imputed strategy, bounding rectangle, replaces NAN values with converted values.

ⓘ	Categorical values take long time to fit.

Imputation Using K-NN

The k-nearest neighbors is an algorithm that is used for simple classification purposes. The algorithm uses "feature similarity" to predict the values of any new data points.

The idea is impute array with a passed in initial impute function (mean impute) and then use the resulting complete array to construct a KDTree.[97] Use this KDTree to compute nearest neighbors. After finding "k"-nearest neighbors, take the weighted average of them.

K-D tree is a partitioning data structure that arrange points in k-dimensional space. K-D tree performs analysis in memory.

Let's see some example code using Impyute library which provides a simple and easy way to use KNN for imputation:

Software code for this model can be found on GitHub link: Bahamas_RuralAgGrowthkNN (Jupyter notebook code).

[97] Source code for impyute.imputation.cs.fast_knn – https://impyute.readthedocs.io/en/master/_modules/impyute/imputation/cs/fast_knn.html

Step 1 and 2 are the same: Load Bahamas Dataset and creation of Bahamas DF.
Step 3:

Apply kNN = 1. That is, average (by distance) of nearest neighbor 1 is fills all null cells with the nearest cell value.

```
import sys
from impyute.imputation.cs import fast_knn
from math import sqrt
import random
import numpy as np

sys.setrecursionlimit(100000) #Increase the recursion limit of the OS

data = bahamasDF.to_numpy()
# Weighted average (by distance) of nearest 1 neighbour
print('\naverage (by distance) of nearest 1 neighbour')
imputed_bahamasDF_knn1_array =fast_knn(data, k=1)
print(imputed_bahamasDF_knn1_array)
```

Output:

```
average (by distance) of nearest 1 neighbour
[[ 1.99000000e+03  4.66931912e+00  2.01620000e+01  7.36780000e+01
   6.67560000e+01 -3.32065650e+00 -1.77523004e+00]
 [ 1.99100000e+03  7.11468120e+00  1.99330000e+01  7.40250000e+01
   6.70370000e+01 -5.93839822e+00  2.41910537e+01]
 [ 1.99200000e+03  5.73819886e+00  1.97070000e+01  7.43380000e+01
   6.72930000e+01 -5.64076660e+00 -1.36249147e+01]]
```

Step 4: Convert kNN array back into Dataframe:

```
imputed_kNNtraindf = pd.DataFrame(imputed_bahamasDF_knn1_array, columns = ['Year','Inflation, consumer prices (annual %)','Rural population (% of total population)','Life expectancy at birth, female (years)',
          'Life expectancy at birth, male (years)','GDP per capita growth (annual %)','Agriculture, forestry, and fishing, value added (annual % growth)'])
imputed_kNNtraindf['Year'] = imputed_kNNtraindf['Year'].astype(int)
imputed_kNNtraindf.tail(10)
```

Output:

	Year	Inflation, consumer prices (annual %)	Rural population (% of total population)	Life expectancy at birth, female (years)	Life expectancy at birth, male (years)	GDP per capita growth (annual %)	Agriculture, forestry, and fishing, value added (annual % growth)
20	2010	1.344027	17.573	74.614	69.619	0.011342	1.256281
21	2011	3.198781	17.524	74.670	69.799	-0.683537	-7.775021
22	2012	1.973368	17.468	74.771	70.014	1.950585	3.139013
23	2013	1.034670	17.404	74.914	70.260	-3.628949	-18.608696
24	2014	47.775963	17.332	75.093	70.528	0.109573	-5.982906
25	2015	-30.243160	17.254	75.296	70.802	-0.726544	0.681818
26	2016	-0.330038	17.168	75.509	71.064	0.444174	5.191874
27	2017	1.499926	17.075	75.718	71.300	2.081111	-5.257511
28	2018	2.267515	16.975	75.911	71.499	1.989701	-10.645527
29	2019	2.491235	16.868	75.911	71.499	0.219422	-11.913815

For K = 2 and higher, the weighted average values of the data are as below. As you can see the weighted average values are computed for K is greater than 1.

```
# Weighted average of nearest 2 neighbour
print('\nWeighted average of nearest 2 neighbour')
imputed_bahamasDF_knn2_array =fast_knn(data, k=2)
print(imputed_bahamasDF_knn2_array)

# Weighted average of nearest 2 neighbour
print('\nWeighted average of nearest 3 neighbour')
imputed_bahamasDF_knn3_array =fast_knn(data, k=3)
print(imputed_bahamasDF_knn3_array)
```

Output:

K = 2.

	Year	Inflation, consumer prices (annual %)	Rural population (% of total population)	Life expectancy at birth, female (years)	Life expectancy at birth, male (years)	GDP per capita growth (annual %)	Agriculture, forestry, and fishing, value added (annual % growth)
20	2010	1.344027	17.573	74.614000	69.619000	0.011342	1.256281
21	2011	3.198781	17.524	74.670000	69.799000	-0.683537	-7.775021
22	2012	1.973368	17.468	74.771000	70.014000	1.950585	3.139013
23	2013	1.034670	17.404	74.914000	70.260000	-3.628949	-18.608696
24	2014	47.775963	17.332	75.093000	70.528000	0.109573	-5.982906
25	2015	-30.243160	17.254	75.296000	70.802000	-0.726544	0.681818
26	2016	-0.330038	17.168	75.509000	71.064000	0.444174	5.191874
27	2017	1.499926	17.075	75.718000	71.300000	2.081111	-5.257511
28	2018	2.267515	16.975	75.911000	71.499000	1.989701	-10.645527
29	2019	2.491235	16.868	75.779436	71.363346	0.219422	-11.913815

K = 3.

	Year	Inflation, consumer prices (annual %)	Rural population (% of total population)	Life expectancy at birth, female (years)	Life expectancy at birth, male (years)	GDP per capita growth (annual %)	Agriculture, forestry, and fishing, value added (annual % growth)
20	2010	1.344027	17.573	74.614000	69.619000	0.011342	1.256281
21	2011	3.198781	17.524	74.670000	69.799000	-0.683537	-7.775021
22	2012	1.973368	17.468	74.771000	70.014000	1.950585	3.139013
23	2013	1.034670	17.404	74.914000	70.260000	-3.628949	-18.608696
24	2014	47.775963	17.332	75.093000	70.528000	0.109573	-5.982906
25	2015	-30.243160	17.254	75.296000	70.802000	-0.726544	0.681818
26	2016	-0.330038	17.168	75.509000	71.064000	0.444174	5.191874
27	2017	1.499926	17.075	75.718000	71.300000	2.081111	-5.257511
28	2018	2.267515	16.975	75.911000	71.499000	1.989701	-10.645527
29	2019	2.491235	16.868	75.281415	70.661118	0.219422	-11.913815

K-NN imputation is much more accurate than Simple Imputer with mean/median or frequent imputation strategies. Nonetheless, computation expensive as k-NN does in memory data wrangling and training. If dataset has quite higher outliers, the performance and accuracy of k-NN is affected immensely. Needless to mention, expense would be much higher as compared to datasets with least outliers.

Imputation Using Multivariate Imputation by Chained Equation (MICE)

"Multivariate Imputation using Chained Equations" or "Multiple Imputation by Chained Equations" (MICE) imputation account for the process that created missing data, preserve the relations in the data, a stark contrast to what constant or mean/median simple imputer, and preserves the uncertainty about relations. Please find main steps used in MICE[98]: MICE perform various regressions to predict missing values from the dataset [27]. The type of regression is based on the type of missing column. With a multiple imputation method, each variable with missing data is modeled conditionally using the other variables in the data before filling in the missing values. The MICE,[99] sometimes called "fully conditional specification" or "sequential regression multiple imputation," has emerged in the statistical literature as one principled method of addressing missing data [28]. Creating multiple imputations, as opposed to single imputations, accounts for the statistical uncertainty in the imputations. In addition, the chained equations approach is very flexible and can handle variables of varying types (e.g., continuous, or binary) as well as complexities such as bounds or survey skip patterns.

The leftmost side of the picture indicates that the analysis starts with an observed, incomplete dataset Y_{obs}. The issue we cannot estimate Q from Y_{obs} without unrealistic assumption. Multiple imputation is a general framework that several imputed versions of the data by replacing the missing values by plausible data values. These plausible values are drawn from a distribution specifically modeled for each missing entry. Above figure portrays m = 3 imputed datasets $Y^{(1)}$ …, $Y^{(3)}$. The three imputed sets are identical for the non-missing data but differ in the imputed values. The magnitude of these differences reflects our uncertainty about what value to impute (mids store the imputed value). The second step is to estimate Q value for each imputed dataset (with.mids). The last step is to estimate variable. Numerous regression models are run in such a way that the variable with missing data is modeled depending on other variables in the dataset.[100] Each variable is modeled taking into account the type of the variable (please see Fig. 2.26).

For example, logistic regression is used to model binary variables, whereas predictive mean matching is used for continuous variables [26, 28].

- Step 1: The first step involves replacing every missing data with the mean of the observed values for the variable, which acts as a placeholder.
- Step 2: The second step involves setting these mean imputations back to "missing."
- Step 3: In the third step, the observed values of a variable (e.g., "x") are regressed on the other variables such that "x" is the dependent variable and the rest are independent variables.

---

[98] MICE – https://www.jstatsoft.org/article/view/v045i03/v45i03.pdf

[99] Multiple imputation by chained equations: what is it and how does it work? – https://www.ncbi.nlm.nih.gov/pmc/articles/PMC3074241/

[100] MICE vs PPCA: Missing data imputation in healthcare – https://www.sciencedirect.com/science/article/pii/S2352914819302783#bib13

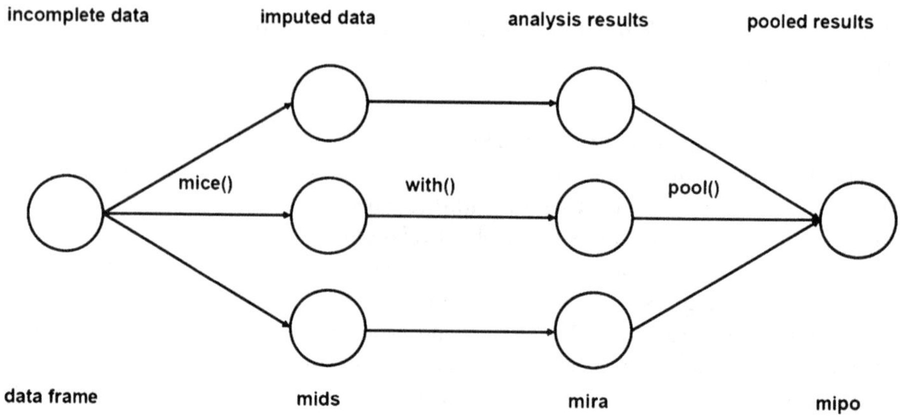

Fig. 2.26 MICE [28]

- Step 4: The fourth step involves replacing the missing values with the predictions derived from the regression model. This imputed value would then be part of the independent variables along with observed values for other variables. Steps "2" to "4" are then repeated for each variable that has missing values, constituting one "iteration." After one iteration, all missing values are replaced by the regressed predictions related to the observed data. The imputed values are replaced after every iteration, and the number of iterations may vary.

Software code for this model can be found on GitHub link: Bahamas_RuralAgGrowthMICE (Jupyter notebook code).

Step 1: Load Bahamas Dataset:

```
from sklearn.linear_model import LinearRegression
from sklearn.model_selection import StratifiedKFold
from sklearn.metrics import mean_squared_error
from math import sqrt
import random
import numpy as np
random.seed(0)

#Fetching the dataset
import pandas as pd
bahamasDF = pd.read_csv('Bahamas.csv')

bahamasDF.tail(10)
```

Output:

	Year	Inflation, consumer prices (annual %)	Rural population (% of total population)	Life expectancy at birth, female (years)	Life expectancy at birth, male (years)	GDP per capita growth (annual %)	Agriculture, forestry, and fishing, value added (annual % growth)
20	2010	1.344027	17.573	74.614	69.619	0.011342	1.256281
21	2011	3.198781	17.524	74.67	69.799	-0.683537	-7.775021
22	2012	1.973368	17.468	74.771	70.014	1.950585	3.139013
23	2013	1.034670	17.404	74.914	70.26	-3.628949	-18.608696
24	2014	47.775963	17.332	75.093	70.528	0.109573	-5.982906
25	2015	-30.243160	17.254	75.296	70.802	-0.726544	0.681818
26	2016	-0.330038	17.168	75.509	71.064	0.444174	5.191874
27	2017	1.499926	17.075	75.718	71.3	2.081111	-5.257511
28	2018	2.267515	16.975	75.911	71.499	1.989701	-10.645527
29	2019	2.491235	16.868	..	..	0.219422	-11.913815

Step 2: Load entire dataset:

```
bahamasDF = bahamasDF[['Year','Inflation, consumer prices (annual %)','Rural population (% of total
population)','Life expectancy at birth, female (years)',
          'Life expectancy at birth, male (years)','GDP per capita growth (annual %)','Agriculture,
forestry, and fishing, value added (annual % growth)']]
bahamasDF.tail(10)
```

Output:

	Year	Inflation, consumer prices (annual %)	Rural population (% of total population)	Life expectancy at birth, female (years)	Life expectancy at birth, male (years)	GDP per capita growth (annual %)	Agriculture, forestry, and fishing, value added (annual % growth)
20	2010	1.344027	17.573	74.614	69.619	0.011342	1.256281
21	2011	3.198781	17.524	74.67	69.799	-0.683537	-7.775021
22	2012	1.973368	17.468	74.771	70.014	1.950585	3.139013
23	2013	1.034670	17.404	74.914	70.26	-3.628949	-18.608696
24	2014	47.775963	17.332	75.093	70.528	0.109573	-5.982906
25	2015	-30.243160	17.254	75.296	70.802	-0.726544	0.681818
26	2016	-0.330038	17.168	75.509	71.064	0.444174	5.191874
27	2017	1.499926	17.075	75.718	71.3	2.081111	-5.257511
28	2018	2.267515	16.975	75.911	71.499	1.989701	-10.645527
29	2019	2.491235	16.868	..	..	0.219422	-11.913815

Step 3: replace ".." with NAN values:

```
bahamasDF['Life expectancy at birth, female (years)']=bahamasDF['Life expectancy at birth, female
(years)'].replace('..', np.nan)
bahamasDF['Life expectancy at birth, male (years)']=bahamasDF['Life expectancy at birth, male
(years)'].replace('..', np.nan)
```

	Year	Inflation, consumer prices (annual %)	Rural population (% of total population)	Life expectancy at birth, female (years)	Life expectancy at birth, male (years)	GDP per capita growth (annual %)
20	2010	1.344027	17.573	74.614	69.619	0.011342
21	2011	3.198781	17.524	74.67	69.799	-0.683537
22	2012	1.973368	17.468	74.771	70.014	1.950585
23	2013	1.034670	17.404	74.914	70.26	-3.628949
24	2014	47.775963	17.332	75.093	70.528	0.109573
25	2015	-30.243160	17.254	75.296	70.802	-0.726544
26	2016	-0.330038	17.168	75.509	71.064	0.444174
27	2017	1.499926	17.075	75.718	71.3	2.081111
28	2018	2.267515	16.975	75.911	71.499	1.989701
29	2019	2.491235	16.868	NaN	NaN	0.219422

Step 4: Convert Life Expectancy Male/Female Columns to FLOAT:

```
bahamasDF['Life expectancy at birth, female (years)'] = bahamasDF['Life expectancy at birth, female (years)'].astype(float)
bahamasDF['Life expectancy at birth, male (years)'] = bahamasDF['Life expectancy at birth, male (years)'].astype(float)
```

Step 5: Apply the Multivariate Imputation by Chained Equation (MICE):

```
imputed_bahamasDF_array = mice(bahamasDF.values)
print(imputed_bahamasDF_array)
```

Output:

```
[[ 1.99000000e+03 4.66931912e+00 2.01620000e+01 7.36780000e+01
   6.67560000e+01 -3.32065650e+00 -1.77523004e+00]
 [ 1.99100000e+03 7.11468120e+00 1.99330000e+01 7.40250000e+01
   6.70370000e+01 -5.93839822e+00 2.41910537e+01]
 [ 1.99200000e+03 5.73819886e+00 1.97070000e+01 7.43380000e+01
   6.72930000e+01 -5.64076660e+00 -1.36249147e+01]
 [ 1.99300000e+03 2.72289767e+00 1.94820000e+01 7.46080000e+01
   6.75220000e+01 -1.55722196e+00 1.03794899e+01]
 [ 1.99400000e+03 1.39933424e+00 1.92600000e+01 7.48310000e+01
   6.77230000e+01 1.36262048e+00 1.04104377e+00]]
```

Step 6: Convert Imputed array to Dataframe:

```
imputed_MICEtraindf = pd.DataFrame(imputed_bahamasDF_array, columns = ['Year','Inflation, consumer prices (annual %)','Rural population (% of total population)','Life expectancy at birth, female (years)',
        'Life expectancy at birth, male (years)','GDP per capita growth (annual %)','Agriculture, forestry, and fishing, value added (annual % growth)'])


imputed_MICEtraindf['Year'] = imputed_MICEtraindf['Year'].astype(int)
imputed_MICEtraindf
```

Output:

	Year	Inflation, consumer prices (annual %)	Rural population (% of total population)	Life expectancy at birth, female (years)	Life expectancy at birth, male (years)	GDP per capita growth (annual %)	Agriculture, forestry, and fishing, value added (annual % growth)
20	2010	1.344027	17.573	74.614000	69.619000	0.011342	1.256281
21	2011	3.198781	17.524	74.670000	69.799000	-0.683537	-7.775021
22	2012	1.973368	17.468	74.771000	70.014000	1.950585	3.139013
23	2013	1.034670	17.404	74.914000	70.260000	-3.628949	-18.608696
24	2014	47.775963	17.332	75.093000	70.528000	0.109573	-5.982906
25	2015	-30.243160	17.254	75.296000	70.802000	-0.726544	0.681818
26	2016	-0.330038	17.168	75.509000	71.064000	0.444174	5.191874
27	2017	1.499926	17.075	75.718000	71.300000	2.081111	-5.257511
28	2018	2.267515	16.975	75.911000	71.499000	1.989701	-10.645527
29	2019	2.491235	16.868	71.871676	69.262625	0.219422	-11.913815

After reading this chapter, you should be able to answer data engineering and feature engineering frameworks-related processing ML datasets. You should be able to answer Knowledge Discovery in Databases (KDD), Data Mining Reference Architecture, exploratory data analysis (EDA), and missing values (SMOTE, ADASYN) strategies, impute strategies, and, finally, able to programmatically develop and run models to address EDA, missing values, and imputation strategies.

References

1. Han, J., Kamber, M., Pei, J., Mining, D.: Concepts and Techniques, 3rd edn. Morgan Kaufmann (2011) ISBN-10: 9780123814791
2. Camp, K.M.: The relationship between crude oil prices and export prices of major agricultural commodities, 8(7), https://www.bls.gov/opub/btn/volume-8/the-relationship-between-crude-oil-and-export-prices-of-major-agricultural-commodities.htm?view_full (2019, April), Accessed 18 Sept 2020
3. Darr, T., Gribsons, G.: How U.S. exports are faring in the World. https://www.bls.gov/opub/mlr/1985/10/art2full.pdf (1985, October). Accessed 12 Dec 2020
4. Pratt, S.: Wheat exports to U.S. fall by half. https://www.producer.com/markets/wheat-exports-to-u-s-fall-by-half/ (2019, October 10), Accessed 8 Jan 2020
5. Ranganathan, J., Waite, R., Searchinger, T., Hanson, C.: How to Sustainably Feed 10 Billion People by 2050, in 21 Charts. https://www.wri.org/blog/2018/12/how-sustainably-feed-10-billion-people-2050-21-charts (2018, December 5). Accessed 18 Sept 2019
6. Vuppalapati, C.: Building Enterprise IoT Applications, 1st edn. CRC Press (2019) ISBN-13: 978-0367173852
7. Sakamoto, T., Gitelson, A.A., Arkebauer, T.J.: MODIS-based corn grain yield estimation model incorporating crop phenology information. Rem. Sens. Environ. **131**, 215–231 (2013). https://doi.org/10.1016/j.rse.2012.12.017. ISSN 0034-4257
8. Semega, J., Kollar, M., Shrider, E.A., Creamer, J.: Income and Poverty in the United States: 2019, Report Number P60-270, https://www.census.gov/data/tables/2020/demo/income-poverty/p60-270.html (2020, September). Accessed 8 Jan 2021
9. Nseir, A., White, T.: Family-owned farms account for 96% of U.S. farms, according to the Census of Agriculture Typology Report. https://www.nass.usda.gov/Newsroom/2021/01-22-2021.php (2021, Jan 22). Accessed 5 Feb 2021
10. Venancio, L.P., Filgueiras, R., Mantovani, E.C., et al.: Impact of drought associated with high temperatures on Coffea canephora plantations: a case study in Espírito Santo State, Brazil. Sci. Rep. **10**, 19719 (2020). https://doi.org/10.1038/s41598-020-76713-y
11. NIST/SEMATECH e-Handbook of Statistical Methods, http://www.itl.nist.gov/div898/handbook/ (April, 2012) Accessed 18 Sept 2018
12. Liu, G., Li, Y., Migliaccio, K.W., Ouyang, Y., Alva, A.K.: Identification of factors most important for ammonia emission from fertilized soils for potato production using principal component analysis. J. Sust. Water. Sci. Manag. **1**(1), 21–30 (2011)
13. Aguiar, R.: An overview of model explain ability in modern machine learning. https://towardsdatascience.com/an-overview-of-model-explainability-in-modern-machine-learning-fc0f22c8c29a (2019, Dec 4). Accessed 30 Apr 2020
14. Turek, M.: Explainable Artificial Intelligence (XAI), https://www.darpa.mil/program/explainable-artificial-intelligence (2020). Accessed 20 Dec 2020
15. Lipovetsky, S., Conklin, M.: Analysis of regression in game theory approach. Appl. Stoch. Model. Bus. Ind. **17**(4), 319–330 (2001)

16. Ribeiro, M.T., Singh, S., Guestrin, C.: Why should I trust you? Explaining the predictions of any classifier. Proceedings of the 22nd ACM SIGKDD international conference on knowledge discovery and data mining. ACM (2016)
17. Brownlee, J.: SMOTE for Imbalanced Classification with Python. https://machinelearningmastery.com/smote-oversampling-for-imbalanced-classification/ (2021, March 17). 25 Mar 2021
18. Chemchem, A., Alin, F., Krajecki, M.: Combining SMOTE sampling and machine learning for forecasting wheat yields in France. International Conference on Artificial Intelligence and Knowledge Engineering (AIKE), Cagliari, Italy (2019). https://doi.org/10.1109/AIKE.2019.00010ff.ffhal-02523637f
19. de Brix, D.: Oversampling SMOTE and ADASYN. https://www.kaggle.com/ruzarx/oversampling-smote-and-adasy (2019, March). Accessed Mar 2021
20. Chawla, N.V., Bowyer, K.W., Hall, L.O., Philip Kegelmeyer, W.: SMOTE: Synthetic minority over-sampling technique. J. Artif. Int. Res. **16**(1), 321–357 (2002, January)
21. Veenadhari, S., Mishra, B., Singh, C.D.: Soybean productivity modelling using decision tree algorithms. Int. J. Comp. Appl. **27**(7), 11–15 (2011, August)
22. Sarkar, D.: Continuous Numeric Data Strategies for working with continuous, numerical data, https://towardsdatascience.com/understanding-feature-engineering-part-1-continuous-numeric-data-da4e47099a7b (2018, Jan 4). Accessed 8 Jan 2021
23. Box, G.E.P., Cox, D.R.: An analysis of transformations. J. Royal Stat. Soc. B Methodol. **26**(2), 211–252 (1964) JSTOR, www.jstor.org/stable/2984418. Accessed 30 Jan 2021
24. Hussan, A.: Project: Gold vs Silver Price. http://rstudio-pubs-static.s3.amazonaws.com/368841_224fe698bd7c45d8a5dd9a6b639988fc.html (2021). Accessed 30 Jan 2021
25. Schafer, J.L., Graham, J.W.: Missing data: our view of the state of the art. Psychol. Methods. **7**(2), 147–177 (2002, Jun)
26. Hegde, H., Shimpi, N., Panny, A., Glurich, I., Christie, P., Acharya, A.: MICE vs PPCA: Missing data imputation in healthcare. Info Med Unlock. **17**, 100275 (2019). https://doi.org/10.1016/j.imu.2019.100275. ISSN 2352-9148. Accessed 8 Nov 2020
27. van Buuren, S., Groothuis-Oudshoorn, K.: mice: Multivariate Imputation by Chained Equations in R. J. Stat. Soft. [Online]. **45**(3), 1–67 (2011) Web. 27 Mar. 2021
28. Azur, M.J., Stuart, E.A., Frangakis, C., Leaf, P.J.: Multiple imputation by chained equations: what is it and how does it work? https://www.ncbi.nlm.nih.gov/pmc/articles/PMC3074241/ (2011, Mar). Accessed May 2020

Section II
GDP and Commodity Markets

Chapter 3
Agricultural Economy and ML Models

This chapter covers:
- Global agriculture landscape.
- Agriculture farm size, scale classification, and ML algorithms.
- Agricultural economics.
- Data sources.
- Macroeconomy and gross domestic product (GDP).
- GDP calculation and machine learning models.
- Predicting real GDP growth.
- Challenges.

The chapter starts with the introduction of global agriculture landscape and farm size economy across the world. Next, it introduces agricultural economics and influence of macrolevel economic indicators on the local agriculture. Next, study of econometrics-related agricultural models is introduced. Finally, it concludes with the challenges that agriculture faces, especially for small-scale farmers and the role of machine learning in addressing such challenges, as well as the opportunities machine learning provides to improve the future of small-scale farming.

Global Agriculture Landscape

Agriculture is one of the oldest and most noble professions. Although agriculture has no single or simple origin,[1] the domestication of animals and agriculture is assumed to have started 12,000 years ago.[2] The development of agriculture about

[1] How Agriculture and Domestication Began – https://www.britannica.com/topic/agriculture/How-agriculture-and-domestication-began

[2] Where Did Agriculture Begin? Oh Boy, It's Complicated – https://www.npr.org/sections/thesalt/2016/07/15/485722228/where-did-agriculture-begin-oh-boy-its-complicated#:~:text=Sometime%20around%2012%2C000%20years%20ago,like%20goats%20and%20wild%20oxen.

C. Vuppalapati, *Machine Learning and Artificial Intelligence for Agricultural Economics*, International Series in Operations Research & Management Science 314,
https://doi.org/10.1007/978-3-030-77485-1_3

12,000 years ago changed the way we, humans, lived. Thanks to agriculture, we switched from nomadic hunter-gatherer lifestyles to permanent settlements and farming.[3] In the present day, agriculture is one of the largest worldwide employers and one of the biggest industries; in 2018, agriculture contributed to 4% (see Fig. 3.1) of the global gross domestic product (GDP), and, in some developing countries, it can account for more than 25% of the GDP[4] [1]. Globally, food and agriculture business form a $5 trillion industry that is growing [2], with the largest agricultural output, in 2018, being from China ($978.5 billion), India ($397 billion), the Unites States ($178.6 billion), Indonesia ($133.5 billion), and Nigeria ($84.2 billion).[5] The countries that developed with agriculture include China, India, the United States, the Republic of Indonesia, Brazil, Liberia, and Somalia [3].

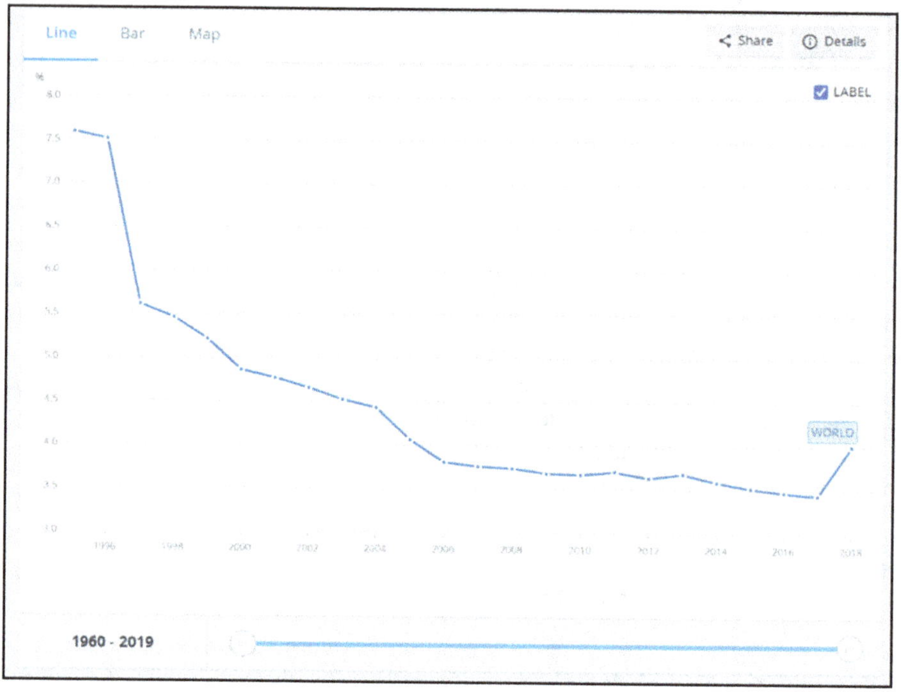

Fig. 3.1 Agriculture, forestry, and fishing, value added (% of GDP): Source – The World Bank. https://data.worldbank.org/indicator/NV.AGR.TOTL.ZS

[3] The Development of Agriculture – https://www.nationalgeographic.org/article/development-agriculture/

[4] Agriculture and Food Industries – https://www.worldbank.org/en/topic/agriculture/overview

[5] Largest Agricultural Output – https://worldinfigures.com/rankings/index/103

The Unites States of America

In the United States, agriculture, food, and related industries contributed $1.109 trillion to the gross domestic product (GDP) in 2019,[6] a 5.2% share [4]. The output of America's farms contributed $136.1 billion of this sum—about 0.6% of the GDP. The overall contribution of agriculture to GDP is larger than 0.6%[7] because sectors related to agriculture rely on agricultural inputs to contribute added value to the economy [5]. In the Unites States, 22.2 million full-time and part-time jobs were related to agricultural and food sector, in 2019 (please see Fig. 3.2). Agriculture provides 10.9% of the total US employment. Direct on-farm employment accounted for about 2.6 million of these jobs, or 1.3% of US employment [4].

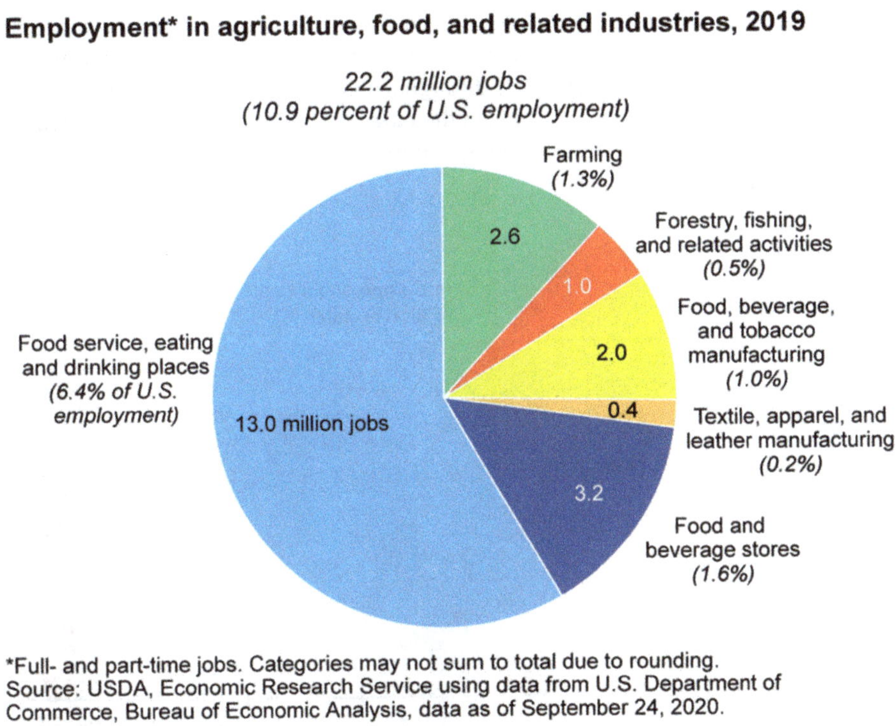

Fig. 3.2 Employment in Agriculture [4]

[6] Ag and Food Sector and the Economy – https://www.ers.usda.gov/data-products/ag-and-food-statistics-charting-the-essentials/ag-and-food-sectors-and-the-economy/

[7] What is agriculture's share of the overall US economy? – https://www.ers.usda.gov/data-products/chart-gallery/gallery/chart-detail/?chartId=58270

Agricultural exports contribute 23% of overall exports in the Unites States [3].

European Union

In Europe, agriculture is the major employment provider with about 10.5 million agricultural holdings in 2016. Despite this, farm numbers, ironically, have been in steep decline for many years. Most of the EU's farms are small in nature (please see Fig. 3.3); two-thirds were less than 5 hectares in size in 2016.[8] These small farms play an important role in reducing the risk of rural poverty, by providing additional income and food. Europe's larger farms (those over 100 hectares in size) constitute the majority of all farmlands (52.7% of 173 million hectares of land); however the majority of agriculture employment is through small farms [6]. In 2016, around 9.7 million people worked in agriculture, and the bulk (96%) of them are family owned. Additionally, in 2016, Europe had more specialist crop farms (52.5%) than livestock farms (25.1%).

In Europe, small farms cultivate around 116.9 million hectares [6].

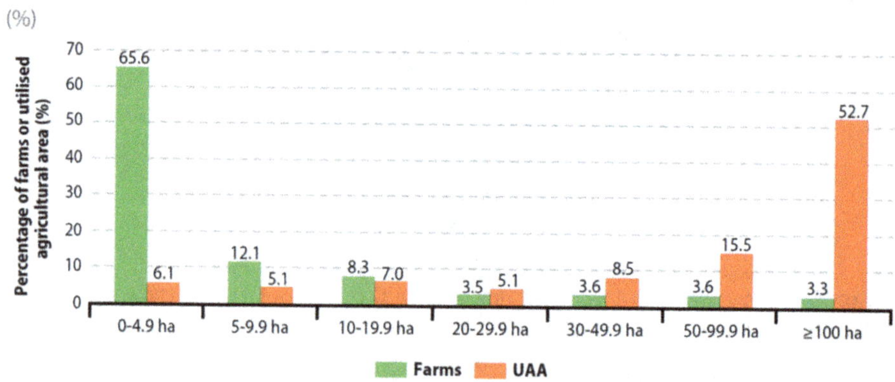

Note: there are some differences in the threshold applied by some Member States, often to exclude the very smallest agricultural holdings which together contribute 2% or less to the total UAA excluding common land, and 2% or less to the total number of farm livestock units.

Fig. 3.3 Distribution of EU farms and utilized agricultural area according to farm size, 2016 [6]

---

[8]Agriculture, Forestry, and fishery statistics – 2018 – https://ec.europa.eu/eurostat/documents/3217494/9455154/KS-FK-18-001-EN-N.pdf/a9ddd7db-c40c-48c9-8ed5-a8a90f4faa3f

Asia

In Asia, about 39 of its countries are contributing to most of its agricultural produce, and a are source of food for 99% of its population. The agriculture sector plays an important role in enhancing[9] the standard of life for people in these countries [7]. Like Europe, agriculture and animal husbandry are the major sources of income for the people in Asia. It is fundamentally true that animal resource development activities play an important role in countries with an agrarian economy. Agriculture is important for all countries of Asia and the Pacific, with more than 2.2 billion people relying[10] on agriculture for their livelihood [8]. Small and marginal farmers hold 86% of the agricultural landholding in India.[11] The economies of China, India, and the Republic of Indonesia grow with agriculture. Notable Asian countries with agricultural economies:

Country	Notes
China	Agriculture contributes to 12% of the GDP. 50% of workforce depend on agriculture.
India	Agriculture contributes to 18% of GDP. Over 58% of rural households depend on the agriculture.
Indonesia	14% of GDP is contributed by the agriculture. Indonesia has the highest per capita price production in the world [3].

Cash crops[12] (1964–1966 and 1970–1972): cotton, jute, and sugarcane [9]

Africa

Agriculture is by far the single most important economic activity in Africa.[13] It provides employment for about two-thirds of the continent's working population and for each country contributes an average of 30–60% of gross domestic product and

[9] Statistical Evaluation of agricultural development in Asian countries – http://www.stats.gov.cn/english/icas/papers/P020071017429375003907.pdf

[10] Building Climate Resilience in the Agriculture Sector of Asia and the Pacific – https://www.adb.org/sites/default/files/publication/27531/building-climate-resilience-agriculture-sector.pdf

[11] Agricultural Ordinances 2020 – https://www.prsindia.org/sites/default/files/bill_files/Legislative%20Brief%20-%20Agriculture%20Ordinances%2C%202020.pdf

[12] Parikh, Ashok, 1982. "An econometric model of the agricultural sector of the Indian economy" – https://doi.org/10.1016/0161-8938(82)90027-8

[13] Africa – Agriculture – https://www.britannica.com/place/Africa/Agriculture

about 30% of the value of exports. While Africa holds more than 60% of the world's arable land,[14] the continent's share in global agricultural production remains low. Vast areas of land are uncultivated, and productivity is lower than in much of the rest of the world. Like Europe and Asia, the vast majority of farming is working on small-scale farms[15] that currently produce around 90% of all output [10]. Family farming accounts for 90% of agricultural activity in Africa.[16] Despite the increased production, domestic demand in these states is set to continue to outstrip supply, with populations expected to double by 2050. Therefore, meeting consumption demand will continue to depend on global markets and imports over the medium term. However, despite the challenges faced, prospects for Africa's agricultural sector are relatively positive [11]. UN institutions expect cultivated areas to expand and farmers to increase productivity, through greater use of technology and improved inputs. According to the "Agriculture Outlook 2018–2027" report from the Organization for Economic Co-operation and Development (OECD) and the UN Food and Agriculture Organization, the sector will undergo robust growth, with crop production in sub-Saharan Africa projected to rise 30% between 2018 and 2028 [12]. Moreover, greater access to innovative technologies is expected to support the development and application of smart and precision farming techniques.[17]

Most farmers engage in subsistence agriculture, which can prevent them from making a profit and investing in their farms. Several initiatives have been designed to empower farmers, particularly women and youth, and help them turn their farming into sustainable businesses [12].

The Latin America and Caribbean

The Latin America and Caribbean (LAC) region covers more than 2 billion ha[18] and encompasses 34 countries with a total estimated population in 2018 of 657 million—a low average population density of 0.34 persons per hectare [13]. Latin America and Caribbean (LAC) accounts for about a quarter of global exports in agricultural and fisheries products. Agriculture is the most important sector for the

[14] Arable Land – https://www.sciencedirect.com/topics/earth-and-planetary-sciences/arable-land

[15] Agriculture in Africa 2019: Special Report – https://oxfordbusinessgroup.com/blog/souhir-mzali/focus-reports/agriculture-africa-2019-special-report

[16] Agriculture in Africa 2019 – https://oxfordbusinessgroup.com/sites/default/files/blog/specialreports/949525/africa_2019_special_report.pdf

[17] OECD-FAO Agricultural Outlook 2018–2027 – http://www.fao.org/3/i9166en/I9166EN.PDF

[18] LATIN AMERICAN AGRICULTURE: PROSPECTS AND CHALLENGES – http://www.fao.org/3/CA4076EN/CA4076EN_Chapter2_Latin_American_Agriculture.pdf

economy across LAC, accounting for an average of 4.7% of GDP in 2015–2017. The region has positioned itself as a leading exporter of agricultural products. Among countries in the LAC, Brazil is the largest agricultural and food exporter (USD 79.3 billion in 2017) in the region, followed by Argentina, (USD 35.0 billion), Mexico (USD 32.5 billion), Chile (USD 17 billion), Ecuador (USD 10.4 billion), and Peru (USD 8.8 billion). Nevertheless, the region also contains one of the second largest importer of wheat, Mexico [13].

Agriculture is the largest employer in the LAC region. In 2018, 14.1% of total labor force in the LAC region was employed in agriculture. Countries such as Bolivia, Ecuador, Guatemala, Honduras, Haiti, Guatemala, Nicaragua, and Peru employed more than a quarter of its labor force in the agricultural sector (World Bank, 2019). The LAC region experiences high incidence of poverty and extreme poverty in rural areas (48.6% and 22.5%, respectively) since 2017 and widening gap between rural and urban poor with high undernourished people in the region (39.3 million). Finally, food security continues to be a concern in this region, with many households unable to afford the food they need.

Agriculture has served as a buffer during recessionary periods in Latin America [13].

Agriculture is the largest employer in the developing countries (please see Fig. 3.4). Given sheer number, agricultural development is one of the most powerful tools to end extreme poverty, boost shared prosperity, and feed a projected 9.7 billion people by 2050 [1]. Unlike other sectors, the growth in the agriculture sector is two to four times more effective in raising incomes among the poorest. Analyses in 2016 found that 65% of poor working adults[19] made a living through agriculture [14].

In summary, countries[20] that are the most economically dependent (% of GDP) on agriculture (please see Table 3.1) are Sierra Leone, Guinea-Bissau, Chad, Niger, Mali, Liberia, Kenya, Sudan, Ethiopia, and Central Africa Republic,[21] and the countries that are the least[22] economically dependent on agriculture are Macau, Singapore, Hong Kong, Luxembourg, Qatar, Bahrain, Kuwait, Isle of Man, Curacao, Cayman Islands, Andorra, Belgium, and the United Kingdom.

[19] Who Are the Poor in the Developing World – http://documents1.worldbank.org/curated/en/187011475416542282/pdf/WPS7844.pdf

[20] The World-fact Book – https://www.cia.gov/library/publications/the-world-factbook/fields/214.html

[21] Most economically dependent on agriculture – https://worldinfigures.com/rankings/index/104

[22] Least economically dependent on agriculture – https://worldinfigures.com/rankings/index/105

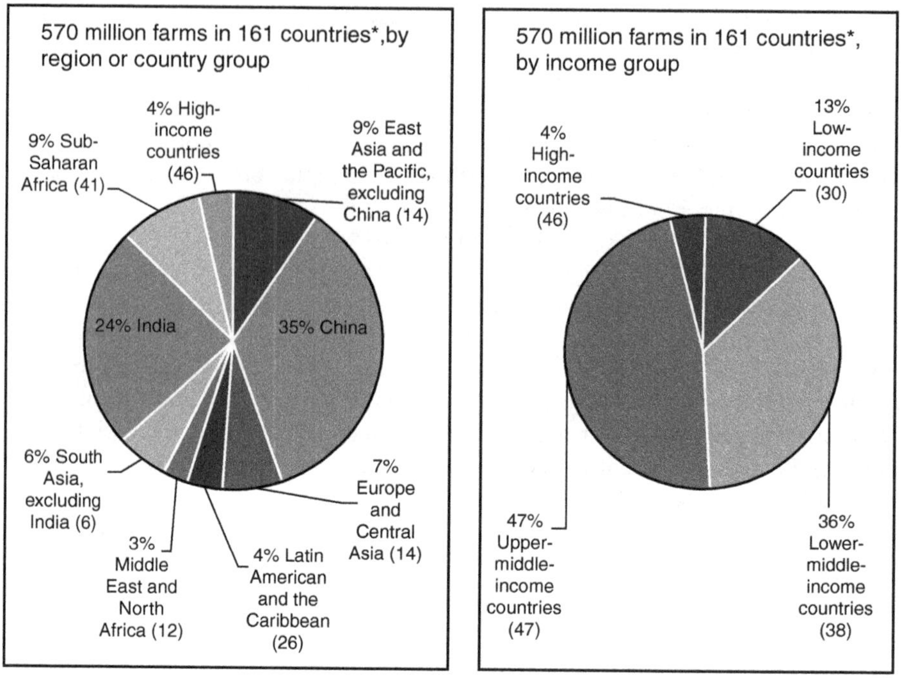

Fig. 3.4 Farms in 161 countries [14]

Real-time GDP prediction, i.e., "nowcasting[23]" or "GDPNow,[24]" is essential for farmers to assess macrolevel economy trends and its influence on their farm produce, price, and finances.

Agriculture Farm Sizes: Scale Classification

Based on the size and output, agricultural farms, in general, are classified in three categories: small-scale, medium-scale, and large-scale farms. In Europe and Africa, however, small-scale farms are further divided into subsistence small scale and small-scale farms with the distinction that subsistence farming performed to feed farmer families. Most of the worldwide agriculture is performed by small-scale farmers who operate as a family owned and highly fragmented. Younger generation

[23] Federal Reserve Bank of New York – https://www.newyorkfed.org/research/policy/nowcast/methodology.html

[24] Federal Reserve Bank of Atlanta – https://www.frbatlanta.org/cqer/research/gdpnow#:~:text=The%20GDPNow%20model%20estimate%20for,11.0%20percent%20on%20December%2016

Table 3.1 Countries and dependence on agriculture (Ag)

Most dependent Ag. (% of GDP), 2018		Least dependent Ag.(%of GDP), 2018[a]	
Sierra Leone	58.9	Macau	0.0
Guinea-Bissau	47.5	Singapore	0.0
Chad	44.9	Hong Kong	01
Niger	39.2	Luxembourg	0.2
Mali	38.7	Qatar	0.2
Liberia	37.3	Bahrain	0.3
Kenya	34.2	Kuwait	0.4
Sudan	31.5	Isle of Man	0.4
Ethiopia	31.2	Curacao	0.4
Central African Rep.	31.2	Cayman Islands	0.4
Burundi	29.0	Andorra	0.5
Rwanda	29.0	Belgium	0.5
Uzbekistan	28.8	United Kingdom	0.6
Tanzania	28.7	United Arab Emirates	0.7
Burkina Faso	28.0	Switzerland	0.7

[a]Countries with least dependent on Agriculture – https://worldinfigures.com/rankings/index/105

of these farms, in general youth, migrate to urban areas. Ironically, in search for industrial employment and onus of maintaining the farms fall onto old and seniors. The small farmers make big contribution to agriculture and dairy production[25] in the world [15]. Unlike the dairy farms of the west, milk originates in highly decentralized villages with the help of small farmers who own three to five cattle, and they bring milk twice a day to milk collection centers to get paid.[26] Simply put, the livelihoods of roughly 2 billion people (26.7% of the world population) of small farmers in the world depend on agriculture and the climate change adversely impacting their survival. Agriculture is the largest employer with majority of small-scale farmers depending on their livelihood [16].

25 Small Farmers Make Big Contributions: Dairy Production in India – https://ucanr.edu/blogs/blogcore/postdetail.cfm?postnum=23184

26 Strategies for the Bottom of the Pyramid: Creating Sustainable Development – http://pdf.wri.org/2001summit_hartarticle.pdf

Economies of size[27] and relative cost efficiencies are associated with farm sizes. As farm size increases in most field crop regions, per unit cost declines at first, and they are relatively constant. Medium-size farms achieve most available technical cost efficiencies. Societies would likely benefit little in terms of lower food costs if these mid-size farms further expanded. Farmers expand to increase income rather than to reduce per unit cost. In essence, the goal of expansion is not to achieve cost synergies but to sustain.

North America

Based on the size and economic activities, agriculture farms in the United States fall into three categories (please see Fig. 3.5). Small farms are rapidly disappearing, and large farms are major source of agriculture activity. 97% of US farms are family owned.[28]

Europe

In Europe, farms are classified in four categories: very small, small, medium, large, and very large farms (please see Fig. 3.6). Very small farms (semi-substance) focus on growing a high portion food to feed farmers and their families. The size of farms

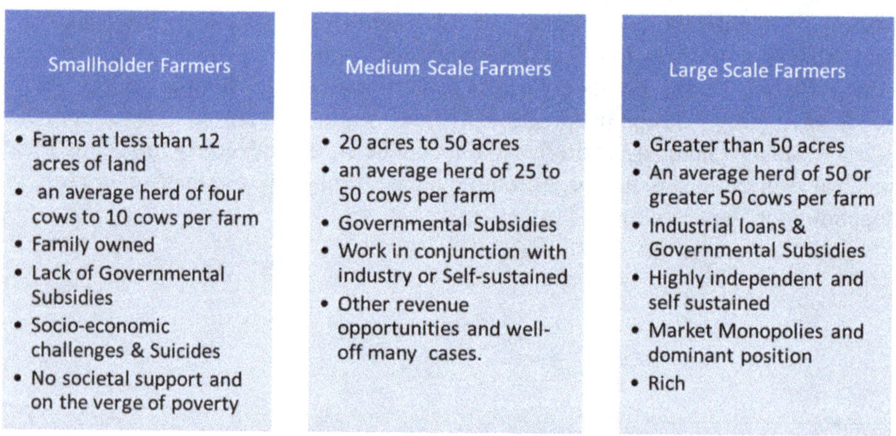

Fig. 3.5 Farm size classification in the United States

[27] Economies of size in U.S. Field crop Faming – https://naldc.nal.usda.gov/download/CAT87202149/PDF

[28] Family Farms – https://texasfarmbureau.org/percent-of-us-farms-are-familyowned/

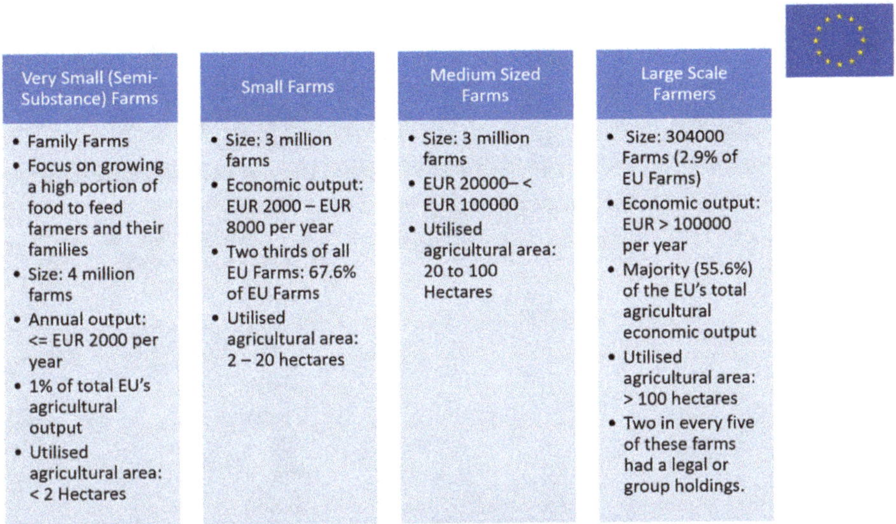

Fig. 3.6 Farm classification in EU

is around four million (in 2016), and their annual economic output is below EUR 2000 per year.[29] Small farms, on the other hand, generate EUR 2000 to 8000 per year. Both very small and small farms represent 67.6% EU farms[30] (67.6% of 173 million hectares = **116.9 million hectares or 288.8 million acres almost equal to China's expected to harvest 288.8 million**[31]) [17, 18].

Asia

In Asia, especially India, China, Bangladesh, Pakistan, and Indonesia, the 90% of farms are small and family owned (please see Fig. 3.7).

Though machine learning and data science services are required across all farm sizes, the large farms have built in information communication technology (ICT) infrastructure and can afford to have sophisticated ML algorithms deployed on farm premises.[32] A good example would be computer vision technology (CVT),[33] sensor connected edge ML applications, and web mining systems [19].

Some of the popular ML algorithms include:

[29] Agriculture, forestry, and fishing, value added (% of GDP) https://data.worldbank.org/indicator/NV.AGR.TOTL.ZS

[30] Small and large farms in the EU – statistics from the farm structure survey – https://ec.europa.eu/eurostat/statistics-explained/pdfscache/54736.pdf

[31] China's Acreage Gap – https://aei.ag/2018/10/01/chinas-acreage-gap/

[32] SmartHerd management: A microservices-based fog computing–assisted IoT platform towards data-driven smart dairy farming – https://onlinelibrary.wiley.com/doi/full/10.1002/spe.2704

[33] Machine Learning in Agriculture: A Review – https://www.mdpi.com/1424-8220/18/8/2674/pdf

Crop	Farm Size	ML Algorithm
Wheat[a]	Large	Artificial Neural Networks Multi-layer perceptron (MLP) Time Series
Tomato	Large	Clustering
Green Citrus[b]	Large	Support Vector Machine (SVM)
Rice[c]	Large	SVM Ensemble Learning (EL) Random Forest (RF) Time Series
Dairy	Medium and Large	Sensors & Edge ML Decision Trees Kalman Filter Regression SVM Time Series
Soybeans	Large	ANN[d]
Corn	Large	ANN[e]
Coffee[f]	Medium/ Large	SVM
Cotton[g]	Large	SVM
Commodity Pricing	Large	Pricing Spread & Time Series

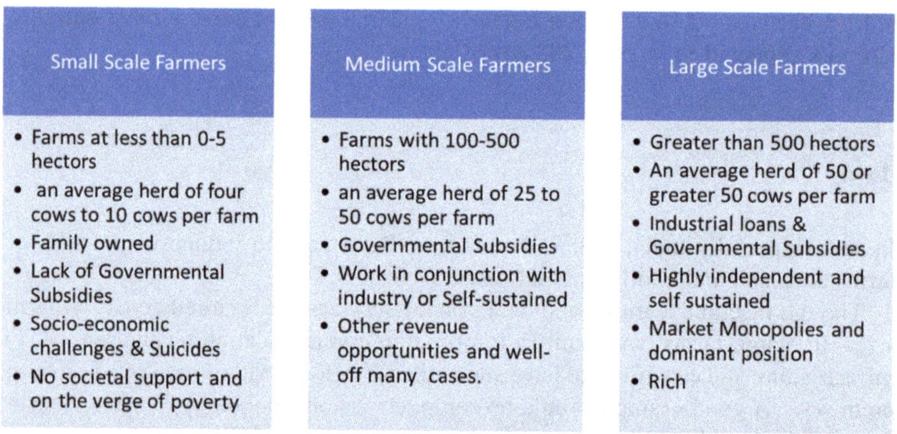

Fig. 3.7 Farm size classification in Asia

[a]Pantazi, X.-E.; Moshou, D.; Alexandridis, T.K.; Whetton, R.L.; Mouazen, A.M. Wheat yield prediction using machine learning and advanced sensing techniques. Comput. Electron. Agric. 2016, 121, 57–65,

[b]Sengupta, S.; Lee, W.S. Identification and determination of the number of immature green citrus fruit in a canopy under different ambient light conditions. Biosyst. Eng. 2014, 117, 51–61,

[c]Su, Y.; Xu, H.; Yan, L. Support vector machine-based open crop model (SBOCM): Case of rice production in China. Saudi J. Biol. Sci. 2017, 24, 537–547,

[d]Kaul M, Hill L, Charles Walthall R (2005) Artificial neural networks for corn and soybean yield prediction,

[e]Evaluating machine learning algorithms for predicting maize yield under conservation agriculture in Eastern and Southern Africa – https://link.springer.com/article/10.1007/s42452-020-2711-6#:~:text=Machine%20learning%20techniques%20available%20for,is%20a%20critical%20first%20step.

[f]Ramos, P.J.; Prieto, F.A.; Montoya, E.C.; Oliveros, C.E. Automatic fruit count on coffee branches using computer vision. Comput. Electron. Agric. 2017, 137, 9–22,

[g]Zhang, M.; Li, C.; Yang, F. Classification of foreign matter embedded inside cotton lint using short wave infrared (SWIR) hyperspectral transmittance imaging. Comput. Electron. Agric. 2017, 139, 75–90.

Across the different crops, some common ML algorithms include feature reduction techniques (principal component analysis), time series, regression, and clustering techniques.

As you can see from the above, small farmers form major workforce across the agricultural industry, and they make major economic force in terms of spending money and improving overall rural and national economy. Profit maximization of crop yields is possible through predicting crop prices in advance and to ensure farmers receive maximum price on their crops. High crop prices translated quickly into needed income for farmers. This extra income meant that farmers could buy new equipment, more food, clothing, and so on. The net *multiplier* effect is increase in GDP, and it underscores the fact that the GDP as a measure of economic well-being.[34] In-time crop price prediction would have a huge impact on overall economy of the world [20].

To augment small-scale farmers with information and data services that enable them better plan crops with higher market price yields and reduce the risk of economic uncertainty and market shocks, we need machine learning-enabled agricultural economic information services that provide real-time gauge into commodity price market and that provide advanced prognostic view to course correct, if any, unintended consequences. The goal is to put more money into the hands of small-scale farmers and reduce their overall agricultural debt and thus make them more prosperous.

"Indian Agrarian Crisis: Farmer,[35] the most endangered species" [21]

[34] GDP as a measure of an economic well-being – https://www.brookings.edu/wp-content/uploads/2018/08/WP43-8.23.18.pdf

[35] Farmers' suicide: Across culture – https://www.ncbi.nlm.nih.gov/pmc/articles/PMC2802368/

Agricultural Economics

Agricultural economics[36] is much more than a farm business. It is greater than animal husbandry and dairy farming. Agricultural economics involves the study of farming system, environmental and natural resources, local and international economics, law, policy, technological innovations, and precision devices.[37] Put it simply, agricultural economics studies the linkage between macroeconomic variables (please see Fig. 3.8) and its influence on the productivity and profitability of the farmers [22].

Traditionally, econometric techniques played a key role in the development of agricultural economics. With a wide variety of reasonably competitive product markets and abundant public data (statistical purposes and time series), agriculture presented a ready field of application for various statistical regression techniques as they were developed.[38] Nonetheless, econometric models apply statistical techniques with historical and national level key metrics [23].

Fig. 3.8 Macroeconomic View

[36] Agricultural Economics – https://www.uky.edu/academics/bachelors/agricultural-economics

[37] Strengthening agribusiness linkages with small-scale farmers – http://www.fao.org/3/y6001e/y6001e00.htm#Contents

[38] MATHEMATICAL MODELING IN AGRICULTURAL ECONOMICS – http://www.eolss.net/Sample-Chapters/C02/E6-154-12-00.pdf

> Macroeconomics is the study of the major economic totals, or aggregates. The "Big Three" concepts of macroeconomics involve the study of unemployment, inflation rate, and economic growth.[39]

Advanced Analytics

With the advent of big data and machine learning techniques, agricultural economics now can be benefited from data science to predict more accurately the dynamic nature of interdependent economic markets and its influences. As global economies have become more interdependent, large-scale events (see Fig. 3.14) introduced volatility and uncertainty to global markets affecting industries and people across the world. Over time, the frequency of these events has accelerated. This is particularly important to farm employees, especially small farmers as they're highly susceptible to changes in the economic markets[40,41,42,43,44,45] [24–27]. Data mining techniques, machine learning feature reduction techniques such as principal component analysis (PCA), clustering techniques, and regression machine learning models have confirmed statistically significant co-movements among these time series (57 US macroeconomic indicators (18 leading, 30 coincidental, and 9 lagging) and 5 other trade/money indexes) and identify noteworthy economic events. Machine learning techniques can study these complex nexus relationships to better predict economic conditions[46] to help small-scale farmers.

---

39 Macroeconomics 12th Edition by Robert Gordon, ISBN-13: 978-0138014919, Pearson; 12th edition (April 8, 2011)

40 GDP Now: A Model for GDP "Nowcasting" – https://www.frbatlanta.org/-/media/documents/research/publications/wp/2014/wp1407.pdf

41 Concepts and Methods of the U.S. National Income and Product Accounts – https://www.bea.gov/system/files/2019-12/All-Chapters.pdf

42 Now Casting Report – https://www.newyorkfed.org/research/policy/nowcast/methodology.html

43 GDP Now – Federal Reserve Bank of Atlanta – https://www.frbatlanta.org/cqer/research/gdpnow#:~:text=The%20GDPNow%20model%20estimate%20for,11.0%20percent%20on%20December%2016.

44 Nowcasting New Zealand GDP Using Machine Learning Algorithms – https://cama.crawford.anu.edu.au/sites/default/files/publication/cama_crawford_anu_edu_au/2018-09/47_2018_richardson_mulder_vehbi.pdf

45 Predicting surprises to GDP: a comparison of econometric and machine learning techniques – https://dspace.mit.edu/bitstream/handle/1721.1/109649/987002546-MIT.pdf?sequence=1&isAllowed=y

46 Relationship between Macroeconomic Indicators and Economic Cycles in U.S – https://www.ncbi.nlm.nih.gov/pmc/articles/PMC7242465/#MOESM1

Potential GDP or full employment GDP (Y^P) is the level of output if all resources (land, labor, and capital) are fully employed. Actual GDP is usually below potential GDP. Business cycles: recessions are defined to be a decline in GDP that is sustained at least two quarters. Rising GDP is a recovery or boom period. The output gap is the amount by which actual GDP is less than potential GDP. The output ratio is actual GDP/potential GDP.

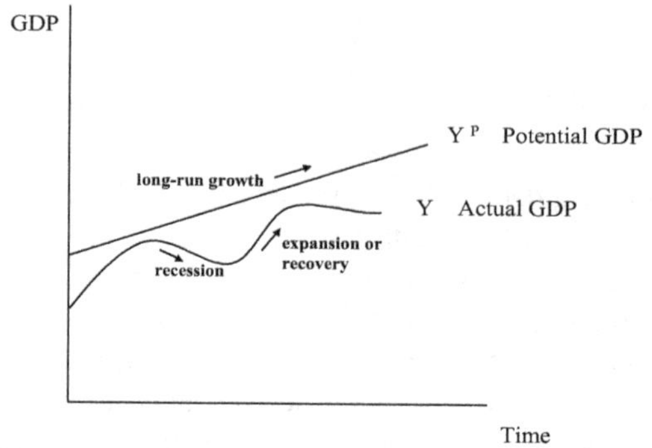

Data Units and Frequencies

Second, macroeconomic indicators (most of them sourced from FRED) exhibit different units and data collection frequencies. Seasonally adjusted indicators given by the FRED Economic Data are used if available. Time series mining of macroeconomic indicators for the absence of seasonality is verified to better predictability and coverability of the variables on the economic activity.

In the above charts, Time Series Decomposition Power BI Utility[47] does the calculation of seasonality and trend. The decomposition of time series is a statistical method that splits a time series into several components, each representing one of the underlying processes. There are three components that are typically of interest: trend, seasonality, and noise. Time series decomposition is an essential analytics tool to understand time series components and to improve a forecast.

For clustering time series signals and markers, the symbolic representation of time series and storing those symbols in persistent data store are performed. Generally, the time series signals are compared one need to have a common frequency of measurement (i.e., the time between observations). With that in place, identify a common model that would reasonably describe each series separately. This might be an ARIMA model or a multiply trended regression model with possible level shifts or a composite model integrating both memory (ARIMA) and dummy variables. This common model could be estimated globally and separately for each of the two series, and then one could construct an F test to test the hypothesis of a common set of parameters.

[47] Understand the time series components using "Seasonal and Trend decomposition using Loess" – https://appsource.microsoft.com/en-us/product/power-bi-visuals/wa104380897?tab=overview

Private Housing Build Permits[48]	Average Weekly Hours of Production and Nonsupervisory Employees, Manufacturing[49]
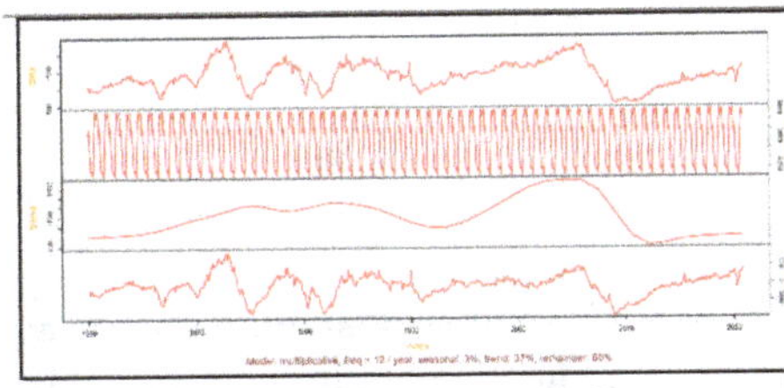	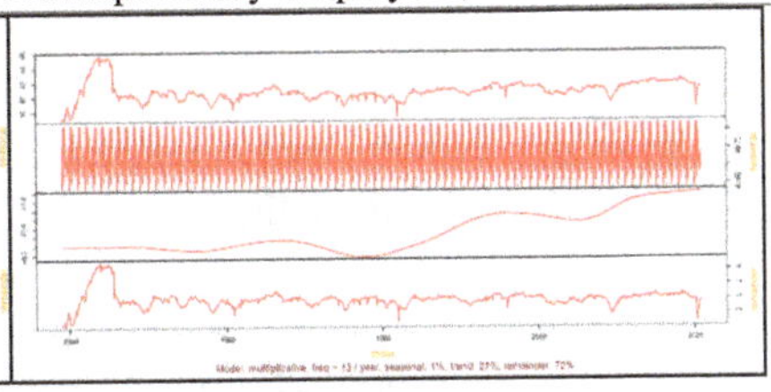
Units: Thousands of Units, Seasonally Adjusted Annual Rate Frequency: Monthly Seasonal: 1%	Units: Hours, Seasonally Adjusted Frequency: Monthly Seasonal: 3%
4-Week Moving Average of Continued Claims (Insured Unemployment)[50]	US ISM Manufacturing New Orders Index[51]
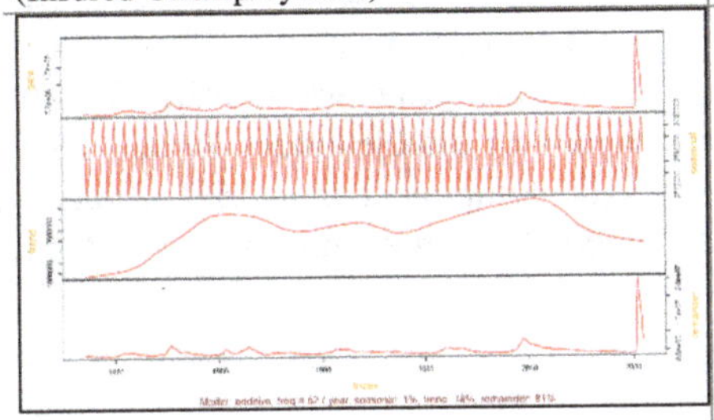	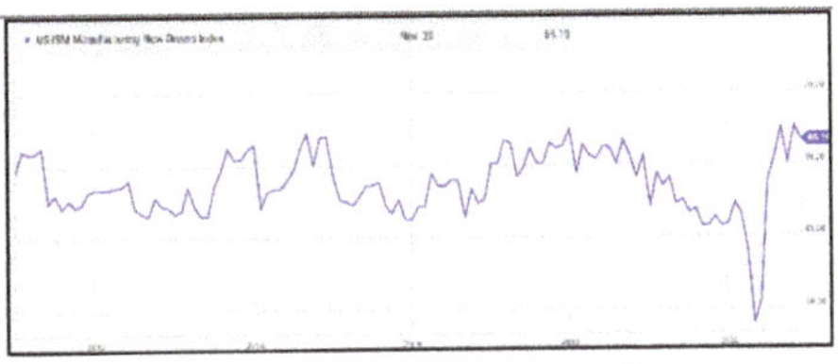
Units: Number, Seasonally Adjusted Frequency: Weekly, Ending Saturday Seasonal: 1%	Units: Number, Seasonally Adjusted Frequency: Monthly Seasonal: 1%
New Private Housing Units Authorized by Building Permits – in Structures with 5 Units (PERMIT5)[52]	Leading Index for the United States (USSLIND)[53]
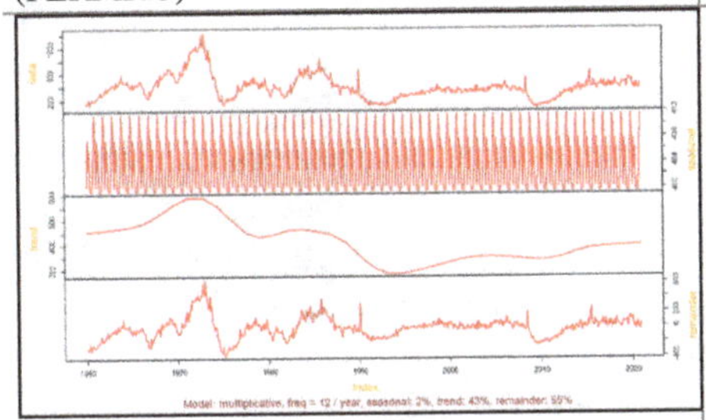	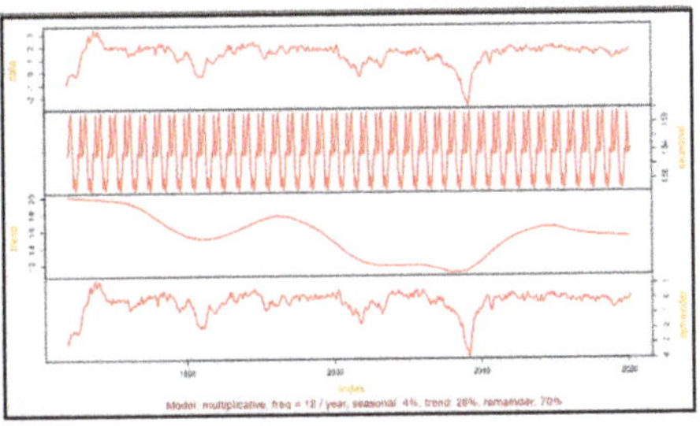
Units: Thousands of Units, Seasonally Adjusted Annual Rate Frequency: Monthly Seasonal: 2%	Units: Percent, Seasonally Adjusted Frequency: Monthly Seasonal: 4%

[48] New Private Housing Units Authorized by Building Permits – https://fred.stlouisfed.org/series/PERMIT

[49] Average Weekly Hours of Production and Nonsupervisory Employees, Manufacturing (AWHMAN) – https://fred.stlouisfed.org/series/AWHMAN

[50] 4-Week Moving Average of Continued Claims (Insured Unemployment) (CC4WSA) – https://fred.stlouisfed.org/series/CC4WSA

[51] ISM New Orders – https://ycharts.com/indicators/us_ism_manufacturing_new_orders_index

[52] New Private Housing Units Authorized by Building Permits – in Structures with 5 Units – https://fred.stlouisfed.org/series/PERMIT5

[53] Leading Index for the United States (USSLIND) – https://fred.stlouisfed.org/series/USSLIND

(For all other economic indicators, please refer to Appendix D.)

Machine Learning Modeling

Time series decomposition can be performed using Python programming interface. Time series is the representation of series of data points ordered in time. In a time series, usually time is an independent variable, and the target variable is the one for which we would like to forecast. In our case, for example, it is "Average Weekly Hours of Production and Nonsupervisory Employees, Manufacturing (AWHMAN[54])."

Average weekly hours relate to the average hours per worker for which pay was received and is different from standard or scheduled hours. Factors such as unpaid absenteeism, labor turnover, part-time work, and stoppages cause average weekly hours to be lower than scheduled hours of work for an establishment. Group averages further reflect changes in the workweek of component industries. Average weekly hours are the total weekly hours divided by the employees paid for those hours.

Source of Data: US Bureau of Labor Statistics.
Release: Employment Situation.
Units: Hours, Seasonally Adjusted.
Frequency: Monthly.

Average Weekly Hours of Production and Nonsupervisory Employees, Manufacturing (AWHMAN)
Production and related employees include working supervisors and all nonsupervisory employees (including group leaders and trainees) engaged in fabricating, processing, assembling, inspecting, receiving, storing, handling, packing, warehousing, shipping, trucking, hauling, maintenance, repair, janitorial, guard services, product development, auxiliary production for plant's own use (e.g., power plant), recordkeeping, and other services closely associated with the above production operations.

Nonsupervisory employees include those individuals in private, service-providing industries who are not above the working-supervisor level. This group includes individuals such as office and clerical workers, repairers, salespersons, operators, drivers, physicians, lawyers, accountants, nurses, social workers, research aides, teachers, drafters, photographers, beauticians, musicians, restaurant workers, custodial workers, attendants, line installers and repairers, laborers, janitors, guards, and other employees at similar occupational levels whose services are closely associated with those of the employees listed.

[54] AWHMAN – https://fred.stlouisfed.org/series/AWHMAN

Time series factors to be detected:
Stationary.
Seasonal.
Autocorrelation.

The following code loads "Average Weekly Hours of Production and Nonsupervisory Employees, Manufacturing" FRED data – January 01, 1939, to October 2020:

Software code for this model can be found on GitHub Link: ts_analysis_awhman.ipynb (Jupyter Notebook Code)

Step 1: Load Libraries:

```
import logging
import pandas as pd
import numpy as np
from numpy import random
import nltk
from sklearn.model_selection import train_test_split
from sklearn.feature_extraction.text import CountVectorizer, TfidfVectorizer
from sklearn.metrics import accuracy_score, confusion_matrix
import matplotlib.pyplot as plt
from nltk.corpus import stopwords
import re
from bs4 import BeautifulSoup
import pandas as pd
import matplotlib.pyplot as plt
import matplotlib as mpl
import numpy as np
import scipy.stats as spstats
import seaborn as sns


%matplotlib inline
mpl.style.reload_library()
mpl.style.use('classic')
mpl.rcParams['figure.facecolor'] = (1, 1, 1, 0)
mpl.rcParams['figure.figsize'] = [6.0, 4.0]
mpl.rcParams['figure.dpi'] = 100

sns.set()
```

Step 2: Load AWHMAN Dataset:

```
DATAPATH = 'AWHMAN.csv'
data = pd.read_csv(DATAPATH, index_col=['DATE'], parse_dates=['DATE'])
data.head(10)
```

Output:

	AWHMAN
DATE	
1939-01-01	37.1
1939-02-01	37.2
1939-03-01	37.3
1939-04-01	37.3
1939-05-01	37.0
1939-06-01	37.4
1939-07-01	37.3
1939-08-01	38.1
1939-09-01	38.1
1939-10-01	39.1

The following code creates time series plot (please see Fig. 3.9):

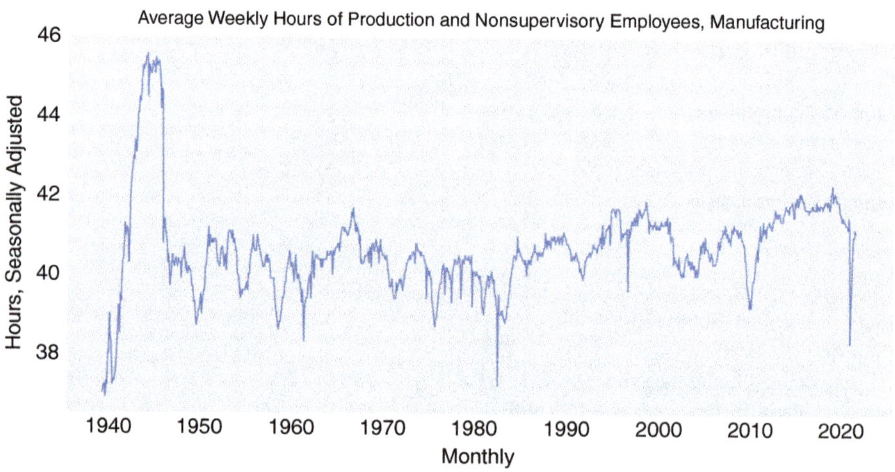

Fig. 3.9 Average weekly hours

```
plt.figure(figsize=(17, 8))
plt.plot(data.AWHMAN)
plt.title('Average Weekly Hours of Production and Nonsupervisory Employees, Manufacturing')
plt.ylabel('Hours, Seasonally Adjusted')
plt.xlabel('Monthly')
plt.grid(False)
plt.show()
```

Moving Averages

The moving average model is probably the naivest approach to time series modeling. This model simply states that the next observation is the mean of all past observations. The purpose of the moving averages is to find the trend in the data (please see Fig. 3.10).

```
def plot_moving_average(series, window, plot_intervals=False, scale=1.96):

    rolling_mean = series.rolling(window=window).mean()

    plt.figure(figsize=(17,8))
    plt.title('Moving average\n window size = {}'.format(window))
    plt.plot(rolling_mean, 'g', label='Rolling mean trend')

    #Plot confidence intervals for smoothed values
    if plot_intervals:
        mae = mean_absolute_error(series[window:], rolling_mean[window:])
        deviation = np.std(series[window:] - rolling_mean[window:])
        lower_bound = rolling_mean - (mae + scale * deviation)
        upper_bound = rolling_mean + (mae + scale * deviation)
        plt.plot(upper_bound, 'r--', label='Upper bound / Lower bound')
        plt.plot(lower_bound, 'r--')

    plt.plot(series[window:], label='Actual values')
    plt.legend(loc='best')
    plt.grid(True)

#Smooth by the previous 12 months (by year)
plot_moving_average(data.AWHMAN, 12)
```

The moving average can be used to identify interesting trends in the data.[55] We can define a window (in the above chart, we have used 12 month window) to apply the moving average model to smooth the time series and highlight different trends [28]. Please note the longer the window, the smoother the trend will be.

---

[55] The Complete Guide to Time Series Analysis and Forecasting – https://towardsdatascience.com/the-complete-guide-to-time-series-analysis-and-forecasting-70d476bfe775

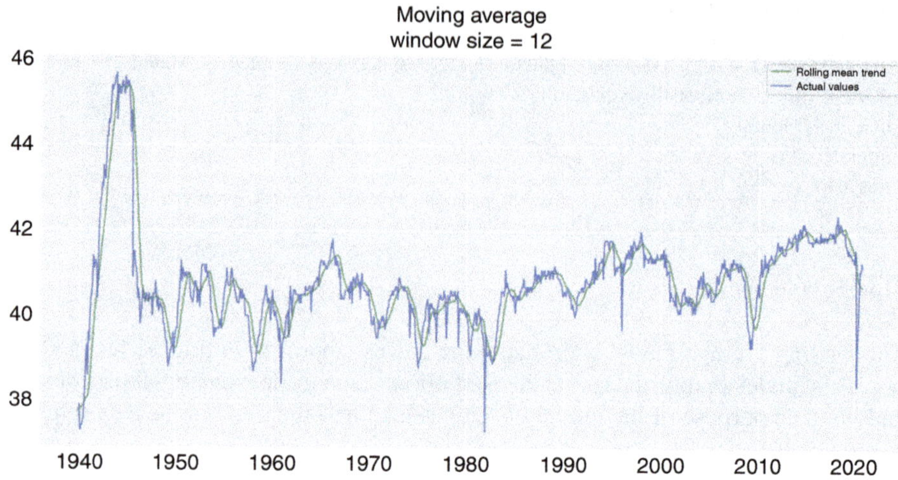

Fig. 3.10 Moving averages (window size =12)

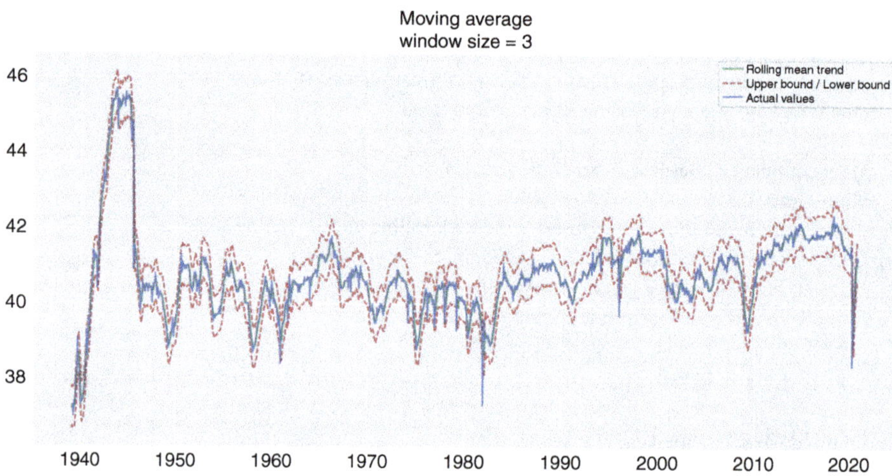

Fig. 3.11 Moving averages (window size = 3)

In the following code, we have used 3 months' moving averages (please see Fig. 3.11):

```
#Smooth by previous quarter (3 months)
plot_moving_average(data.AWHMAN, 3, plot_intervals=True)
```

Stationary

In a stationary time series, the statistical properties of time series do not change over time.[56] In other words, the statistical properties (mean, variance, and covariance) are constant and independent of time. Dickey-Fuller test helps to determine a time series is stationary [29].

Ideally, we would like to possess a stationary statistic for modeling. Of course, not all of them are stationary, but we will make different transformations to form them stationary.

```
def tsplot(y, lags=None, figsize=(12, 7), syle='bmh'):

    if not isinstance(y, pd.Series):
        y = pd.Series(y)

    with plt.style.context(style='bmh'):
        fig = plt.figure(figsize=figsize)
        layout = (2,2)
        ts_ax = plt.subplot2grid(layout, (0,0), colspan=2)
        acf_ax = plt.subplot2grid(layout, (1,0))
        pacf_ax = plt.subplot2grid(layout, (1,1))

        y.plot(ax=ts_ax)
        p_value = sm.tsa.stattools.adfuller(y)[1]
        ts_ax.set_title('Time Series Analysis Plots\n Dickey-Fuller: p={0:.5f}'.format(p_value))
        smt.graphics.plot_acf(y, lags=lags, ax=acf_ax)
        smt.graphics.plot_pacf(y, lags=lags, ax=pacf_ax)
        plt.tight_layout()

tsplot(data.AWHMAN, lags=12)
```

How to Test if a Process Is Stationary

You may have noticed within the title of the plot above Dickey-Fuller. This is often the statistical test that we run to work out if a statistic is stationary or not. The intuition behind a unit root test is that it *determines how strongly a time series is defined by a trend.*

There are several unit root tests, and the augmented Dickey-Fuller may be one of the more widely used (please see Fig. 3.12). It uses an autoregressive model and optimizes an information criterion across multiple different lag values.

The null hypothesis of the test is that the time series can be represented by a unit root that it is not stationary (has *some time-dependent structure*). The alternate hypothesis (rejecting the null hypothesis) is that the time series is stationary.

[56] Introduction to Time Series Analysis – https://itl.nist.gov/div898/handbook/pmc/section4/pmc4.htm

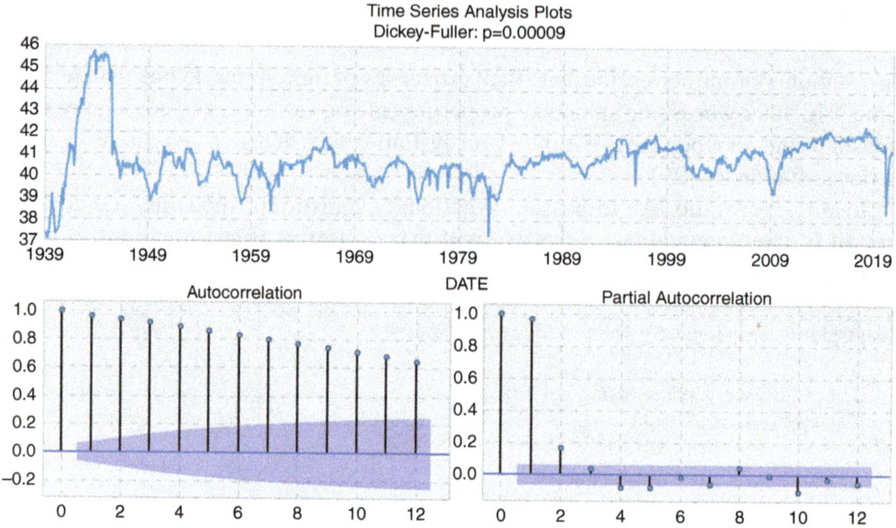

Fig. 3.12 Time series analysis plot

- Null Hypothesis (H0): If failed to be rejected, it suggests the time series has a unit root, meaning it is non-stationary. It has some time-dependent structure.
- Alternate Hypothesis (H1): The null hypothesis is rejected; it suggests the time series does not have a unit root, meaning it is stationary. It does not have time-dependent structure.

We interpret this result using the p-value from the test. A p-value below a threshold (such as 5% or 1%) suggests we reject the null hypothesis (stationary); otherwise a p-value above the threshold suggests we fail to reject the null hypothesis (non-stationary).

- p-value >0.05: Fail to reject the null hypothesis (H0), the data has a unit root and is non-stationary.
- p-value <= 0.05: Reject the null hypothesis (H0), the data does not have a unit root and is stationary.

Running the "Average Weekly Hours of Production and Nonsupervisory Employees, Manufacturing (AWHMAN)" dataset prints the test statistic value of −4.685582. The more negative this statistic, the more likely we are to reject the null hypothesis (we have a stationary dataset).

As part of the output, we get a lookup table to help determine the ADF statistic. We can see that our statistic value of −4.685582 is less than the value of −3.437 at 1%. This suggests that we can reject the null hypothesis with a significance level of less than 1%.

```
from pandas import read_csv
from statsmodels.tsa.stattools import adfuller
series = read_csv(DATAPATH, header=0, index_col=0, squeeze=True)
X = series.values
print(X)
result = adfuller(X)
print()
print('ADF Statistic: %f' % result[0])
print('p-value: %f' % result[1])
print('Critical Values:')
for key, value in result[4].items():
	print('\t%s: %.3f' % (key, value))
```

Output:

```
ADF Statistic: -4.685582
p-value: 0.000090
Critical Values:
	1%: -3.437
	5%: -2.865
	10%: -2.568
```

Rejecting the null hypothesis[57] means that the process has no unit root and in turn that the time series is stationary or does not have time-dependent structure [30].

Let's log transform the dataset again to make the distribution of values more linear and better meet the expectations of this statistical test.

```
from pandas import read_csv
from statsmodels.tsa.stattools import adfuller
from numpy import log
series = read_csv(DATAPATH, header=0, index_col=0, squeeze=True)
X = series.values
X = log(X)
result = adfuller(X)
print('ADF Statistic: %f' % result[0])
print('p-value: %f' % result[1])
for key, value in result[4].items():
	print('\t%s: %.3f' % (key, value))
```

Output:

Running the dataset shows a negative value for the ADF statistic. We can also see that the value is smaller than the critical values (1%), again, meaning that we can reject the null hypothesis and in turn that the time series is stationary.

[57] How to Check if Time Series Data is Stationary with Python – https://machinelearningmastery.com/time-series-data-stationary-python/

```
ADF Statistic: -4.731345
p-value: 0.000073
      1%: -3.437
      5%: -2.865
```

Data drift and/or time series difference between two signals could be identified using a cumulative distribution function (CDF) measures such as Wasserstein distance[58] and probability distribution function (PDF) such as Kullback-Leibler divergence[59] (KL divergence).

Commodity Pricing Signals

Third, agricultural commodities (corn, soybeans, and wheat) make up a large portion of staple food and animal feed consumed worldwide.[60] The United States is the largest exporter of some of the staple commodities (36% of global corn, share close to 40% of global soybean, 16% of world wheat production). The export prices of these commodities are more volatile than overall US export prices [31]. This volatility is demonstrated in Fig. 3.12, which tracks price changes over time for corn,[61] soybeans,[62] and wheat[63] exports and crude oil imports[64] (please see Fig. 3.13). During the period shown in the chart, the graphs for export corn, soybeans, and wheat reveal prices subject to volatility—in other words, recording large price changes on 1- or 12-month bases.

This volatility occurs because agricultural goods prices are susceptible to weather events, crude oil price changes, and the value of the US dollar. In addition, prices of some of these commodities (corn, soybeans, and wheat) generally trend together because the crops share common price-determining factors such as substitutability, demand, biofuels, the value of the US dollar, weather, and crude oil. The increased link[65] between energy and non-energy commodity prices, strong demand by

[58] scipy.stats.wasserstein_distance – https://docs.scipy.org/doc/scipy/reference/generated/scipy.stats.wasserstein_distance.html

[59] scipy.special.kl_div – https://docs.scipy.org/doc/scipy/reference/generated/scipy.special.kl_div.html?highlight=kullback%20leibler%20divergence

[60] The relationship between crude oil prices and export prices of major agricultural commodities – https://www.bls.gov/opub/btn/volume-8/the-relationship-between-crude-oil-and-export-prices-of-major-agricultural-commodities.htm?view_full

[61] Soybean Export Price Index – https://beta.bls.gov/dataViewer/view/timeseries/EIUIQ00100

[62] Corn export – https://beta.bls.gov/dataViewer/view/timeseries/WPS012202

[63] Wheat export – https://beta.bls.gov/dataViewer/view/timeseries/EIUIQ00000

[64] Crude oil imports – https://beta.bls.gov/dataViewer/view/timeseries/EIUIR10000

[65] MACROECONOMIC ENVIRONMENT, COMMODITY MARKETS: A LONGER TERM OUTLOOK – http://www.fao.org/3/a-ak967e.pdf

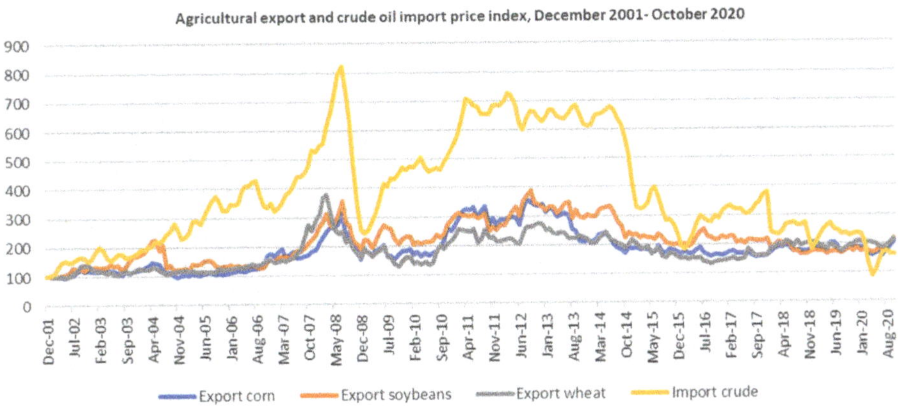

Fig. 3.13 Agricultural export and oil import price index

developing countries and changing weather patterns will be the dominant forces that are likely to shape developments in commodity markets [32]. Application of data, time series, and signal mining of common price determinants of the commodities would provide prognostic and/or predictive insight to farmers to achieve profit maximization. The commodities prices have direct influence[66] on farm profitability and sustainability of small farmer and machine learning algorithms would enable farmers to sustain pricing shocks [33].

Economics of Rural America: three major shifts have taken place in the rural America:
- Ethanol-based crops.
- Biofuels.
- Livestock.

Information Services and ICT

Fourth, dissecting the interplay of macrolevel economics, commodity prices, demand factors, environmental factors, transportation and storage factors, and weather patterns is complex. Enabling the small-scale farmer with this information on a constant basis would provide a better planning tool and information services and enable small-scale farmer to join information and ICT. Application of advanced data science techniques, natural language processing, and econometric models to deliver easily digestible form would bring big data revolution to small-scale farmer.

[66] The interactions between agricultural commodity and oil prices: an empirical analysis – https://www.agriculturejournals.cz/publicFiles/162049.pdf

Government Ordinances

Finally, many national governments across the world, historically, have supported minimum support price (MSP) or minimum guarantee prices (MGP) for farm produce [34]—a pricing system that have protected national farmers from the free-market fluctuations and market shocks for decades[67,68,69,70] [35]. A notable MSP system includes Dairy Forward Pricing Program in the United States, Milk and Dairy Pricing in European Union, MSP in India, Minimum Guarantee Prices in Brazil,[71] Minimum Support Prices in China,[72] Minimum Price Policy in Indonesia,[73] and many others. Ironically, decommissioning of the MSP or MGP have taken foothold, exposing small farmers to the uncertainties of commodity and price markets.[74] With these two seismic shifts, it is pragmatic and prudent to have macroeconomic view to small-scale farmers in real time to prevent[75] them in the turmoil of the market[76,77,78] [36–40].

> Time series is one of the most important datasets that provides historical perspective of agriculture and potentially contains keys to climate change.

---

[67] Farm bills: Are India's new reforms a 'death warrant'for farmers? – https://www.bbc.com/news/world-asia-india-54233080

[68] EU Milk and Dairy Productions – https://ec.europa.eu/info/food-farming-fisheries/animals-and-animal-products/animal-products/milk-and-dairy-products_en

[69] Agricultural Policy of the United States – https://en.wikipedia.org/wiki/Agricultural_policy_of_the_United_States

[70] The 2018 Farm Bill (P.L. 115–334): Summary and Side-by-Side Comparison – https://crsreports.congress.gov/product/pdf/R/R45525

[71] Brazil sets new minimum prices – http://www.fao.org/giews/food-prices/food-policies/detail/en/c/1185343/

[72] China Announces 2020 Wheat Minimum support Prices at 2019 Level – https://apps.fas.usda.gov/newgainapi/api/Report/DownloadReportByFileName?fileName=China%20Announces%202020%20Wheat%20Minimum%20Support%20Price%20at%202019%20Level_Beijing_China%20-%20Peoples%20Republic%20of_12-03-2019

[73] Sup. – https://www.oecd-ilibrary.org/sites/9e2cf2f4-en/index.html?itemId=/content/component/9e2cf2f4-en

[74] 'If open markets were so benevolent, farmers in US or Europe- http://timesofindia.indiatimes.com/articleshow/78328364.cms?utm_source=contentofinterest&utm_medium=text&utm_campaign=cppst

[75] US Farm Bankruptcies Reach Eight-Year High – https://www.forbes.com/sites/niallmccarthy/2020/02/10/us-farm-bankruptcies-reach-eight-year-high-infographic/?sh=77b2189f7de0

[76] There Is Turmoil in Agriculture. Here's What Investors Need to Know. – https://www.barrons.com/articles/agriculture-farmers-investors-china-trade-latin-america-brazil-argentina-tyson-corteva-fmc-wasde-report-51566336014

[77] As small US farms face crisis, Trump's trade aid flowed to corporations – https://www.cnbc.com/2020/09/02/as-small-us-farms-face-crisis-trumps-trade-aid-flowed-to-corporations.html

[78] More Farmers Declare Bankruptcy Despite Record Levels of Federal Aid – https://www.wsj.com/articles/more-farmers-declare-bankruptcy-despite-record-levels-of-federal-aid-11596706201

How farms respond in times of uncertainty is as varied as the farms themselves. Many small farms will go out of business, and mid/large farms slow down operations to operate more cost-effectively. Across this entire spectrum, the ability to cost-effectively adapt and scale to rapidly changing conditions and manage the underlying risk is paramount and data science and ML will proactively help farmers to manage effectively.

The landscape of agricultural economics, now, centered around the data. The data has become new tool to design new agricultural techniques, formulate pricing and competitive strategies, harvest techniques for creating sustainable agricultural practices for the future, and understand the influence of climate change on the future of the agriculture. The data for building predictive and prognostic analytical models come from different national and international datasets.

Machine learning continues to find applications in all data-centric disciplines, and macroeconomics is no exception. This trend is unlikely to abate, and all macroeconomic forecasting problems have the potential to benefit from the application of machine learning prediction techniques. In this paper, we have added to the body of evidence which shows, quite clearly, that machine learning algorithms are a highly effective set of tools for making macroeconomic predictions. While the random forest algorithm applied in this paper is effective, it is likely that other machine learning methods would also perform well on these data. One avenue of future work for this prediction problem would be to explore different machine learning[79] and AI algorithms, as well as to conduct a deeper study of the optimal hyperparameter configurations of those algorithms [41].

Data

Data is central to the sustainable, precision, and smart agriculture. In the era of the big data and artificial intelligence, data plays not only competitive advantage but also pivotal to our future sustainability.

Data	Source	
US Bureau of Economic Analysis (BEA)[80]	The Bureau of Economic Analysis (BEA) of the US Department of Commerce is a US government agency that provides official macroeconomic and industry statistics and most notably reports about the gross domestic product (GDP) of the United States and its various units—states, cities/towns/townships/villages/counties and metropolitan areas. They also provide information about personal income, corporate profits, and government spending in their national income and product accounts (NIPAs).	

[79] Improvements to Consumption Prediction: Machine Learning Methods and Novel Features – https://scholar.smu.edu/cgi/viewcontent.cgi?article=1055&context=datasciencereview

[80] The Bureau of Economic Analysis of the United States – https://www.bea.gov/

The US Census Bureau, officially the Bureau of the Census[81]	The Census Bureau is dedicated to providing current facts and figures about America's people, places, and economy.	
The Bureau of Labor Statistics is a unit of the United States Department of Labor.[82]	The Bureau of Labor Statistics is the principal fact-finding agency for the US government in the broad field of labor economics and statistics and serves as a principal agency of the US Federal Statistical System.	
US Department of Agriculture[83]	The US Department of Agriculture, also known as the Agriculture Department, is the US federal executive department responsible for developing and executing federal laws related to farming, forestry, rural economic development, and food.	
The Federal Reserve Bank of Atlanta	The Federal Reserve Bank of Atlanta is part of the central bank of the United States. The Federal Reserve System—the Fed, as it is often called—consists of 12 Reserve Banks located around the country and the Board of Governors in Washington, D.C.	
Economic Research-Federal Reserve Bank of St. Louis[84]	Providing free US and international economic data, graphs and other data-related tools, plus quality research from St. Louis Fed economists.	
The World Bank[85]	The World Bank is an international financial institution that provides loans and grants to the governments of low- and middle-income countries for the purpose of pursuing capital projects.	
Food and Agriculture Organization[86]	The Food and Agriculture Organization of the United Nations is a specialized agency of the United Nations that leads international efforts to defeat hunger and improve nutrition and food security.	
S&P 500 Index[87]	The S&P 500, or simply the S&P, is a stock market index that measures the stock performance of 500 large companies listed on stock exchanges in the United States. It is one of the most followed equity indices.	

[81] The Census Bureau – https://www.census.gov/

[82] The Bureau of Labor Statistics – https://www.bls.gov/

[83] The USDA – https://www.usda.gov/

[84] FRED – https://fred.stlouisfed.org/

[85] The World Bank – https://data.worldbank.org/

[86] FAO- http://www.fao.org/home/en/

[87] S&P 500 Index – https://finance.yahoo.com/quote/%5EGSPC/

The International Monetary Fund[88,89]	The International Monetary Fund (IMF) is an international organization, headquartered in Washington, D.C., consisting of 190 countries working to foster global monetary cooperation, secure financial stability, facilitate international trade, promote high employment and sustainable economic growth, and reduce poverty. Inflation Rate[90]	
Organization for Economic Co-operation and Development (OECD) Data[91]	The Organization for Economic Co-operation and Development is an intergovernmental economic organization with 37 member countries, founded in 1961 to stimulate economic progress and world trade.	
BLS Beta Labs[92]	BLS Beta Labs is part of the US Bureau of Labor Statistics.	

Macroeconomy and GDP

Macroeconomy is the branch of economics that studies the behavior and performance of an economy. Macroeconomy focuses on the aggregate changes in the economy such as unemployment, growth, gross domestic product (GDP), and inflation. Generally, national central banks and federal government set the macroeconomic policies. Changes in the wider economy and macroeconomic policies affect the performance of agriculture sector.[93]

The US Bureau of Economic Analysis (BEA) defines GDP as the market value of the goods, services, and structures produced by the economy in each period. GDP is equal to the sum of personal consumption expenditures, gross private domestic investment, net exports of goods and services, and government consumption expenditures and gross investment. GDP is also equal to the sum of value added by industry across all industries [24].

Conceptually, the GDP can be arrived at by three means (please see Fig. 3.15): as the sum of goods and services sold to final users, as the sum of income payments and other costs incurred in the production of goods and services, and as the sum of value added at each stage of production [24].

[88] IMF GDP Data – https://www.imf.org/external/datamapper/NGDP_RPCH@WEO/OEMDC/ADVEC/WEOWORLD

[89] IMF Dataset – https://www.imf.org/external/datamapper/datasets/WEO

[90] CPI – https://www.imf.org/external/datamapper/PCPIPCH@WEO/OEMDC/ADVEC/WEOWORLD

[91] OECD Net Export Data – International trade – Trade in goods and services forecast – OECD Data

[92] BLS Beta Labs – https://beta.bls.gov/dataQuery/find?st=0&r=20&more=0&q=export+corn

[93] Global Development Bank – Macroeconomy – http://www.fao.org/3/i2490e/i2490e01c.pdf

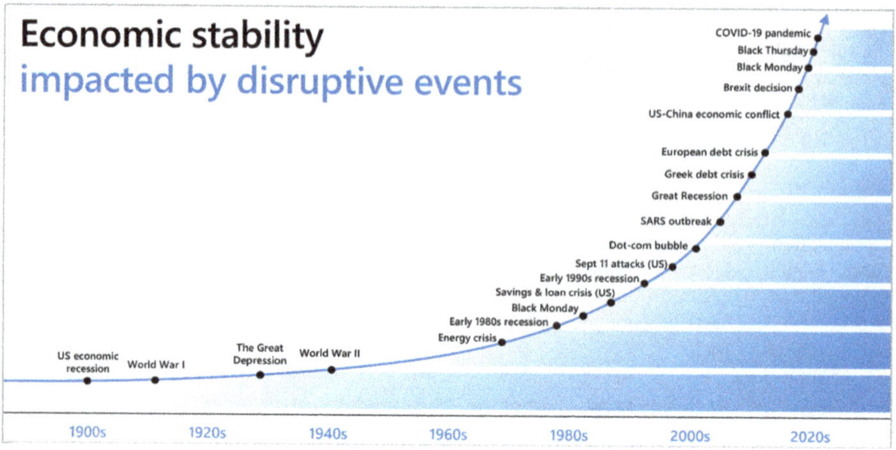

Fig. 3.14 Economic events

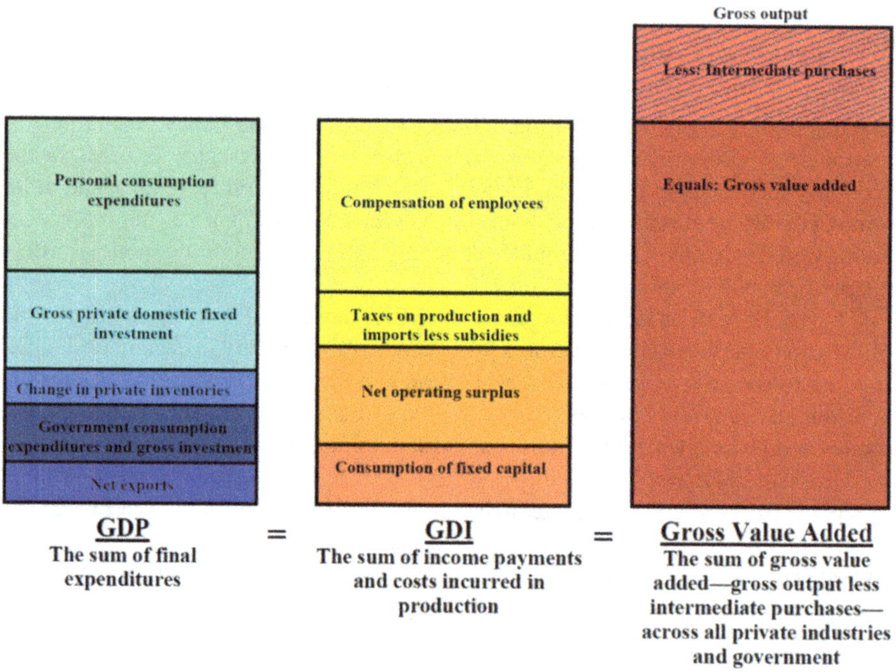

Fig. 3.15 GDP calculations [24]

GDP Method: As the Sum of Goods and Services Sold to Final Users

This measure, known as the expenditures approach, is used to identify the goods and services purchased by persons, businesses, governments, and foreigners. It is calculated by summing the following final expenditures components:

$$GDP = \text{Personal Consumption Expenditures} + \text{Government Expenditures} + \text{Investment} + \text{Net Exports}$$

Personal consumption expenditures (PCE) which measure the value of the goods and services purchased by, or on the behalf of, persons. The persons include households, nonprofit institutions that primarily serve households, private noninsured welfare funds, and private trust funds. The largest component of GDP is personal consumption expenditures (PCE), which accounted[94] for 70% of the total US GDP [42] (please see Fig. 3.16).

The Bureau of Economic Analysis (BEA) defines PCE as, "… the value of the goods and services purchased by, or on the behalf of, U.S. residents." At the national level, BEA publishes annual, quarterly, and monthly reports containing estimates of PCE. The goods included in the PCE calculation comprise both durable and non-durable goods.[95] Durable goods include motor vehicles, furniture and household equipment, and recreational goods and vehicles. Non-durable goods include food and beverages, clothing and footwear, and fuel and energy. The services included in the PCE calculation include housing and utilities, healthcare, transportation services, recreation services, food services and accommodations, financial services and insurance, and other miscellaneous services. In simple terms, it is a measure of consumption economy.

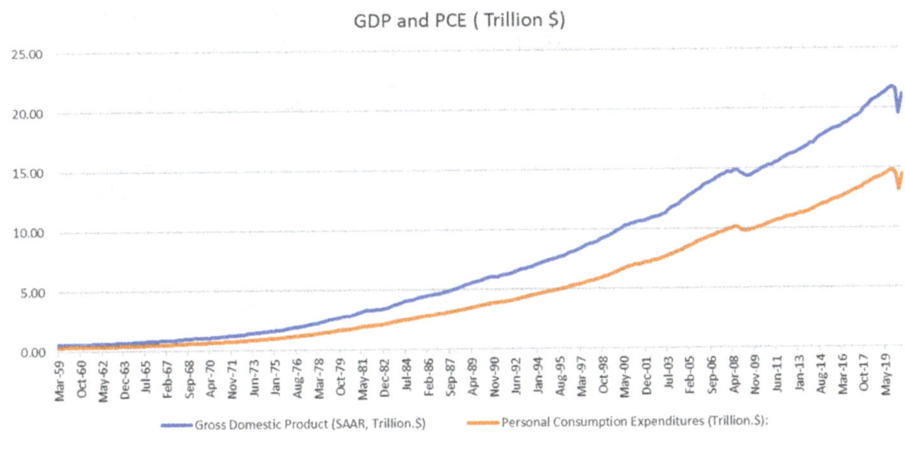

Fig. 3.16 GDP and PCE

[94] The Components of GDP – https://www.thebalance.com/components-of-gdp-explanation-formula-and-chart-3306015#:~:text=1.-,Personal%20Consumption%20Expenditures,figures%20reported%20are%20real%20GDP

[95] Improvement to Consumption Prediction: Machine Learning Methods and Novel Features-https://scholar.smu.edu/cgi/viewcontent.cgi?article=1055&context=datasciencereview

Vector autoregression (VAR) and random forest regression algorithms are very useful in predicting PCE. Additionally, natural language processing (NLP) techniques enable to identify sentiment of consumers!

Please find components of 2017 GDP[96]:

Exports	Mar-17	Jun-17	Sep-17	Dec-17
Exports of goods (SAAR, Mil.$)	$315,324	$312,833	$320,881	$338,007
Exports of services (SAAR, Mil.$)	$812,713	$828,190	$836,668	$853,592
Total exports	$2,326,357	$2,333,107	$2,370,118	$2,468,658

Imports	Mar-17	Jun-17	Sep-17	Dec-17
Imports of goods (SAAR, Mil.$)	$2,334,343	$2,342,088	$2,351,425	$2,479,653
Imports of services (SAAR, Mil.$)	$535,576	$550,540	$562,274	$564,476
Total imports	$2,869,919	$2,892,628	$2,913,699	$3,044,129

Net exports (exports—imports)(Mil.$)	−$543,562	−$559,521	−$543,581	−$575,471

Personal consumption expenditures (PCE)	Mar-17	Jun-17	Sep-17	Dec-17
PCE: Durable goods (SAAR, Mil.$)	$ 1,383,173	$1,394,003	$1,412,984	$1,452,553
PCE: Nondurable goods (SAAR, Mil.$)	$2,724,895	$2,736,303	$2,764,310	$2,820,799
PCE: Services (SAAR, Mil.$)	$9,045,123	$9,110,961	$9,193,627	$9,322,684
Total PCE	$13,153,191	$13,241,267	$13,370,921	$13,596,036

Components of GDP	2017 amount ($USD trillions)	% of total
Net exports	−$0.55	−2.81%
Government spending	$3.40	17.4%
Private investment	$3.34	17.4%
Personal consumption expenditures	$13.34	68.30%
Total GDP	$19.53	100%

[96] GDP Tracking Model And Forecasts: https://www.frbatlanta.org/-/media/Documents/cqer/researchcq/gdpnow/GDPTrackingModelDataAndForecasts.xlsx?la=en

The following variables are used to compute GDP using PCE:

Variable	Source	Frequency	Ticker	Units
Exports of Goods (SAAR, Mil.$)[a]	BEA	Quarterly	XMX_USNA	Billions of Dollars, Seasonally Adjusted Annual Rate
Exports of Services (SAAR, Mil.$)[b]	BEA	Quarterly	XSX_USNA	Billions of Chained 2012 Dollars, Seasonally Adjusted Annual Rate
Imports of Goods (SAAR, Mil.$)[c]	BEA	Quarterly	MMX_USNA	Billions of Chained 2012 Dollars, Seasonally Adjusted Annual Rate
Imports of Services (SAAR, Mil.$)	BEA	Quarterly	MMX_USNA	Billions of Chained 2012 Dollars, Seasonally Adjusted Annual Rate
Government Consumption Expenditures & Gross Investment (SAAR, Mil.$)[d]	BEA	Quarterly	MSX_USNA	Billions of Dollars, Seasonally Adjusted Annual Rate
Private Fixed Investment (SAAR, Mil.$)[e]	BEA	Quarterly	FX_USNA	Billions of Chained 2012 Dollars, Seasonally Adjusted Annual Rate
Personal Consumption Expenditures: Durable Goods (SAAR, Mil.$)[f]	BEA	Monthly	CDX_USNA	Billions of Dollars, Seasonally Adjusted Annual Rate
Personal Consumption Expenditures: Nondurable Goods (SAAR, Mil.$)[g]	BEA	Monthly	CNX_USNA	Billions of Dollars, Seasonally Adjusted Annual Rate
Personal Consumption Expenditures: Services (SAAR, Mil.$)[h]	BEA	Quarterly	CSX_USNA	Billions of Dollars, Seasonally Adjusted Annual Rate

[a]Exports of Goods and Services (EXPGS) – https://fred.stlouisfed.org/series/EXPGS

[b]Real Exports of Goods and Services (EXPGSC1) – https://fred.stlouisfed.org/series/EXPGSC1

[c]Real imports of goods and services (IMPGSC1) – https://fred.stlouisfed.org/series/IMPGSC1

[d]Government Consumption Expenditures and Gross Investment (GCE) – https://fred.stlouisfed.org/series/GCE

[e]Real Gross Private Domestic Investment: Fixed Investment: Nonresidential: Equipment (Y033RX1Q020SBEA) – https://fred.stlouisfed.org/series/Y033RX1Q020SBEA

[f]Personal Consumption Expenditures: Durable Goods (PCEDG) – https://fred.stlouisfed.org/series/PCEDG
[g]Personal Consumption Expenditures: Nondurable Goods (PCEND) – https://fred.stlouisfed.org/series/PCEND
[h]Personal Consumption Expenditures: Services (PCESV) – https://fred.stlouisfed.org/series/PCESV

The following table lists some of the key PCE components of the GDP[97]:

Real PCE [a]	Nominal PCE [b]
Core retail goods ex foods services	Core retail goods ex foods services
New Motor Vehicles	New Motor Vehicles
Net Purchases of Used Motor Vehicles	Net Purchases of Used Motor Vehicles
Gasoline & Other Energy Goods	Gasoline & Other Energy Goods

[a]GDP Tracking Model and Forecasts XLSX – https://www.frbatlanta.org/-/media/Documents/cqer/researchcq/gdpnow/GDPTrackingModelDataAndForecasts.xlsx?la=en

[b]GDP Tracking Model and Forecasts XLSX – https://www.frbatlanta.org/-/media/Documents/cqer/researchcq/gdpnow/GDPTrackingModelDataAndForecasts.xlsx?la=en

PCE as a Leading Indicator—Market Signal Mining

Signal mining is an important technique to understand and infer the overall market movements. For instance, PCE constitutes around a 70% of the GDP, and it is prudent to data mine the PCE and/or its core constituents. As of ending of year 2020, consumer cut back spending (FRED—Advance Monthly Sales for Retail and Food Services[98]) during December, the third straight month of decline.[99]

[97] GDP NOW XLSX – https://www.frbatlanta.org/-/media/Documents/cqer/researchcq/gdpnow/GDPTrackingModelDataAndForecasts.xlsx?la=en

[98] Advance Monthly Sales for Retail and Food Services – https://fred.stlouisfed.org/series/RSAFS

[99] US Retail Sales Fell 0.7% in December as Covid-19 Cases Rose – https://www.wsj.com/articles/us-economy-december-retail-sales-coronavirus-recovery-11610676272

PCE as a Leading Indicator - Market Signal Mining

Signal mining is an important technique to understand and infer the overall market movements. For instance, PCE constitutes around a 70% of the GDP and it is prudent to data mine the PCE and/or its core constituents. As of ending of year 2020, consumer cut back spending (FRED - Advance Monthly Sales for Retail and Food Services[116]) during December, the third straight month of decline[117].

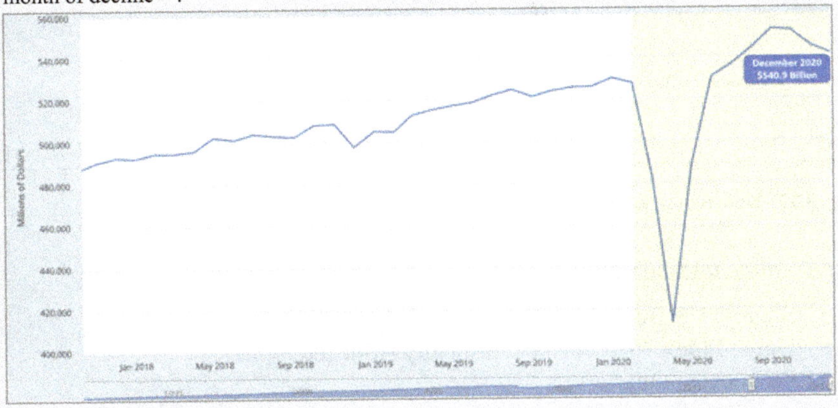

Units: Millions of Dollars, Seasonally Adjusted

Units: Millions of Dollars, Seasonally Adjusted.

New Motor Vehicles[100]

The PCE components of GDP are time series data points. These data points are used in econometric and machine learning calculations. In the below graph (please see Fig. 3.17), the new motor sales data presented from January 1959 to October 2020. Although the magnitude of numbers change is of the interest, the important factor to be analyzed is the scale of shifts during economic volatility.

For instance,

1. During the 1980s early recession the drop of 30%.
2. The early 1990s recession and the 9-11 attack.
3. During the great recession, 2008.
4. COVID-19 crisis. And the results confirm Federal Reserve Bank of St. Louis Economic Research that auto industry was one of the hardest hit industries during the recession. During 2018[101] recession, New vehicle sales fell nearly 40% [43].

---

[100] US Bureau of Economic Analysis, Personal consumption expenditures: New motor vehicles (chain-type price index) [AB67RG3Q086SBEA], retrieved from FRED, Federal Reserve Bank of St. Louis; https://fred.stlouisfed.org/series/AB67RG3Q086SBEA, December 23, 2020.

[101] Auto Sales and the 2007–09 Recession – https://research.stlouisfed.org/publications/economic-synopses/2019/07/05/auto-sales-and-the-2007-09-recession

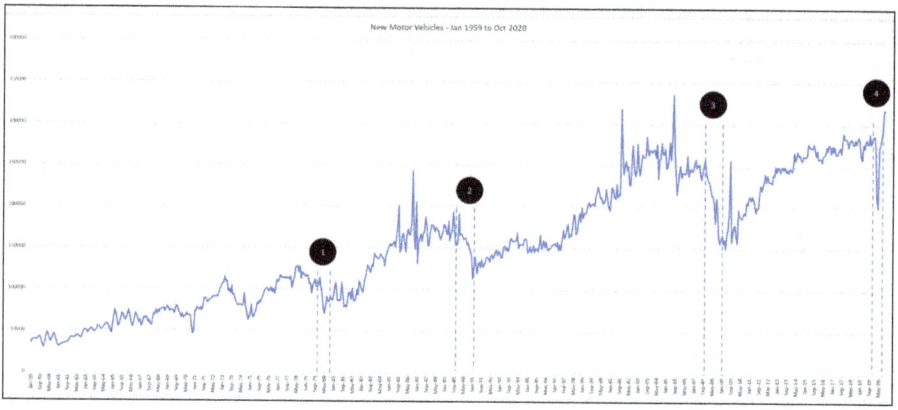

Fig. 3.17 New motor vehicles

Data Source and Attributes:

- Source: US Bureau of Economic Analysis.[102]
- Units: Index 2012 = 100, Seasonally Adjusted.
- Frequency: Quarterly.

From econometric model point of view [44], the New Motor Vehicles is calculated as[103]:

New Motor Vechicles=Latest number predicted with growth in "Light Weight Vehicle Sales {Autos+Light Trucks} (SAAR, Mil.Units)" -- TLVAR@USECON when available.

The machine learning algorithms, importantly, can codify the time series signature and help to cluster such signatures in database to predict and provide a prognostic view of the events.

Application of time series (TS) analysis on New Motor Vehicles data provides above time series analysis (please see Fig. 3.18):

- Time series model: Multiplicative.
- Frequency: 12 data points per year.
- Seasonal: 1%.
- Trend: 76%.
- Remainder: 23%.

[102] Personal consumption expenditures: New motor vehicles (chain-type price index) – https://fred.stlouisfed.org/series/AB67RG3Q086SBEA

[103] Current Quarter GDP Tracking – https://macroblog.typepad.com/files/111020CurrentQuarterGDPTrackingDescription.pdf

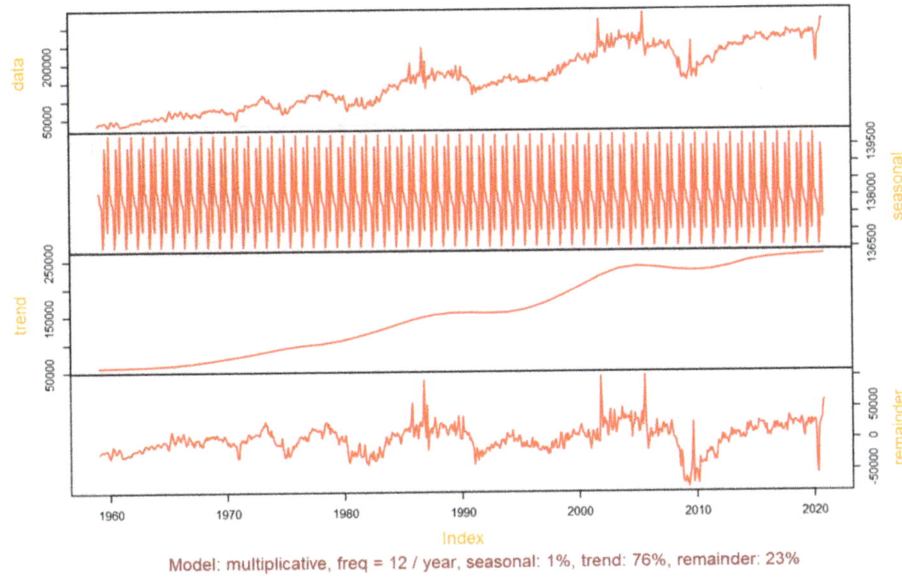

Fig. 3.18 New Motor Vehicles time series

Net Purchase of Used Motor Vehicles

Unlike any other industry, the unique importance of the automobile industry to the wider US economy[104] is quite evident. After the purchase of a house, a car is the next biggest purchase a household usually makes [45]. The volatility experienced in new car markets also had flow on effects on the used car market (please see Fig. 3.19). Cutbacks in consumer spending saw used car demand fall, while business fleets decided to extend their leases to forego new capital expenditure.[105] This is particularly true during COVID-19 [46] (please see Fig. 3.20).

Data Source and Attributes:

- Source: U.S. Bureau of Economic Analysis.[106]
- Units: Billions of Dollars, Seasonally Adjusted Annual Rate.
- Frequency: Quarterly.
- BEA Account Code: A137RC.

[104] Why Did Car Sales Drop So Dramatically During the Financial Crisis? – https://insight.kellogg.northwestern.edu/article/was-the-government-bailout-of-car-financers-necessary

[105] What the Coronavirus means for car sales – https://datiuminsights.com.au/what-the-coronavirus-means-for-car-sales/

[106] Personal consumption expenditures: Net purchases of used autos – https://fred.stlouisfed.org/series/A137RC1Q027SBEA

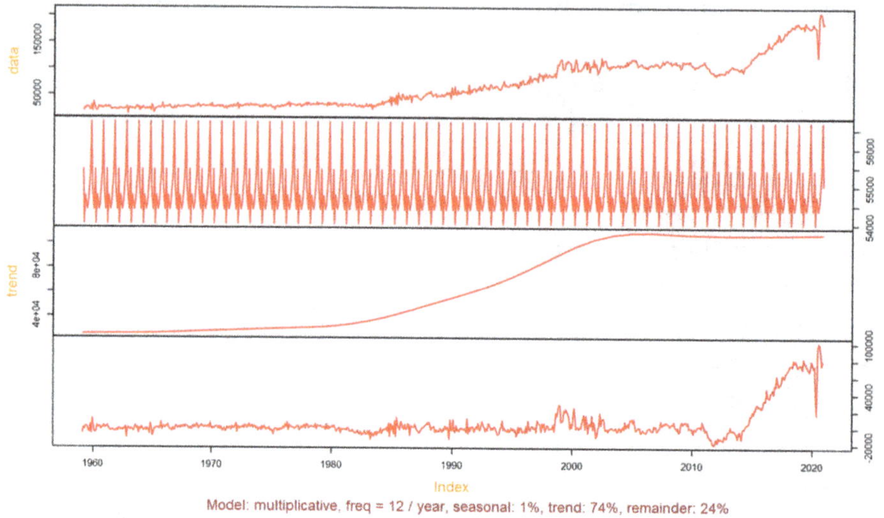

Fig. 3.19 Used vehicles

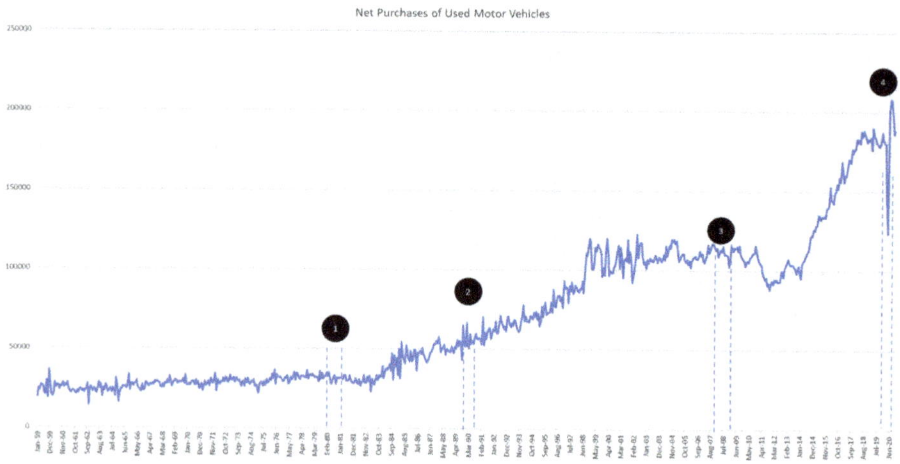

Fig. 3.20 Used vehicles sales during COVID-19

Application of time series analysis on New Motor Vehicles data provides above time series analysis.

- Time series model: Multiplicative.
- Frequency: 12 data points per year.
- Seasonal: 1%.
- Trend: 74%.
- Remainder: 24%.

The time series for new car sales and used car sales is different from trend and other TS properties. Nevertheless, we can witness the reduction of sales for both in economic contractions and shock!

Purchased Meals and Beverages

For purchased meals and beverages[107] (excluding school sales), food services sales from the economic census are allocated to PCE by type of eating place.

- Source: US Bureau of Economic Analysis.[108]
- Units: Billions of Dollars, Not Seasonally Adjusted.
- Frequency: Annual.

As indicated in Fig. 3.21, COVID-19 has disproportionately affected units of purchased means and beverages. Prices for perishable foods with a short shelf life were strongly affected, and disruptions in dairy and egg production and distribution caused the greatest price volatility in all BLS price indexes. The shift away from institutional and restaurant food consumption toward food at home consumption created short-term disruptions and shocks in the economy, particularly for perishable foods. These shocks rippled through the economy and affected the prices consumers pay in the grocery store.[109]

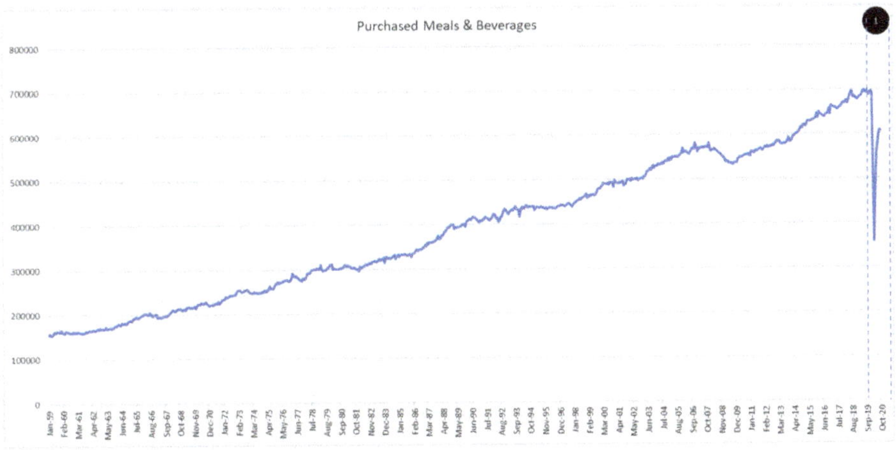

Fig. 3.21 Purchased meals and beverages

[107] Personal consumption expenditures: Nondurable goods: Food and beverages purchased for off-premises consumption/Gross Domestic Product – https://fred.stlouisfed.org/graph/?g=79tc

[108] Personal consumption expenditures: Services: Purchased meals and beverages – https://fred.stlouisfed.org/series/DPMBRC1A027NBEA

[109] The impact of the COVID-19 pandemic on food price indexes and data collection – https://www.bls.gov/opub/mlr/2020/article/the-impact-of-the-covid-19-pandemic-on-food-price-indexes-and-data-collection.htm

Econometric Models

PCE plays an important role in calculating the GDP. The current models for predicting personal consumption expenditures (PCE) employ statistical techniques and rely upon traditional economic features. For instance, the abovementioned key attributes enable to calculate the GDP, and most of the calculations are econometric! The sources for the data are listed in Table 3.2.

> Predicting surprise to GDP[110] could be achieved both from econometrics and machine learning techniques [47]. For instance, **ordinary least squares (OLS) and logistic regression** are common in econometric literature, whereas **random forest and neural network** machine learning algorithms most seen in engineering settings are capable of predicting changes to GDP.

Machine Learning

There are several techniques to predict GDP growth, for instance, using artificial neural networks[111] [48], in addition, other leading indicators that help to understand the status of the economy[112] and provide good glimpse on where we're on the business cycle [49].

Table 3.2 Data releases used for GDP calculation[a]

Primary data release	Source	GDP component forecasted directly from release
BEA unit auto sales	Bureau of Economic Analysis	Real PCE New Motor Vehicles, change in private inventories
Retail trade and inventories	US Census Bureau	Real PCE (retail control) spending, change in private inventories
Consumer price index	Bureau of Labor Statistics	Real PCE (retail control) spending, change in private inventories
Personal income and outlays	Bureau of Economic Analysis	Real PCE

[a]GDPNow: A Model for GDP "Nowcasting" – https://www.frbatlanta.org/research/publications/wp/2014/07.aspx

[110] Predicting surprises to GDP: a comparison of econometric and machine learning techniques – https://dspace.mit.edu/handle/1721.1/109649

[111] Forecasting GDP Growth Using Artificial Neural Networks – https://www.bankofcanada.ca/wp-content/uploads/2010/05/wp99-3.pdf

[112] Forecasting Inflation and GDP Growth: Automatic Leading Indicator (ALI) Method versus Macro Econometric Structural Models (MESMs) – https://www.adb.org/sites/default/files/publication/29861/tn-18-forecasting-inflation-gdp-growth.pdf

Predicting Real GDP Growth

The goal is to predict real GDP growth using leading economic indicators. Economists and financial analysts use leading indicators to help understand the future state of the economy. It helps pinpoint where we will be on the business cycle curve. Below are the leading [50] indicators[113]:

- Average Weekly Initial Claims for Unemployment Insurance[114]: An initial claim is a claim filed by an unemployed individual after a separation from an employer. The claim requests a determination of basic eligibility for the Unemployment Insurance Program. A good measure of initial layoffs and rehiring.
  - Source: US Employment and Training Administration.
  - Units: Number, Seasonally Adjusted.
  - Frequency: Weekly, ending Saturday.
- Manufacturing Average Hours Worked[115] (Quarterly % Change): Hours data for the labor productivity and cost measures include hours worked for all persons working in the sector-wage and salary workers, the self-employed, and unpaid family workers. The primary source of hours and employment data is the BLS Current Employment Statistics (CES) program, which provides monthly survey data on the number of jobs held by wage and salary workers in nonfarm establishments, counting a person who is employed by two or more establishments at each place of employment. Businesses are more likely to cut overtime before layoffs and increase overtime before rehiring; therefore, this metric moves first before the business cycle turns.
  - Source: US Bureau of Labor Statistics.
  - Units: Percent Change from Quarter 1 Year Ago, Seasonally Adjusted.
  - Frequency: Quarterly.
- ISM Manufacturers' New Orders[116]: Nondefense Capital Goods Excluding Aircraft—monthly new orders offer first signal of movement in industrial sector and good proxy for business expectations.
  - Source: US Census Bureau.
  - Units: Millions of Dollars, Seasonally Adjusted.
  - Frequency: Monthly.

---

[113] Predicting real GDP Growth – https://towardsdatascience.com/predicting-real-gdp-growth-85f34fdca97a

[114] 4-Week Moving Average of Initial Claims (IC4WSA) – https://fred.stlouisfed.org/series/IC4WSA

[115] Productivity and Costs: https://www.bls.gov/news.release/pdf/prod2.pdf

[116] Manufacturers' New Orders: Durable Goods (DGORDER) – https://fred.stlouisfed.org/series/DGORDER

- Building Permits for Private Housing Units: It is a data attribute that provides New Private Housing Units Authorized by Building Permits and clearly foretells new construction activity.
  - Source: US Census Bureau.
  - Units: Thousands of Units, Seasonally Adjusted Annual Rate.
  - Frequency: Monthly.
- S&P 500 Index[117]: It provides early indications of movement in the economic business cycles.
- Interest Rate Spread Between 10-Year Treasury Yield and Federal Funds Rate[118]: Series is calculated as the spread between 10-Year Treasury Constant Maturity (BC_10YEAR) and Effective Federal Funds Rate. Wider spreads anticipate economic upswings and narrower spreads anticipate downturns.
  - Source: Federal Reserve Bank of St. Louis.
  - Units: Percent, Not Seasonally Adjusted.
  - Frequency: Daily.
- Average Consumer Expectations/Confidence[119]: if consumers are confident, spending will increase. Consumption makes of 66% of the US economy.
  - Source: University of Michigan.
  - Units: Index 1966: Q1 = 100, Not Seasonally Adjusted.
  - Frequency: Monthly.

Data

All data points are quarterly aggregates retrieved from St. Louis FRED website. The data is from 1947 to 2020 Q1 with 293 records.

Feature Engineering

- New variables created:
  - Quarter: created from date.
  - S&P 500_pct: Replaced actual values with percent change of S&P 500, because over time S&P 500 values increase.

[117] S&P Index – https://finance.yahoo.com/quote/%5EGSPC/history?p=%5EGSPC

[118] 10-Year Treasury Constant Maturity Minus Federal Funds Rate (T10YFF) – https://fred.stlouisfed.org/series/T10YFF

[119] University of Michigan: Consumer Sentiment (UMCSENT) – https://fred.stlouisfed.org/series/UMCSENT

Target

The target variable is the annualized real GDP growth. This value is derived from GDPNow Dataset.

Machine Learning Models (ML)

Software code for this model can be found on GitHub Link: GDPPrediction.ipynb (Jupyter Notebook Code).

We have used the following ML models:

- Elastic net regression model because it does a better job dealing with highly correlated independent variables (see Fig. 3.22).

```
#Separate Features and target
X_Target = dataframe3['GDP_Change'] # the
feature_df= dataframe3.loc[:, dataframe3.columns != 'GDP_Change']
#Split the data
X_train, X_test, y_train, y_test = train_test_split(feature_df, X_Target, test_size=0.3,
random_state=0)
alphas = [0.0001, 0.001, 0.01, 0.1, 0.3, 0.5, 0.7, 1]
elastic_cv=ElasticNetCV(alphas=alphas, cv=5)
model = elastic_cv.fit(X_train, y_train)
ypred = model.predict(X_test)
score = model.score(X_test, y_test)
mse = mean_squared_error(y_test, ypred)
print("R2:{0:.4f}, MSE:{1:.4f}, RMSE:{2:.4f}"
    .format(score, mse, np.sqrt(mse)))
```

Output:

```
R2:0.4323, MSE:0.0008, RMSE:0.0275
```

Mean squared error was quite low, but root mean squared error was larger because it penalizes the larger errors. These large errors may be due to the volatility in the market resulting from unforeseen events. The R-squared of 0.4323 reveals that the leading indicators only explain 43% of variance in GDP growth. Thus, using only the leading indicators is not enough to forecast GDP growth. To have a better idea of where the economy is heading, we need to look at more metrics such as coincident indicators.

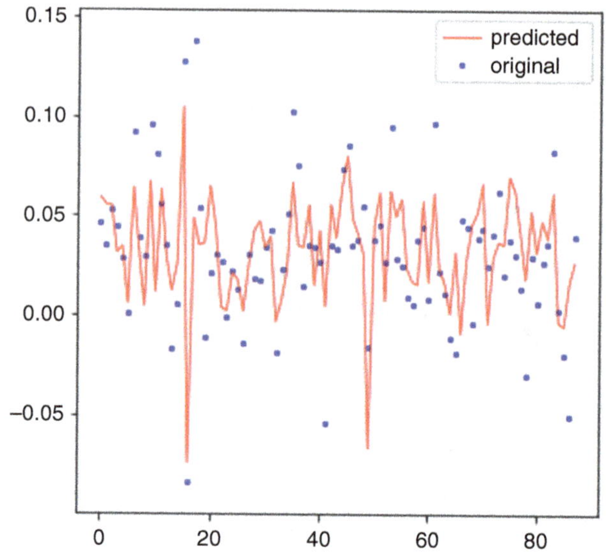

Fig. 3.22 Original and predicted values

Coincident Indicators

Coincident indicators measure the current state of the economy. So, by knowing where we are on the business cycle, we can foretell where we are heading.

I added the following metrics:

- Change in Monthly Unemployment[120] (quarterly average): The unemployment rate represents the number of unemployed as a percentage of the labor force. Labor force data are restricted to people 16 years of age and older, who currently reside in 1 of the 50 states or the District of Columbia, who do not reside in institutions (e.g., penal and mental facilities, homes for the aged), and who are not on active duty in the Armed Forces.
  - Source: US Bureau of Labor Statistics.
  - Units: Percent, Seasonally Adjusted.
  - Frequency: Monthly.
- Personal Consumption Expenditures Excluding Food and Energy (Chain-Type Price Index)[121]: price change from a year ago.
  - Source: US Bureau of Economic Analysis.
  - Units: Index 2012 = 100, Seasonally Adjusted.
  - Frequency: Monthly.

[120] Unemployment Rate (UNRATE) – https://fred.stlouisfed.org/series/UNRATE

[121] Personal Consumption Expenditures Excluding Food and Energy (Chain-Type Price Index) (PCEPILFE) – https://fred.stlouisfed.org/series/PCEPILFE

- Monthly Supply of House[122]: The months' supply is the ratio of houses for sale to houses sold. This statistic provides an indication of the size of the for-sale inventory in relation to the number of houses currently being sold. The months' supply indicates how long the current for-sale inventory would last given the current sales rate if no additional new houses were built.
    - Source: US Census Bureau.
    - Units: Months' Supply, Seasonally Adjusted.
    - Frequency: Monthly.
- Mortgage Debt Service Payments as a Percent of Disposable Personal Income[123], price change from a year ago.
    - Source: Board of Governors of the Federal Reserve System (US).
    - Units: Percent, Seasonally Adjusted.
    - Frequency: Quarterly.

The new elastic net regression model had slightly lower errors and an increased in R-squared. This model explains about 65% of variability in GDP growth. Using both coincident and leading indicators paints a better picture of where we're headed [50].

```
R2:0.6477, MSE:0.0006, RMSE:0.0235
```

> A coincident indicator is an economic statistical indicator that changes (more or less) simultaneously with general economic conditions and therefore reflects the status of the economy. Personal income, GDP, and monthly supply of houses are examples of coincident indicator.

In summary, the developments in the macroeconomy can have significant consequences for agriculture. Key factors linking agriculture to the US and global macroeconomy are[124]:

- Exchange rates.
- International trade.
- Foreign and domestic income.
- Employment, interest rates.
- Energy costs.

---

[122] Monthly Supply of Houses in the United States (MSACSR) – https://fred.stlouisfed.org/series/MSACSR

[123] Mortgage Debt Service Payments as a Percent of Disposable Personal Income (MDSP) – https://fred.stlouisfed.org/series/MDSP

[124] Macroeconomics and agriculture – https://www.ers.usda.gov/topics/international-markets-us-trade/macroeconomics-agriculture/#:~:text=Key%20factors%20linking%20agriculture%20

International and domestic macroeconomic changes can cause major shifts in the values of these indicators, altering a country's agricultural prices, production, consumption, and trade. Changing interest rates influence capital investments, land values, and storage, while inflation affects real interest rates, as well as input, commodity, and land prices.[125] The total demand (or consumption of food grains) is determined by food grain prices and total disposable income, while the previous year's consumption does not have an important role (due to collinearity) [9]. Both price and income elasticity coefficients are of correct signs, and income elasticity seems to be nearer unity for food grains in a country line India [9].

There are several techniques to predict GDP growth. Traditional econometric models and novel machine learning models.

Coincident Economic Activity Index for the United States[126]: FRED tracks coincident economic activity index for the US economy. The Coincident Economic Activity Index includes four indicators: nonfarm payroll employment, the unemployment rate, average hours worked in manufacturing, and wages and salaries. The trend for each state's index is set to match the trend for gross state product.

The index is a monthly indicator with units Index 2007 = 100, Seasonally Adjusted and source from Federal Reserve Bank of Philadelphia.

Challenges

Despite the popular image of farming as a peaceful and healthy way of life, agriculture has its own challenges. Firstly, farming is a very tough, back breaking hard industry. Farming is sensitive to local resource changes to global macrolevel economic factors. This is true as size of the farms get smaller. Secondly, farming has several uncertain and threat vectors. Threat vectors include change in people consumption patterns, changes in people's caloric intake, changes in weather and climate change, change in commodity markets, tariff[127] battles [51], change in trade policies, changes in oil and crude oils, changes in currency valuations, new government regulations and ordinances,[128] changes in stock market, changes in family, and

to,interest%20rates%2C%20and%20energy%20costs.&text = ERS%20provides%20a%20range%20of,relating%20to%20macroeconomics%20and%20agriculture.

[125] Macroeconomy – http://www.fao.org/3/i2490e/i2490e01c.pdf

[126] Coincident Economic Activity Index for the United States – https://fred.stlouisfed.org/series/USPHCI

[127] On America's Farms, Worries About Regulation, Hope for Trade Deals Under Biden – https://www.wsj.com/articles/on-americas-farms-worries-about-regulation-hope-for-trade-deals-under-biden-11604917800?page=1

[128] Agriculture Ordinance 2020 – https://www.prsindia.org/sites/default/files/bill_files/Legislative%20Brief%20-%20Agriculture%20Ordinances%2C%202020.pdf

life-changing events, especially for small-scale farmers. Thirdly, the availability of financial resources and investment capital is very critical in farming. Selling farm produce at the right pricing, after a long and tough agriculture season, is critical and stark makes difference between survival or debt stricken sinking enterprise[129] [21, 52]. Fourthly, the availability of human resources is a great challenge and poses a huge constraint to the sustainability of faming, especially small-scale farming. Despite improvement in productivity and automation, it is still a challenge to get resources for farming. For instance, total global employment (please see Table 3.3) in agriculture was 43.68% in 1991 and has fallen to 26.49% in 2020. At the same time, the employment in services (34.50% in 1991 and 50.52% in 2020) and the manufacturing industry (21.82% in 1991 to 23.08% in 2020) have grown, respectively. Finally, farming, like any other entrepreneurial pursuit, is filled with uncertainties: the probability of failure is very high and the probability of success is stacked against farmers. To be afloat, the emphasis to profit and loss is key and as stringent and rigorous as any other industry or enterprise.

For proper and effective agricultural development, it is desirable that crop productions and animal husbandry activities should prosper together in the country.[130]

Decline in Agriculture Output and Increase in Rural Poverty

Agricultural outputs are declining worldwide (Asia, Europe, America, LAC, and other parts). One of the reasons for this is the availability of human resources. For instance, employment in agriculture had peaked at 43.6% in 1991 and has since fallen to 26.49% in 2020. The shift in employment to services (34.50% in 1991, 50.42% in 2020) and industry (21.81% in 1991, 23.08% in 2020) in urban settings is most noticeable across the world (please see Table 3.2). Most notably, the service industry absorbed a major chunk! This is true in Western parts of the world, but developing economies still exhibit the same scale of productivity as in the early part of the century [53].

The World Bank Policy Research Working Paper (WPS7844) coins a new demographic profile of extreme and moderate poverty, defined as those living on less than \$1.90 and between \$1.90 and \$3.10 per day in 2013, based on household survey data from 89 developing countries (please see Fig. 3.23). The face of poverty is primarily rural and young; 80% of the extreme poor and 75% of the moderate poor live in rural areas [14].

In the developing countries, the face of poverty is primarily rural and young.

[129] They're Trying to Wipe Us Off the Map.' Small American Farmers Are Nearing Extinction - https://time.com/5736789/small-american-farmers-debt-crisis-extinction/

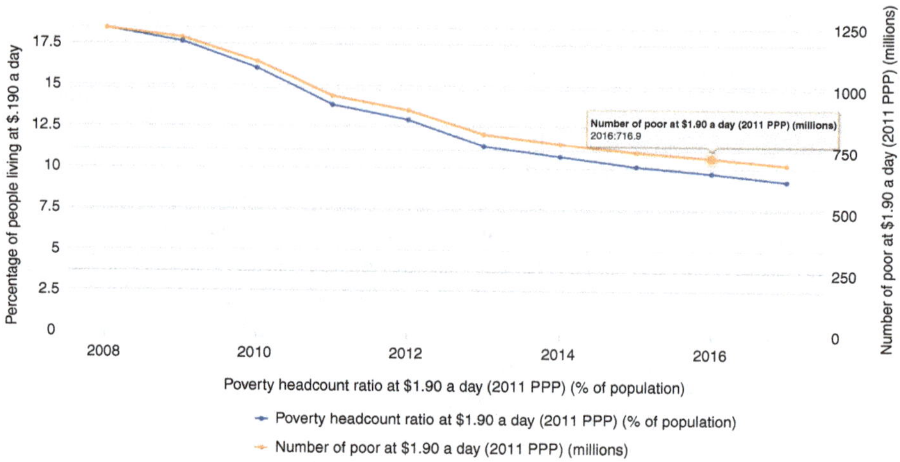

Fig. 3.23 Global poverty indicator

Global Poverty Indicators[131]:

- Poverty headcount ratio at $1.90 a day (% of population).
  - 35.9% 1990 9.2% 2017
- Number of poor at $1.90 a day (2011 purchasing power parity (PPP)).
  - 1.9120 billion 1990 689 million 2017.

The results reinforce the central importance of households in rural areas and those containing large numbers of children in efforts to reduce extreme poverty and are consistent with increased educational attainment and urbanization hastening poverty reduction.

Rural populations across the world experience more economic stress and less economic opportunities (please see Fig. 3.24). The gap between rural and urban indicators of economic well-being persists.[132]

Farmers' Suicides in Different Cultures

Suicide among farmers is a global phenomenon. Farmers are exposed to a high rate of stress due to the isolating, financially precarious and physically demanding nature of their work.[133] The physical stressors, availability of resources, and hazards

[130] Statistical Evaluation of agricultural development in Asian countries – http://www.stats.gov.cn/english/icas/papers/P020071017429375003907.pdf

[131] http://povertydata.worldbank.org/poverty/home/ Access Date: October 26, 2020, 12:35 AM

[132] Selected Charts from Ag and Food Statistics: Charting the Essentials, February 2020 – https://www.ers.usda.gov/publications/pub-details/?pubid=96956

[133] Quiet Epidemic of Suicide Claims France's Farmers – https://www.nytimes.com/2017/08/20/world/europe/france-farm-suicide.html

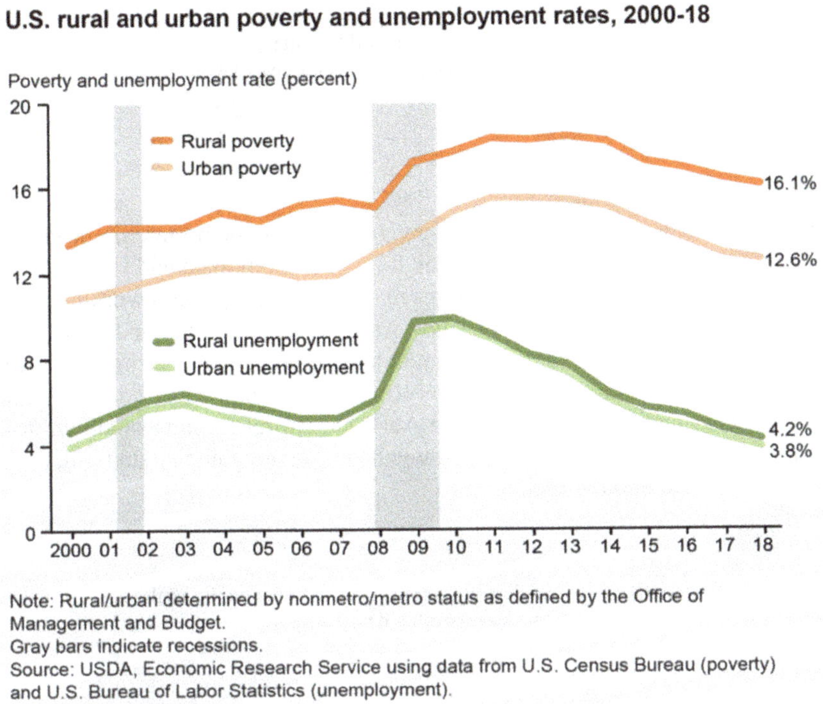

Fig. 3.24 US rural and urban poverty and unemployment rates [54]

in the farm environment are compounded by fluctuations in commodities markets (economic recessions and shocks), climate change, international politics and regulatory framework, and economic dynamics of managing farm business. Farmers operate in the context of declining trends of trade for agricultural produce, limited availability of farm labor, financial capital, availability of ICT, information services, volatile commodity markets, limited availability of off-farm employment, growing cost of machinery and production, and loss of farm or livelihood due to crop failures. Economic concerns and government bureaucracy have been consistently identified as a major cause of stress and a contributor to suicide. Farming is one of the most dangerous industries and is associated with a higher suicide rate than the general population [21]. The mortality rates of French farmers by suicide are 20% higher than that of the general population [55].[134] Stress is a major killer, and natural calamities are driving the farmers to the edge. For example, farmers in Southern Africa are committing suicide as droughts reach crisis levels.[135]

---

[134] A French Farmer Commits Suicide Every 2 Days, Study Says – https://www.huffpost.com/entry/french-farmer-suicide-study_n_4085319

[135] Farmers are committing suicide as drought in southern Africa reaches crisis levels – https://nypost.com/2019/11/21/farmers-are-committing-suicide-as-drought-in-southern-africa-reaches-crisis-levels/

"Farmers are often reluctant to talk about their difficulties, and it is hard for them to imagine doing anything else. They have a strong culture of labor and effort, and they're not used to complaining.
Véronique Louazel[136]"

The findings on farmer suicides are corroborated by the research. For instance, the HUNT study[137] performed in Norway found that both male and female farmers had higher levels of depression symptoms than those of the general working population. The hypothesis for the HUNT study was that working in an industry that has been under long-term structural and economic pressure may be detrimental to mental health. A similar study conducted by Rural and Remote Health in Australia in 2014, "Suicide and accidental death in Australia's rural farming communities: A review of the literature," found that farmers had a suicide rate one-and-a-half times that of the general population.[138]

According to the survey conducted by the French national public health agency, one farmer commits suicide almost every other day, especially men aged 45–54.

Agriculture provides livelihoods for about half of the Indian population, most of whom are smallholder farmers, yet most of the government agricultural subsidies are used by medium- and large-scale farmers. Parallel to India's tremendous successes in the modernization of agriculture, smallholder farmers have been marginalized. The average debt of a farming household has risen fivefold in a decade, while increases in farm incomes have not kept up, and more than 300,000 Indian farmers have committed suicide since 1995.[139]

Over-regulation "will kill us in the end"—French farmer[140]

[136] Quiet Epidemic of Suicide Claims France's Farmers – https://www.nytimes.com/2017/08/20/world/europe/france-farm-suicide.html

[137] Anxiety and Depression Symptoms Among Farmers: The HUNT Study, Norway – https://www.tandfonline.com/doi/full/10.1080/1059924X.2015.1106375

[138] Suicide and accidental death in Australia's rural farming communities: a review of the literature – https://www.rrh.org.au/journal/article/2517

[139] How digital innovation is transforming agriculture: Lessons from India – https://www.mckinsey.com/industries/agriculture/our-insights/how-digital-innovation-is-transforming-agriculture-lessons-from-india

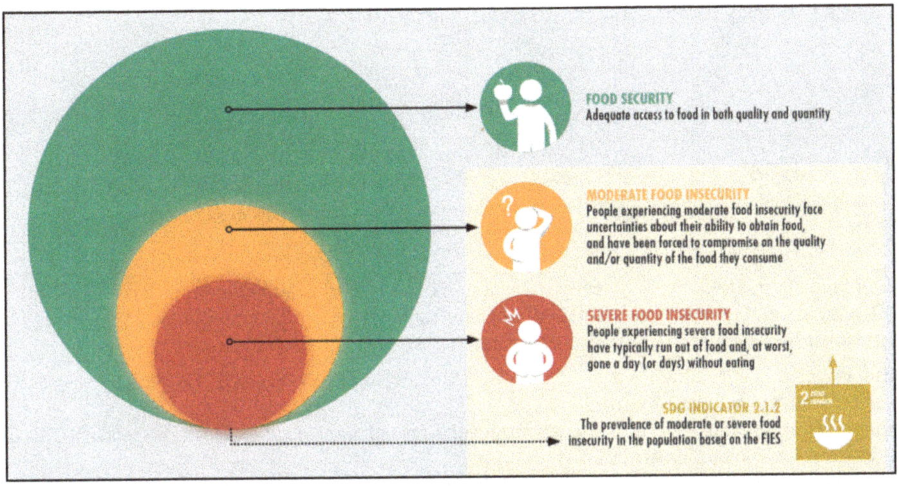

Fig. 3.25 FAO, IFAD, UNICEF, WFP and WHO. 2019. The State of Food Security and Nutrition in the World 2019. Safeguarding against economic slowdowns and downturns. Rome, FAO

Food Security

Food security is one of the biggest issues that humanity is facing. For instance, Asia-Pacific is home to over half of all people worldwide who do not obtain sufficient dietary energy to maintain normal, active, and healthy lives. It is perplexing that Asia-Pacific, having one of the fastest-growing GDP globally, of which is predicted to overtake the combined worldwide GDP by 2020,[141] fails to assure the dietary needs of more than half of its very young children (aged 6–23 months) and unable to meet minimum standards of diversity, leading to micronutrient deficiencies that affect child development and therefore the potential of future generations (Fig. 3.25).[142]

To feed the people of the world and to overcome food security issues, we need more optimized production with the help of technology. Machine learning and AI farm technology can yield new growth[143] and ensure food security to future generations.

Finally, farmers incur huge debt, bank loans, and harvesting difficulties when planting crop, nevertheless diligently waiting for returns. In the end of agriculture

[140] Secret anguish of the French farmer – https://www.politico.eu/article/brussels-economy-agriculture-secret-anguish-of-the-french-farmer/

[141] In 2020 Asia will have the world's largest GDP. Here's what that means – https://www.weforum.org/agenda/2019/12/asia-economic-growth/#:~:text=In%202020%20Asia's%20GDP%20will,class%20entering%20the%20global%20economy.

[142] Asia and Pacific Regional Overview of Food Security and Nutrition – http://www.fao.org/3/ca7062en/ca7062en.pdf

[143] Agriculture's Connected future: How technology can yield new growth – https://www.mckinsey.com/industries/agriculture/our-insights/agricultures-connected-future-how-technology-can-yield-new-growth

Table 3.3 Employment in agriculture, service, and industry

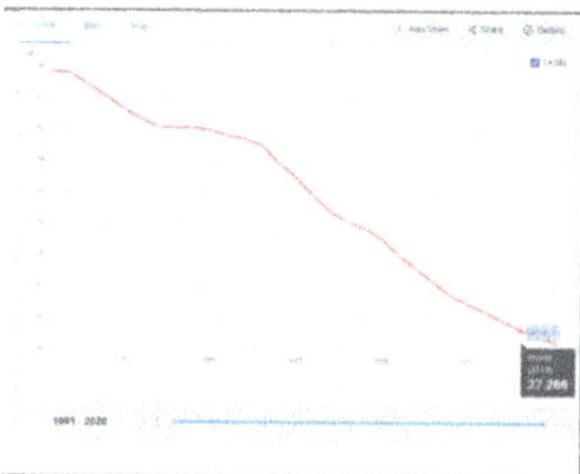	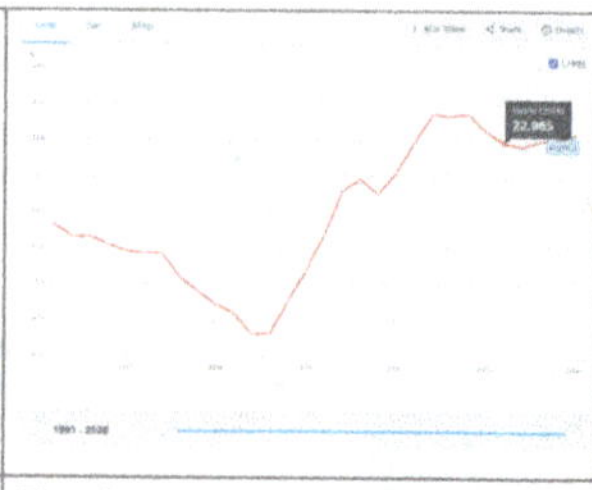	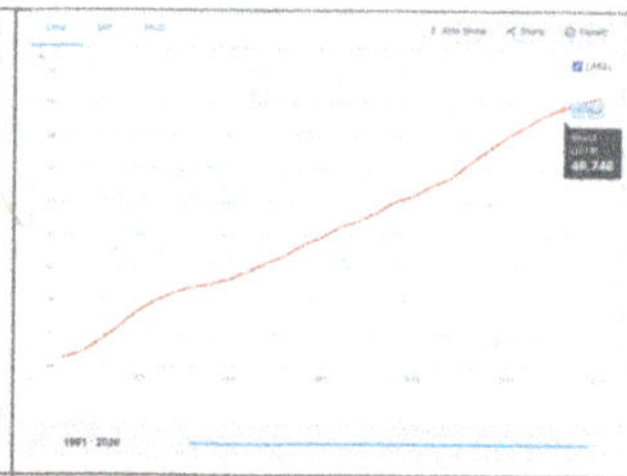
Employment in agriculture (% of total employment) (modeled ILO estimate)[a]	Employment in industry (% of total employment) (modeled ILO estimate)[b]	Employment in services (% of total employment) (modeled ILO estimate)[c]

[a]Employment in agriculture – https://data.worldbank.org/indicator/SL.AGR.EMPL.ZS?end=2020&start=1991&view=chart&year=2020

[b]Employment in Industry – https://data.worldbank.org/indicator/SL.IND.EMPL.ZS?end=2020&start=1991&view=chart&year=2020

[c]Employment in Services – https://data.worldbank.org/indicator/SL.SRV.EMPL.ZS?end=2020&start=1991&view=chart&year=2020

season, crop revenues keep them afloat and allow them to plan for the next season's harvest. The bottom line is, agriculture is woven with uncertainties, difficulties, heart pains, and risks. The purpose of machine learning is to reduce the risks that farmers face from macro- and microeconomic environment changes. For instance, if a GDP model forecasts a lower economic trend, a plan-b should be baked in. In this case, the actionable insight to the small farmer would be to reduce plant harvest capacities, thus lowering overall economic risk exposure, such as falling crop prices.

"As farmers go, so goes the nation" clearly connects the role of higher crop prices to the recovery of the United from the Great Depression in the spring of 1933.[144]

Data science infused machine learning analysis of macroeconomic factors, enabling farmers to achieve crop profit maximization with the intent to sustain farming through uncertainties posed by dynamics of life and markets!

The need for Macrolevel Insightful Informational Services to manage microlevel farm outcomes for small farmers is the need of the day!

[144]As farmers go, so goes the nation How did agriculture help the US recover from the Great Depression in the spring of 1933? – https://www.aeaweb.org/research/farmer-recovery-great-depression

Chapter Summary
After reading the chapter you should comfortably answer:

- Global agriculture landscape.
- Agriculture farms and classification.
- Farm size and appropriate machine learning algorithm.
- Macroeconomics and GDP.
- Estimate GDP using open-source data sources.
- Machine learning algorithms in agriculture.
- Small-scale farmer challenges and need to have information services.

References

1. World Bank: Agriculture and Food. https://www.worldbank.org/en/topic/agriculture/overview (2020, September). Accessed 22 Sept 2020
2. Goedde, L., Horii, M., Sanghvi, S.: Global agriculture's many opportunities. https://www.mckinsey.com/~/media/McKinsey/Industries/Private%20Equity%20and%20Principal%20Investors/Our%20Insights/Global%20agricultures%20many%20opportunities/Global%20agricultures%20many%20opportunities.ashx (2015). Accessed 22 Nov 2020
3. Aririahu, C.: Top Seven Countries That Developed With Agriculture. https://www.afterschoolafrica.com/47413/top-seven-countries-that-developed-with-agriculture/ (2020, Apr 8). Accessed 20 Dec 2020
4. Kassel, K., Martin, A.: Ag and Food Sectors and the Economy. https://www.ers.usda.gov/data-products/ag-and-food-statistics-charting-the-essentials/ag-and-food-sectors-and-the-economy/ (2020, December 16). Accessed 24 Dec 2020
5. Zahniser, S.: What is agriculture's share of the overall U.S. economy? https://www.ers.usda.gov/data-products/chart-gallery/gallery/chart-detail/?chartId=58270 (2020, October 15). Accessed 20 Dec 2020
6. Wirtz, C.: Agriculture, forestry and fishery statistics 2018 Edition, eurostat. ISBN 978-92-79-94758-2 ISSN 1977-2262, https://ec.europa.eu/eurostat/documents/3217494/9455154/KS-FK-18-001-EN-N.pdf/a9ddd7db-c40c-48c9-8ed5-a8a90f4faa3f (2018, November)
7. Bhatia, V.K., Rai, S.C.: Statistical Evaluation of Agricultural Development in Asian Countries, http://www.stats.gov.cn/english/icas/papers/P020071017429375003907.pdf (2020). Accessed 18 Sept 2020
8. Yao, X.: Building Climate Resilience in the Agriculture Sector in Asia and the Pacific, Asian Development Bank, ISBN 978-971-561-827-4 (2009)
9. Parikh, A.: An econometric model of the agricultural sector of the Indian economy. J. Policy Model. **4**(3), 395–411 (1982) https://www.sciencedirect.com/science/article/abs/pii/0161893882900278?via%3Dihub
10. Mzali, S.: Agriculture in Africa 2019: Special Report, https://oxfordbusinessgroup.com/blog/souhir-mzali/focus-reports/agriculture-africa-2019-special-report (2019, 27 Jun). Access 18 Sept 2020
11. Cornock, O.: The Report Africa 2019. https://oxfordbusinessgroup.com/sites/default/files/blog/specialreports/949525/africa_2019_special_report.pdf (2019). Accessed 18 Sept 2020
12. Gurría, A., da Silva, J.G.: OECD-FAO Agricultural Outlook 2018–2027 – Special focus: Middle East and North Africa, ISBN 978-92-64-06203-0 (PDF), http://www.fao.org/3/i9166en/I9166EN.PDF (2018)

13. OECD/FAO: Latin American Agriculture: Prospects and Challenges, in OECD-FAO Agricultural Outlook 2019–2028, OECD Publishing, Paris, https://doi.org/10.1787/b2b742eb-en (2019)
14. Castañeda, A., Doan, D., Newhouse, D., Nguyen, M.C., Uematsu, H., Azevedo, J.P.: Policy Research Working Paper 7844 – WPS7844 – Who Are the Poor in the Developing World, October 2016, World Bank Group, Poverty and Shared Prosperity Report 2016: Taking on Inequality, http://documents1.worldbank.org/curated/en/187011475416542282/pdf/WPS7844.pdf (2016, October)
15. Rao, D.R., Satyanarayan, K.: Small Farmers Make Big Contributions: Dairy Production in India. https://ucanr.edu/blogs/blogcore/postdetail.cfm?postnum=23184 (2017, February 7). Accessed 18 Sept 2019
16. Prahalad, C.K., Hart, S.L.: Strategies for the Bottom of the Pyramid: Creating Sustainable Development. http://pdf.wri.org/2001summit_hartarticle.pdf (1999, August). Accessed 22 Nov 2019
17. Eurostat: Small and large farms in the EU – statistics from the farm structure survey. https://ec.europa.eu/eurostat/statistics-explained/pdfscache/54736.pdf (2018, Sept 28). Accessed 15 Nov 2019
18. Widmar, D.: China's Acreage Gap. https://aei.ag/2018/10/01/chinas-acreage-gap/ (2018, October 1). Accessed 15 Nov 2019
19. Taneja, M., Jalodia, N., Byabazaire, J., Davy, A., Olariu, C.: SmartHerd management: A microservices-based fog computing–assisted IoT platform towards data-driven smart dairy farming. https://doi.org/10.1002/spe.2704 (2019, 16 May). Accessed 6 Aug 2019
20. Dynan, K., Sheiner, L.: GDP as a Measure of Economic Well-being. https://www.brookings.edu/wp-content/uploads/2018/08/WP43-8.23.18.pdf (2018, August 24). Accessed 10 Dec 2019
21. Behere, P.B., Bhise, M.C.: Farmers' suicide: Across culture. Indian J. Psych. **51**(4), 242–243 (2009). https://doi.org/10.4103/0019-5545.58286
22. Santacoloma, P., Suárez, R., Riveros, H.: Strengthening agribusiness linkages with small-scale farmers – Case studies in Latin America and the Caribbean. http://www.fao.org/3/y6001e/y6001e00.htm#Contents (2005). Accessed 7 Mar 2018
23. Just, R.E.: Mathematical Models In Economics. Vol. II, Mathematical Modeling in Agricultural Economics, http://www.eolss.net/sample-chapters/c02/E6-154-12-00.pdf (2020). Accessed 20 Dec 2020
24. Bureau of Economic Analysis (BEA): Concepts and Methods of the U.S. National Income and Product Account. https://www.bea.gov/system/files/2019-12/All-Chapters.pdf (2019, November). Accessed 10 Dec 2020
25. Richardson, A., van Florenstein Mulder, T., Vehbi, T.: Nowcasting New Zealand GDP Using Machine Learning Algorithms, Centre for Applied Macroeconomic Analysis (CAMA) Working Paper 47/2018. https://cama.crawford.anu.edu.au/sites/default/files/publication/cama_crawford_anu_edu_au/2018-09/47_2018_richardson_mulder_vehbi.pdf (2018, September). Accessed 14 Nov 2020
26. Rajkumar, V.: Predicting Surprises to GDP: A Comparison of Econometric and Machine Learning Techniques. https://dspace.mit.edu/bitstream/handle/1721.1/109649/987002546-MIT.pdf?sequence=1&isAllowed=y (2017, February). Accessed 18 Sept 2020
27. Iyetomi, H., et al.: Relationship between Macroeconomic Indicators and Economic Cycles in U.S. Sci. Rep. **10**(1), 8420 (2020). https://doi.org/10.1038/s41598-020-65002-3
28. Peixeiro, M.: The Complete Guide to Time Series Analysis and Forecasting. https://towardsdatascience.com/the-complete-guide-to-time-series-analysis-and-forecasting-70d476bfe775 (2019, Aug 7). Accessed 2 May 2020
29. NIST/SEMATECH: e-Handbook of Statistical Methods. http://www.itl.nist.gov/div898/handbook/ (2012, April). Accessed 7 Feb 2019

30. Brownlee, J.: How to Check if Time Series Data is Stationary with Python. https://machinelearningmastery.com/time-series-data-stationary-python/ (2016, December 30; Last Updated on 2020, August 15) Accessed 18 Sept 2020
31. Camp, K.M.: The relationship between crude oil prices and export prices of major agricultural commodities. https://www.bls.gov/opub/btn/volume-8/the-relationship-between-crude-oil-and-export-prices-of-major-agricultural-commodities.htm?view_full (2019, April 11). Accessed 25 Dec 2020
32. van der Mensbrugghe, D., Osorio-Rodarte, I., Burns, A., Baffes, J.: Macroeconomic Environment, Commodity Markets: A Longer Term Outlook – Expert Meeting on How to feed the World in 2050. http://www.fao.org/3/a-ak967e.pdf (2020). Accessed 1 Dec 2020
33. Kapusuzoglu, A., Ulusoy, M.K.: The interactions between agricultural commodity and oil prices: An empirical analysis. Agric. Econ Czech. **61**(9), 410–421 (2015). https://doi.org/10.17221/231/2014-AGRICECON
34. Inamdar, N.: Farm bills: Are India's new reforms a 'death warrant' for farmers? https://www.bbc.com/news/world-asia-india-54233080 (2020, September 23). Accessed 2 Dec 2020
35. Mark, A.: McMinimy, The 2018 Farm Bill (P.L. 115-334): Summary and Side-by-Side Comparison. https://crsreports.congress.gov/product/pdf/R/R45525 (2019, February 22). Accessed July 2020
36. Mcleod, L., Mcgrath, C.: China Announces 2020 Wheat Minimum Support Price at 2019 Level, https://apps.fas.usda.gov/newgainapi/api/Report/DownloadReportByFileName?fileName=China%20Announces%202020%20Wheat%20Minimum%20Support%20Price%20at%202019%20Level_Beijing_China%20-%20Peoples%20Republic%20of_12-03-2019 (2019, December 6). Accessed 10 Dec 2020
37. McCarthy, N.: U.S. Farm Bankruptcies Reach Eight-Year High [Infographic]. https://www.forbes.com/sites/niallmccarthy/2020/02/10/us-farm-bankruptcies-reach-eight-year-high-infographic/?sh=77b2189f7de0 (2020, Feb 10). Accessed 16 May 2020
38. Root, A.: There Is Turmoil in Agriculture. Here's What Investors Need to Know. https://www.barrons.com/articles/agriculture-farmers-investors-china-trade-latin-america-brazil-argentina-tyson-corteva-fmc-wasde-report-51566336014 (2019, Aug. 21). Accessed June 2020
39. Heeb, G.: As small U.S. farms face crisis, Trump's trade aid flowed to corporations. https://www.cnbc.com/2020/09/02/as-small-us-farms-face-crisis-trumps-trade-aid-flowed-to-corporations.html (2020, Sept 2; Updated 2020, Sept 3). Accessed Dec 2020
40. Newman, J.: More Farmers Declare Bankruptcy Despite Record Levels of Federal Aid. https://www.wsj.com/articles/more-farmers-declare-bankruptcy-despite-record-levels-of-federal-aid-11596706201 (2020, Aug. 6). Accessed 20 Dec 2020
41. Kinskey, I., Oswald, G., McCann, C., Finch, T., Tanaydin, A.: Improvements to Consumption Prediction: Machine Learning Methods and Novel Features. SMU Data Sci. Rev. **1**(4) (2018) Article 3. Available at https://scholar.smu.edu/datasciencereview/vol1/iss4/3
42. Amadeo, K., Barnier, B.: Components of GDP Explained, https://www.thebalance.com/components-of-gdp-explanation-formula-and-chart-3306015#:~:text=1.-,Personal%20Consumption%20Expenditures,figures%20reported%20are%20real%20GDP (2020, June 26). Accessed 20 Dec 2020
43. Dupor, B.: Auto Sales and the 2007-09 Recession, Economic Synopses, No. 16. https://doi.org/10.20955/es.2019.16 (2019)
44. Higgins, P.: Nowcasting GDP growth, https://macroblog.typepad.com/files/111020CurrentQuarterGDPTrackingDescription.pdf (2020, November 10). Accessed 22 Nov 2020
45. Benmelech, E., Meisenzahl, R.R., Ramcharan, R.: Why Did Car Sales Drop So Dramatically During the Financial Crisis? https://insight.kellogg.northwestern.edu/article/was-the-government-bailout-of-car-financers-necessary (2016, APR 4) Accessed 14 Nov 2020
46. Ahmed, T.: What the Coronavirus means for car sales. https://datiuminsights.com.au/what-the-coronavirus-means-for-car-sales/ (2020, March 12). Accessed 10 Dec 2020
47. Rajkumar, V.: Predicting surprises to GDP: a comparison of econometric and machine learning techniques. https://dspace.mit.edu/handle/1721.1/109649 (2017). Accessed 18 Sept2020

48. Tkacz, G., Hu, S.: Forecasting GDP Growth Using Artificial Neural Networks – Working Paper 99-3/Document de travail 99-3 https://www.bankofcanada.ca/wp-content/uploads/2010/05/wp99-3.pdf (2010, May). ISSN 1192-5434, ISBN 0-662-27537-3. Accessed 7 Mar 2020
49. Qin, D., Cagas, M.A., Ducanes, G., Magtibay-Ramos, N., Quising, P.: Forecasting Inflation And Gdp Growth: Automatic Leading Indicator (ALI) Method Versus Macro Econometric Structural Models (MESMS). https://www.adb.org/sites/default/files/publication/29861/tn-18-forecasting-inflation-gdp-growth.pdf (2006, July). Access 25 Mar 2020
50. Rana, J.: Predicting Real GDP Growth How well do economic leading indicators predict the direction of real GDP growth? https://towardsdatascience.com/predicting-real-gdp-growth-85f34fdca97a (2020, May 21). Accessed 7 Oct 2020
51. Bunge, J.: On America's Farms, Worries About Regulation, Hope for Trade Deals Under Biden, https://www.wsj.com/articles/on-americas-farms-worries-about-regulation-hope-for-trade-deals-under-biden-11604917800?page=1 (2020, Nov. 9). Accessed 10 Dec 2020
52. Semuels, A.: They're Trying to Wipe Us Off the Map. Small American Farmers Are Nearing Extinction, https://time.com/5736789/small-american-farmers-debt-crisis-extinction/ (2019, November 27). Accessed 20 Aug 2020
53. Roser, M.: Employment in Agriculture. Published online at OurWorldInData.org. Retrieved from https://ourworldindata.org/employment-in-agriculture [Online Resource] (2013)
54. Kassel, K., Martin, A.: Selected Charts from Ag and Food Statistics: Charting the Essentials, February 2020. https://www.ers.usda.gov/publications/pub-details/?pubid=96956 (2020, February). Accessed 18 Sept 2020
55. Michalopoulos, S.: One French farmer commits suicide every two days, survey says https://www.euractiv.com/section/agriculture-food/news/one-french-farmer-commits-suicide-every-two-days-survey-says/ (2018, Oct 16). Accessed 10 Mar 2019

Chapter 4
Commodity Markets: Machine Learning Techniques

This chapter covers:

- Demand and supply framework.
- Stocks to use ratio and importance to pricing of commodities.
- Machine learning and time series characteristics.
- Vector autoregression (VAR) model.
- Granger's causality.
- Durbin-Watson statistic.
- Commodity gold price ML modeling (VAR).
- Fertilizers, crude oil, and agricultural commodity model.
- World Bank Commodity Market Data—Pink Sheet.
- Fertilizer price predict ML model.

In this chapter, machine learning techniques are introduced to perform historical commodity analysis to derive leading economic macro and micro indicators that would provide a comprehensive market view to small farmers. As part of the chapter, demand and supply framework is introduced. Next, commodity stocks to use ratio indictor are introduced and the impact of stocks to use ratio on the pricing model developed. Time series techniques are introduced to validate trend, seasonality, and stationarity of the time series. Next, the time series feature causation and correlation models and vector autoregression (VAR) are introduced. Finally, the chapter presents two time series-based use cases: predicting gold commodity prices and worldwide study of fertilizer price predict.

C. Vuppalapati, *Machine Learning and Artificial Intelligence for Agricultural Economics*, International Series in Operations Research & Management Science 314,
https://doi.org/10.1007/978-3-030-77485-1_4

Commodities

Forecasting commodity prices play an important role in terms of decision-making of forecasted planted/harvested acreage of crops and financial well-being of small farmers. Expected agricultural commodity prices can influence production decisions of farmers and ranchers on planted/harvested acreage of crops or inventory of livestock and, thus, affect the supply of agricultural commodities[1] [1].

Commodity price changes also affect farms' financial well-being, for example, sustained periods of low commodity prices reduce farm revenues and cause farmers to increasingly rely on credit, making them vulnerable to higher interest rates and other changes to economic conditions. Sustained periods of high commodity prices can contribute to increases in farm revenues and farm operator resilience to changes in economic conditions. Changes to commodity prices also have implications for food security: sustained low prices increase consumers' ability to purchase adequate quantities of food, while sustained high prices decrease their food security, particularly in developing countries.

Commodities pricing, generally, is influenced by supply and demand side of economics that contribute to higher or lower agricultural commodity prices and estimates (please see Fig. 4.1) their effects on commodity markets in terms of trade. Gross domestic product (GDP), economic outlook, employment, and PCE contribute to the demand of agriculture. Planted acreage, harvested for a particular grain (million acres), production (million bushels), and yield per harvested acre (bushels per acre) are supply side of commodity. Other factors such as weather, oil, and climate change can affect both demand and supply side of the commodities. A better understanding of these factors and their impacts can inform and enhance public and private decision-making on issues relating to agricultural markets.

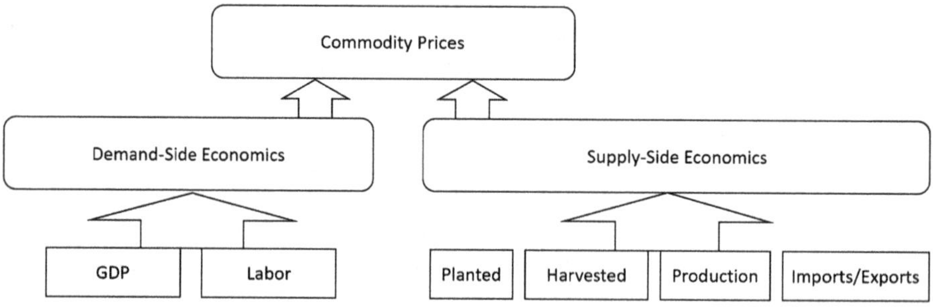

Fig. 4.1 Commodity framework

[1] Factors Contributing to Changes in Agricultural Commodity Prices and Trade for the United States and the World – https://ageconsearch.umn.edu/record/301181/files/err-272.pdf

For instance, sugar production in the United States and Mexico is expected to be significantly lower for 2019/2020 in Mexico—the United States' largest foreign sugar supplier—as drought conditions in several key sugarcane-producing regions are expected to reduce output considerably.[2]

Machine Learning and Commodity Prices

The behavior of agricultural product prices is sufficiently unusual and requires special treatment.[3] Agricultural Commodity Markets are sensitive to macroeconomic environment,[4] oil prices,[5] demand/supply, consumer tastes/preferences, adverse climatic conditions, biofuels, stock to use ratios, dollar exchange rates, speculation,[6] food storage, speculative activity, financial markets, fertilizers,[7] trade restrictions, wealth of nations, and other economic conditions. Economic growth, money supply, weather,[8] and inflation have a positive impact on the commodity prices (bubbles[9]), while effect of interest rate is negative [2–7].

Predicting commodity prices is a challenge and includes data science and econometrics overlay with macroeconomic factors, governmental regulations, subsidies, import/export, production, and demand[10] [8]. The application of econometric analysis, advanced statistics techniques, and machine learning and artificial intelligence techniques provide understanding of price transmission, correlation, and causality.

Commodity price prediction is vital to local and global economy. It would better regulate commodity pricing to consumers (please see Fig. 4.2) and helps small farmers to have a sustained agriculture. Inherently, commodity prices data is a time

---

[2] Weather sours US and Mexican sugar production in 2019/20 – https://www.ers.usda.gov/data-products/charts-of-note/charts-of-note/?page=1&topicId=2608b4e1-fded-4f6e-9855-9e78584d678a

[3] Agricultural product prices by William G. Tomek and Kenneth L. Robinson – ASIN: 0801440939, Cornell University Press; fourth edition (April 10, 2003)

[4] MACROECONOMIC ENVIRONMENT, COMMODITY MARKETS: A LONGER-TERM OUTLOOK – http://www.fao.org/3/a-ak967e.pdf

[5] The interactions between agricultural commodity and oil prices: an empirical analysis – https://www.agriculturejournals.cz/publicFiles/162049.pdf

[6] Did speculation affect world Rice prices? – http://www.fao.org/3/a-ak232e.pdf

[7] Agricultural Fertilizer – https://cfpub.epa.gov/roe/indicator_pdf.cfm?i=55#:~:text=The%20three%20major%20types%20of,can%20also%20occur%20as%20nitrate

[8] US Crop Prices Are Rising, and China Is Buying – https://www.wsj.com/articles/u-s-crop-prices-are-rising-and-china-is-buying-11605522607

[9] Commodity Price Bubbles and Macroeconomics – https://ageconsearch.umn.edu/record/235068/files/Commodity%20Price%20Bubbles%20and%20Macroeconomics.pdf

[10] ROBUST U.S. ECONOMY, HIGHER COMMODITY PRICES IN 2021, SAYS USDA – https://www.agriculture.com/news/business/robust-us-economy-higher-commodity-prices-in-2021-says-usda

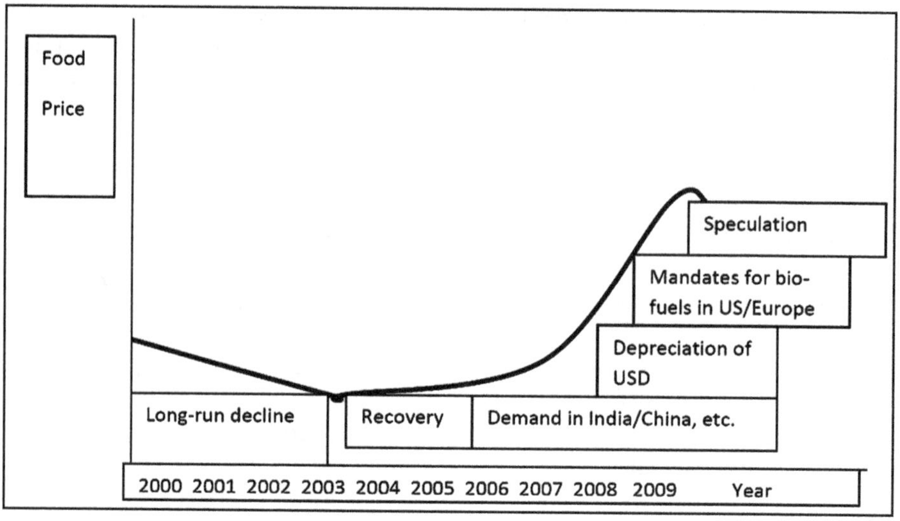

Fig. 4.2 Demand Trend

series data. Regression models, Zero-Inflated Poisson Regression Models,[11] Time Series AUF, and other ML models help to predict commodity prices [7]. Before applying principles of machine learning to understand and predict the demand, here are some basic macroeconomic principles of commodity supply and demand curves.

> Demand/supply vs. speculative effects of financial investors: Whether price movements in commodity markets are determined by changes in physical supply and demand fundamentals or by the speculative effects of financial investors seems to find some element of agreement on one point: The need for increased transparency and improved information on future markets and physical commodity markets[12] [9].

Demand and Supply

Demand: the quantity of good that consumers and households purchase at various prices. As price rises, people are willing and able to purchase less, so the curve is downward sloping (D line) (please see Fig. 4.3).

Supply: the quantity of a good that the firm willing and able to sell at various prices. As price rises, firms are willing to sell more so the supply curve is upward trending (the S line).

[11] Commodity Price Bubbles and Macroeconomics – https://ageconsearch.umn.edu/record/235068/files/Commodity%20Price%20Bubbles%20and%20Macroeconomics.pdf

[12] AN OVERVIEW OF MAJOR SOURCES OF DATA AND ANALYSES RELATING TO PHYSICAL FUNDAMENTALS IN INTERNATIONAL COMMODITY MARKETS – https://unctad.org/system/files/official-document/osgdp20112_en.pdf

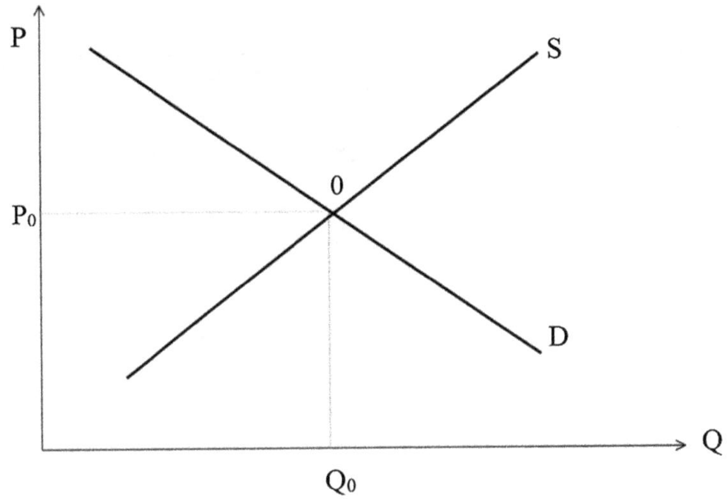

Fig. 4.3 Demand and supply

Stocks to Use Ratio

The stocks to use[13] ratio is a convenient measure of supply and demand interrelationships of commodities. The stocks to use ratio indicates the level of carryover stock for any given commodity as a percentage of the total demand or use. The mathematical formula for this relationship is as follows (please see Eq. 4.1):

$$\text{Stocks to Use} = \frac{(\text{Beggining Stock} + \text{Total Production} - \text{Total Use})}{\text{Total Use}} \tag{4.1}$$

We can fairly say: sum of *Begging Stock* and *Total Production* and subtracting *Total Use* is nothing but carryover stock. Put it simply (please see Eq. 4.2):

$$\text{Stocks to Use ratio} = \frac{\text{Carryover}}{\text{Total Use}} \times 100 \tag{4.2}$$

For instance, Table 4.1 has the stocks to use calculation for the hard red winter wheat: supply and disappearance (million bushels)[14]: Ending stocks is equal to total supply minus total domestic use. Stocks to use is the ratio of ending stocks divided by total disappearance. As it can be observed the inverse relationship between ending stock to price of the commodity.

[13] Stock to Use – https://futures.tradingcharts.com/learning/stocks_to_use.html

[14] Wheat Data (WheatYearbookTable07-Full)– https://www.ers.usda.gov/data-products/wheat-data/

Table 4.1 Hard red winter wheat

	Supply				Disappearance							
					Domestic use							
Mkt year 1/	Beginning stocks	Production	Imports 2/	Total supply 3/	Food use	Seed use	Feed and residual use	Total domestic use 3/	Exports	Total disappearance 3/	Ending stocks	Stocks to Use
1984/85	745	1,251	--	1,996	275	47	242	564	715	1,279	717	56.08%
1985/86	717	1,230	--	1,947	281	43	221	545	393	938	1,009	107.56%
1986/87	1,009	1,017	--	2,026	276	39	309	624	429	1,053	973	92.38%
1987/88	973	1,019	--	1,992	290	39	195	524	901	1,425	567	39.77%
1988/89	567	882	--	1,449	330	44	133	507	639	1,146	302	26.38%
1989/90	302	711	--	1,013	303	41	95	439	359	798	215	26.93%
1990/91	215	1,196	--	1,411	321	38	321	681	369	1,050	360	34.29%
1991/92	360	901	--	1,261	338	40	129	507	559	1,067	194	18.23%
1992/93	194	967	--	1,162	343	40	111	494	464	958	204	21.26%
1993/94	204	1,066	3.53	1,273	384	39	137	560	486	1,046	227	21.74%
1994/95	227	971	3.38	1,202	359	38	190	586	422	1,008	194	19.21%
1995/96	194	825	.29	1,019	346	40	96	481	384	865	154	17.81%
1996/97	154	759	.28	914	320	38	127	485	286	771	143	18.54%
1997/98	143	1,098	.63	1,242	381	36	156	573	362	935	307	32.79%
1998/99	307	1,179	.90	1,487	387	35	186	608	444	1,052	435	41.37%
1999/00	435	1,051	.11	1,486	386	34	132	552	476	1,028	458	44.55%
2000/01	458	846	.05	1,304	375	32	93	500	393	893	411	46.03%
2001/02	411	766	.81	1,178	364	34	68	465	349	815	363	44.58%
2002/03	363	620	.28	984	377	37	74	488	308	795	188	23.67%
2003/04	188	1,071	.38	1,260	378	35	109	522	510	1,033	227	21.99%
2004/05	227	857	.66	1,084	382	33	87	502	389	891	193	21.65%
2005/06	193	930	.43	1,123	370	33	78	481	428	909	215	23.64%
2006/07	215	682	1.06	898	366	37	50	453	280	733	165	22.48%
2007/08	165	956	1.09	1,121	397	35	15	448	536	984	138	13.98%
2008/09	138	1,046	1.51	1,185	385	36	63	483	447	930	254	27.35%
2009/10	254	926	1.56	1,182	361	32	34	427	370	797	385	48.32%
2010/11	385	1,006	.90	1,392	359	32	-3	389	617	1,005	387	38.44%
2011/12	387	783	.48	1,170	404	33	19	455	397	853	317	37.19%
2012/13	317	998	17.67	1,333	404	33	171	608	382	990	343	34.63%
2013/14	343	747	18.71	1,109	370	34	22	426	446	872	237	27.15%
2014/15	237	739	9.85	985	370	33	17	419	272	692	294	42.48%
2015/16	294	830	6.20	1,130	391	30	37	458	227	685	446	65.05%
2016/17	446	1,082	5.05	1,533	385	26	80	490	453	943	589	62.47%
2017/18	589	750	6.75	1,346	392	26	-25	392	373	765	581	75.92%
2018/19	581	662	4.87	1,248	384	25	-8	400	332	732	516	70.44%
2019/20	516	845	1.89	1,363	378	24	78	480	376	856	506	59.13%
2020/21	506	659	2.0	1,167	375	25	35	435	370	805	362	44.97%

For instance, year 2007/2008 has ended with the lowest stocks to use (13.98%). Reflection of it can be seen on subsequent year price 2008/2009 (price jumped to \$403[15]) (please see Fig. 4.4). The same can be observed in 1995–1996.

The stocks to use ratio is a useful indicator. Nonetheless, it won't provide a price guidance to all the commodities. For instance, corn price[16] does not exhibit the same elastic behavior. What is implied in the case of corn price is that stocks to use is not a good indicator of market price predictor [11].

Agricultural grain price increase (please see Fig. 4.5) can be witnessed as a leading indicator by looking into stock to use ratio of major grains. The stock to use[17]

[15] HRW Wheat Prices – https://www.indexmundi.com/commodities/?commodity=wheat&months=360

[16] Irwin, S. and D. Good. "The Relationship between Stocks-to-Use and Soybean Prices Revisited." farmdoc daily (5):89, Department of Agricultural and Consumer Economics, University of Illinois at Urbana-Champaign, May 14, 2015. – https://farmdocdaily.illinois.edu/2015/05/relationship-between-stock-to-use-and-soybean-prices.html

[17] Stocks to Use ratios and prices as indicators of vulnerability to spikes in global cereal markets – https://are.berkeley.edu/~bwright/Wright/Publications_files/Stocks%20to%20use.pdf

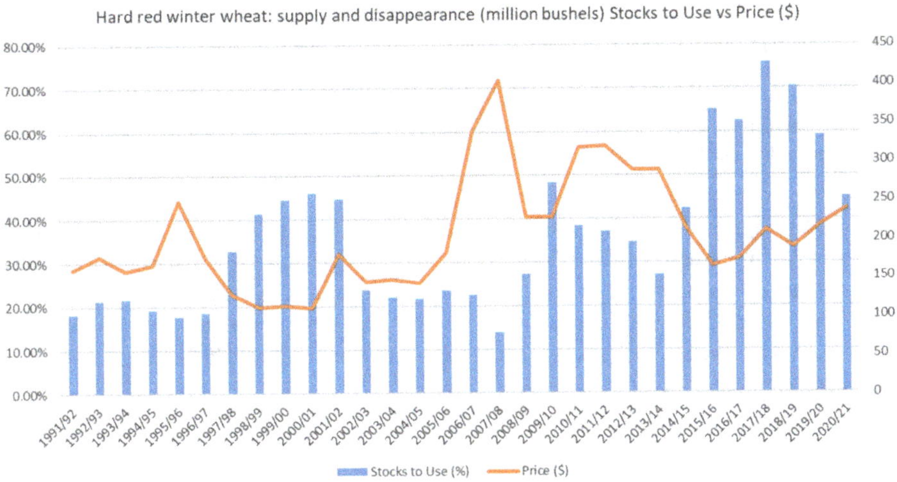

Fig. 4.4 Stocks to use ratio

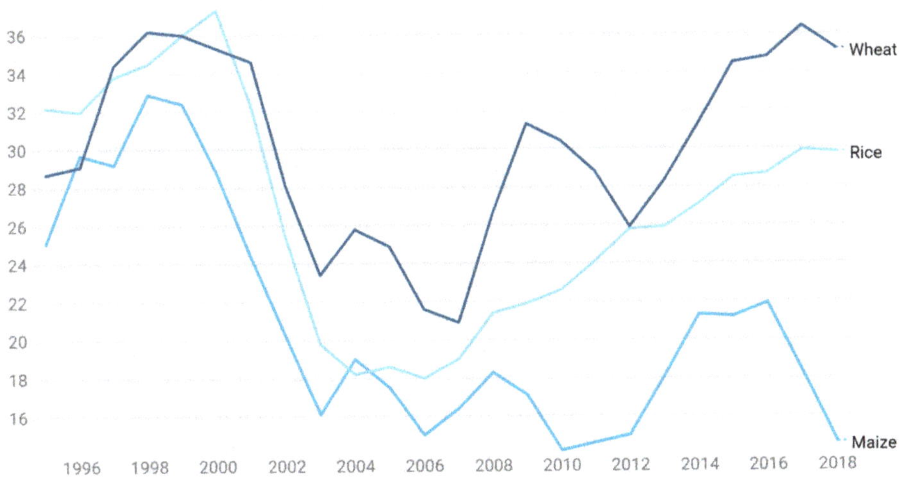

Fig. 4.5 Stocks to use—commodities

ratios (a measure of supply relative to demand) of two of the three main grains [12], which have been on the rise for several years, reversed course. Tightening supply conditions due to lower plantings in North America (a result of persistently low grain prices) and weather problems in South America were the main reasons for this.[18] Grain stock to use ratios: slipping.

[18] Why commodity prices are rising, in nine charts – https://blogs.worldbank.org/developmenttalk/why-commodity-prices-are-rising-nine-charts

Price forecasting for crops based on yield rates that help predict total volumes produced are invaluable in defining pricing strategies for a given crop[19] [13]. Understanding yield rates and quality levels of crops help agricultural firms, co-ops, and farmers better negotiate for the best possible price for their harvests. Considering the total demand for a given crop to determine if the price elasticity curve for a given crop is inelastic, unitary, or highly elastic defines what the pricing strategy will be. Knowing this data alone saves agricultural businesses millions of dollars a year in lost revenue!

Commodity pricing follows pricing strategies of any financial securities. A security or investment is a tradable financial asset. Examples are stocks, bonds, mutual funds, shares, transferable shares, investment contracts, voting-trust certificates, certificate of deposits, calls, and options. Other financial arrangements (OFAs) are loans, material leases, mortgages, bank accounts, credit cards, brokerage accounts, and insurance policies.

Predicting Commodity Prices: Gold Price Predictability

A commodity is usually defined as a raw material used in the productive process or consumed on its own. Gold is often hailed as a hedge against inflation - increasing in value as the purchasing power of the dollar declines. Gold is bought by countries and individuals as hedge against any future risk. Liquidity and portability of gold are very high. Oil, on other hand, is very essential in the modern economy. Oil fuels economy. Everything depends on oil; it heats our homes, runs our cars and machinery, and produces our electricity which in turn runs our computers, the Internet, entertainment, lights, and HVAC. This is one of the reasons why both gold and oil have garnered major attention from academics, investors, and government agencies alike. These two products are known for their substantial influence on global economy. The gold prices are closely related with other commodities. A hike in oil prices will have positive impact on gold prices and vice versa. Historically, we have seen that, when there is a hike in equities, gold prices goes down. Un this Use case, Time series[20]-based causality test can identify and predict the relationship between gold-silver and oil price commodities. Time series was analyzed using vector autoregression[21] (VAR) [14, 15].

[19] 10 Ways AI Has The Potential To Improve Agriculture In 2021 – https://www.forbes.com/sites/louiscolumbus/2021/02/17/10-ways-ai-has-the-potential-to-improve-agriculture-in-2021/?sh=7f611e297f3b

[20] Handbook of Economic Forecasting – by Graham Elliot and Allan Timmermann Elsevier Science and Technology Books, ISBN-13: 978-0444627322, North Holland; 1st edition (September 9, 2013)

[21] Vector Autoregression (VAR) – Comprehensive Guide with Examples in Python – 811

Machine Learning Techniques

Time series (TS) is the representation of series of data points ordered in time, for instance, the value of stock over a period. In TS, usually time is an independent variable and the target variable; value of stock, for example, is the one for which we would like to forecast.[22] The target variable could be sales of the next 6 months, the demand, price of a commodity, or account receivable or payable amount of a corporate account. Depending on the frequency,[23] a time series can be of yearly, quarterly, monthly, etc. Time series forecasting is one of the important areas of machine learning. This is very important when it comes to prediction problems that involve a time component [16, 17].

There are two things which time series make different from the regular regression problem. First one is time-dependent. In linear regression models, observations are independent, but, in this case, observations depend on time. And there can be seasonality trends, where variations are specific to a particular time frame.

The time series can represent a single value or scalar over a period, univariate time series, or represents more than one variable, multivariate. The term "univariate time series" refers to a time series that consists of single (scalar) observations recorded sequentially over equal time increments. Multivariate analysis is a branch of statistics concerned with the analysis of multiple measurements, made on one or several samples of individuals. In multivariate time series, the goal is to identify and model the influence of one time series model on the others.

- Univariate time series forecasting: only two variables in which one is time and the other is the field to forecast.
- Multivariate time series forecasting: contain multiple variables keeping one variable as time and others will be multiple in parameters.

Time series has the following important characteristics (please see Fig. 4.6):

- Trend—The trend is showing the general tendency of the data to increase or decrease with time.
- Seasonality—Seasonality in a time series is a regular pattern of changes that repeats over S time periods, where S defines the number of periods until the pattern repeats.
- Stationary—Stationarity means that the statistical properties of a time series which are mean, variance, and covariance do not change over time.
- In the first plot,[24] mean varies (increases) with time which results in an upward trend (please see Fig. 4.6).

[22] Introduction to Time Series Analysis – https://itl.nist.gov/div898/handbook/pmc/section4/pmc4.htm

[23] Introduction to Time Series forecasting – https://towardsdatascience.com/introduction-to-time-series-forecasting-7e03c4bd83e0

[24] Time Series plot – https://cdn.analyticsvidhya.com/wp-content/uploads/2018/09/ns5-e1536673990684.png

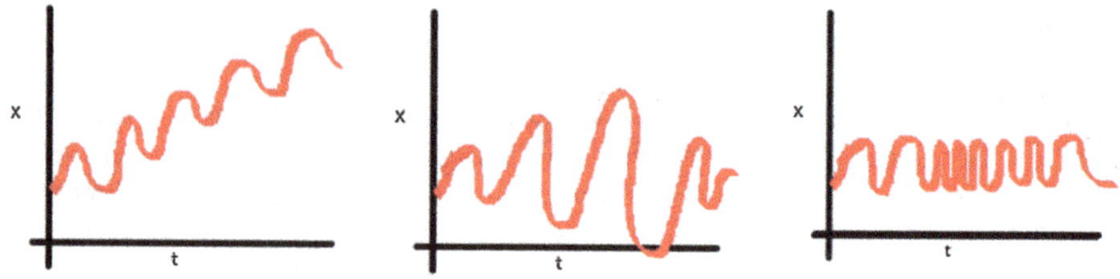

Fig. 4.6 Time series characteristics

- In the second plot, no trend in the series, but the variance of the series is a vary over time.
- In the third plot, the spread becomes closer as the time increases, which means that the covariance is varying over time.

In the above diagram,[25] all three properties are constant with time which stationary time series looks like (please see Fig. 4.7). We can use several methods to identify whether the time series is stationary or not.

- Visual test which identifies the series simply by looking at each plot.
- ADF (augmented Dickey-Fuller) test which used to determine the presence of unit root in the series.
- KPSS (Kwiatkowski-Phillips-Schmidt-Shin) test.

In the following dataset, the goal was to predict the price of gold commodity. The price of gold commodity depends upon several macrolevel parameters as well the perceive of risk from the market investors. During the economic stress times, the price of gold goes higher as many investors find comfort in investing in gold to reduce their risk exposure. The price of gold in true sense is a multivariate time series in that the gold price at any point is dependent upon the market condition, the inflation, the adjacent commodity to gold, and silver price. To predict the gold price, vector autoregression (VAR) is a forecasting algorithm used as it can find price of gold when two or more time series influence each other. That is, the relationship between the time series involved is bi-directional.

Let's understand how a multivariate time series is formulated. Below are the simple K-equations of multivariate time series where each equation is a lag of the other series. X is the exogenous (please see Eq. 4.3) series here. The objective is to see if the series is affected by its own past and the past of the other series:

$$\begin{aligned} Y_{1,t} &= f_1\left(y_{1,t-1}\ldots, y_{k,t-1}\ \ldots, y_{1,t-p}\ \ldots, y_{k,t-p}\ \ldots, X_{t-1}, X_{t-2}\ldots\right) \\ Y_{K,t} &= f_k\left(y_{1,t-1}\ldots, y_{k,t-1}\ \ldots, y_{1,t-p}\ \ldots, y_{k,t-p}\ \ldots, X_{t-1}, X_{t-2}\ldots\right) \end{aligned} \tag{4.3}$$

[25] Stationary–https://cdn.analyticsvidhya.com/wp-content/uploads/2018/09/ns6-e1536674898419.png

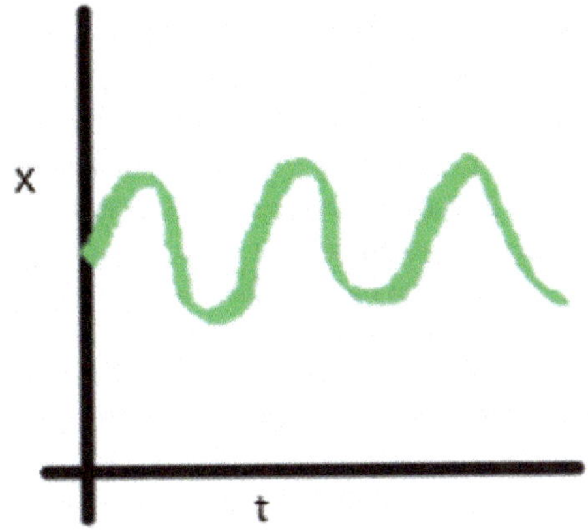

Fig. 4.7 Stationary

In the VAR model, each variable is modeled as a linear combination of past values of itself and the past values of other variables in the system.[26] [15] Since you have multiple time series that influence each other, it is modeled as a system of equations with one equation per variable (time series).

Now, let's explain Eq. 4.1 in terms of autoregressive (AR) model (please see Fig. 4.4):

$$Y_1 = \alpha + \beta_{1\,Yt-1} + \beta_2\, Y_{t-2} + \ldots. + \beta_p\, Y_{t-p} + \epsilon_t \tag{4.4}$$

where α is the intercept, a constant, and β_1, β_2, … and β_p are the coefficients of the lags of Y till order p. Order "p" means, up to p-lags of Y is used, and they are the predictors in the equation. The ϵ{t} is the error, which is considered as white noise.

In the VAR model, each variable is modeled as a linear combination of past values of itself and the past values of other variables in the system. Since you have multiple time series that influence each other, it is modeled as a system of equations with one equation per variable (time series).

That is, if you have time series 5 {Gold Price, Silver Price, S&P Index, Oil Price, and Interest Rate} time series that influence each other, we will have a system of 5 equations.

Let's suppose, you have two variables (time series) Y1 and Y2, and you need to forecast the values of these variables at time (t).

To calculate Y1(t), VAR will use the past values of both Y1 and Y2. Likewise, to compute Y2(t), the past values of both Y1 and Y2 can be used.

For example, the system of equations for a VAR(1) model with two time series (variables "Y1" and "Y2") is as follows (please see Eq. 4.5):

[26] Time Series – Vector Autoregression (VAR) – Comprehensive Guide with Examples in Python – Vector Autoregression (VAR) – Comprehensive Guide with Examples in Python – ML+ (machinelearningplus.com)

$$\begin{aligned} Y_{1,t} &= \alpha_1 + \beta_{11,1} Y_{1,t-1} + \beta_{12,1} Y_{2,t-1} + \epsilon_{1,t} \\ Y_{2,t} &= \alpha_2 + \beta_{21,1} Y_{1,t-1} + \beta_{22,1} Y_{2,t-1} + \epsilon_{2,t} \end{aligned} \quad (4.5)$$

where, $Y_{1,t-1}$ and $Y_{2,t-1}$ are the first lag of time series Y_1 and Y_2, respectively. The above equation is referred to as a VAR(1) model, because each equation is of order 1, that is, it contains up to one lag of each of the predictors (Y_1 and Y_2). Since the Y terms in the equations are, the Y's are considered as endogenous variables, rather than as exogenous predictors.

Likewise, the second-order VAR(2) model for two variables would include up to two lags for each variable (Y_1 and Y_2) (please see Eq. 4.6).

$$\begin{aligned} Y_{1,t} &= \alpha_1 + \beta_{11,1} Y_{1,t-1} + \beta_{12,1} Y_{2,t-1} + \beta_{11,2} Y_{1,t-2} + \beta_{12,2} Y_{2,t-2} + \epsilon_{1,t} \\ Y_{2,t} &= \alpha_2 + \beta_{21,1} Y_{1,t-1} + \beta_{22,1} Y_{2,t-1} + \beta_{21,2} Y_{1,t-2} + \beta_{22,2} Y_{2,t-2} + \epsilon_{2,t} \end{aligned} \quad (4.6)$$

Now, let's see VAR(2) model for {Gold Price (Y_G), Silver Price (Y_S), S&P Index (Y_{SP}), Oil Price (Y_O), and Interest Rate (Y_{IR})} (please see Eq. 4.7).

$$\begin{aligned}
&\mathbf{Y}_{\mathbf{G,t}} = \alpha_1 + \beta_{\text{Gold},1} Y_{G,t-1} + \beta_{G-S,1} Y_{S,t-1} + \beta_{G-SP,1} Y_{SP,t-1} + \beta_{G-O,1} Y_{O,t-1} \\
&+ \beta_{G-IR,1} Y_{IR,t-1} + \beta_{\text{Gold},2} Y_{G,t-2} + \beta_{G-S,2} Y_{S,t-2} + \beta_{G-SP,2} Y_{SP,t-2} + \beta_{G-O,2} Y_{O,t-2} \\
&+ \beta_{G-IR,2} Y_{IR,t-2} + \epsilon_{G,t} \\
&\mathbf{Y}_{\mathbf{S,t}} = \alpha_1 + \beta_{So;ver,1} Y_{S,t-1} + \beta_{S-G,1} Y_{G,t-1} + \beta_{S-SP,1} Y_{SP,t-1} + \beta_{S-O,1} Y_{O,t-1} \\
&+ \beta_{S-IR,1} Y_{IR,t-1} + \beta_{\text{Silver},2} Y_{S,t-2} + \beta_{S-G,2} Y_{G,t-2} + \beta_{S-SP,2} Y_{SP,t-2} + \beta_{S-O,2} Y_{O,t-2} \\
&+ \beta_{S-IR,2} Y_{IR,t-2} + \epsilon_{S,t} \\
&\mathbf{Y}_{\mathbf{SP,t}} = \alpha_1 + \beta_{\text{S\&PIndex},1} Y_{SP,t-1} + \beta_{SP-G,1} Y_{G,t-1} + \beta_{SP-S,1} Y_{S,t-1} + \beta_{SP-O,1} Y_{O,t-1} \\
&+ \beta_{SP-IR,1} Y_{IR,t-1} + \beta_{\text{S\&PIndex},2} Y_{SP,t-2} + \beta_{SP-G,2} Y_{G,t-2} + \beta_{SP-S,2} Y_{S,t-2} \\
&+ \beta_{SP-O,2} Y_{O,t-2} + \beta_{SP-IR,2} Y_{IR,t-2} + \epsilon_{SP,t} \\
&\mathbf{Y}_{\mathbf{O,t}} = \alpha_1 + \beta_{\text{Oil},1} Y_{O,t-1} + \beta_{O-G,1} Y_{G,t-1} + \beta_{O-S,1} Y_{S,t-1} + \beta_{O-SP,1} Y_{SP,t-1} + \beta_{O-IR,1} \\
&Y_{IR,t-1} + \beta_{\text{Oil},2} Y_{O,t-2} + \beta_{O-G,2} Y_{G,t-2} + \beta_{O-S,2} Y_{S,t-2} + \beta_{O-IR,2} Y_{SP,t-2} + \beta_{O-IR,2} Y_{IR,t-2} + \epsilon_{O,t} \\
&\mathbf{Y}_{\mathbf{IR,t}} = \alpha_1 + \beta_{\text{InterestRate},1} Y_{IR,t-1} + \beta_{IR-G,1} Y_{G,t-1} + \beta_{IR-S,1} Y_{S,t-1} + \beta_{IR-SP,1} Y_{SP,t-1} + \\
&\beta_{IR-O,1} Y_{O,t-1} + \beta_{\text{InterestRate},2} Y_{IR,t-2} + \beta_{IR-G,2} Y_{G,t-2} + \beta_{IR-S,2} Y_{S,t-2} + \beta_{IR-SP,2} Y_{SP,t-2} \\
&+ \beta_{IR-O,2} Y_{O,t-2} + \epsilon_{IR,t}
\end{aligned} \quad (4.7)$$

As you increase the number of cc, the system of equations become larger.

An endogenous variable is a variable in a statistical model that's changed or determined by its relationship with other variables within the model. in other words, an **endogenous variable is synonymous with a dependent variable**, meaning it correlates with other factors within the system being studied. In an economic model, an **exogenous variable** is one whose value is **determined outside the model** and is imposed on the model, and an exogenous change is a change in an exogenous variable. In contrast, an endogenous variable is a variable whose value is determined by the model.

Data Sources

A hike in oil prices will have positive impact on gold prices and vice versa. Historically, when there is a hike in equities, gold prices goes down.[27] [18]

> Crude futures tend to be supported during periods of high inflation, while gold is traditionally used as a hedge against inflation. This positive correlation has often meant higher oil prices have coincided with higher gold prices, although one does not directly impact the other.[28] more than 60% of the time, gold and crude oil have a direct relationship. This means that when gold prices rise, crude oil prices also tend to rise and vice versa[29] [19] [20].

Data	Source	
Gold[30] [21]	London bullion market association (LBMA): The London Gold fixing companies set the prices for gold that are globally considered as the international standard for pricing of gold. The gold price in London is set twice a day by five LBMA[31] market makers who comprise the London Gold market fixing limited (LGMFL).	
Silver	The London silver fixing companies set the prices for gold that are globally considered as the international standard for pricing of silver. The silver price in London is set once a day by three LBMA market makers who comprise the London silver market fixing limited (LSMFL).[32]	
Brent and WTI spot prices[33]	Oil prices are collected from Europe Brent and WTI (Western Texas intermediate) spot prices (annual/monthly/weekly/daily) from EIA US (Energy Information Administration). Data series is available through the EIA open data API Europe Brent spot Price FOB (dollars per barrel)—From may 20, 1987, till today. Please note: The US Energy Information Administration (EIA) is a principal agency of the US Federal Statistical System responsible for collecting, analyzing, and disseminating energy information to promote sound policy making, efficient markets, and public understanding of energy and its interaction with the economy and the environment. EIA programs cover data on coal, petroleum, natural gas, electric, renewable, and nuclear energy. EIA is part of the US Department of Energy.	

[27] Time Series Forecasting using Granger's Causality and Vector Auto-regressive Mode – https://towardsdatascience.com/granger-causality-and-vector-auto-regressive-model-for-time-series-forecasting-3226a64889a6

[28] Where's crude heading? Morgan Stanley says to watch the oil-gold ratio – https://www.cnbc.com/2020/07/14/oil-and-gold-prices-morgan-stanley-says-to-watch-the-oil-gold-ratio.html

[29] Relationship Between Gold and Oil Prices: Why There's a Big Disconnection in 2020 – https://ca.finance.yahoo.com/news/relationship-between-gold-oil-prices-153048345.html

[30] Project: Gold vs Silver Price – http://rstudio-pubs-static.s3.amazonaws.com/368841_224fe698bd7c45d8a5dd9a6b639988fc.html

[31] London Bullion Market Association (LBMA) – https://www.quandl.com/data/LBMA/GOLD-Gold-Price-London-Fixing

[32] SilverPrice:LondonFixing–https://www.quandl.com/data/LBMA/SILVER-Silver-Price-London-Fixing

[33] Oil Source: https://datahub.io/core/oil-prices/r/brent-daily.csv

Data	Source	
Economic research -Federal Reserve Bank of St. Louis[34]	Providing free US and international economic data, graphs, and other data-related tools, plus quality research from St. Louis Fed's economists. Trade weighted US Dollar index: Broad, goods and services (DTWEXBGS)[35]	ECONOMIC RESEARCH
S&P 500 index[36]	The S&P 500, or simply the S&P, is a stock market index that measures the stock performance of 500 large companies listed on stock exchanges in the United States. It is one of the most followed equity indices.	STANDARD &POOR'S 500
Economic research -Federal Reserve Bank of St. Louis[37]	Interest rate: 10-year Treasury constant maturity rate (DGS10)[38]	ECONOMIC RESEARCH

Exploratory Data Analysis

Exploratory data analysis (EDA) is a very common technique applied to understand the data and feature engineering. For time series data, EDA has one more extra component—to identify the stationarity of time series. Stationarity implies the statistical properties of time series are relative constant or change over a time, additionally, to identify causality[39, 40] in the time series variables [18, 23].

Before deep diving on the EDA, let's perform eyeball the time series:

Gold Price Time Series Data.

The London Bullion Market Association (LBMA) Gold Price (please see Fig. 4.8) data contains time series values from the 1970s onward. The gold price time series is multiplicative, frequency of 253 data points per year, seasonality 1%, trend of 76% (as you can see the increase in gold price), and remainder of 24%.

> In the multiplicative model, the original time series is expressed as the product of trend, seasonal, and irregular components.[41]

[34] FRED – https://fred.stlouisfed.org/

[35] Trade Weighted USD – https://fred.stlouisfed.org/series/DTWEXBGS

[36] S&P 500 (Yahoo) – https://finance.yahoo.com/quote/%5EGSPC/history?period1=-1325635200&period2=1611360000&interval=1d&filter=history&frequency=1d&includeAdjustedClose=true

[37] FRED – https://fred.stlouisfed.org/

[38] 10-Year Treasury Constant Maturity Rate (DGS10) – https://fred.stlouisfed.org/series/DGS10

[39] Testing for Granger Causality Using Python – https://rishi-a.github.io/2020/05/25/granger-causality.html

[40] Time Series Forecasting using Granger's Causality and Vector Auto-regressive Model – https://towardsdatascience.com/granger-causality-and-vector-auto-regressive-model-for-time-series-forecasting-3226a64889a6

[41] Time Series Analysis: the basics – https://www.abs.gov.au/websitedbs/d3310114.nsf/home/time+series+analysis:+the+basics#:~:text=In%20the%20multiplicative%20model%2C%20

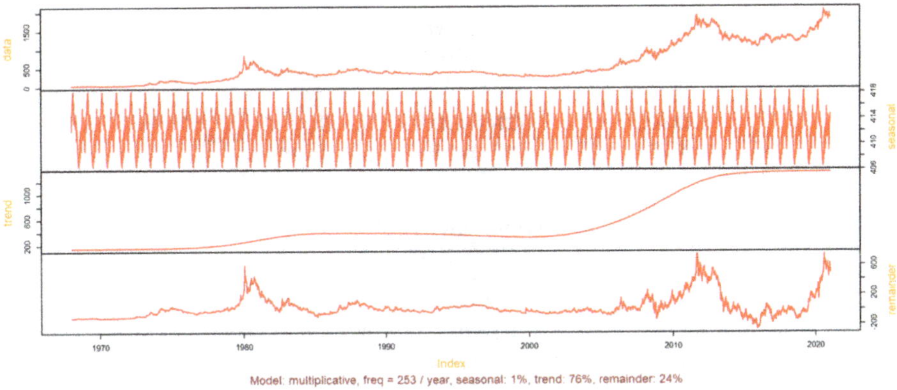

Fig. 4.8 Gold commodity time series

Gold is precious commodity not only from individual's perception of value of the commodity but also as a hedge against inflation and risk for many corporation, governments, and cooperatives. Gold plays an important role in the commodity weights and index calculation issues by the World Bank's monthly Pink Sheet data[42] (March 2021 Pink[43] sheet data). Figure 4.9 shows the weights assigned to gold and the index impact of it, as a reference, on the pricing of others. At the energy level, crude oil and natural gas have higher price index over coal. Non-energy schema, agriculture leads followed by food and metals.

Please note, the higher the weight, the more influence that commodity would have. For instance, crude oil will influence prices of all most all essential commodities as well as precious metals and even agricultural fertilizers.

Silver Price Time Series Data.

The silver price time series (please see Fig. 4.10), very the same as gold time series, is multiplicative, frequency of 253 data points per year, seasonality 1%, trend of 60% (as you can see the increase in silver price), and remainder of 39%.

Oil Price Time Series Data.

The Europe Brent and WTI (Western Texas Intermediate) Spot Oil Price (please see Fig. 4.11) data contains time series values from the 1990s onward. The oil price time series is multiplicative, frequency of 254 data points per year, seasonality 2%, trend of 75% (as you can see Brent Crude Oil Spot Price is at a current level of 55.68, up from 55.66 the previous market day and down from 62.11 1 year ago—peaked in 2008 because of large cuts in production. However, because of the

---

the,trend%2C%20seasonal%20and%20irregular%20components.&text=Under%20this%20model%2C%20the%20trend,unitless%20factors%2C%20distributed%20around%201

[42] Pink Sheet Data – https://www.worldbank.org/en/research/commodity-markets

[43] The World Bank – Pink Sheet March 2021 – https://thedocs.worldbank.org/en/doc/5d903e848db1d1b83e0ec8f744e55570-0350012021/original/CMO-Pink-Sheet-March-2021.pdf

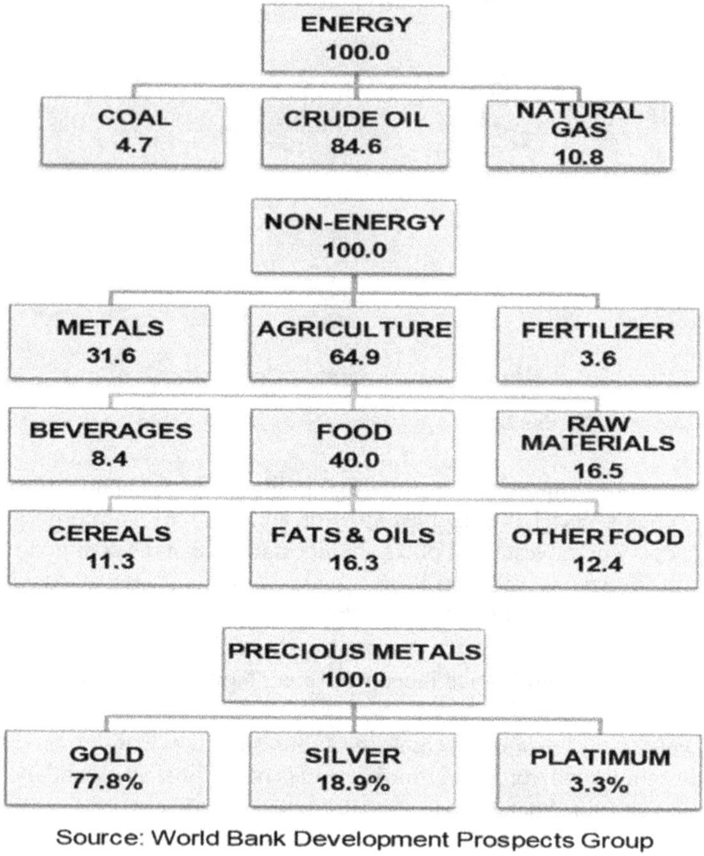

Fig. 4.9 The World Bank Commodity Price Index - Groups and Weights

financial crisis and an abrupt loss of demand for oil globally, the price of Brent Crude fell as much at 70% off highs in January of 2009.[44]), and remainder of 23%.

S&P 500 Index.

The S&P 500 Index time series values from the 1940s onward (please see Fig. 4.12). The S&P time series is multiplicative, frequency of 251 data points per year, seasonality 0%, trend of 92% (the index is in a rising trend channel in the medium long term. Signal increasing optimism among investors and indicates continued rise. There is no resistance in the price chart and further rise is indicated.[45]), and remainder of 8%.

[44] Brent Crude Oil Spot Price – https://ycharts.com/indicators/brent_crude_oil_spot_price#:~:text=One%20of%20the%20most%20notable,of%20large%20cuts%20in%20production.

[45] S&P500 Index – https://www.investtech.com/main/market.php?CompanyID=99200005

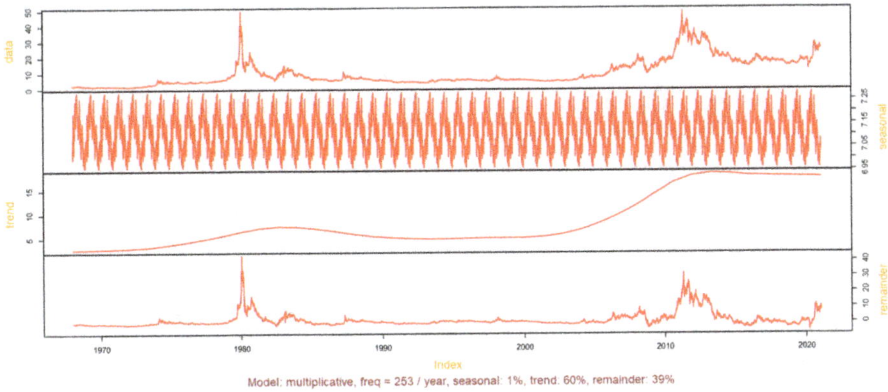

Fig. 4.10 Silver commodity time series

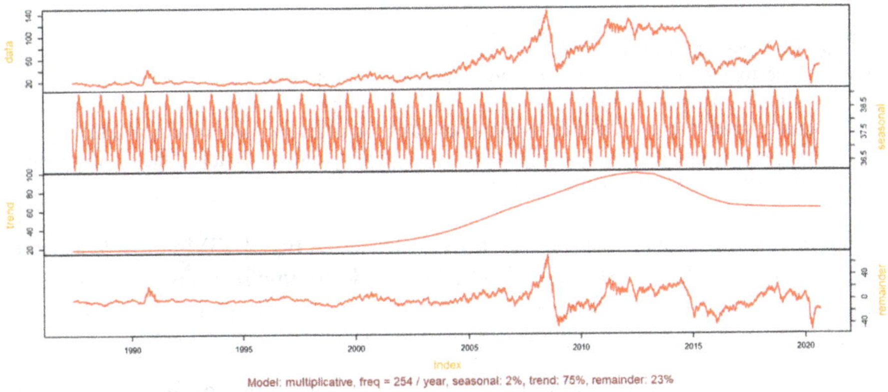

Fig. 4.11 Brent Gas time series

10-Year Treasury Constant Maturity Rate (DGS10)

The US Treasury yield is an important macroeconomic indicator that provides a great insight into market. The higher treasury yields indicate that the market returns are competing bond investment. On the other hand, the lower treasury yield indicates the investors are parking, a factor determined by the risk in the equities[46] or other high return markets, their money on more stable investment, albeit a low return [23].

[46] US Stocks Waver After Tech Selloff – https://www.wsj.com/articles/global-stock-markets-dow-update-02-26-2021-11614314418?page=1

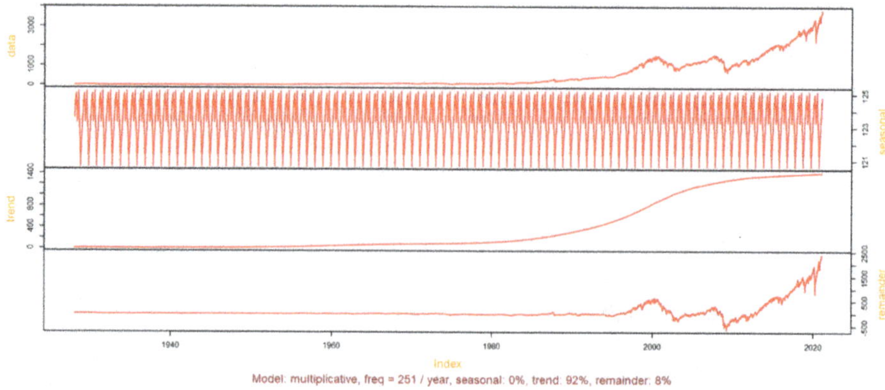

Fig. 4.12 S&P 500 Index

US Treasury yields steady

Investors tend to sell treasury bonds when they expect faster market growth and inflation, which lowers the value of bonds' fixed payments and can prompt the Fed to raise interest rates.[47] Investor's optimism has been lifted recently, February 2021, by improving economic data, the promise of more government spending and the expanding distribution of coronavirus vaccines.

The 10-Year Treasury Constant Maturity (please see Fig. 4.13) Interest Rate time series values from the 1960s onward. The Interest Rate time series is multiplicative, frequency of 250 data points per year, seasonality 2%, trend of 71%, and remainder of 27%. The most notable observations on the data:

- Peak value: The Treasury completed an extraordinary week of financing yesterday with the sale of $4.5 billion of 1-year bills at a near-record interest rate of 14.54%[48] [24].
- On June 1, 2012, the 10-year Treasury rate fell to its lowest point since the early 1800s.[49] It hit an intra-day low of 1.442%. Investors worried about the eurozone debt crisis and a poor jobs report. On July 25, 2012, it closed at 1.43, the lowest point in 200 years[50] [25, 26].

With time series values varying from dates, peaks, and troughs, apply exploratory data analysis (EDA) to unleash the causality relationships among prices and indexes.

---

47 US Treasury Yields Steady – https://www.wsj.com/articles/u-s-treasury-yields-steady-11614359571?page=1

48 TREASURY BILL RATE AT 14.54% – https://www.nytimes.com/1981/08/08/business/treasury-bill-rate-at-14.54.html

49 TREASURIES-Weak jobs data knock U.S. yields lower – https://www.reuters.com/article/markets-usa-bonds-idUSL1E8H1A3Z20120601

50 10-Year Treasury Note and How It Works – https://www.thebalance.com/10-year-treasury-note-3305795

Data Model Diagram

All the data sources in this experiment (Economic Research, S&P 500 Index, LBMA, and Brent) are connected using Date (please see Fig. 4.14). That is, Date is combined key to prepare the VAR model.

Model Development

The following steps help to develop VAR model for multivariate time series (please Fig. 4.15).

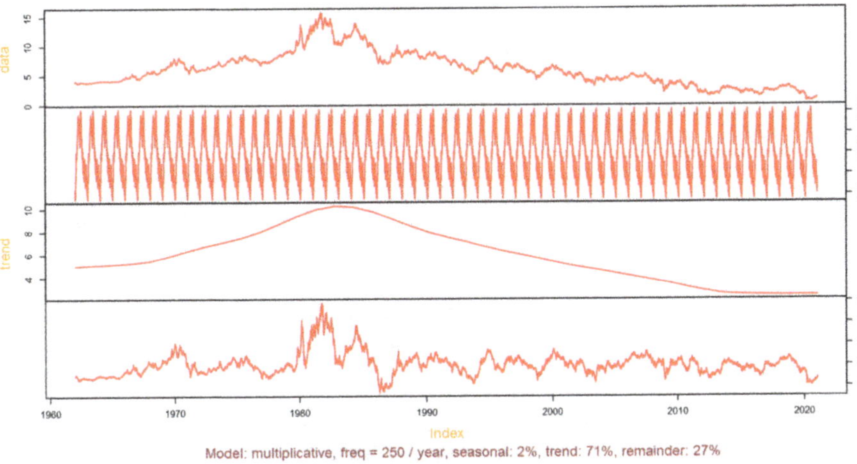

Fig. 4.13 10-year treasury

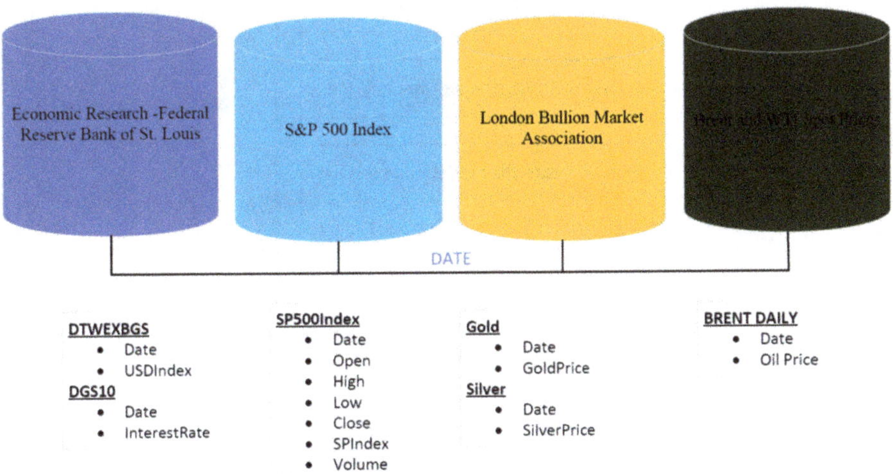

Fig. 4.14 Data Model

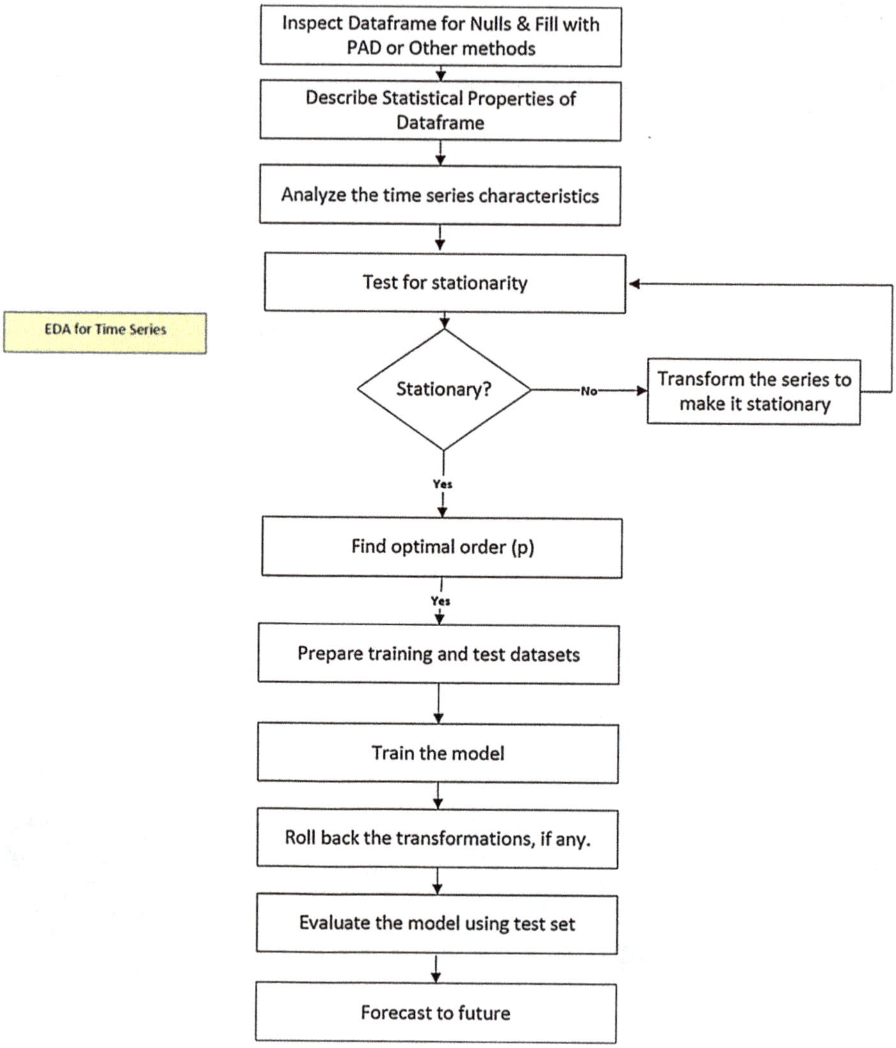
EDA for Time Series
Inspect Dataframe for Nulls & Fill with PAD or Other methods
Describe Statistical Properties of Dataframe
Analyze the time series characteristics
Test for stationarity
Stationary?
No
Transform the series to make it stationary
Yes
Find optimal order (p)
Yes
Prepare training and test datasets
Train the model
Roll back the transformations, if any.
Evaluate the model using test set
Forecast to future

Fig. 4.15 EDA

Step 1: Load required Python libraries.

```
import logging
import pandas as pd
import numpy as np
from numpy import random
import nltk
from sklearn.model_selection import train_test_split
from sklearn.feature_extraction.text import CountVectorizer, TfidfVectorizer
from sklearn.metrics import accuracy_score, confusion_matrix
import matplotlib.pyplot as plt
from nltk.corpus import stopwords
import re
from bs4 import BeautifulSoup
import pandas as pd
import matplotlib.pyplot as plt
import matplotlib as mpl
import numpy as np
import scipy.stats as spstats
import seaborn as sns

%matplotlib inline
mpl.style.reload_library()
mpl.style.use('classic')
mpl.rcParams['figure.facecolor'] = (1, 1, 1, 0)
mpl.rcParams['figure.figsize'] = [6.0, 4.0]
mpl.rcParams['figure.dpi'] = 100
```

Software code for this model can be found on GitHub Link: VARSilverGoldCommodityPricingPredict.Ipynb (Jupyter Notebook)

Step 2: Load Gold dataset: in the below code, load LBMA gold data into Python Data Frame and set Date as an index column. The purpose is to align all the datasets to a common date index column.

```
goldDF = pd.read_csv("LBMA-GOLD-USD.csv",parse_dates=['Date'], dayfirst=True)
goldDF = goldDF.set_index('Date')
goldDF.head()
```

Output:

Date	GoldPrice
2021-01-22	1853.60
2021-01-21	1867.65
2021-01-20	1854.60
2021-01-19	1843.10
2021-01-18	1833.95

Sometimes, data contains nulls, mixed integers, alpha numeric, and, plainly put, load errors. To confirm the data loaded correctly, check Data Frame Information:

```
goldDF.info()
```

Output:

```
<class 'pandas.core.frame.DataFrame'>
DatetimeIndex: 13412 entries, 2021-01-22 to 1968-02-01
Data columns (total 1 columns):
 #   Column     Non-Null Count  Dtype
---  ------     --------------  -----
 0   GoldPrice  13411 non-null  float64
dtypes: float64(1)
memory usage: 209.6 KB
```

Plot the gold price value:

```
fig, ax = plt.subplots()
goldDF['GoldPrice'].hist(color='#A9C5D3', edgecolor='black',
                         grid=False)
ax.set_title('Gold Price', fontsize=12)
ax.set_xlabel('Gold Price', fontsize=12)
ax.set_ylabel('Frequency', fontsize=12)
```

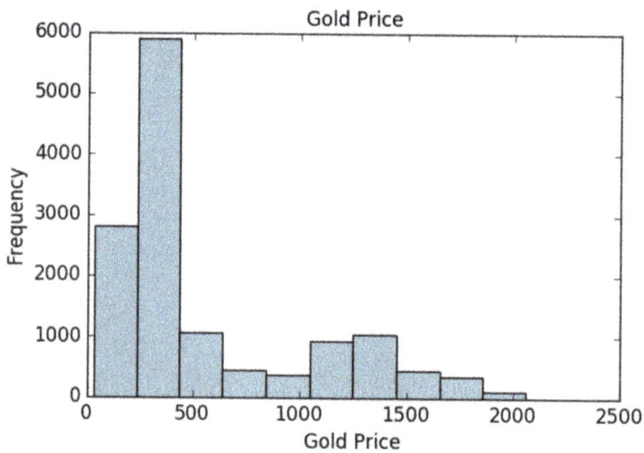

Gold price is heavily skewed toward left. To get more clear view, divide gold price into five quantiles.

```
goldDF['GoldPrice_bin_round'] = np.array(np.floor(
                    np.array(goldDF['GoldPrice']) / 10.))
goldDF[['GoldPrice', 'GoldPrice_bin_round']].iloc[1071:2076]

quantile_list = [0, .25, .5, .75, 1.]
quantiles = goldDF['GoldPrice'].quantile(quantile_list)
quantiles
```

Plot the bins and quantiles.

```
fig, ax = plt.subplots()
goldDF['GoldPrice'].hist(bins=30, color='#A9C5D3',
                    edgecolor='black', grid=False)
for quantile in quantiles:
   qvl = plt.axvline(quantile, color='r')
ax.legend([qvl], ['Quantiles'], fontsize=10)
ax.set_title('Gold Price Histogram with Quantiles',
         fontsize=12)
ax.set_xlabel('Gold Price', fontsize=12)
ax.set_ylabel('Frequency', fontsize=12)
```

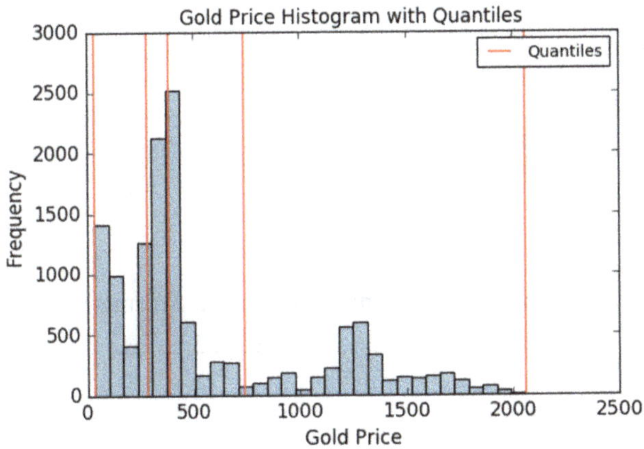

Gold price is left skewed and it's not normal.
To test normality, let's perform the following extra test:

```
goldDF['GoldPrice_log'] = np.log((1+ goldDF['GoldPrice']))
goldDF[[ 'GoldPrice', 'GoldPrice_log']].iloc[4:9]
```

	GoldPrice	GoldPrice_log
Date		
2021-01-18	1833.95	7.514773
2021-01-15	1853.85	7.525559
2021-01-14	1840.25	7.518200
2021-01-13	1852.40	7.524777
2021-12-01	1861.85	7.529863

Plot log shift of gold price:

```
goldprice_log_mean = np.round(np.mean(goldDF['GoldPrice_log']), 2)
fig, ax = plt.subplots()
goldDF['GoldPrice_log'].hist(bins=30, color='#A9C5D3',
                             edgecolor='black', grid=False)
plt.axvline(goldprice_log_mean, color='r')
ax.set_title('Gold Price Histogram after Log Transform',
         fontsize=12)
ax.set_xlabel('Gold Price (log scale)', fontsize=12)
ax.set_ylabel('Frequency', fontsize=12)
ax.text(11.5, 450, r'$\mu$='+str(goldprice_log_mean), fontsize=10)
```

```
Text(11.5, 450, '$\\mu$=5.93')
```

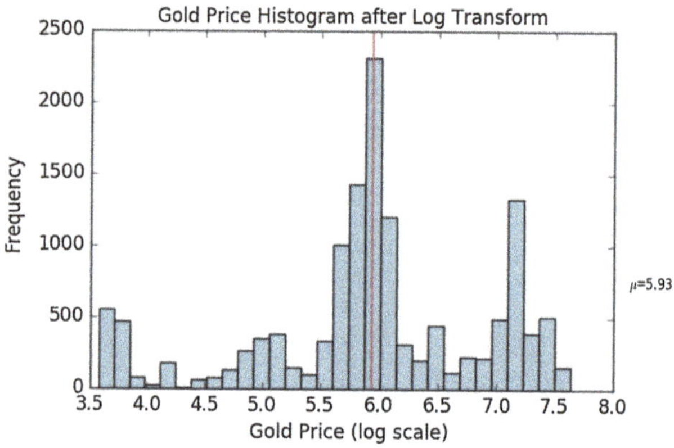

Another way of checking for normality is applying Box-Cox transformation.

```
goldprice_boxcox_mean = np.round(
                   np.mean(
                    goldDF['goldprice_boxcox_lambda_opt']),2)
fig, ax = plt.subplots()
goldDF['goldprice_boxcox_lambda_opt'].hist(bins=30,
                  color='#A9C5D3', edgecolor='black', grid=False)
plt.axvline(goldprice_boxcox_mean, color='r')
ax.set_title('Gold Price after Box–Cox Transform',
         fontsize=12)
ax.set_xlabel('Gold Price (Box–Cox transform)', fontsize=12)
ax.set_ylabel('Frequency', fontsize=12)
ax.text(24, 450, r'$\mu$='+str(goldprice_boxcox_mean), fontsize=10)
```

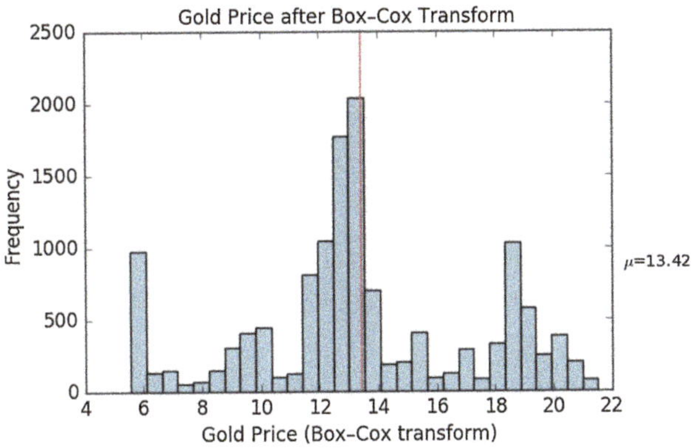

Step 3: Load silver commodity price data into data frame.

```
#Silver price data
silverDF = pd.read_csv("LBMA-SILVER_USD.csv",parse_dates=['Date'], dayfirst=True)
silverDF = silverDF.set_index('Date')
silverDF.head()
```

Output:

Date	SilverPrice
2021-01-22	25.320
2021-01-21	25.865
2021-01-20	25.295
2021-01-19	25.295
2021-01-18	24.865

To validate silver price data loaded as a numeric column values, seek silver data frame information:

```
silverDF.info()
```

Output of the above code:

```
<class 'pandas.core.frame.DataFrame'>
DatetimeIndex: 13422 entries, 2021-01-22 to 1968-02-01
Data columns (total 1 columns):
 #   Column       Non-Null Count  Dtype
---  ------       --------------  -----
 0   SilverPrice  13403 non-null  float64
dtypes: float64(1)
memory usage: 209.7 KB
```

```
sns.displot(silverDF, x="SilverPrice")
```

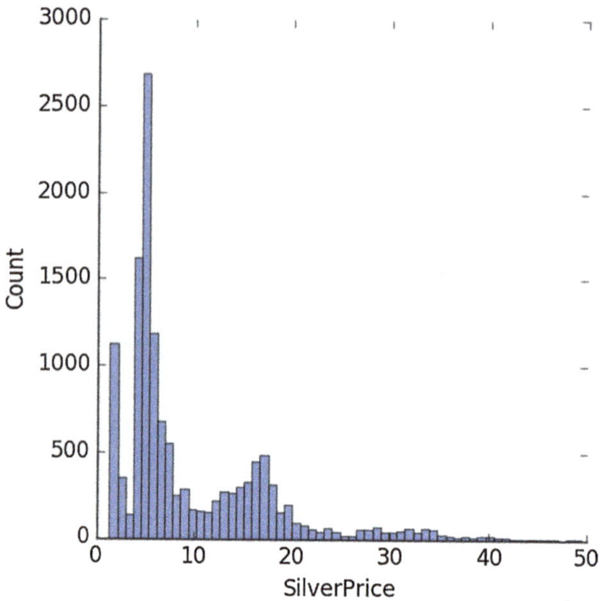

Step 4: load USD Index, Interest Rate, S&P 500, and Oil Prices historical datasets.

```
# USD Index

usdindexDF = pd.read_csv("DTWEXBGS.csv",parse_dates=['DATE'], dayfirst=True)
usdindexDF = usdindexDF.set_index('DATE')

# Interest
# 10-Year Treasury Constant Maturity Rate (DGS10)
# https://fred.stlouisfed.org/series/DGS10
interestRateDF = pd.read_csv("DGS10.csv",parse_dates=['DATE'], dayfirst=True)
interestRateDF = interestRateDF.set_index('DATE')

# S&P Index
spindex500Index = pd.read_csv("SP500Index.csv",parse_dates=['Date'], dayfirst=True)
spindex500Index = spindex500Index.set_index('Date')

#corn price data
oilPricesDF = pd.read_csv("brent-daily_csv.csv",parse_dates=['Date'], dayfirst=True)
oilPricesDF = oilPricesDF.set_index('Date')
```

After completion of Step 4, all the required datasets are available to perform causality test.

Step 5: Concatenate Gold Prices, Silver Price, Oil, USD Index, S&P 500, and Treasury Interest into one single data frame.

```
dataset = pd.concat([goldDF,silverDF,oilPricesDF,usdindexDF,interestRateDF,spindex500Index],
axis=1)
print('Number of colums in Dataframe : ', len(dataset.columns))
print('Number of rows in Dataframe : ', len(dataset.index))
print(dataset)
```

Output:

Number of columns in Dataframe: 6.
Number of rows in Dataframe: 23722.

	GoldPrice	SilverPrice	OilPrice	USDIndex	InterestRate	SPIndex
2007-01-01	NaN	NaN	NaN	97.04	NaN	NaN
2007-01-02	652.55	13.58	56.74	97.71	4.84	1445.939941
2007-01-03	673.60	14.29	61.18	97.34	4.56	1403.170044
2007-01-05	677.50	13.45	67.40	95.02	4.64	1486.300049
2007-01-06	662.50	13.53	68.65	94.15	4.95	1536.339966

Verify combine data frame details:

```
dataset.info()

Output:
<class 'pandas.core.frame.DataFrame'>
DatetimeIndex: 23722 entries, 1927-12-30 to 2021-12-01
Data columns (total 6 columns):
 #  Column       Non-Null Count  Dtype
--- ------       --------------  -----
 0  GoldPrice    13411 non-null  float64
 1  SilverPrice  13403 non-null  float64
 2  OilPrice     8449 non-null   float64
 3  USDIndex     3831 non-null   float64
 4  InterestRate 14748 non-null  float64
 5  SPIndex      23376 non-null  float64
dtypes: float64(6)
memory usage: 1.3 MB
```

Please note all price and index columns are non-null and float types. This is required to apply statistical techniques.

> Gold's climb adds momentum to year-end rally[51] [27]
> Gold price reflects investors hedge against market uncertainties and risks. Factors that increase market risk, for instance, surge in COVID-19 infections, can push the gold price toward a higher number. Other macroeconomic factors such as weak US dollar, increase in oil prices, and inflation are great influencers of gold price.

[51] Gold's Climb Adds Momentum to Year-End Rally – https://www.wsj.com/articles/golds-climb-adds-momentum-to-year-end-rally-11609859184?page=1

Step 6: Time Series of Prices (Gold, Silver, Oil, Treasury, and Interest) span different range. Let's fix time series dates:

```
dataset=dataset.loc['20070101':'20200101']
dataset.head()
```

Output:

	GoldPrice	SilverPrice	OilPrice	USDIndex	InterestRate	SPIndex
2007-01-01	NaN	NaN	NaN	97.04	NaN	NaN
2007-01-02	652.55	13.58	56.74	97.71	4.84	1445.939941
2007-01-03	673.60	14.29	61.18	97.34	4.56	1403.170044
2007-01-05	677.50	13.45	67.40	95.02	4.64	1486.300049
2007-01-06	662.50	13.53	68.65	94.15	4.95	1536.339966

Step 7: Fix nulls and missing values:

```
dataset.isnull().sum() ## missing values
```

Output: summary of missing values.

```
GoldPrice      89
SilverPrice    89
OilPrice       90
USDIndex       64
InterestRate  122
SPIndex       103
dtype: int64
```

One way to overcome missing values is filling the values with previous one so that we have continuity of time series, assuming that it won't disturb intrinsic nature of the time series.

```
dataset=dataset.fillna(method='pad')
dataset.isnull().sum() ## missing values
```

The following columns have been fixed with missing values.

```
GoldPrice      1
SilverPrice    1
OilPrice       1
USDIndex       0
InterestRate   1
SPIndex        1
dtype: int64
```

The NaN values in the dataset are filled with previous days' data. The dataset now clean for further analysis.

```
dataset=dataset.dropna()
dataset.head()
```

Output:

	GoldPrice	SilverPrice	OilPrice	USDIndex	InterestRate	SPIndex
2007-01-02	652.55	13.58	56.74	97.71	4.84	1445.939941
2007-01-03	673.60	14.29	61.18	97.34	4.56	1403.170044
2007-01-05	677.50	13.45	67.40	95.02	4.64	1486.300049
2007-01-06	662.50	13.53	68.65	94.15	4.95	1536.339966
2007-01-08	660.25	12.77	77.11	93.42	4.76	1465.810059

Plot the dataset to visualize time series:

```
fig, axes = plt.subplots(nrows=3, ncols=2, dpi=120, figsize=(10,6))
for i, ax in enumerate(axes.flatten()):
 data = dataset[dataset.columns[i]]
 ax.plot(data, color='red', linewidth=1)
 ax.set_title(dataset.columns[i])
 ax.xaxis.set_ticks_position('none')
 ax.yaxis.set_ticks_position('none')
 ax.spines["top"].set_alpha(0)
 ax.tick_params(labelsize=6)
plt.tight_layout();
```

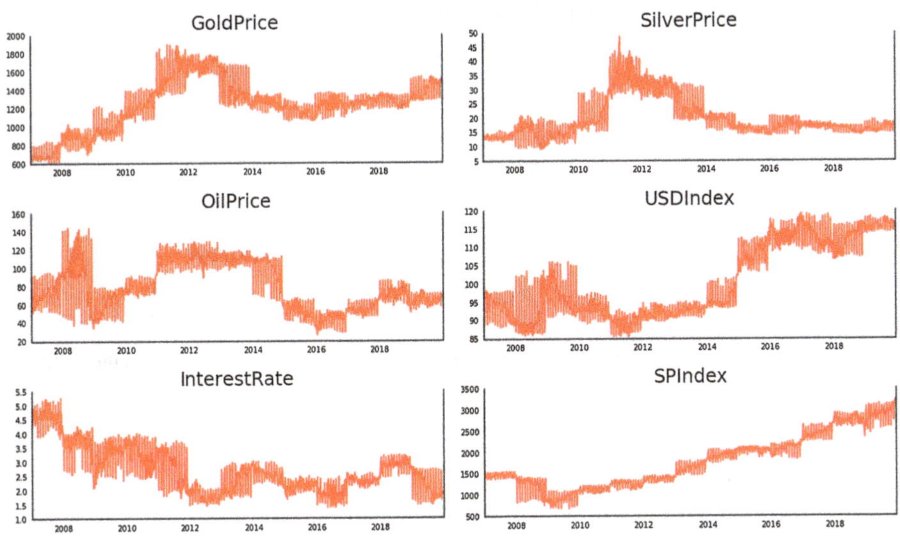

Step 8: To model the data and extract information from the dataset, it is required to check distribution of the data. Normal distribution is required to extract maximum information from the data.

Using SCIPY statistical normaltest,[52] we can identify whether sample differs from a normal distribution. This tests null hypothesis and alternate hypothesis intuition—rejection of null hypothesis implies the data distribution is normal.

API: Scipy.Stats.Normaltest
This function tests the null hypothesis that a sample comes from a normal distribution. It is based on D'Agostino and Pearson's test that combines skew and kurtosis to produce an omnibus test of normality.

Normal distribution test for gold price distribution:

```
from scipy import stats
GoldPrice=dataset.GoldPrice.values
print(GoldPrice)
stat,p = stats.normaltest(GoldPrice)
print("GoldPrice Statistics = %.3f, p=%.3f" % (stat,p))
alpha = 0.05
if p> alpha:
   print('Data looks Gaussian (fail to reject null hypothesis)')
else:
   print('Data looks non-Gaussian (reject null hypothesis)')
```

Output:

```
[ 652.55  673.6  677.5 ... 1510.6  1511.5  1523. ]
GoldPrice Statistics = 29.362, p=0.000
Data looks non-Gaussian (reject null hypothesis)
```

Other price variables normal distribution test:

[52] Normal Test – https://docs.scipy.org/doc/scipy/reference/generated/scipy.stats.normaltest.html

```
from scipy import stats

SilverPrice=dataset.SilverPrice.values
stat,p = stats.normaltest(SilverPrice)
print("SilverPrice Statistics = %.3f, p=%.3f" % (stat,p))
alpha = 0.05
if p> alpha:
   print('Data looks Gaussian (fail to reject null hypothesis)')
else:
   print('Data looks non-Gaussian (reject null hypothesis)')
```

Output:
SilverPrice Statistics = 756.825, p=0.000
Data looks non-Gaussian (reject null hypothesis)

```
from scipy import stats

OilPrice=dataset.OilPrice.values
stat,p = stats.normaltest(OilPrice)

print("OilPrice Statistics = %.3f, p=%.3f" % (stat,p))
alpha = 0.05
if p> alpha:
   print('Data looks Gaussian (fail to reject null hypothesis)')
else:
   print('Data looks non-Gaussian (reject null hypothesis)')
```

Output:
OilPrice Statistics = 2136.097, p=0.000
Data looks non-Gaussian (reject null hypothesis)

```
from scipy import stats

USDIndex=dataset.USDIndex.values
stat,p = stats.normaltest(USDIndex)

print("USDIndex Statistics = %.3f, p=%.3f" %(stat,p))
alpha = 0.05
if p> alpha:
   print('Data looks Gaussian (fail to reject null hypothesis)')
else:
   print('Data looks non-Gaussian (reject null hypothesis)')
```

Output:
USDIndex Statistics = 27596.353, p=0.000
Data looks non-Gaussian (reject null hypothesis)

```
from scipy import stats

InterestRate=dataset.InterestRate.values

stat,p = stats.normaltest(InterestRate)

print("InterestRate Statistics = %.3f, p=%.3f" % (stat,p))
alpha = 0.05
if p> alpha:
   print('Data looks Gaussian (fail to reject null hypothesis)')
else:
   print('Data looks non-Gaussian (reject null hypothesis)')
```

Output:
InterestRate Statistics = 273.748, p=0.000
Data looks non-Gaussian (reject null hypothesis)

Observe: p-values are zero!

Step 8b: Take dataset statistics—that is, describe Dataframe.

```
dataset.describe()
```

	GoldPrice	SilverPrice	OilPrice	USDIndex	InterestRate	SPIndex
count	3374.000000	3374.000000	3374.000000	3374.000000	3374.000000	3374.000000
mean	1232.024244	19.577390	79.095261	100.353159	2.740498	1787.430482
std	271.274404	6.932474	25.781097	10.145543	0.848580	616.190485
min	608.300000	8.880000	26.010000	85.490000	1.370000	676.530029
25%	1104.900000	15.260000	57.922500	92.110000	2.080000	1306.522461
50%	1252.950000	17.130000	74.510000	95.270000	2.580000	1641.630005
75%	1357.875000	20.638750	106.567500	110.990000	3.297500	2167.332458
max	1896.500000	48.700000	143.950000	119.210000	5.260000	3240.020020

The gold price has total 3374 records with the highest gold price is \$1896, and the lowest in the dataset is \$608.30 with mean of \$1232.02. The quantile percentages 25%, 50%, and 75% prices are \$1104.9, \$1252.95, and \$1357.8, respectively.

Step 8c: Perform kurtosis to verify the deviation of tail of the distribution from normal distribution. Kindly note, kurtosis is a statistical measure that defines how heavily the tails of a distribution differ from the tails of a normal distribution. The kurtosis of the normal distribution is zero.[53] The distribution with a higher kurtosis has a heavier tail. The zero valued kurtosis of the normal distribution in Fisher's definition can serve as a reference point. In other words, kurtosis identifies whether the tails of a given distribution contain extreme values.

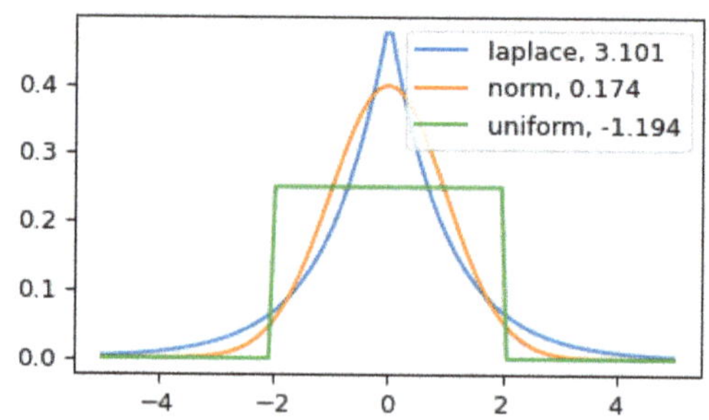

[53] Kurtosis – https://docs.scipy.org/doc/scipy/reference/generated/scipy.stats.kurtosis.html

Note: The Laplace distribution has a heavier tail than the normal distribution. The uniform distribution (which has negative kurtosis) has the thinnest tail.

Step 8c: check the skewness of data distribution. For normal distribution, the skewness should be zero.[54]

```
import matplotlib.pyplot as plt
import scipy.stats as stats
from scipy.stats import kurtosis

dataset.GoldPrice.plot(kind = 'density')

print('Gold: Kurtosis of normal distribution: {}'.format(stats.kurtosis(dataset.GoldPrice)))
print('Gold:Skewness of normal distribution: {}'.format(stats.skew(dataset.GoldPrice)))
```

Output:

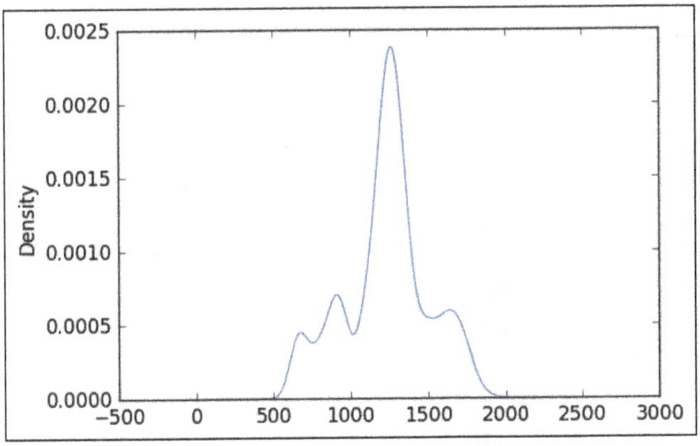


```
Gold: Kurtosis of normal distribution: -0.18547338284799597
Gold:Skewness of normal distribution: -0.207167414848137
```

The above two distributions give us some intuition about the distribution of our data[55] [28]. A value close to 0 for kurtosis indicates a normal distribution where asymmetrical nature is signified by a value between −0.5 and +0.5 for skewness. The tails are heavier for kurtosis greater than 0 and vice versa. Moderate skewness refers to the value between −1 and −0.5 or 0.5 and 1.

[54] scipy.stats.skew – https://docs.scipy.org/doc/scipy/reference/generated/scipy.stats.skew.html

[55] Gold and silver manipulation: What can be empirically verified? – https://brianmlucey.files.wordpress.com/2011/05/goldsilvermanipulation.pdf

Silver Price:

```
import matplotlib.pyplot as plt
import scipy.stats as stats
from scipy.stats import kurtosis

dataset.SilverPrice.plot(kind = 'density')

print('Silver: Kurtosis of normal distribution: {}'.format(stats.kurtosis(dataset.SilverPrice)))
print('Silver:Skewness of normal distribution: {}'.format(stats.skew(dataset.SilverPrice)))
```

Output:

Silver: Kurtosis of normal distribution—1.3132692116867828.
Silver: Skewness of normal distribution—1.4241832959005907.

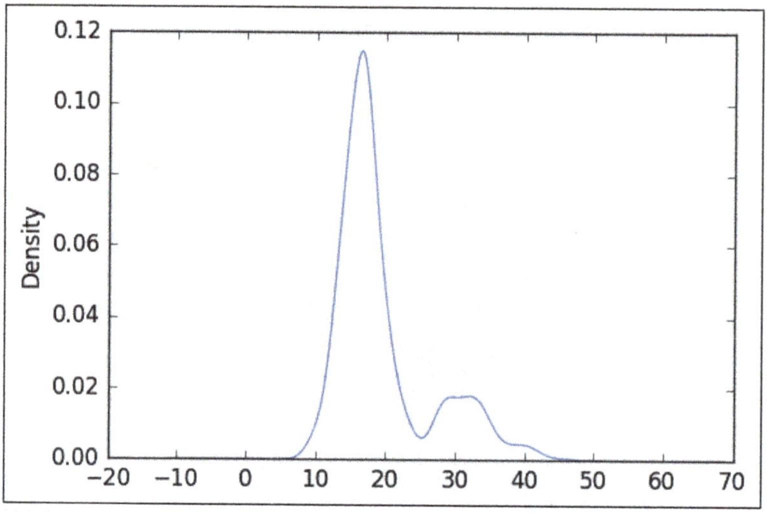

Oil Price: exhibited two peaks.

```
import matplotlib.pyplot as plt
import scipy.stats as stats
from scipy.stats import kurtosis

dataset.OilPrice.plot(kind = 'density')

print('Oil: Kurtosis of normal distribution: {}'.format(stats.kurtosis(dataset.OilPrice)))
print('Oil:Skewness of normal distribution: {}'.format(stats.skew(dataset.OilPrice)))
```

Output:

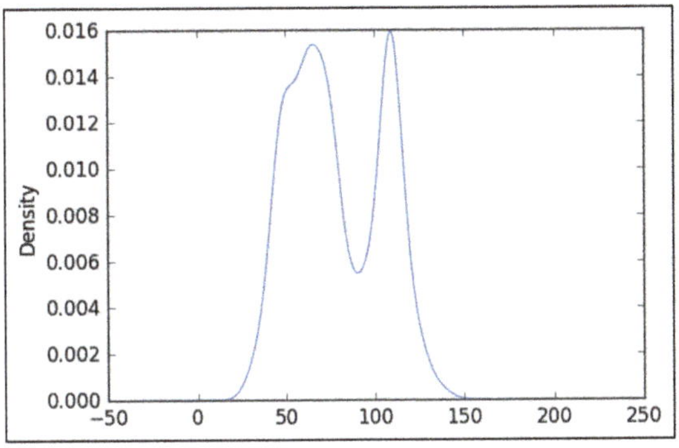

USD Index:

```
import matplotlib.pyplot as plt
import scipy.stats as stats
from scipy.stats import kurtosis

dataset.USDIndex.plot(kind = 'density')

print('USDIndex: Kurtosis of normal distribution: {}'.format(stats.kurtosis(dataset.USDIndex)))
print('USDIndex:Skewness of normal distribution: {}'.format(stats.skew(dataset.USDIndex)))
```

Output:

USD Index: Kurtosis of normal distribution—−1.4333432455556974.
USD Index: Skewness of normal distribution—0.3784173557673985.

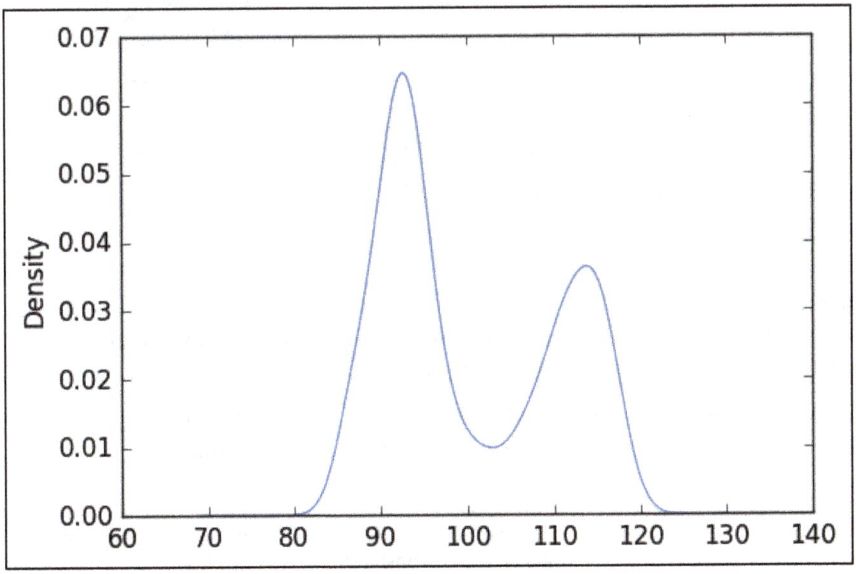

Step 8d: Plot the distribution:

```
plt.figure(figsize=(14,6))
plt.subplot(1,2,1)
dataset['GoldPrice'].hist(bins=50)
plt.title('Gold')
plt.subplot(1,2,2)
stats.probplot(dataset['GoldPrice'], plot=plt);
dataset.GoldPrice.describe().T
```

Output:

```
count     3374.000000
mean      1232.024244
std        271.274404
min        608.300000
25%       1104.900000
50%       1252.950000
75%       1357.875000
max       1896.500000
Name: GoldPrice, dtype: float64
```

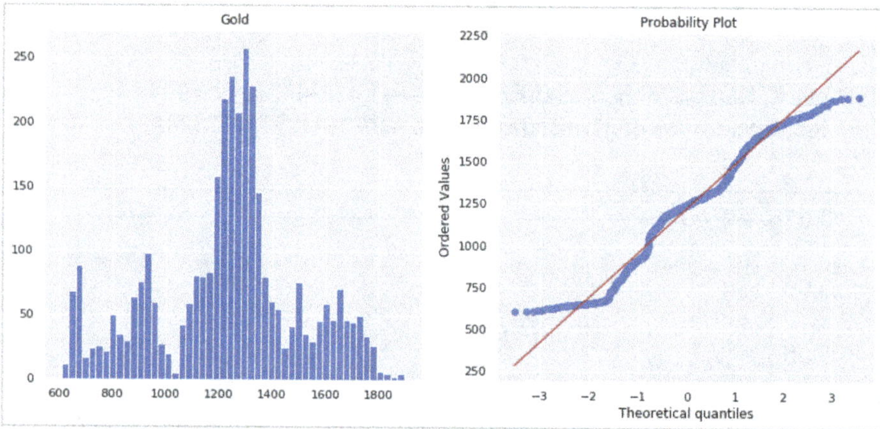

Step 9: Data Correlation.

The statistical relationship between two variables is referred to as their correlation.[56] Correlation could be positive, zero, or negative [29].

- Positive Correlation: both variables change in the same direction.
- Neutral Correlation: No relationship in the change of the variable.

[56] How to Calculate Correlation Between Variables in Python – https://machinelearningmastery.com/how-to-use-correlation-to-understand-the-relationship-between-variables/

- Negative Correlation: Variables change in the opposite direction.

Data Correlation: perform data correlation to see the correlation of different prices in the dataset. Prices in the dataset could be related in several ways:

- Price value of a variable could cause or depend on the other variables.
- One price tightly associated with the other (Gold Price to Silver Price or SP Index to USD Index).
- Two variables could depend on third variable.

```
import matplotlib.pyplot as plt
import pandas as pd
import numpy as np
import seaborn as sns; sns.set()
corr = dataset.corr()

fig, ax = plt.subplots()
sns.heatmap(corr,xticklabels=corr.columns.values, yticklabels=corr.columns.values,
annot=True,annot_kws={'size':12})
heat_map=plt.gcf()
heat_map.set_size_inches(10,6)
plt.xticks(fontsize=10)
plt.yticks(fontsize=10)
plt.show()
```

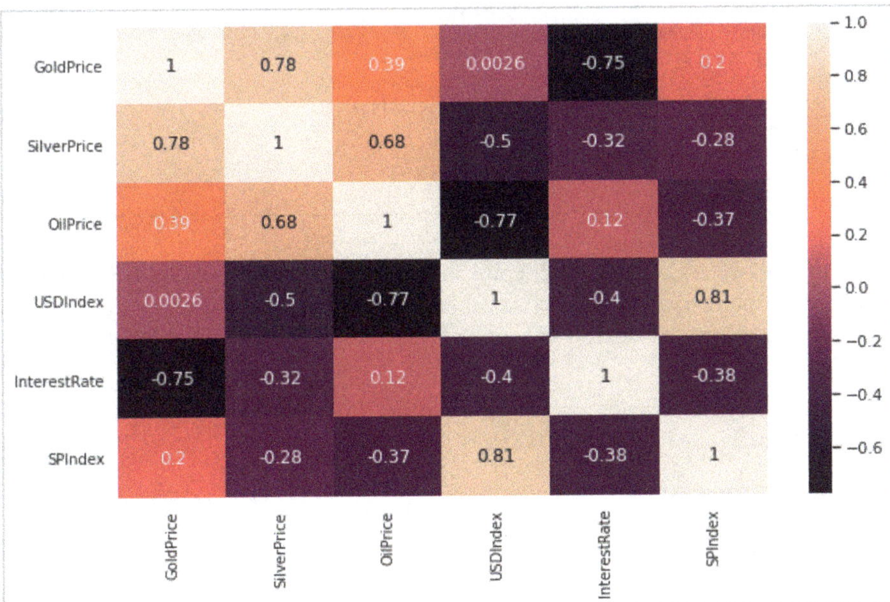

Step 10: Autocorrelation or serial correlation.

In Step 8, Dataset Price and Index columns are tested for normal distribution. Autocorrection is the process of calculating correlation of a time series variable with its previous time steps, called lags.[57] Because the correlation of the time series observations is calculated with values of the same series at previous times, this is called a serial correlation, or an autocorrelation. A plot of the autocorrelation of a time series by lag is called the AutoCorrelation Function, or the acronym ACF. This plot is sometimes called a correlogram or an autocorrelation plot [30].

Autocorrelation plots are a commonly used tool for checking randomness in a dataset[58] [16]. This randomness is ascertained by computing autocorrelations for data values at varying time lags. If random, such autocorrelations should be near zero for all time-lag separations. If non-random, then one or more of the autocorrelations will be significantly non-zero.

Autocorrelation plot function: Statsmodels.Graphics.Tsaplots.plot_acf[59]
Plot the autocorrelation function. Plot lags on the horizontal and the correlations on vertical axis.

Given dataset columns are normally distributed, Pearson's correlation coefficient is applied to summarize the correlation between the Gold, Silver, Oil, USD Index, Interest, and SPIndex variables.
Note: Autocorrelation or serial correlation can be a significant problem in analyzing historical data if we do not know how to look out for it

```
import pandas as pd
import matplotlib.pyplot as plt
import statsmodels.api as sm

# plots the autocorrelation plots for each stock's price at 50 lags
for i in dataset:
   sm.graphics.tsa.plot_acf(dataset[i], lags = 50)
   plt.title('ACF for %s' % i)
plt.show()
```

Output: above code creates a 2D plot showing the lag values of Gold, Silver, Oil, USD Index, Interest, and SPIndex variables along the x-axis and the correlation on the y-axis between −1 and 1.

[57] A Gentle Introduction to Autocorrelation and Partial Autocorrelation – https://machinelearningmastery.com/gentle-introduction-autocorrelation-partial-autocorrelation/

[58] Autocorrelation Plot – https://itl.nist.gov/div898/handbook/eda/section3/autocopl.htm

[59] PLOT_ACF: https://www.statsmodels.org/devel/generated/statsmodels.graphics.tsaplots.plot_acf.html

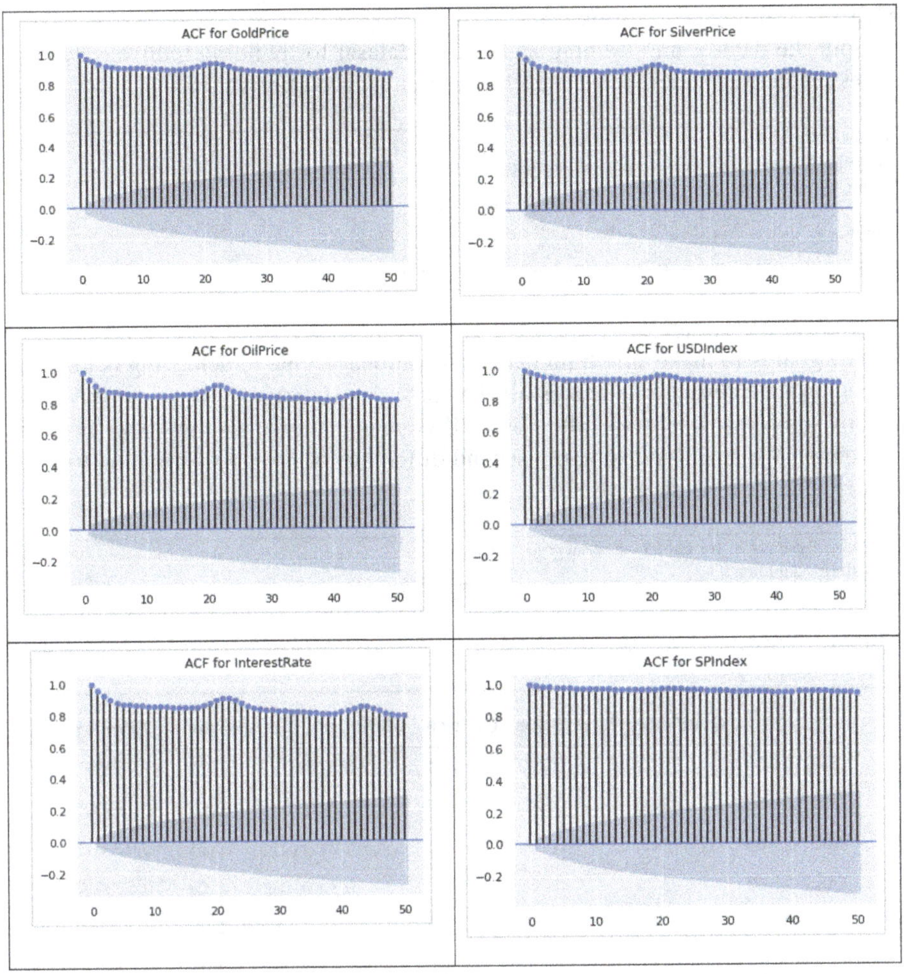

Higher value of autocorrelation implies non-random series. The usage of autocorrelation plots provides autoregression and moving average intuitions.

Please note[60]: If random, such autocorrelations should be near zero for all time-lag separations. If non-random, then one or more of the autocorrelations will be significantly non-zero [16].

[60] Autocorrelation Function – https://itl.nist.gov/div898/handbook/eda/section3/autocopl.htm

Step 11: Training and Testing Data.

Split the dataset into training and testing dataset for building multivariate time series predictive model. Test Dataset is split on lagged observations.

```
n_obs=15
X_train, X_test = dataset[0:-n_obs], dataset[-n_obs:]
print(X_train.shape, X_test.shape)

Output:
(3359, 6) (15, 6)
```

Step 12: Transformation.

The goal is to make sure train dataset is stationary. One of achieving is first differencing on training data makes all the six series stationary. However, this is an iterative process where we after first differencing, the series may still be non-stationary. We shall have to apply second difference or log transformation to standardize the series in such cases.

```
transform_data = X_train.diff().dropna()
transform_data.head()
```

Output:

	GoldPrice	SilverPrice	OilPrice	USDIndex	InterestRate	SPIndex
2007-01-03	21.05	0.710	4.44	-0.37	-0.28	-42.769897
2007-01-05	3.90	-0.840	6.22	-2.32	0.08	83.130005
2007-01-06	-15.00	0.080	1.25	-0.87	0.31	50.039917
2007-01-08	-2.25	-0.760	8.46	-0.73	-0.19	-70.529907
2007-01-10	85.00	1.005	1.22	-2.51	-0.20	81.229980

```
transform_data.describe()
```

Output:

	GoldPrice	SilverPrice	OilPrice	USDIndex	InterestRate	SPIndex
count	3358.000000	3358.000000	3358.000000	3358.000000	3358.000000	3358.000000
mean	0.252948	0.001005	0.000116	0.005688	-0.000950	0.427862
std	60.076368	1.897170	7.959643	1.856990	0.238154	71.064877
min	-463.500000	-19.490000	-44.610000	-16.310000	-1.290000	-484.299805
25%	-12.337500	-0.365000	-1.300000	-0.330000	-0.060000	-12.225067
50%	0.000000	0.000000	0.020000	0.000000	0.000000	1.385071
75%	13.000000	0.323750	1.617500	0.350000	0.050000	20.914948
max	450.500000	11.360000	86.070000	13.160000	1.570000	600.039978

Step 12: Stationarity Check.

In a stationary time series, the statistical properties of time series do not change over time. In other words, the statistical properties (mean, variance, and covariance) are constant and independent of time. In order to test stationarity, augmented Dickey-Fuller test is performed.

Dicker-Fuller test is used to check stationarity of a time series. DF tests the null hypothesis that a unit root is present in time series data. A unit root is a stochastic trend in a time series, sometimes called a "random walk with drift." If a time series has a unit root, it shows systematic pattern that is unpredictable. The augmented Dickey-Fuller test can be used to test for a unit root in a univariate process in the presence of serial correlation. The null hypothesis of the Augmented Dickey-Fuller is that there is a unit root, with the alternative that there is no unit root. If the p-value is above a critical size, then we cannot reject that there is a unit root. The p-values are obtained through regression surface approximation from MacKinnon 1994, but using the updated 2010 tables. If the p-value is close to significant, then the critical values should be used to judge whether to reject the null.

When you run the test, you'll get an ADF value and a p-value. The ADF number should be a negative number and the p-value should be beneath a certain threshold value (e.g., 1% or 5%, etc.) for a confidence level. For this example, we'll use 5% (or 95% confidence level), so if the p-value is greater than 0.05, then we say we fail to reject the null hypothesis because the data has a unit root and is non-stationary. If the p-value is less than or equal to 0.05, we can say we reject the null hypothesis because the data does not have a unit root and is stationary.

```
import statsmodels.tsa.stattools as sm
def augmented_dickey_fuller_statistics(time_series):
    result = sm.adfuller(time_series.values, autolag='AIC')
    print('ADF Statistic: %f' % result[0])
    print('p-value: %f' % result[1])
    print('Critical Values:')
    for key, value in result[4].items():
        print('\t%s: %.3f' % (key, value))
```

Calling ADF Test on 6 series:

```
print('Augmented Dickey-Fuller Test: Gold Price Time Series')
augmented_dickey_fuller_statistics(transform_data['GoldPrice'])
print('Augmented Dickey-Fuller Test: Silver Price Time Series')
augmented_dickey_fuller_statistics(transform_data['SilverPrice'])
print('Augmented Dickey-Fuller Test: Oil Price Time Series')
augmented_dickey_fuller_statistics(transform_data['OilPrice'])
print('Augmented Dickey-Fuller Test: Interest_rate Time Series')
augmented_dickey_fuller_statistics(transform_data['InterestRate'])
print('Augmented Dickey-Fuller Test: Stock_Index Time Series')
augmented_dickey_fuller_statistics(transform_data['SPIndex'])
print('Augmented Dickey-Fuller Test: USD_Index Time Series')
augmented_dickey_fuller_statistics(transform_data['USDIndex'])
```

Output:

```
Augmented Dickey-Fuller Test: Gold Price Time Series
ADF Statistic: -17.557047
p-value: 0.000000
Critical Values:
	1%: -3.432
	5%: -2.862
	10%: -2.567
Augmented Dickey-Fuller Test: Silver Price Time Series
ADF Statistic: -15.942107
p-value: 0.000000
Critical Values:
	1%: -3.432
	5%: -2.862
	10%: -2.567
Augmented Dickey-Fuller Test: Oil Price Time Series
ADF Statistic: -16.914470
p-value: 0.000000
Critical Values:
	1%: -3.432
	5%: -2.862
	10%: -2.567
Augmented Dickey-Fuller Test: Interest_rate Time Series
ADF Statistic: -18.651821
p-value: 0.000000
Critical Values:
	1%: -3.432
	5%: -2.862
	10%: -2.567
Augmented Dickey-Fuller Test: Stock_Index Time Series
ADF Statistic: -18.505585
p-value: 0.000000
Critical Values:
	1%: -3.432
	5%: -2.862
	10%: -2.567
Augmented Dickey-Fuller Test: USD_Index Time Series
ADF Statistic: -20.017560
p-value: 0.000000
Critical Values:
	1%: -3.432
	5%: -2.862
	10%: -2.567
```

All 6 series are stationary as p-value is less than 0.05. Plot the series:

```
fig, axes = plt.subplots(nrows=3, ncols=2, dpi=120, figsize=(10,6))
for i, ax in enumerate(axes.flatten()):
    d = transform_data[transform_data.columns[i]]
    ax.plot(d, color='red', linewidth=1)
    # Decorations
    ax.set_title(dataset.columns[i])
    ax.xaxis.set_ticks_position('none')
    ax.yaxis.set_ticks_position('none')
    ax.spines['top'].set_alpha(0)
    ax.tick_params(labelsize=6)

plt.tight_layout();
```

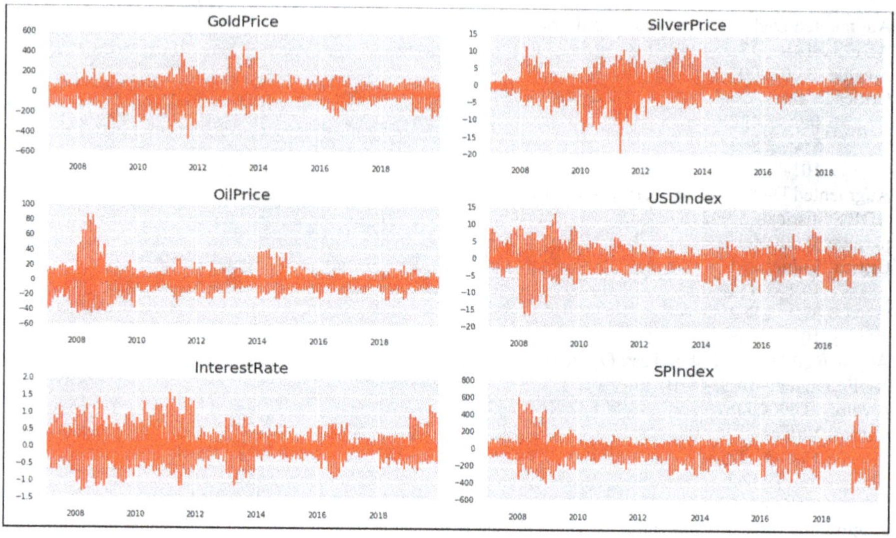

Step 13: Testing causation using Granger's causality test.

Causation of all six series (that is multivariate variables influence each other) can be tested using Granger's causality test. Granger's causality tests the null hypothesis that the coefficients of past values in the regression equation is zero. In simpler terms, the past values of time series (x) do not cause the other series (y). So, if the p-value obtained from the test is lesser than the significance level of 0.05, then, you can safely reject the null hypothesis. This has been performed on original dataset.

```
import statsmodels.tsa.stattools as sm
maxlag=12
test = 'ssr-chi2test'
def grangers_causality_matrix(X_train, variables, test = 'ssr_chi2test', verbose=False):
    dataset = pd.DataFrame(np.zeros((len(variables), len(variables))), columns=variables,
index=variables)
    for c in dataset.columns:
        for r in dataset.index:
            test_result = sm.grangercausalitytests(X_train[[r,c]], maxlag=maxlag, verbose=False)
            p_values = [round(test_result[i+1][0][test][1],4) for i in range(maxlag)]
            if verbose: print(f'Y = {r}, X = {c}, P Values = {p_values}')
            min_p_value = np.min(p_values)
            dataset.loc[r,c] = min_p_value
    dataset.columns = [var + '_x' for var in variables]
    dataset.index = [var + '_y' for var in variables]
    return dataset
```

Calling Granger causality test[61]: [22].

```
grangers_causality_matrix(dataset, variables = dataset.columns)
```

[61] Testing for Granger Causality Using Python – https://rishi-a.github.io/2020/05/25/granger-causality.html

Output:

	GoldPrice_x	SilverPrice_x	OilPrice_x	USDIndex_x	InterestRate_x	SPIndex_x
GoldPrice_y	1.0000	0.0022	0.0209	0.0153	0.0000	0.1617
SilverPrice_y	0.2459	1.0000	0.0002	0.0000	0.0003	0.0022
OilPrice_y	0.0026	0.0000	1.0000	0.0000	0.0712	0.0000
USDIndex_y	0.0009	0.1539	0.0000	1.0000	0.0435	0.0000
InterestRate_y	0.0000	0.0007	0.0000	0.0000	1.0000	0.0000
SPIndex_y	0.0119	0.0000	0.0000	0.0000	0.0000	1.0000

Test for Granger non-causality as follows.[62] Test the hypothesis that the coefficients of (only) the first p lagged values of X are zero in the Y equation, using a standard Wald test. Then do the same thing for the coefficients of the lagged values of Y in the X equation. For instance, if we take the value 0. 0.0022 in (row 1, column 2), it refers to the p-value of the Granger's Causality test for SilverPrice_x causing GoldPrice_y. The 0.2459 in (row 2, column 1) refers to the p-value of GoldPrice_y causing SilverPrice_x and so on. Except for SPIndex and GoldPrice, Silver and Gold, and Silver and USD, all the other variables (time series) in the system are interchangeably causing each other. This justifies the VAR modeling approach for this system of multi time-series to forecast.

Step 14: VAR model.

VAR requires stationarity of the series which means the mean to the series do not change over time (we can find this out from the plot drawn next to augmented Dickey-Fuller test). Perform VAR model on training set and then used the fitted model to forecast the next 15 observations. These forecasts will be compared against the actual present in test data. The maximum lag (15) to identify the required lags for VAR model.

```
import numpy as np
import pandas
import statsmodels.api as sm
from statsmodels.tsa.api import VAR

mod = VAR(transform_data)
res = mod.fit(maxlags=15, ic='aic')
print(res.summary())
```

Output:

The biggest correlations are 0.77 (Silver & Gold) and −0.63 (Oil & USD); however, there are small enough to ignore in this case.

---

[62] Testing for Granger Causality – https://davegiles.blogspot.com/2011/04/testing-for-granger-causality.html

```
  Summary of Regression Results
==================================
Model:                         VAR
Method:                        OLS
Date:           Sun, 24, Jan, 2021
Time:                     06:06:03
--------------------------------------------------------------------
No. of Equations:        6.00000    BIC:                   18.1108
Nobs:                    3343.00    HQIC:                  17.4694
Log likelihood:         -56518.0    FPE:               2.70241e+07
AIC:                     17.1122    Det(Omega_mle):    2.30021e+07
--------------------------------------------------------------------

Correlation matrix of residuals
              GoldPrice  SilverPrice  OilPrice  USDIndex  InterestRate   SPIndex
GoldPrice      1.000000     0.778871  0.327238 -0.445650     -0.419671  0.052984
SilverPrice    0.778871     1.000000  0.456898 -0.485468     -0.106323  0.125395
OilPrice       0.327238     0.456898  1.000000 -0.638941      0.303663  0.415947
USDIndex      -0.445650    -0.485468 -0.638941  1.000000     -0.100569 -0.378803
InterestRate  -0.419671    -0.106323  0.303663 -0.100569      1.000000  0.202174
SPIndex        0.052984     0.125395  0.415947 -0.378803      0.202174  1.000000
```

Residual values:

```
y_fitted = res.fittedvalues
y_fitted
```

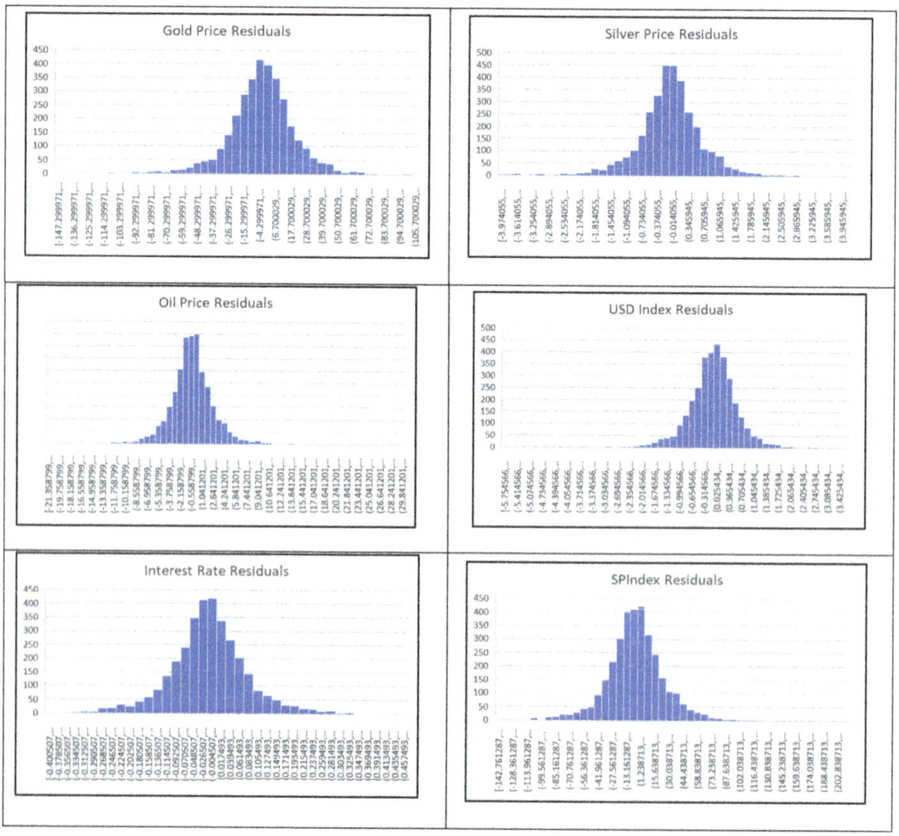

Residual plot:

```
y_fitted = res.fittedvalues
residuals = res.resid
plt.figure(figsize = (15,5))
plt.plot(residuals, label='resid')
plt.plot(y_fitted, label='VAR prediction')
plt.xlabel('Date')
plt.xticks(rotation=45)
plt.ylabel('Residuals')
plt.grid(True)
```

Output:

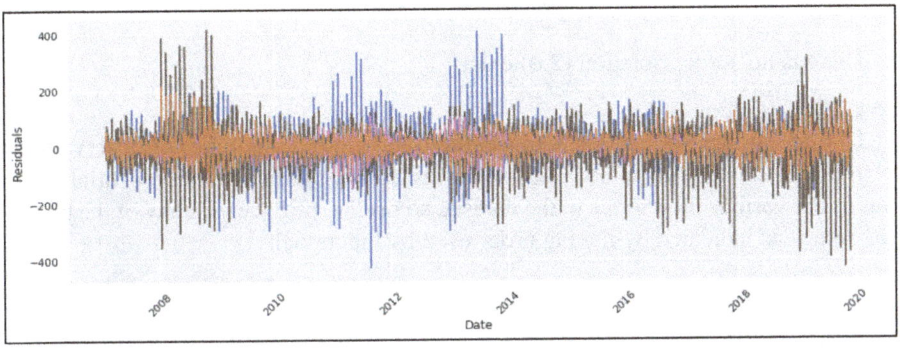

Residual plot looks normal with constant mean throughout apart from some large fluctuation during 2009, 2011, 2014, etc.

> Causation models[63] are at the center of solving agriculture challenges: The agribusiness production cycles, environmental sustainability, climate change, and food security are current demands that challenge food producers, distributors, and consumers. This line of study will focus on cause-and-effect models for agricultural production chains, especially for small producers. The goal will be to use advanced probabilistic models for decision-making based on cause and effect, addressing many sources of concern, such as food and water waste. [31]

Step 15: Durbin-Watson statistic.

Durbin-Watson statistic is related to auto correlation. The Durbin-Watson statistic will always have a value between 0 and 4. A value of 2.0 means that there is no autocorrelation detected in the sample. Values from 0 to less than 2 indicate positive autocorrelation, and values from 2 to 4 indicate negative autocorrelation. A rule of thumb is that test statistic values in the range of 1.5 to 2.5 are relatively normal. Any value outside this range could be a cause for concern.

[63] C4AI – A New Center for Artificial Intelligence to accelerate science and innovation in Brazil- https://www.ibm.com/blogs/research/2020/10/c4ai-a-new-center-for-artificial-intelligence-to-accelerate-science-and-innovation-in-brazil/

```
from statsmodels.stats.stattools import durbin_watson
out = durbin_watson(res.resid)
for col,val in zip(transform_data.columns, out):
    print((col), ":", round(val,2))
```

Output:

```
GoldPrice : 2.02
SilverPrice : 2.01
OilPrice : 2.01
USDIndex : 2.01
InterestRate : 2.01
SPIndex : 2.03
```

There is no autocorrelation (2.0) exist.

Step 15: Prediction.

To forecast, the VAR model expects up to the lag order number of observations from the past data. This is because, the terms in the VAR model are essentially the lags of the various time series in the dataset, so need to provide as many of the previous values as indicated by the lag order used by the model.

```
# Get the lag order
lag_order = res.k_ar
print(lag_order)
# Input data for forecasting
input_data = transform_data.values[-lag_order:]
print(input_data)
# forecasting
pred = res.forecast(y=input_data, steps=n_obs)
pred = (pd.DataFrame(pred, index=X_test.index, columns=X_test.columns + '_pred'))
print(pred)
```

Output:

```
            GoldPrice_pred  SilverPrice_pred  OilPrice_pred  USDIndex_pred  \
2019-12-09        4.177465          0.002858      -0.868513       0.599603
2019-12-11      -21.039277         -0.231194       3.767794      -0.568016
2019-12-12      -18.731486          0.047470       2.397978      -0.332362
2019-12-13      -19.368022         -0.159041      -0.376208       0.045862
2019-12-16       -5.652834         -0.041259      -0.078600      -0.003112
2019-12-17       -6.652626         -0.117316      -0.378357       0.082016
2019-12-18       -7.450345         -0.000987       0.605729      -0.170279
2019-12-19      -16.401983         -0.603018      -0.399840       0.378818
2019-12-20       17.122682          0.734008       1.940819      -0.476212
2019-12-23       20.204580          0.176870       0.774359      -0.413320
2019-12-24       -5.735587         -0.110948       0.142006       0.028089
2019-12-26       -3.422185         -0.027944      -0.444365       0.007159
2019-12-27       -4.644322         -0.151591       0.606357      -0.102356
2019-12-30      -11.107344         -0.249916      -0.959227       0.130709
2019-12-31        3.068490         -0.061377      -1.137009       0.147184
```

Invert the Transformation to Get the Real Forecast.

The forecasts are generated, but it is on the scale of the training data used by the model. So, to bring it back up to its original scale, we need to de-difference it.

The way to convert the differencing is to add these differences consecutively to the base number. An easy way to do it is to first determine the cumulative sum at index and then add it to the base number.

This process can be reversed by adding the observation at the prior time step to the difference value. Inverted(ts) = differenced(ts) + observation(ts−1).

```
# inverting transformation
def invert_transformation(X_train, pred):
    forecast = pred.copy()
    columns = X_train.columns
    for col in columns:
        forecast[str(col)+'_pred'] = X_train[col].iloc[-1] + forecast[str(col)+'_pred'].cumsum()
        return forecast

output = invert_transformation(X_train, pred)
## https://stackoverflow.com/questions/20461165/how-to-convert-index-of-a-pandas-dataframe-into-a-column
output['Date'] = output.index

print(output)
```

Output:

```
index GoldPrice_pred GoldPrice accuracy
14 2019-12-31     1426.32   1523.00  93.65%
13 2019-12-30     1423.25   1511.50  94.16%
12 2019-12-27     1434.36   1510.60  94.95%
11 2019-12-26     1439.00   1490.85  96.52%
10 2019-12-24     1442.42   1490.85  96.75%
9 2019-12-23      1448.16   1483.95  97.59%
8 2019-12-20      1427.95   1476.90  96.69%
7 2019-12-19      1410.83   1474.40  95.69%
6 2019-12-18      1427.23   1478.90  96.51%
5 2019-12-17      1434.68   1478.40  97.04%
4 2019-12-16      1441.34   1477.40  97.56%
3 2019-12-13      1446.99   1470.60  98.39%
2 2019-12-12      1466.36   1474.70  99.43%
1 2019-12-11      1485.09   1455.00 102.07%
0 2019-12-09      1506.13   1502.95 100.21%
```

The VAR model is a popular tool for the purpose of predicting joint dynamics of multiple time series based on linear functions of past observations. More analysis, e.g., impulse response (IRF) and forecast error variance decomposition (FEVD), can also be done along with VAR for assessing the impacts of shock from one asset on another to assess the impacts of shock from one asset on another. You can download the code and run using FeatureEngineeringSilverGoldAgCommodities Jupyter notebook.

Fertilizers, Crude Oil, and Agricultural Commodity Model

Fertilizer is still a significant expense for most grain farms. Fertilizer expense as a percentage of crop expenses varies across continents and countries, for example, high cost of farm inputs—fertilizer is the leading cause of major debt faced by small farms in India[64,65] [32]. Costs of agricultural fertilizers have a huge influence on overall yield and production of maize in Malawi[66] [33]. Predicting cost of fertilizer would have a lifesaving impact and have a multiplier effect on developing economies. One it will reduce the debt of small farmers. In a 2018 study conducted by National Bank for Agriculture and Rural Development in India, it showed that 52.5% of all the agricultural households were indebted, as per prevailing banking rules that do not allow farmers who have running loans to borrow more credit. Hence, indebted families in need of credit often push female farmers to take loans, pushing them deeper into the vicious debt cycle that is now so common across this western Indian region and many developing countries where farms have plunged into cyclic debt due to cost of fertilizers and low yield crops.[67] Data released earlier this year showed that India had 93 million microfinance accounts, most of whom were women in self-help groups, a rise of 22% from the previous year. Microfinance lending rates are much lesser than those quoted by private money lenders, who thrive in these areas.

Finding the lowest price fertilizer combination that meets the nutrient requirements of a field can sometimes be difficult. Agronomists give recommendations for the pounds of N, P, and K required, while the available fertilizers have different percentages of a given nutrient and may contain two or more nutrients. The only 100% accurate way to find the lowest price fertilizer combination is to use a linear programming approach.[68] solver-based tool offers the best pricing for N, P, and K combination[69]:

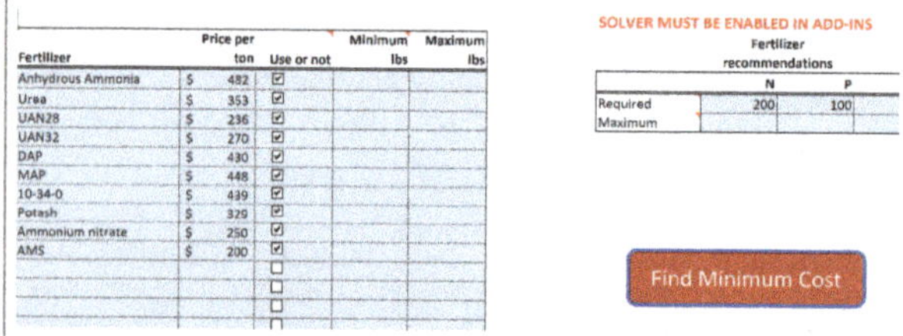

Fertilizer	Price per ton	Use or not	Minimum lbs	Maximum lbs
Anhydrous Ammonia	$ 482	☑		
Urea	$ 353	☑		
UAN28	$ 236	☑		
UAN32	$ 270	☑		
DAP	$ 430	☑		
MAP	$ 448	☑		
10-34-0	$ 439	☑		
Potash	$ 329	☑		
Ammonium nitrate	$ 250	☑		
AMS	$ 200	☑		
		☐		
		☐		
		☐		

SOLVER MUST BE ENABLED IN ADD-INS

Fertilizer recommendations

	N	P
Required	200	100
Maximum		

Find Minimum Cost

[64] FAO In India – http://www.fao.org/india/fao-in-india/india-at-a-glance/en/#:~:text=Agriculture%2C%20with%20its%20allied%20sectors,275%20million%20tonnes%20(MT).

[65] "Free" Fertilizer is Saving Rural Farmers – https://www.permaculturenews.org/2017/10/02/free-fertilizer-saving-rural-farmers/

[66] Agricultural household effects of fertilizer price changes for smallholder farmers in central Malawi – https://www.sciencedirect.com/science/article/pii/S0308521X16304942

[67] As debt grows, more Indian women farmers taking their lives – https://www.aljazeera.com/features/2019/11/5/as-debt-grows-more-indian-women-farmers-taking-their-lives#:~:text=A%202018%20study%20by%20the,loans%20to%20borrow%20more%20credit.

[68] Fertilizer Cost Tool – https://www.agmanager.info/ksu-fertilizer-tool

[69] Solver Tool – https://www.agmanager.info/sites/default/files/spreadsheets/KSU_FertilizerTool.xlsm

Common fertilizers used in agriculture as per the World Bank Pink Sheet includes[70]:

- DAP (diammonium phosphate), spot, f.o.b. US Gulf.
- Phosphate rock, f.o.b. North Africa.
- Potassium chloride (muriate of potash), f.o.b. Vancouver.
- TSP (triple superphosphate), spot, import US Gulf.
- Urea, (Ukraine), f.o.b. Black Sea.

Urea is the top fertilizer used across the world with consumption around 113,122.89 tons per year.

Top Fertilizers in Agricultural Use Case

Data Sources: Fertilizer by Product.[71]

Data Fields

Countries: Select all	Elements: Agricultural use case	Items: All
Years: All		

Please select the following (see Fig. 4.16) on FAO Stats Database.[72]
Urea and NPK Fertilizer are top 10 agricultural use case (please see Fig. 4.17).

Top 10 Countries: Urea Agricultural Use Case

Top countries that use Urea (please see Fig. 4.18): India (around 26 million tons), the United States (around 5.7 million tons), Pakistan (5.4 million tons), Indonesia (around 4.9 million), and Egypt (3.2 million tons) (please see Fig. 4.19 for World Wide Usage).

Please find US consumption of plant nutrients (please see Table 4.2).[73]

---

[70] The World Bank Pink Sheet – https://thedocs.worldbank.org/en/doc/5d903e848db1d1b83e0ec8f744e55570-0350012021/original/CMO-Pink-Sheet-March-2021.pdf

[71] Fertilizer By Product – http://www.fao.org/faostat/en/#data/RFB

[72] FAO Stats Database – http://www.fao.org/faostat/en/#data/RFB

[73] US consumption of plant nutrients – Table 1 – https://www.ers.usda.gov/webdocs/DataFiles/50341/fertilizeruse.xls?v=5103

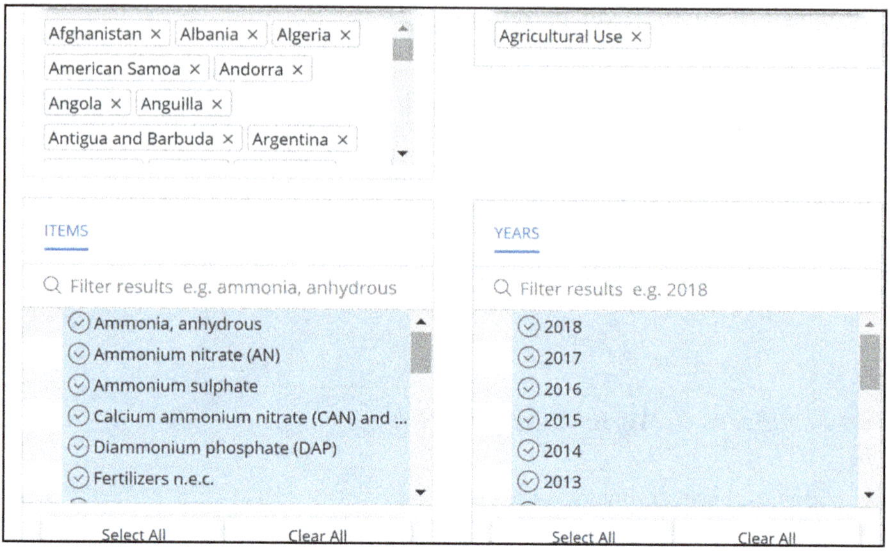

Fig. 4.16 FAO settings

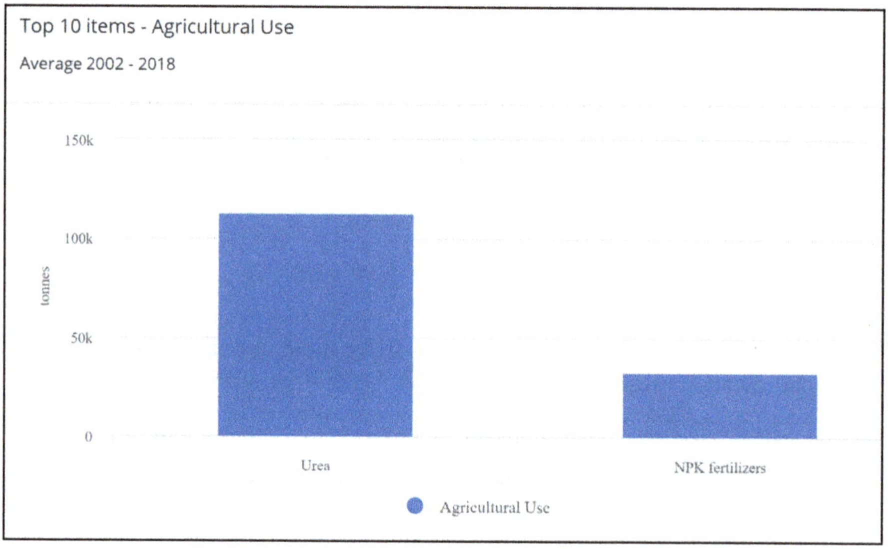

Fig. 4.17 Top 10 Items

In the United States, the higher dollar per acre fertilizer cost can be attributed to three factors (please see Fig. 4.20). The first is higher fertilizer prices. As will be shown later, the fertilizer price is closed tied to the price of oil, and this period saw higher oil prices. The second is a shift in the crop mix to more corn. Corn requires a high level of nitrogen fertilizer which means this crop is more expensive to grow.

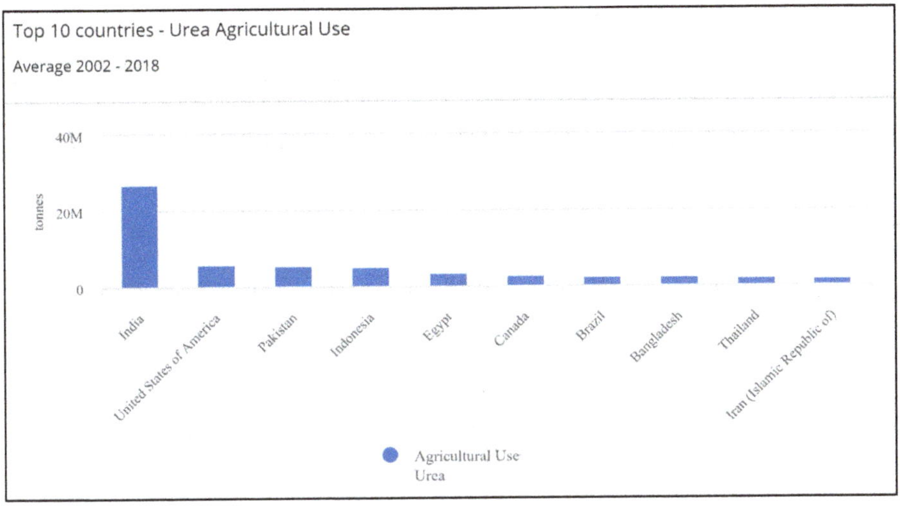

Fig. 4.18 Top 10 countries - Urea Use

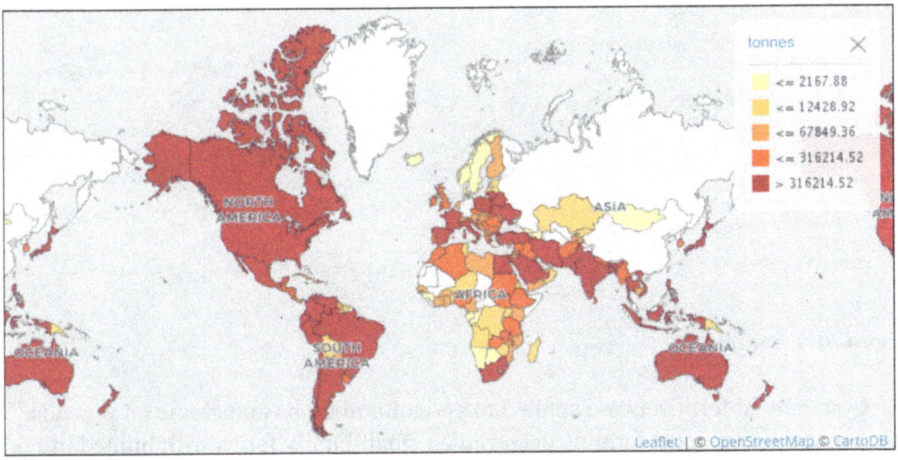

Fig. 4.19 Worldwide Urea Usage

Third, the higher grain prices leading up to 2012 meant that it was profitable for farmers to fertilize at higher rates. Since 2012, fertilizer costs per acre have declined for the same reasons they increased. Even with this recent decline, fertilizer costs per acre are nearly double what they were in the nineties (in real dollars)[74] [35].

[74] Predicting Fertilizer Prices – https://www.agmanager.info/sites/default/files/pdf/2019.4.pdf

Table 4.2 US fertilizer consumption

Year ending June 30	Primary nutrient content 1/ Nitrogen (N)	Phosphate (P_2O_5)	Potash (K_2O)	Total	Share of total Nitrogen	Phosphate	Potash
	------------------1,000 nutrient short tons--------------				----------------Percent----------------		
1960	2,738	2,572	2,153	7,464	37	34	29
1961	3,031	2,645	2,169	7,844	39	34	28
1962	3,370	2,807	2,271	8,448	40	33	27
1963	3,929	3,073	2,503	9,505	41	32	26
1964	4,353	3,378	2,730	10,460	42	32	26
1965	4,639	3,512	2,835	10,985	42	32	26
1966	5,326	3,897	3,221	12,445	43	31	26
1967	6,027	4,305	3,642	13,974	43	31	26
1968	6,788	4,453	3,793	15,034	45	30	25
1969	6,958	4,666	3,892	15,515	45	30	25
1970	7,459	4,574	4,036	16,068	46	28	25
1971	8,134	4,803	4,231	17,168	47	28	25

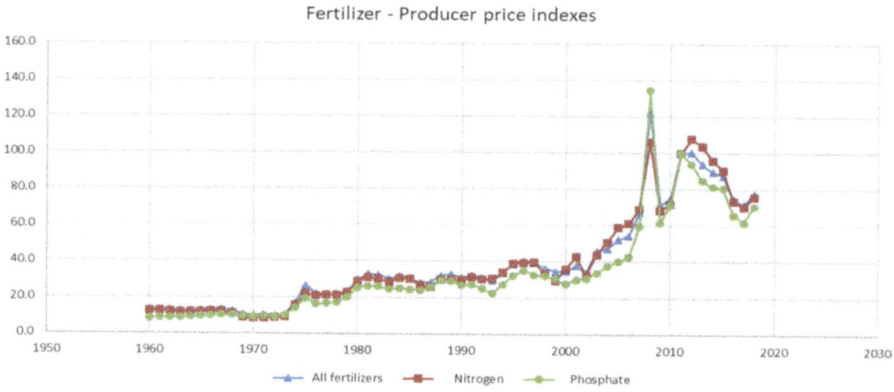

Fig. 4.20 Fertilizer PPI

Commercial fertilizers are applied to agricultural crops to increase crop yields.[75] Before the 1950s, most farming occurred on small family farms with limited use of chemicals. The shift since then to larger corporate farms has coincided with the use of chemical fertilizers in modern agricultural practices. Internationally, the use of fertilizer has exploded and has huge influence on the agricultural outcome. Additionally, through data science and empirical analysis, it can be shown that the crude oil prices affect agricultural commodities and fertilizer prices[76,77] [35]

[75] Agricultural Fertilizers – https://cfpub.epa.gov/roe/indicator_pdf.cfm?i=55#:~:text=The%20three%20major%20types%20of,can%20also%20occur%20as%20nitrate

[76] The relationship between agricultural commodity prices, crude oil prices and US dollar exchange rates: a panel VAR approach and causality analysis – https://www.tandfonline.com/doi/abs/10.1080/02692171.2014.1001325?scroll=top&needAccess=true&journalCode=cira20

[77] Evaluating the impact of rising fertilizer prices on crop yields – https://onlinelibrary.wiley.com/doi/10.1111/agec.12161

With profitability for grain farms limited by low grain prices, producers need to manage their expenses very closely if they want to be profitable. Fertilizer is a good candidate for analysis given that fertilizer is a major expense item for every crop farm. If farmers could predict fertilizer prices 6 months or more in advance, they could purchase their fertilizer at the low-price points of the year and adjust their crop mix to account for years where fertilizer was either higher or lower than normal. *Also, the ability to predict fertilizer prices would help farmers with their planning as they work with lenders to obtain operating loans.*

The three major types of commercial fertilizer used in the United States are nitrogen, phosphate, and potash. The same is true for the world market also. The following analysis shows the relationship between major agricultural commodities, crude oil, and fertilizers.

Data Variables

The data collected from World Bank Commodities Price Data (The Pink Sheet)[78,79] [36]. The CMO Historical Data Monthly Excel file.[80]

Data	Source	Variable
Crude oil, average	Crude oil, average - nominal US dollars	CRUDE_ PETRO
Natural gas, US	Natural gas (U.S.), spot price at Henry hub, Louisiana	NGAS_EUR
Soybeans	Soybeans, U.S. No. 2 yellow meal, CIF Rotterdam beginning December 2007; previously US origin, nearest forward. Soybeans ($/mt)	SOYBEANS
Maize	Maize (U.S.), no. 2, yellow, f.o.b. US Gulf ports ($/mt)	MAIZE
Sorghum	Sorghum (US), no. 2 milo yellow, Texas export bids for grain delivered to export elevators, rail-truck, f.o.b. gulf ports. ($/mt)	SORGHUM
Rice, Thai 5%	Rice (Thailand), 5% broken, white rice (WR), milled, indicative price based on weekly surveys of export transactions, government standard, f.o.b. Bangkok. ($/mt)	RICE_05
Wheat (U.S.), no. 2, soft red winter	Wheat (U.S.), no. 2, soft red winter, export price delivered at the US gulf port for prompt or 30 days shipment. ($/mt)	WHEAT_US_ SRW
Wheat (U.S.), no. 1, hard red winter	Wheat (U.S.), no. 1, hard red winter, ordinary protein, export price delivered at the US gulf port for prompt or 30 days shipment. ($/mt)	WHEAT_US_ HRW
Cotton	Cotton (cotton outlook "CotlookA index"), middling 1–3/32 inch, traded in Far East, C/F beginning 2006; previously northern Europe, c.i.f. ($/kg)	COTTON_A_ INDX
Phosphate rock	Phosphate rock, f.o.b. North Africa ($/mt)	PHOSROCK

[78] Commodity Markets – https://www.worldbank.org/en/research/commodity-markets

[79] World Bank Commodities Price Data (The Pink Sheet) – http://pubdocs.worldbank.org/en/843201604424898761/CMO-Pink-Sheet-November-2020.pdf

[80] Pink Sheet Monthly Data – http://pubdocs.worldbank.org/en/561011486076393416/CMO-Historical-Data-Monthly.xlsx

Data	Source	Variable
Diammonium phosphate	DAP (diammonium phosphate), spot, f.o.b. US gulf ($/mt)	DAP
Triple superphosphate	TSP (triple superphosphate), spot, import US Gulf ($/mt)	TSP
Urea	Urea, (Ukraine), f.o.b. Black Sea ($/mt)	UREA_EE_BULK
Potassium chloride	Potassium chloride (muriate of potash), f.o.b. Vancouver ($/mt)	POTASH

The prices of essential commodities of the CMO Historical Data Monthly (please see Fig. 4.21). As can be seen from the data, commodity markets have experienced turbulent times during the first decade of this millennium, reflected in the increasing volatility of commodity prices. Having registered a long decline until the end of the 1990s, commodity prices started to rise in 2002–2003. The price increases accelerated in 2004–2005, and they peaked in 2008. Commodity prices then plunged during the second half of 2008 with the eruption of the global financial crisis, only to recover by mid-2009 and increase rapidly again in 2010 and early 2011. In May 2011 commodity prices were suffering sharp downward corrections[81] [9].

US Commodity Prices: Producer Price Index

Producer price index is a classical time series data that measures the average change in selling prices received by domestic producers of goods and services. Producer price index[82] (PPI) is measure of prices from producer point of view. The consumer

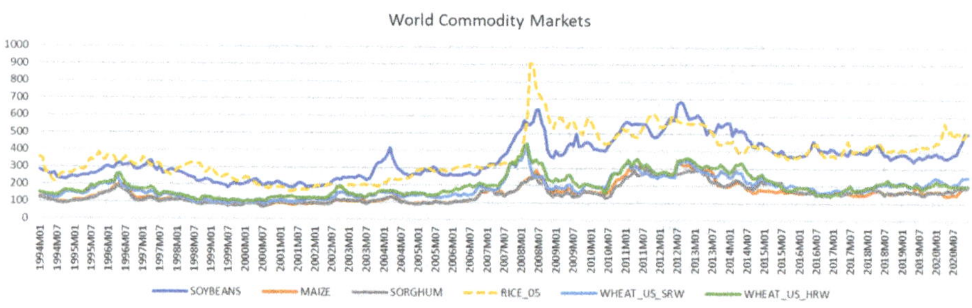

Fig. 4.21 World commodities prices

[81] AN OVERVIEW OF MAJOR SOURCES OF DATA AND ANALYSES RELATING TO PHYSICAL FUNDAMENTALS IN INTERNATIONAL COMMODITY MARKETS – https://unctad.org/system/files/official-document/osgdp20112_en.pdf

[82] PPI – https://www.bls.gov/ppi/ppiover.htm#:~:text=The%20Producer%20Price%20Index%20(PPI,the%20perspective%20of%20the%20seller.

price index (CPI) is measure of price change from consumers point of view. The governmental regulations, government subsidies, sales and excise taxes, and distribution costs make the difference between PPI and CPI [37].

Both PPI and CPI are major economic indicators—the PPI is used to deflate revenue to measure real growth in output, while the CPI is used to adjust income and expenditures for changes in the cost of living.[83] In brief, the CPI includes services, imports, and sales taxes, whereas the PPI excludes them; distribution costs are included in CPI prices, while PPI prices include only producers' costs; and finally, the PPI includes capital equipment, while the CPI does not. The PPI serves as a leading indicator for the CPI, so when producers face input inflation, the increases in their production costs are passed onto the retailers and consumers. The PPI also serves as a true measure of output in that it is not affected by consumer demand.[84] The following are PPI for essential US commodities (please see Fig. 4.22):

- Corn.[85]
- Soybeans.[86]
- Wheat.[87]
- Rice.[88]

Data Sources

The GIEWS[89] Food Price Monitoring and Analysis (FPMA) Tool (please see Fig. 4.23) provides quick access to up-to-date consumer price data in over 90 countries, as well as about 90 international price series.

[83] How Does the Producer Price Index Differ from the Consumer Price Index? – https://www.bls.gov/ppi/ppivcpi.pdf

[84] CPI vs. PPI? – https://www.investopedia.com/ask/answers/011915/what-difference-between-consumer-price-index-cpi-and-producer-price-index-ppi.asp

[85] Producer Price Index by Commodity: Farm Products: Corn – https://fred.stlouisfed.org/series/WPS012202

[86] Producer Price Index by Commodity: Farm Products: Soybeans (WPU01830131) – https://fred.stlouisfed.org/series/WPU01830131

[87] Producer Price Index by Commodity: Farm Products: Wheat (WPU0121) – https://fred.stlouisfed.org/series/WPU0121

[88] Producer Price Index by Commodity: Farm Products: Rough Rice (WPU01230103) – https://fred.stlouisfed.org/series/WPU01230103

[89] FAO International Price Domain – http://www.fao.org/prices/en/

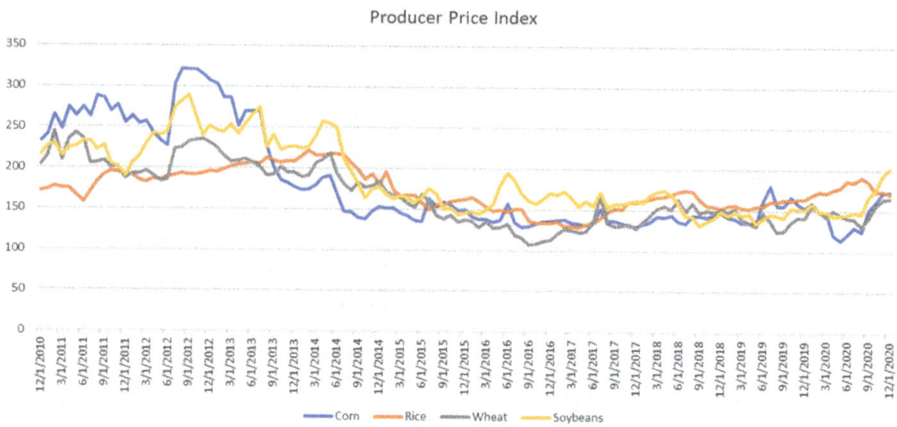

Fig. 4.22 PPI

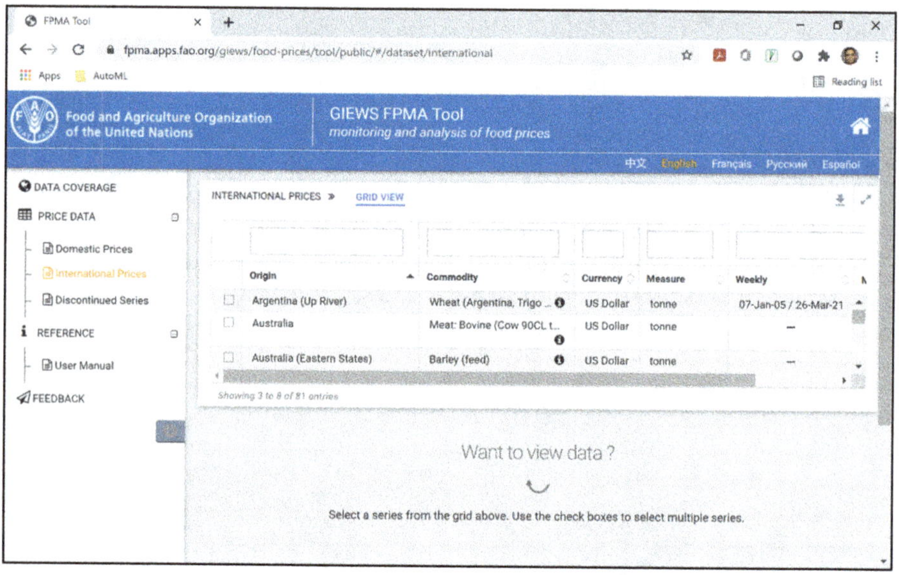

Fig. 4.23 GIEWS

World Bank Commodity Market Data (please see Fig. 4.24) provides comprehensive view of dynamic commodity prices details. We have Pink Sheet data[90] and World Bank Market Data[91] to prepare the model.

---

[90] The World Bank Pink Sheet Data – https://thedocs.worldbank.org/en/doc/5d903e848db1d1b83e0ec8f744e55570-0350012021/original/CMO-Pink-Sheet-March-2021.pdf

[91] World bank Commodities Market Data – https://www.worldbank.org/en/research/commodity-markets

World Bank Commodities Price Data (The Pink Sheet) 2-Mar-2021

			Annual Averages			Quarterly Averages					Monthly Averages		
			Jan-Dec	Jan-Dec	Jan-Dec	Oct-Dec	Jan-Mar	Apr-Jun	Jul-Sep	Oct-Dec	Dec	Jan	Feb
Commodity	Unit		2018	2019	2020	2019	2020	2020	2020	2020	2020	2021	2021
Energy													
Coal, Australia	$/mt	a/	107.0	77.9	60.8	67.5	68.0	54.4	52.1	68.6	83.0	86.8	86.7
Coal, South Africa	$/mt		97.6	71.9	65.7	72.4	76.7	56.9	57.2	71.9	85.2	86.9	82.8
Crude oil, average	$/bbl		68.3	61.4	41.3	60.3	49.1	30.3	42.0	43.6	48.7	53.6	60.5
Crude oil, Brent	$/bbl	a/	71.1	64.0	42.3	62.7	50.5	31.4	42.7	44.5	49.9	54.6	62.0
Crude oil, Dubai	$/bbl	a/	69.2	63.2	42.2	61.4	50.7	31.7	42.5	43.8	49.3	54.2	60.4
Crude oil, WTI	$/bbl	a/	64.8	57.0	39.3	56.9	46.0	27.8	40.9	42.6	47.1	52.1	59.1
Natural gas, Index	2010=100		82.1	61.1	45.5	59.5	44.7	35.9	42.3	59.2	63.9	73.0	96.9
Natural gas, Europe	$/mmbtu	a/	7.68	4.80	3.24	4.94	3.09	1.82	2.87	5.19	5.86	7.27	6.16
Natural gas, U.S.	$/mmbtu	a/	3.16	2.57	2.01	2.41	1.91	1.70	1.99	2.46	2.54	2.67	5.07
Liquefied natural gas, Japan	$/mmbtu	a/	10.67	10.56	8.31	10.03	10.00	9.69	6.67	6.90	7.66	8.91	8.97
Non Energy Commodities													
Agriculture													
Beverages													
Cocoa	$/kg	b/	2.29	2.34	2.37	2.47	2.55	2.27	2.30	2.35	2.41	2.39	2.41
Coffee, Arabica	$/kg	b/	2.93	2.88	3.32	3.12	3.13	3.28	3.50	3.38	3.48	3.54	3.66
Coffee, Robusta	$/kg	b/	1.87	1.62	1.52	1.58	1.51	1.42	1.57	1.56	1.59	1.56	1.61
Tea, average	$/kg		2.85	2.56	2.70	2.59	2.33	2.57	3.09	2.81	2.65	2.68	2.58
Tea, Colombo	$/kg	b/	3.61	3.10	3.40	3.11	3.23	3.57	3.35	3.46	3.41	3.39	3.35
Tea, Kolkata	$/kg	b/	2.36	2.38	2.69	2.37	1.63	2.15	3.97	3.02	2.60	2.62	2.38
Tea, Mombasa	$/kg	b/	2.58	2.21	2.00	2.28	2.14	1.98	1.94	1.97	1.94	2.03	2.02
Food													

Fig. 4.24 World Bank Commodity Pink Data

Globally, most fertilizer prices are significantly affected by the crude oil price,[92] while the volatility of global fertilizer prices and crude oil price from March to December 2008 are higher than in other periods [38].

Software code for this model can be found on GitHub link: FertilizerGrangerCausalityTest.Ipynb (Jupyter notebook).

Model Development

Step 1: Load Python Libraries.

```
import logging
import pandas as pd
import numpy as np
from numpy import random
import nltk
from sklearn.model_selection import train_test_split
from sklearn.feature_extraction.text import CountVectorizer, TfidfVectorizer
from sklearn.metrics import accuracy_score, confusion_matrix
import matplotlib.pyplot as plt
from nltk.corpus import stopwords
import re
from bs4 import BeautifulSoup
%matplotlib inline
```

[92] Modeling the effects of Oil Prices on Fertilizers – https://www.mdpi.com/1911-8074/5/1/78/pdf

Step 2: Load Pink Sheet Commodity Prices.

```
#CMOHistoricalDataMonthly
agriCMOHistoricalDataMonthlyDF =
pd.read_csv("CMOHistoricalDataMonthly.csv",parse_dates=['Date'], dayfirst=True)

agriCMOHistoricalDataMonthlyDF = agriCMOHistoricalDataMonthlyDF.set_index('Date')

agriCMOHistoricalDataMonthlyDF.head()
```

Output:

Date	CRUDE_PETRO	NGAS_EUR	SOYBEANS	MAIZE	SORGHUM	RICE_05	WHEAT_US_SRW	WHEAT_US_HRW	COTTON_A_INDX	PHOSROCK
960M01	1.63	0.4	94.00	45.00	39.00	104.45	140.36	59.89	0.65	13.0
960M02	1.63	0.4	91.00	44.00	39.00	103.54	140.36	60.99	0.65	13.0
960M03	1.63	0.4	92.00	45.00	35.00	103.79	140.36	61.73	0.65	13.0
960M04	1.63	0.4	93.00	45.00	35.00	100.97	140.36	60.99	0.64	13.0
960M05	1.63	0.4	93.00	48.00	35.00	102.15	140.36	57.69	0.65	13.0
960M06	1.63	0.4	91.00	47.00	36.00	103.13	140.36	55.48	0.66	13.0
960M07	1.63	0.4	92.00	47.00	39.90	102.05	140.36	54.75	0.66	13.0
960M08	1.63	0.4	93.00	47.00	39.00	109.71	140.36	55.12	0.65	13.0
960M09	1.63	0.4	92.00	46.00	35.00	116.10	140.36	56.59	0.66	13.0
960M10	1.63	0.4	88.00	42.00	35.00	115.21	140.36	56.95	0.67	13.0
960M11	1.63	0.4	90.00	38.00	35.00	114.43	140.36	57.69	0.67	13.0
960M12	1.63	0.4	93.00	40.00	36.00	112.66	140.36	58.05	0.66	13.0
961M01	1.57	0.4	102.00	43.00	38.00	113.47	140.36	58.79	0.66	13.0
961M02	1.57	0.4	114.00	44.00	39.00	109.52	140.36	59.16	0.68	13.0
961M03	1.57	0.4	121.00	44.00	39.00	117.65	140.36	58.05	0.67	13.0
961M04	1.57	0.4	129.00	43.00	38.00	115.17	140.36	57.32	0.67	13.0

Step 3: Inspect DataFrame.

```
agriCMOHistoricalDataMonthlyDF.info()
```

Output:

```
<class 'pandas.core.frame.DataFrame'>
Index: 732 entries, 1960M01 to 2020M12
Data columns (total 14 columns):
 #   Column         Non-Null Count  Dtype
---  ------         --------------  -----
 0   CRUDE_PETRO    732 non-null    float64
 1   NGAS_EUR       732 non-null    float64
 2   SOYBEANS       732 non-null    float64
 3   MAIZE          732 non-null    float64
 4   SORGHUM        732 non-null    float64
 5   RICE_05        732 non-null    float64
 6   WHEAT_US_SRW   732 non-null    float64
 7   WHEAT_US_HRW   732 non-null    float64
 8   COTTON_A_INDX  732 non-null    float64
 9   PHOSROCK       732 non-null    float64
 10  DAP            732 non-null    float64
 11  TSP            732 non-null    float64
 12  UREA_EE_BULK   732 non-null    float64
 13  POTASH         732 non-null    float64
dtypes: float64(14)
```

Step 3: Load Pink Sheet data from January 1960 through December 2020.

```
dataset=dataset.loc['1960M01':'2020M12']
dataset.head(20)
```

Output:

Date	CRUDE_PETRO	NGAS_EUR	SOYBEANS	MAIZE	SORGHUM	RICE_05	WHEAT_US_SRW	WHEAT_US_HRW	COTTON_A_INDX	PHOSROCK
1960M01	1.63	0.4	94.0	45.0	39.0	104.45	140.36	59.89	0.65	13.0
1960M02	1.63	0.4	91.0	44.0	39.0	103.54	140.36	60.99	0.65	13.0
1960M03	1.63	0.4	92.0	45.0	35.0	103.79	140.36	61.73	0.65	13.0
1960M04	1.63	0.4	93.0	45.0	35.0	100.97	140.36	60.99	0.64	13.0
1960M05	1.63	0.4	93.0	48.0	35.0	102.15	140.36	57.69	0.65	13.0
1960M06	1.63	0.4	91.0	47.0	36.0	103.13	140.36	55.48	0.66	13.0
1960M07	1.63	0.4	92.0	47.0	39.9	102.05	140.36	54.75	0.66	13.0
1960M08	1.63	0.4	93.0	47.0	39.0	109.71	140.36	55.12	0.65	13.0
1960M09	1.63	0.4	92.0	46.0	35.0	116.10	140.36	56.59	0.66	13.0
1960M10	1.63	0.4	88.0	42.0	35.0	115.21	140.36	56.95	0.67	13.0
1960M11	1.63	0.4	90.0	38.0	35.0	114.43	140.36	57.69	0.67	13.0
1960M12	1.63	0.4	93.0	40.0	36.0	112.66	140.36	58.05	0.66	13.0
1961M01	1.57	0.4	102.0	43.0	38.0	113.47	140.36	58.79	0.66	13.0
1961M02	1.57	0.4	114.0	44.0	39.0	109.52	140.36	59.16	0.68	13.0
1961M03	1.57	0.4	121.0	44.0	39.0	117.65	140.36	58.05	0.67	13.0
1961M04	1.57	0.4	129.0	43.0	38.0	115.17	140.36	57.32	0.67	13.0

Step: Describe the dataset.

```
dataset.describe()
```

Output:

As you can see, the total number of records is 475. Crude Petro showed a value of minimum $1.21, December 1970, and maximum of $132.83, July 2008.

	CRUDE_PETRO	NGAS_EUR	SOYBEANS	MAIZE	SORGHUM	RICE_05	WHEAT_US_SRW	WHEAT_US_HRW	COTTON_A_INDX	PHOSROCK
count	475.000000	475.000000	475.000000	475.000000	475.000000	475.000000	475.000000	475.000000	475.000000	475.000000
mean	26.068568	3.579621	263.829642	112.616653	107.726695	288.680105	145.270674	151.453137	1.470253	47.252421
std	20.554709	2.545905	76.404471	32.565015	30.239696	115.542005	37.883467	50.929788	0.352813	55.673769
min	1.210000	0.450000	119.000000	51.330000	49.820000	95.670000	85.300000	52.540000	0.670000	11.000000
25%	13.560000	2.300000	226.500000	95.445000	92.940000	212.370000	128.605000	123.530000	1.240000	31.500000
50%	19.370000	2.820000	252.500000	108.850000	104.960000	271.600000	140.360000	148.810000	1.470000	36.000000
75%	31.435000	4.050000	293.500000	121.245000	116.615000	324.960000	153.260000	169.205000	1.740000	44.000000
max	132.830000	15.930000	637.250000	287.110000	262.190000	907.000000	419.610000	439.720000	2.530000	450.000000

Step: Plot the values of all features.

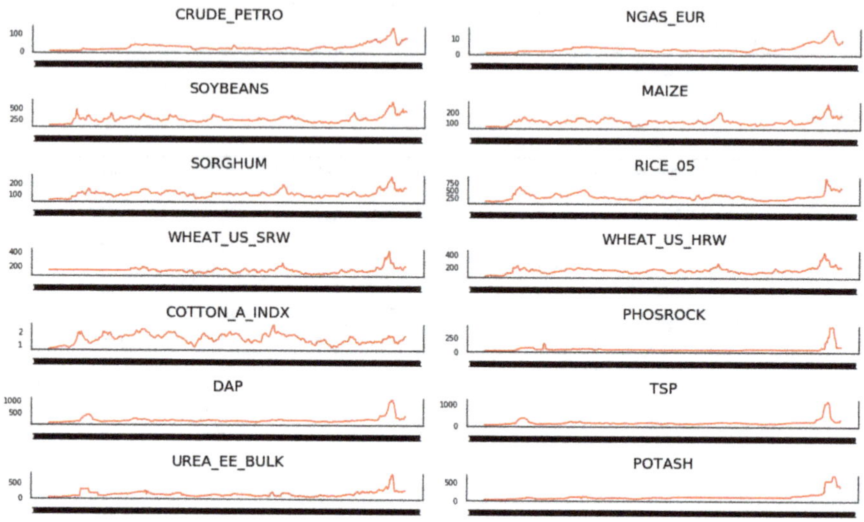

Step: Check normal distribution of the data.

```
from scipy import stats

CRUDE_PETRO=dataset.CRUDE_PETRO.values

stat,p = stats.normaltest(CRUDE_PETRO)

print("CRUDE_PETRO Statistics = %.3f, p=%.3f" % (stat,p))
alpha = 0.05
if p> alpha:
   print('Data looks Gaussian (fail to reject null hypothesis)')
else:
   print('Data looks non-Gaussian (reject null hypothesis)')
```

Output:

```
CRUDE_PETRO Statistics = 215.688, p=0.000
Data looks non-Gaussian (reject null hypothesis)
```

Crude Petro exhibits normal test (p = 0.00).
Similarly, for other features, all are showing normal distribution.

```
from scipy import stats

NGAS_EUR=dataset.NGAS_EUR.values


stat,p = stats.normaltest(NGAS_EUR)

print("NGAS_EUR Statistics = %.3f, p=%.3f" % (stat,p))
alpha = 0.05
if p> alpha:
   print('Data looks Gaussian (fail to reject null hypothesis)')
else:
   print('Data looks non-Gaussian (reject null hypothesis)')
```

Output:

```
NGAS_EUR Statistics = 229.799, p=0.000
Data looks non-Gaussian (reject null hypothesis)
```

All the following features showed normal distribution: SOYBEANS, MAIZE, SORGHUM, RICE_05, WHEAT_US_SRW, WHEAT_US_HRW, PHOSROCK, TSP, and POTASH.

Except Cotton:

```
from scipy import stats

COTTON_A_INDX=dataset.COTTON_A_INDX.values


stat,p = stats.normaltest(COTTON_A_INDX)

print("COTTON_A_INDX Statistics = %.3f, p=%.3f" % (stat,p))
alpha = 0.05
if p> alpha:
   print('Data looks Gaussian (fail to reject null hypothesis)')
else:
   print('Data looks non-Gaussian (reject null hypothesis)')
```

Output:

```
COTTON_A_INDX Statistics = 4.628, p=0.099
Data looks Gaussian (fail to reject null hypothesis)
```

Decline in crude oil prices[93] and sluggish demand have hit the entire cotton value chain, leaving stakeholders complaining about cash crunch as money has been locked up in inventories and prompting yarn manufacturers to cut prices. The monthly yarn price declared at Tirupur has been cut by?10 per kg [39].

---

[93] Cotton hurt by fall in crude, low demand – https://economictimes.indiatimes.com/markets/commodities/news/cotton-hurt-by-fall-in-crude-low-demand/articleshow/67431155.cms?from=mdr

Step: Evaluate skewness and kurtosis of the distributions.

```
print('CRUDE_PETRO: Kurtosis of normal distribution:
{}'.format(stats.kurtosis(dataset.CRUDE_PETRO)))
print('CRUDE_PETRO:Skewness of normal distribution:
{}'.format(stats.skew(dataset.CRUDE_PETRO)))
import matplotlib.pyplot as plt
import scipy.stats as stats
from scipy.stats import kurtosis

dataset.CRUDE_PETRO.plot(kind = 'density')
```

Output:

```
CRUDE_PETRO: Kurtosis of normal distribution: 0.9417868059179275
CRUDE_PETRO:Skewness of normal distribution: 1.2891507804432012
```

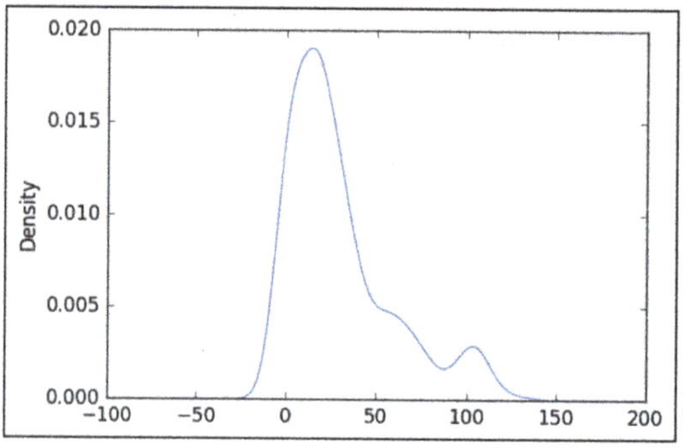

Crude skewed left with tail has multiple changes. It shows that crude prices have variations. Here are the other features:

NGAS_EUR	SOYBEANS
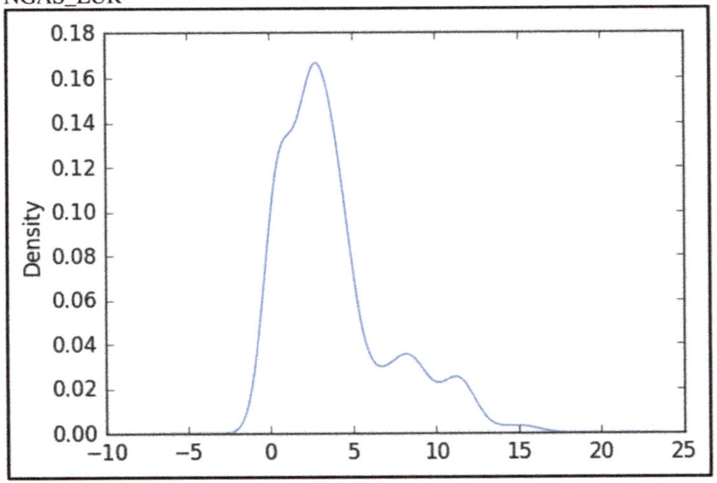	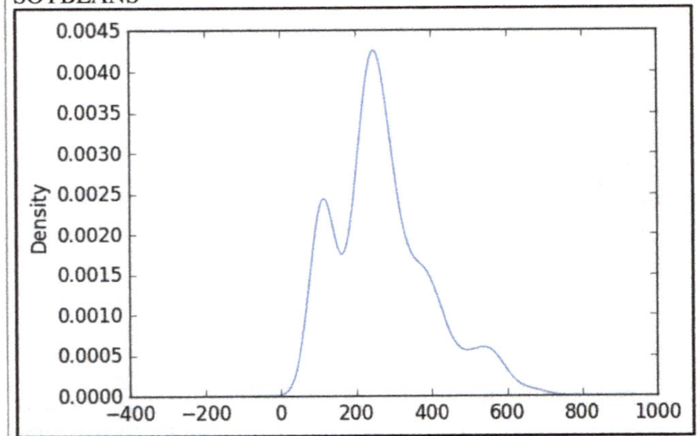
NGAS_EUR: Kurtosis of normal distribution: 1.1282286518507654 NGAS_EUR:Skewness of normal distribution: 1.2680485328680697	SOYBEANS: Kurtosis of normal distribution: 0.31076219905067815 SOYBEANS:Skewness of normal distribution: 0.7460902155459993

MAIZE

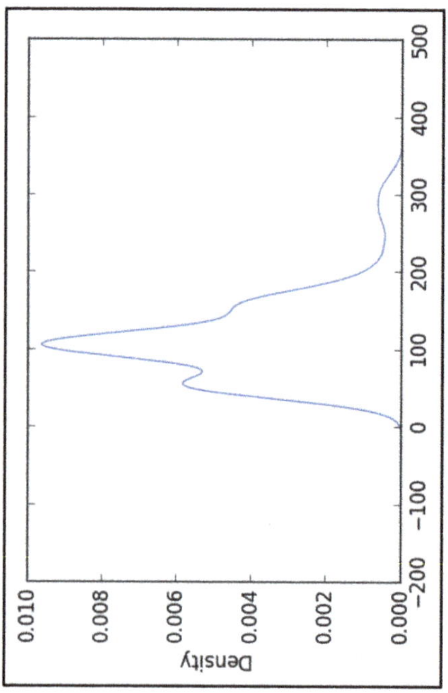

MAIZE: Kurtosis of normal distribution: 2.2327973328318365
MAIZE:Skewness of normal distribution: 1.3262919887175717

SORGHUM

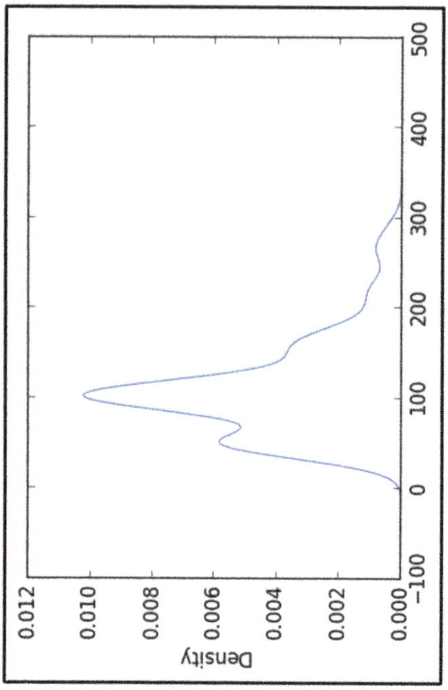

SORGHUM: Kurtosis of normal distribution: 1.1212802163836555
SORGHUM:Skewness of normal distribution: 1.0754413822192457

<table>
<tr>
<td>
RICE_05
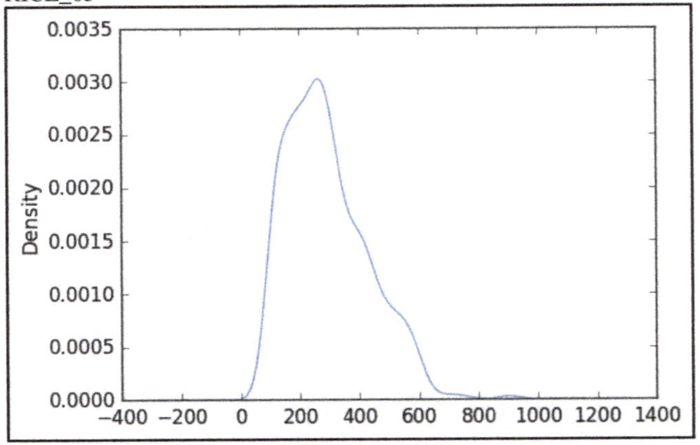

RICE_05: Kurtosis of normal distribution: 0.6034279740095805
RICE_05:Skewness of normal distribution: 0.8153661625248881
</td>
<td>
WHEAT_US_SRW
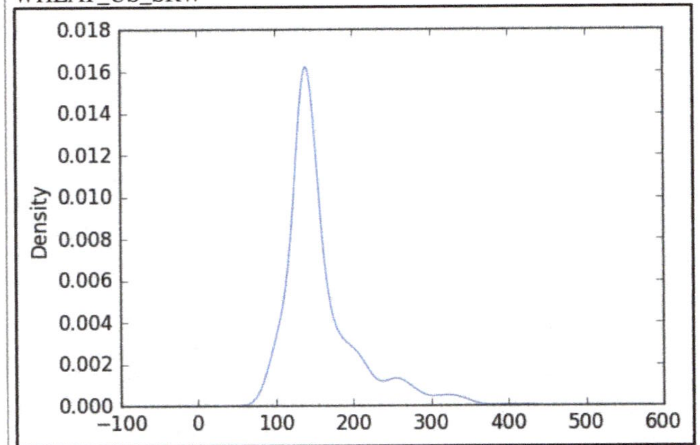

WHEAT_US_SRW: Kurtosis of normal distribution: 3.893584470115263
WHEAT_US_SRW:Skewness of normal distribution: 1.8599749674584594
</td>
</tr>
</table>

COTTON_A_INDX	WHEAT
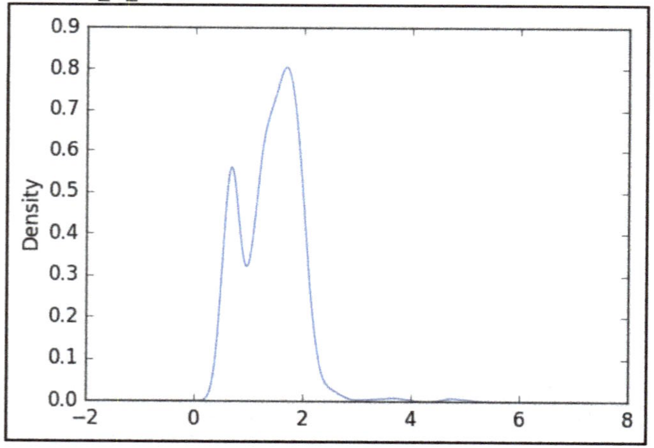	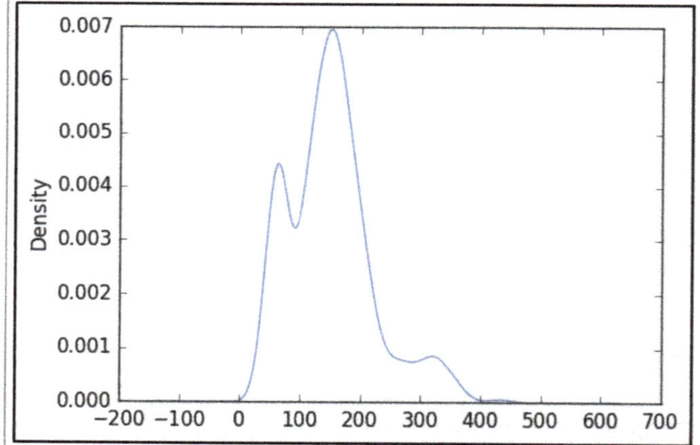
COTTON_A_INDX: Kurtosis of normal distribution: 6.003480419299807 COTTON_A_INDX:Skewness of normal distribution: 1.1728139382221665	WHEAT_US_HRW: Kurtosis of normal distribution: 1.0218529993917569 WHEAT_US_HRW:Skewness of normal distribution: 0.9034087264568631

Fertilizers.

PHOSROCK	DAP
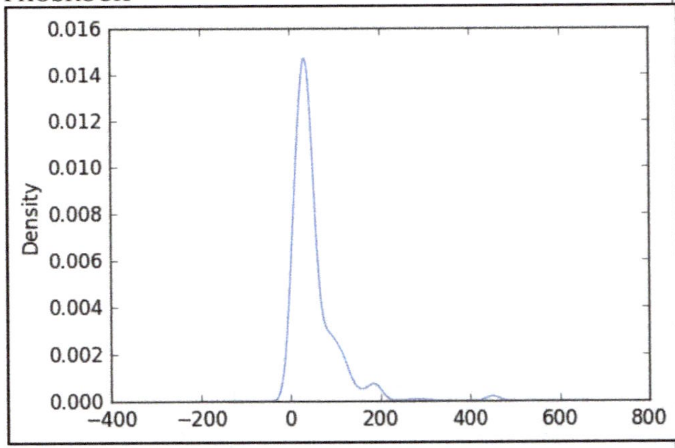	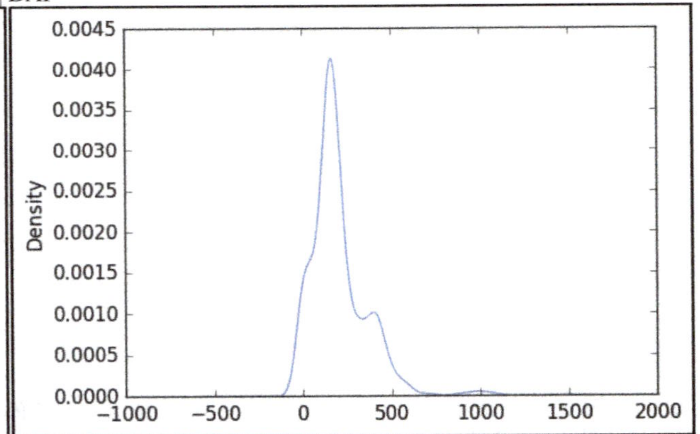
PHOSROCK: Kurtosis of normal distribution: 19.59315549744017 PHOSROCK:Skewness of normal distribution: 3.713790985459138	DAP: Kurtosis of normal distribution: 5.726126211335647 DAP:Skewness of normal distribution: 1.7731086192526118

TSP

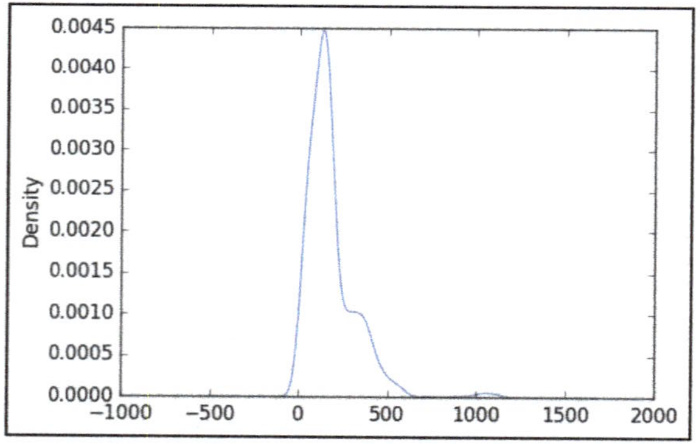

TSP: Kurtosis of normal distribution: 10.811562080492754
TSP:Skewness of normal distribution: 2.622453473991759

UREA

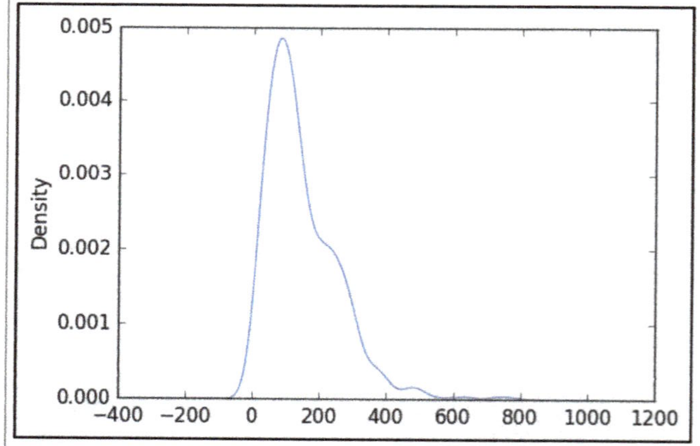

UREA_EE_BULK: Kurtosis of normal distribution: 4.458846679117477
UREA_EE_BULK:Skewness of normal distribution: 1.6542507526335184

Potash 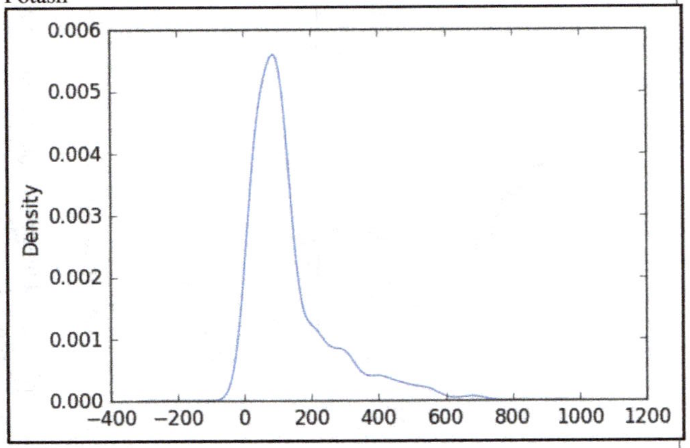 POTASH: Kurtosis of normal distribution: 3.5382448850376838 POTASH:Skewness of normal distribution: 1.8820435856219555	

Urea is one of the most used fertilizers in the world; world's poorest farmers use a nitrogen fertilizer urea. The chemical was initially produced to serve industrial agriculture, but many small-scale farmers were swayed by the fertilizer's promise of increased productivity[94] [32]. However, the fertilizer begins to wreak havoc once absorbed into the soil, destroying the precarious balance of microorganisms the soil needs to provide plants with enough vitamins and minerals. The ecosystem is destroyed. As a result, crops are left vulnerable to disease, produce lower yields, are less nutritious, and even require more water. This kicks off a chain reaction that leads to farmers using more fertilizer, herbicides, and pesticides to remedy these issues. Producers begin investing more money in chemicals[95] and must start purchasing seeds to replant their failing crops—resulting in farmers earning only a 2% profit, intensifying their food insecurity and ongoing poverty.

Step: Probability plot.

```
plt.figure(figsize=(14,6))
plt.subplot(1,2,1)
dataset['CRUDE_PETRO'].hist(bins=50)
plt.title('CRUDE_PETRO')
plt.subplot(1,2,2)
stats.probplot(dataset['CRUDE_PETRO'], plot=plt);
dataset.CRUDE_PETRO.describe().T
```

Output:

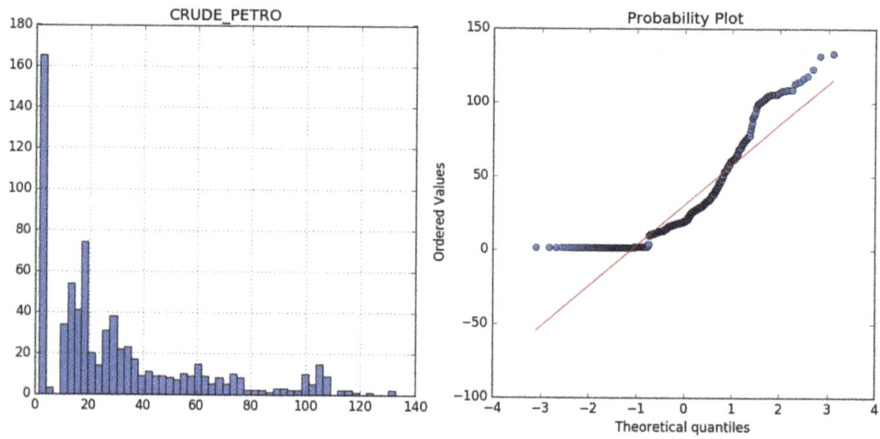

[94] "Free" Fertilizer is Saving Rural Farmers – https://www.permaculturenews.org/2017/10/02/free-fertilizer-saving-rural-farmers/

[95] Simple Approaches to Making and Using Foliar Fertilizers – https://www.permaculturenews.org/2017/09/15/simple-approaches-making-using-foliar-fertilizers/

Step: Correlation.

```
corr
```

Output:

	CRUDE_PETRO	NGAS_EUR	SOYBEANS	MAIZE	SORGHUM	RICE_05	WHEAT_US_SRW	WHEAT_US_HRW	COTTON_A_INDX
CRUDE_PETRO	1.000000	0.944011	0.884810	0.874278	0.882876	0.759408	0.784843	0.881667	0.617591
NGAS_EUR	0.944011	1.000000	0.847135	0.840307	0.848005	0.747379	0.715102	0.862199	0.561009
SOYBEANS	0.884810	0.847135	1.000000	0.941400	0.941896	0.851764	0.791538	0.921522	0.739188
MAIZE	0.874278	0.840307	0.941400	1.000000	0.978896	0.842766	0.819638	0.930020	0.744078
SORGHUM	0.882876	0.848005	0.941896	0.978896	1.000000	0.841890	0.817043	0.934252	0.729728
RICE_05	0.759408	0.747379	0.851764	0.842766	0.841890	1.000000	0.674764	0.838390	0.663193
WHEAT_US_SRW	0.784843	0.715102	0.791538	0.819638	0.817043	0.674764	1.000000	0.831616	0.505756
WHEAT_US_HRW	0.881667	0.862199	0.921522	0.930020	0.934252	0.838390	0.831616	1.000000	0.715699
COTTON_A_INDX	0.617591	0.561009	0.739188	0.744078	0.729728	0.663193	0.505756	0.715699	1.000000
PHOSROCK	0.672479	0.783755	0.687415	0.701810	0.687770	0.715647	0.567893	0.657078	0.399375
DAP	0.855998	0.860595	0.868069	0.850880	0.861241	0.855507	0.727656	0.865629	0.594397
TSP	0.844050	0.843548	0.830268	0.828239	0.827951	0.824578	0.742471	0.838056	0.539334
UREA_EE_BULK	0.879604	0.871613	0.854735	0.855219	0.859178	0.857339	0.734922	0.854686	0.570902
POTASH	0.856832	0.878018	0.829050	0.821054	0.818591	0.796815	0.701735	0.791654	0.488214

As you can see in the correlation values, Fertilizers, UREA_EE_BULK (0.87), POTASH (0.85), DAP (0.855), and TSP (0.844), PHOSROCK (0.67) with CRUDE PETRO. With the Natural Gas also the values of fertilizers show high correlation. One reason could be both CRUDE and Natural Gas also have high correlation—implying one substitutes to others in the manufacturing of fertilizer.

Step: Correlation plot.

```
import matplotlib.pyplot as plt
import pandas as pd
import numpy as np
import seaborn as sns; sns.set()
corr = dataset.corr()

fig, ax = plt.subplots()
sns.heatmap(corr,xticklabels=corr.columns.values,                yticklabels=corr.columns.values,
annot=True,annot_kws={'size':12})
heat_map=plt.gcf()
heat_map.set_size_inches(10,6)
plt.xticks(fontsize=10)
plt.yticks(fontsize=10)
plt.show()
```

Output:

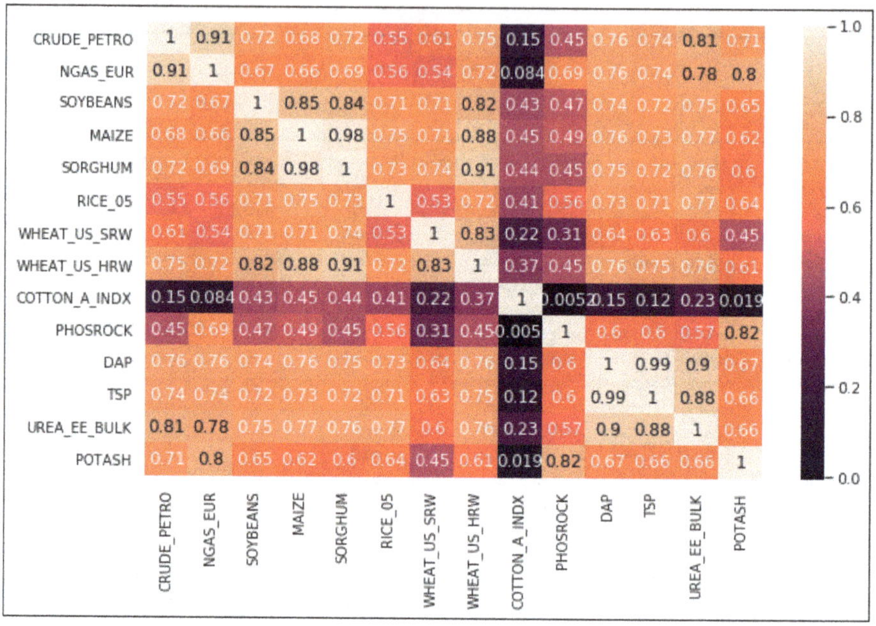

Step 10: Autocorrelation or Serial Correlation.

In Step 8, Dataset Price and Index columns are tested for normal distribution. Autocorrection is the process of calculating correlation of a time series variable with its previous time steps, called lags.[96] Because the correlation of the time series observations is calculated with values of the same series at previous times, this is called a serial correlation, or an autocorrelation. A plot of the autocorrelation of a time series by lag is called the autocorrelation function, or the acronym ACF. This plot is sometimes called a correlogram or an autocorrelation plot [16, 30].

Autocorrelation plots are a commonly used tool for checking randomness in a dataset.[97] This randomness is ascertained by computing autocorrelations for data values at varying time lags. If random, such autocorrelations should be near zero for all time-lag separations. If non-random, then one or more of the autocorrelations will be significantly non-zero.

> Autocorrelation plot function: Statsmodels.Graphics.Tsaplots.plot_acf[98]
> Plot the autocorrelation function. Plots lags on the horizontal and the correlations on vertical axis.

[96] A Gentle Introduction to Autocorrelation and Partial Autocorrelation – https://machinelearningmastery.com/gentle-introduction-autocorrelation-partial-autocorrelation/

[97] Autocorrelation Plot – https://itl.nist.gov/div898/handbook/eda/section3/autocopl.htm

[98] PLOT_ACF: https://www.statsmodels.org/devel/generated/statsmodels.graphics.tsaplots.plot_acf.html

Given dataset columns are normally distributed, Pearson's correlation coefficient is applied to summarize the correlation between the Gold, Silver, Oil, USD Index, Interest, and SPIndex variables.
Note: Autocorrelation or serial correlation can be a significant problem in analyzing historical data if we do not know how to look out for it

```
import pandas as pd
import matplotlib.pyplot as plt
import statsmodels.api as sm

# plots the autocorrelation plots for each stock's price at 50 lags
for i in dataset:
    series = dataset[i]
    sm.graphics.tsa.plot_acf(series)
    plt.title('ACF for %s' % i)
    plt.show()
```

Output: above code creates a 2D plot showing the lag values of Gold, Silver, Oil, USD Index, Interest, and SPIndex variables along the x-axis and the correlation on the y-axis between −1 and 1.

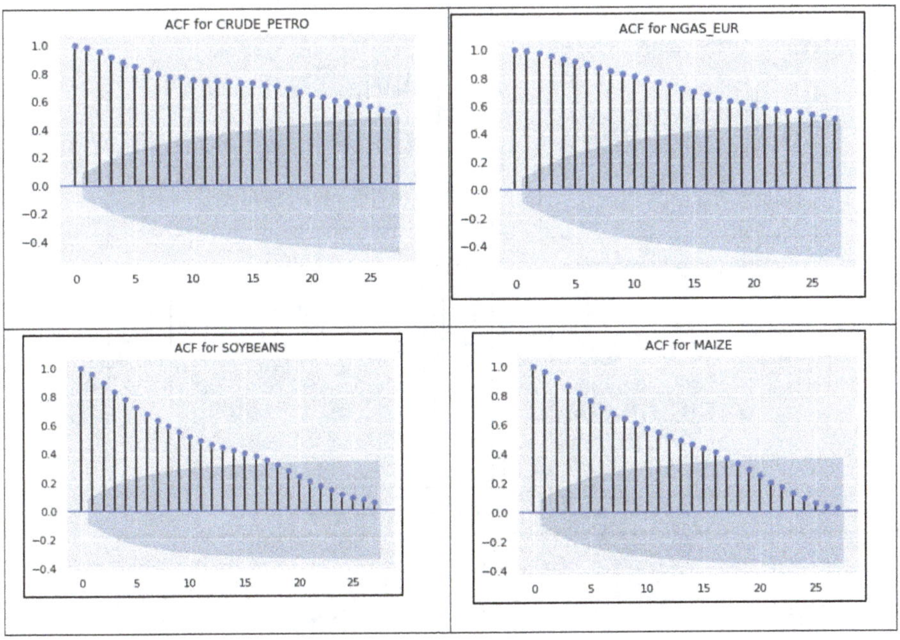

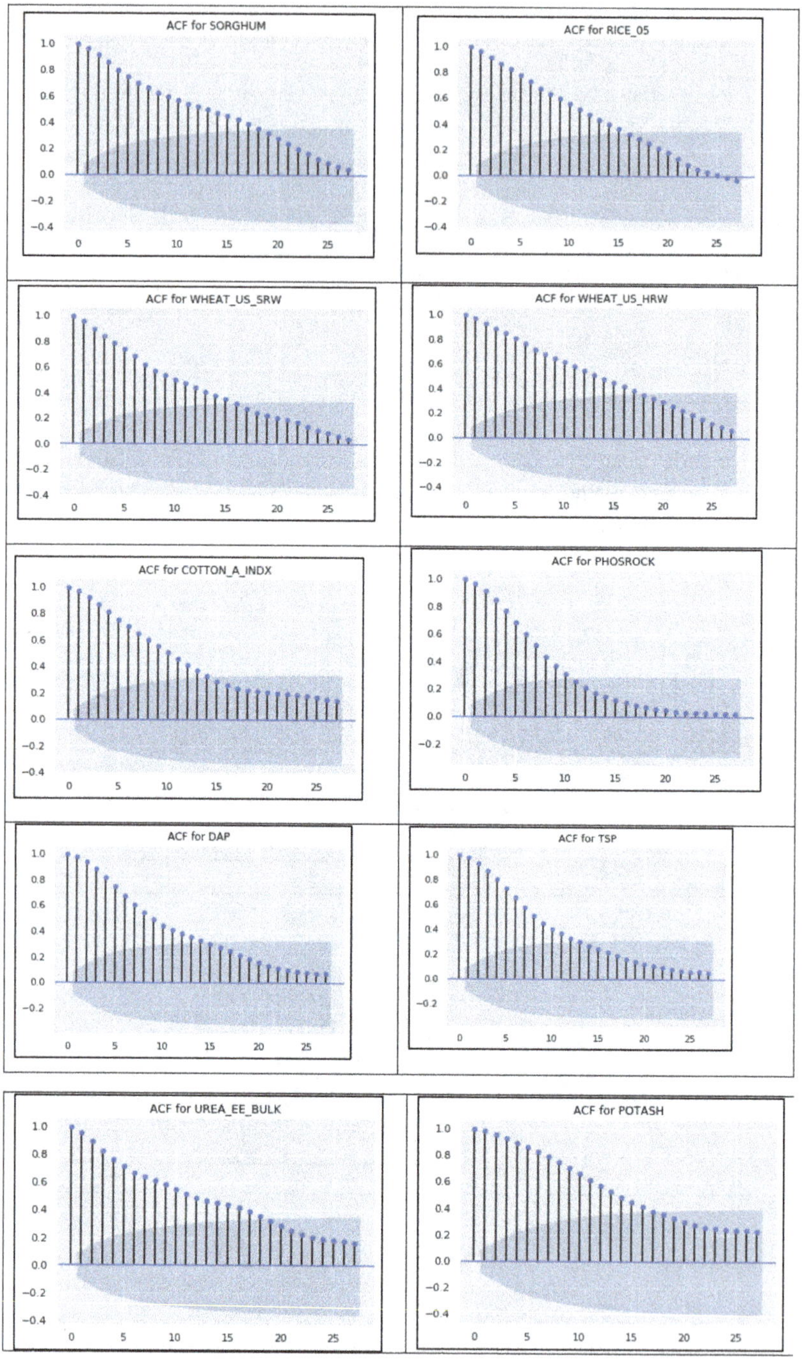
ACF for SORGHUM
ACF for RICE_05
ACF for WHEAT_US_SRW
ACF for WHEAT_US_HRW
ACF for COTTON_A_INDX
ACF for PHOSROCK
ACF for DAP
ACF for TSP
ACF for UREA_EE_BULK
ACF for POTASH

Higher value of autocorrelation implies non-random series. The usage of autocorrelation plots provides autoregression and moving average Intuitions.

Please note[99]: If random, such autocorrelations should be near zero for all time-lag separations. If non-random, then one or more of the autocorrelations will be significantly non-zero [16].

Step 11: Training and Testing Data.

Split the Dataset into training and testing dataset for building multivariate time series predictive model. Test Dataset is split on lagged observations.

```
n_obs=15
X_train, X_test = dataset[0:-n_obs], dataset[-n_obs:]
print(X_train.shape, X_test.shape)

Output:
(460, 14) (15, 14)
```

Step 12: Transformation.

The goal is to make sure train dataset is stationary. One of achieving is first differencing on training data makes all the 14 series stationary. However, this is an iterative process where we after first differencing, the series may still be non-stationary. We shall have to apply second difference or log transformation to standardize the series in such cases.

```
transform_data = X_train.diff().dropna()
transform_data.head()
```

Output:

Date	CRUDE_PETRO	NGAS_EUR	SOYBEANS	MAIZE	SORGHUM	RICE_05	WHEAT_US_SRW	WHEAT_US_HRW	COTTON_A_INDX	PHOSROCK	DAF
1970M08	0.0	0.0	-2.0	0.0	0.0	3.15	0.0	1.47	0.01	0.0	0.(
1970M09	0.0	0.0	-1.0	0.0	0.0	-2.36	0.0	2.21	0.01	0.0	0.(
1970M10	0.0	0.0	4.0	0.0	0.0	-1.28	0.0	2.57	0.01	0.0	0.(
1970M11	0.0	0.0	2.0	0.0	0.0	0.40	0.0	2.20	0.01	0.0	0.(
1970M12	0.0	0.0	-4.0	0.0	0.0	-4.42	0.0	2.21	0.01	0.0	0.(

```
transform_data.describe()
```

[99] Autocorrelation Function – https://itl.nist.gov/div898/handbook/eda/section3/autocopl.htm

Output:

	CRUDE_PETRO	NGAS_EUR	SOYBEANS	MAIZE	SORGHUM	RICE_05	WHEAT_US_SRW	WHEAT_US_HRW	COTTON_A_INDX	PHOSROCK
count	459.000000	459.000000	459.000000	459.000000	459.000000	459.000000	459.000000	459.000000	459.000000	459.000000
mean	0.155730	0.033725	0.605381	0.271590	0.243638	1.055163	0.099259	0.402702	0.001525	0.956427
std	2.802534	0.196885	19.964112	7.440975	7.288852	25.332537	10.713282	11.087762	0.075329	11.787048
min	-26.970000	-1.060000	-140.000000	-50.790000	-52.380000	-144.800000	-96.180000	-77.490000	-0.530000	-79.500000
25%	-0.580000	0.000000	-6.000000	-2.960000	-2.695000	-7.000000	-3.030000	-3.680000	-0.040000	0.000000
50%	0.000000	0.000000	0.000000	0.000000	0.000000	0.250000	0.000000	0.000000	0.000000	0.000000
75%	0.915000	0.020000	7.000000	3.595000	3.655000	7.280000	3.355000	4.040000	0.040000	0.000000
max	13.870000	1.740000	112.000000	43.650000	33.190000	313.000000	69.080000	77.900000	0.250000	172.000000

Step 12: Stationarity Check.

To test stationarity, augmented Dickey-Fuller test is performed. Dicker-Fuller test is used to check stationarity of a time series. DF tests the null hypothesis that a unit root is present in time series data. A unit root is a stochastic trend in a time series, sometimes called a "random walk with drift." If a time series has a unit root, it shows systematic pattern that is unpredictable.

The augmented Dickey-Fuller test can be used to test for a unit root in a univariate process in the presence of serial correlation. The null hypothesis of the Augmented Dickey-Fuller is that there is a unit root, with the alternative that there is no unit root. If the p-value is above a critical size, then we cannot reject that there is a unit root. The p-values are obtained through regression surface approximation from MacKinnon 1994, but using the updated 2010 tables. If the p-value is close to significant, then the critical values should be used to judge whether to reject the null.

When you run the test, you'll get an ADF value and a p-value. The ADF number should be a negative number and the p-value should be beneath a certain threshold value (e.g., 1% or 5%, etc.) for a confidence level. For this example, we'll use 5% (or 95% confidence level), so if the p-value is greater than 0.05, then we say we fail to reject the null hypothesis because the data has a unit root and is non-stationary. If the p-value is less than or equal to 0.05, we can say we reject the null hypothesis because the data does not have a unit root and is stationary.

```
import statsmodels.tsa.stattools as sm
def augmented_dickey_fuller_statistics(time_series):
    result = sm.adfuller(time_series.values, autolag='AIC')
    print('ADF Statistic: %f' % result[0])
    print('p-value: %f' % result[1])
    print('Critical Values:')
    for key, value in result[4].items():
        print('\t%s: %.3f' % (key, value))
```

Calling ADF Test on 14 series:

```
print('Augmented Dickey-Fuller Test: CRUDE_PETRO Time Series')
augmented_dickey_fuller_statistics(transform_data['CRUDE_PETRO'])
print('Augmented Dickey-Fuller Test: NGAS_EUR Time Series')
augmented_dickey_fuller_statistics(transform_data['NGAS_EUR'])
print('Augmented Dickey-Fuller Test: SOYBEANS Time Series')
augmented_dickey_fuller_statistics(transform_data['SOYBEANS'])
print('Augmented Dickey-Fuller Test: MAIZE Time Series')
augmented_dickey_fuller_statistics(transform_data['MAIZE'])
print('Augmented Dickey-Fuller Test: RICE_05 Time Series')
augmented_dickey_fuller_statistics(transform_data['RICE_05'])
print('Augmented Dickey-Fuller Test: WHEAT_US_SRW Time Series')
augmented_dickey_fuller_statistics(transform_data['WHEAT_US_SRW'])
print('Augmented Dickey-Fuller Test: WHEAT_US_HRW Time Series')
augmented_dickey_fuller_statistics(transform_data['WHEAT_US_HRW'])
print('Augmented Dickey-Fuller Test: COTTON_A_INDX Time Series')
augmented_dickey_fuller_statistics(transform_data['COTTON_A_INDX'])
print('Augmented Dickey-Fuller Test: PHOSROCK Time Series')
augmented_dickey_fuller_statistics(transform_data['PHOSROCK'])
print('Augmented Dickey-Fuller Test: DAP Time Series')
augmented_dickey_fuller_statistics(transform_data['DAP'])
print('Augmented Dickey-Fuller Test: TSP Time Series')
augmented_dickey_fuller_statistics(transform_data['TSP'])
print('Augmented Dickey-Fuller Test: UREA_EE_BULK Time Series')
augmented_dickey_fuller_statistics(transform_data['UREA_EE_BULK'])
print('Augmented Dickey-Fuller Test: POTASH Time Series')
augmented_dickey_fuller_statistics(transform_data['POTASH'])
```

Output: ($p < 0.05$).

```
Augmented Dickey-Fuller Test: CRUDE_PETRO Time Series
ADF Statistic: -6.651501
p-value: 0.000000
Critical Values:
        1%: -3.445
        5%: -2.868
        10%: -2.570
Augmented Dickey-Fuller Test: NGAS_EUR Time Series
ADF Statistic: -3.591066
p-value: 0.005934
Critical Values:
        1%: -3.445
        5%: -2.868
        10%: -2.570
```

```
Augmented Dickey-Fuller Test: SOYBEANS Time Series
ADF Statistic: -16.286027
p-value: 0.000000
Critical Values:
        1%: -3.445
        5%: -2.868
        10%: -2.570
Augmented Dickey-Fuller Test: MAIZE Time Series
ADF Statistic: -9.145726
p-value: 0.000000
Critical Values:
        1%: -3.445
        5%: -2.868
        10%: -2.570
Augmented Dickey-Fuller Test: RICE_05 Time Series
ADF Statistic: -13.798323
p-value: 0.000000
Critical Values:
        1%: -3.445
        5%: -2.868
        10%: -2.570
Augmented Dickey-Fuller Test: WHEAT_US_SRW Time Series
ADF Statistic: -6.775981
p-value: 0.000000
Critical Values:
        1%: -3.445
        5%: -2.868
        10%: -2.570
Augmented Dickey-Fuller Test: WHEAT_US_HRW Time Series
ADF Statistic: -8.653836
p-value: 0.000000
Critical Values:
        1%: -3.445
        5%: -2.868
        10%: -2.570
Augmented Dickey-Fuller Test: COTTON_A_INDX Time Series
ADF Statistic: -12.084709
p-value: 0.000000
Critical Values:
        1%: -3.445
        5%: -2.868
        10%: -2.570
Augmented Dickey-Fuller Test: PHOSROCK Time Series
ADF Statistic: 3.408210
p-value: 1.000000
Critical Values:
        1%: -3.445
        5%: -2.868
        10%: -2.570
Augmented Dickey-Fuller Test: DAP Time Series
ADF Statistic: -4.249047
```

```
p-value: 0.000544
Critical Values:
        1%: -3.445
        5%: -2.868
        10%: -2.570
Augmented Dickey-Fuller Test: TSP Time Series
ADF Statistic: -3.809784
p-value: 0.002809
Critical Values:
        1%: -3.445
        5%: -2.868
        10%: -2.570
Augmented Dickey-Fuller Test: UREA_EE_BULK Time Series
ADF Statistic: -10.999361
p-value: 0.000000
Critical Values:
        1%: -3.445
        5%: -2.868
        10%: -2.570
Augmented Dickey-Fuller Test: POTASH Time Series
ADF Statistic: -9.689710
p-value: 0.000000
Critical Values:
        1%: -3.445
        5%: -2.868
        10%: -2.570
```

Except PHOSROCK, all other feature series are stationary as p-value is less than 0.05. Plot the series:

```
'''
Takes long time
'''
fig, axes = plt.subplots(nrows=7, ncols=2, dpi=120, figsize=(10,6))
for i, ax in enumerate(axes.flatten()):
    d = transform_data[transform_data.columns[i]]
    ax.plot(d, color='red', linewidth=1)
    # Decorations
    ax.set_title(dataset.columns[i])
    ax.xaxis.set_ticks_position('none')
    ax.yaxis.set_ticks_position('none')
    ax.spines['top'].set_alpha(0)
    ax.tick_params(labelsize=6)

plt.tight_layout();
```

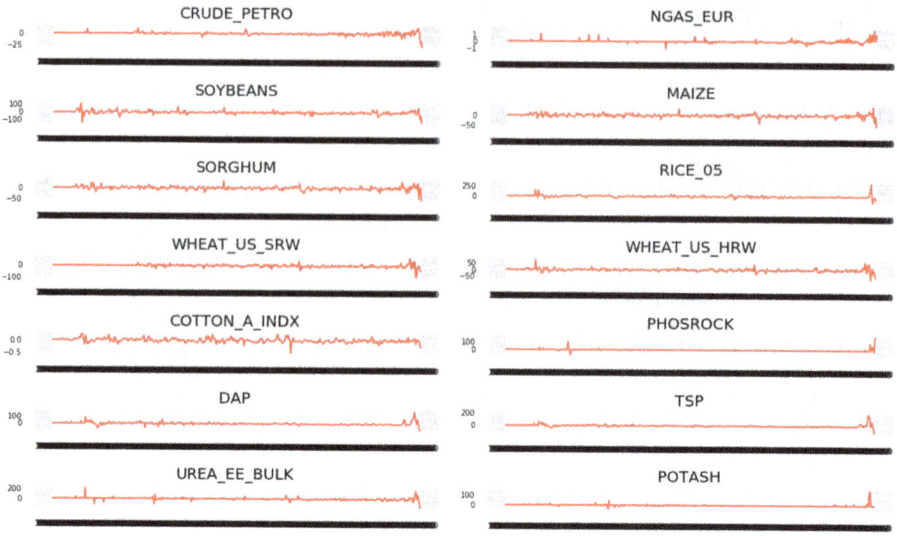

Step 13: Testing Causation using Granger's Causality Test.

Causation of all 13 series (i.e., multivariate variables influence each other) can be tested using Granger's causality test. Granger's causality tests the null hypothesis that the coefficients of past values in the regression equation is zero. In simpler terms, the past values of time series (x) do not cause the other series (y). So, if the p-value obtained from the test is lesser than the significance level of 0.05, then, you can safely reject the null hypothesis. This has been performed on original dataset.

```
import statsmodels.tsa.stattools as sm
maxlag=12
test = 'ssr-chi2test'
def grangers_causality_matrix(X_train, variables, test = 'ssr_chi2test', verbose=False):
    dataset = pd.DataFrame(np.zeros((len(variables), len(variables))), columns=variables,
index=variables)
    for c in dataset.columns:
        for r in dataset.index:
            test_result = sm.grangercausalitytests(X_train[[r,c]], maxlag=maxlag, verbose=False)
            p_values = [round(test_result[i+1][0][test][1],4) for i in range(maxlag)]
            if verbose: print(f'Y = {r}, X = {c}, P Values = {p_values}')
            min_p_value = np.min(p_values)
            dataset.loc[r,c] = min_p_value
    dataset.columns = [var + '_x' for var in variables]
    dataset.index = [var + '_y' for var in variables]
    return dataset
```

Calling Granger causality test[100]: [22].

---

[100] Testing for Granger Causality Using Python – https://rishi-a.github.io/2020/05/25/granger-causality.html

```
grangers_causality_matrix(dataset, variables = dataset.columns)
```

Output:

Test for Granger non-causality as follows.[101] Test the hypothesis that the coefficients of (only) the first p lagged values of X are zero in the Y equation, using a standard Wald test. Then do the same thing for the coefficients of the lagged values of Y in the X equation. For instance, if we take the value 0. 0.0022in (row 1, column 2), it refers to the p-value of the Granger's Causality test for SilverPrice_x causing GoldPrice_y. The 0.2459 in (row 2, column 1) refers to the p-value of GoldPrice_y causing SilverPrice_x and so on. Except for SPIndex and GoldPrice, Silver and Gold, and Silver and USD, all the other variables (time series) in the system are interchangeably causing each other. This justifies the VAR modeling approach for this system of multi-time series to forecast.

Except, COTTON_A_INDX and SOYBEANS, COTTON_A_INDX and SORGHUM, COTTON_A_INDX and WHEAT_US_SRW, COTTON_A_INDX and WHEAT_US_HRW, COTTON_A_INDX and PHOPSROCK, PHOPSROCK and SORGHUM, and PHOPSROCK and COTTON_A_INDX. That is, Cotton and All, Phosprock and Sorghum, and all the other variables (time series) in the system are interchangeably causing each other. This justifies the VAR modeling approach for this system of multi-time series to forecast.

Step 14: VAR Model.

```
import numpy as np
import pandas
import statsmodels.api as sm
from statsmodels.tsa.api import VAR

mod = VAR(transform_data)
res = mod.fit(maxlags=15, ic='aic')
print(res.summary())
```

Output:

The biggest correlations are 0.15 (CRUDE_PETRO & NGAS_EUR) and 0.14 (CRUDE_PETRO & UREA_EE_BULK); however, there are small enough to ignore in this case.

[101] Testing for Granger Causality – https://davegiles.blogspot.com/2011/04/testing-for-granger-causality.html

```
  Summary of Regression Results
==================================
Model:                         VAR
Method:                        OLS
Date:           Sun, 28, Mar, 2021
Time:                     19:31:07
--------------------------------------------------------------------
No. of Equations:         14.0000    BIC:                    66.4171
Nobs:                     445.000    HQIC:                   51.0335
Log likelihood:          -15208.6    FPE:                1.63634e+18
AIC:                      41.0183    Det(Omega_mle):     9.66995e+15
--------------------------------------------------------------------
Results for equation CRUDE_PETRO
====================================================================================
                   coefficient       std. error           t-stat            prob
------------------------------------------------------------------------------------
const                 0.110861         0.101301            1.094           0.274
L1.CRUDE_PETRO        0.164346         0.063246            2.599           0.009
L1.NGAS_EUR          -1.375772         0.901268           -1.526           0.127
```

	CRUDE_PETRO	NGAS_EUR	SOYBEANS	MAIZE	SORGHUM	RICE_05	WHEAT_US_SRW
CRUDE_PETRO	1.000000	0.151729	0.081552	-0.014359	-0.040250	-0.001927	-0.013298
NGAS_EUR	0.151729	1.000000	0.043817	0.023702	0.038340	0.059805	-0.000938
SOYBEANS	0.081552	0.043817	1.000000	0.500918	0.410883	-0.144052	0.181263
MAIZE	-0.014359	0.023702	0.500918	1.000000	0.829285	0.023395	0.276527
SORGHUM	-0.040250	0.038340	0.410883	0.829285	1.000000	-0.015696	0.340964
RICE_05	-0.001927	0.059805	-0.144052	0.023395	-0.015696	1.000000	-0.062557
WHEAT_US_SRW	-0.013298	-0.000938	0.181263	0.276527	0.340964	-0.062557	1.000000
WHEAT_US_HRW	-0.013480	0.026973	0.270257	0.357497	0.456084	-0.053405	0.722635
COTTON_A_INDX	0.085001	0.073200	0.044244	0.033515	0.118284	-0.035743	0.008163
PHOSROCK	0.049068	0.142872	-0.062618	-0.034203	-0.073132	0.070481	-0.089467
DAP	0.067699	0.125243	-0.047840	0.101677	0.091375	0.148455	0.172788
TSP	0.017709	0.022752	-0.066485	0.073193	0.023644	0.244528	0.147322
UREA_EE_BULK	0.143906	0.176861	-0.033694	-0.008852	-0.008960	-0.023203	-0.100842
POTASH	0.051824	0.092136	0.020331	-0.034857	-0.016358	0.058403	-0.262801

	WHEAT_US_HRW	COTTON_A_INDX	PHOSROCK	DAP	TSP	UREA_EE_BULK	POTASH
CRUDE_PETRO	-0.013480	0.085001	0.049068	0.067699	0.017709	0.143906	0.051824
NGAS_EUR	0.026973	0.073200	0.142872	0.125243	0.022752	0.176861	0.092136
SOYBEANS	0.270257	0.044244	-0.062618	-0.047840	-0.066485	-0.033694	0.020331
MAIZE	0.357497	0.033515	-0.034203	0.101677	0.073193	-0.008852	-0.034857
SORGHUM	0.456084	0.118284	-0.073132	0.091375	0.023644	-0.008960	-0.016358
RICE_05	-0.053405	-0.035743	0.070481	0.148455	0.244528	-0.023203	0.058403
WHEAT_US_SRW	0.722635	0.008163	-0.089467	0.172788	0.147322	-0.100842	-0.262801
WHEAT_US_HRW	1.000000	0.057742	-0.068595	0.154623	0.105322	-0.086663	-0.211120
COTTON_A_INDX	0.057742	1.000000	-0.045294	0.077388	-0.087075	0.103948	-0.043603
PHOSROCK	-0.068595	-0.045294	1.000000	0.031650	0.082697	0.167753	0.097705
DAP	0.154623	0.077388	0.031650	1.000000	0.488143	0.222275	-0.020747
TSP	0.105322	-0.087075	0.082697	0.488143	1.000000	0.042454	-0.072386
UREA_EE_BULK	-0.086663	0.103948	0.167753	0.222275	0.042454	1.000000	0.343684
POTASH	-0.211120	-0.043603	0.097705	-0.020747	-0.072386	0.343684	1.000000

Residual Values.

```
y_fitted = res.fittedvalues
y_fitted
```

Residual Plot.

```
y_fitted = res.fittedvalues
residuals = res.resid
plt.figure(figsize = (15,5))
plt.plot(residuals, label='resid')
plt.plot(y_fitted, label='VAR prediction')
plt.xlabel('Date')
plt.xticks(rotation=45)
plt.ylabel('Residuals')
plt.grid(True)
```

Output:

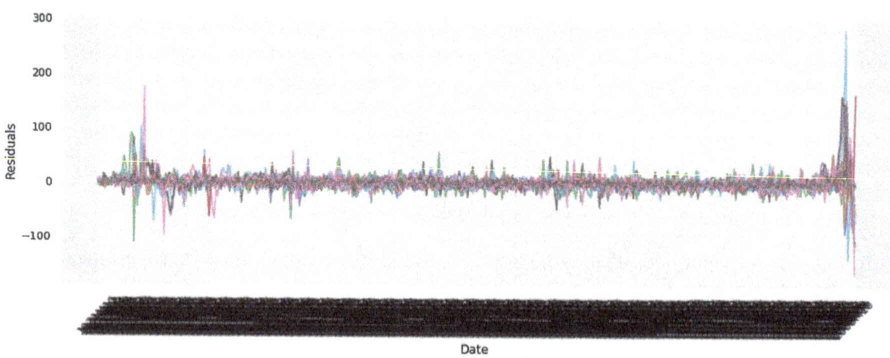

Residual plot looks normal with constant mean throughout apart from some large fluctuation during 1960, 2010, 2014, etc.

Causation models[102] are at the center of solving agriculture challenges: The agribusiness production cycles, environmental sustainability, climate change, and food security are current demands that challenge food producers, distributors, and consumers. This line of study will focus on cause-and-effect models for agricultural production chains, especially for small producers. The goal will be to use advanced probabilistic models for decision-making based on cause and effect, addressing many sources of concern, such as food and water waste. [31]

Step 15: Durbin-Watson statistic.

Durbin-Watson statistic is related to auto correlation. The Durbin-Watson statistic will always have a value between 0 and 4. A value of 2.0 means that there is no autocorrelation detected in the sample. Values from 0 to less than 2 indicate positive autocorrelation and values from 2 to 4 indicate negative autocorrelation. A rule of thumb is that test statistic values in the range of 1.5 to 2.5 are relatively normal. Any value outside this range could be a cause for concern.

```
from statsmodels.stats.stattools import durbin_watson
out = durbin_watson(res.resid)
for col,val in zip(transform_data.columns, out):
    print((col), ":", round(val,2))
```

Output:

```
CRUDE_PETRO : 2.04
NGAS_EUR : 2.01
SOYBEANS : 1.98
MAIZE : 2.02
SORGHUM : 2.04
RICE_05 : 2.01
WHEAT_US_SRW : 2.02
WHEAT_US_HRW : 1.98
COTTON_A_INDX : 1.97
PHOSROCK : 2.02
DAP : 2.02
TSP : 2.05
UREA_EE_BULK : 2.04
POTASH : 2.07
```

There is no autocorrelation (2.0) exist.

Step 15: Prediction.

To forecast, the VAR model expects up to the lag order number of observations from the past data. This is because, the terms in the VAR model are essentially the lags of the various time series in the dataset, so need to provide as many of the previous values as indicated by the lag order used by the model.

[102] C4AI – A New Center for Artificial Intelligence to accelerate science and innovation in Brazil- https://www.ibm.com/blogs/research/2020/10/c4ai-a-new-center-for-artificial-intelligence-to-accelerate-science-and-innovation-in-brazil/

```
# Get the lag order
lag_order = res.k_ar
print(lag_order)
# Input data for forecasting
input_data = transform_data.values[-lag_order:]
print(input_data)
# forecasting
pred = res.forecast(y=input_data, steps=n_obs)
pred = (pd.DataFrame(pred, index=X_test.index, columns=X_test.columns + '_pred'))
print(pred)
```

Output:

```
         CRUDE_PETRO_pred  NGAS_EUR_pred  SOYBEANS_pred  MAIZE_pred  \
Date
2008M11         74.655760      -1.061364     -64.755043  -67.916002
2008M12         78.127235       0.244026      -6.086017   -2.613479
2009M01         73.835742       0.150013      93.762644  -10.077916
2009M02         80.669756      -0.235508     133.718311   24.102612
2009M03         89.438261       0.700204     -25.242655   31.562856
2009M04        102.935299       0.402764    -152.554977   18.915067
2009M05        118.512058      -0.601774    -133.762132  -14.691831
2009M06        116.853798      -0.237653    -187.876966  -14.741226
2009M07        120.574897       0.947157    -108.843358   20.082399
2009M08        113.840758      -0.259518      14.721539   14.541460
2009M09        112.195161      -0.826559      70.271844   28.213955
2009M10        115.878353       1.851453      -9.577520   32.450435
2009M11        114.686216      -0.199584      86.280707   19.200758
2009M12        108.094501      -0.358468     134.664893   -1.351444
2010M01         90.112569       0.739885     123.691271   18.524488

         SORGHUM_pred  RICE_05_pred  WHEAT_US_SRW_pred  WHEAT_US_HRW_pred  \
Date
2008M11    -28.334947   -181.870971          13.388964         -47.113119
2008M12     20.193620   -155.200453          75.757086          66.976083
2009M01      6.428855   -134.080165         126.837292         106.040537
2009M02     36.598296    -90.444142          58.877866          73.685927
2009M03     38.593708    116.784892         -18.763543          46.246697
2009M04     27.736473     59.786752         -69.721766         -21.581973
2009M05     -9.994391    -81.659871         -82.929756         -43.483554
2009M06    -41.696942   -110.435662          37.630161          29.245752
2009M07     -9.208961     95.779299          51.244869          -3.520403
2009M08     -1.196982    178.385708         -78.461825         -61.840698
2009M09     15.366800    108.007670         -57.101227         -30.568545
2009M10     20.026997     -0.677231         -22.385949         -52.905408
2009M11     16.161011    -98.593826          37.222326          41.883765
```

Invert the Transformation to Get the Real Forecast.

The forecasts are generated, but it is on the scale of the training data used by the model. So, to bring it back up to its original scale, we need to de-difference it.

The way to convert the differencing is to add these differences consecutively to the base number. An easy way to do it is to first determine the cumulative sum at index and then add it to the base number.

This process can be reversed by adding the observation at the prior time step to the difference value. Inverted(ts) = differenced(ts) + observation(ts−1).

```
#combining predicted and real data set
combine = pd.concat([output['CRUDE_PETRO_pred'], X_test['CRUDE_PETRO']], axis=1)
combine['accuracy'] = round(combine.apply(lambda row: row.CRUDE_PETRO_pred
/row.CRUDE_PETRO *100, axis = 1),2)
combine['accuracy'] = pd.Series(["{0:.2f}%".format(val) for val in combine['accuracy']],index =
combine.index)
combine = combine.round(decimals=2)
combine = combine.reset_index()
combine = combine.sort_values(by='Date', ascending=False)

print(combine)
```

Output:

```
       Date  CRUDE_PETRO_pred  CRUDE_PETRO accuracy
14  2010M01             90.11        77.12  116.85%
13  2009M12            108.09        74.88  144.36%
12  2009M11            114.69        77.55  147.89%
11  2009M10            115.88        74.08  156.42%
10  2009M09            112.20        68.35  164.15%
9   2009M08            113.84        71.63  158.93%
8   2009M07            120.57        64.67  186.45%
7   2009M06            116.85        69.15  168.99%
6   2009M05            118.51        58.15  203.80%
5   2009M04            102.94        50.28  204.72%
4   2009M03             89.44        46.65  191.72%
3   2009M02             80.67        41.84  192.81%
2   2009M01             73.84        43.86  168.34%
1   2008M12             78.13        41.34  188.99%
0   2008M11             74.66        53.97  138.33%
```

The VAR model is a popular tool for the purpose of predicting joint dynamics of multiple time series based on linear functions of past observations. You can download the code and run using FertilizerGrangerCausalityTest Jupyter notebook.

VAR model is very useful when predicting multivariate time series. Nonetheless, if we wanted to predict price of fertilizer using supply side and demand side influencers, we can use regression or ensemble models.

Software code for this model can be found on GitHub link: UREA_EE_BULK_Fertilizer_PricePredict.Ipynb (Jupyter notebook).

Fertilizer Price Prediction Using Commodity (RICE, SORGHUM, MAIZE, and WHEAT) and Oil Prices

The ML mode to predict fertilizer cost is based on the following demand and supply components[103] [100]:

- Demand Components.
    - The demand for corn (as demand goes high, the need to produce more required and hence to apply more fertilizers to get optimal yield).
    - Demand for wheat.
    - Demand for rice.
    - Soybeans.
- Supply: CRUDE (on supply side).
    - On supply side, consider cost of CRUDE.
    - Natural Gas is complementing of CRUDE – same behavior could be expected from demand swings of Natural Gas.

Data Sources

Data retrieved from the World Bank Commodity Market's Pink Sheet data that gets published every month.[104]

Step 1: Load required Libraries.
The first step in predicting urea fertilizer cost is to load all libraries and Pink Sheet data.

```
# Importing data

CMOHistoricalDataDF = pd.read_csv("CMO-Historical-Data-MonthlyRegressionFinal.csv")

CMOHistoricalDataDF
```

[103] Predicting Fertilizer Prices – https://www.agmanager.info/sites/default/files/pdf/FertilizerProjection_2021-2022.pdf

[104] Pink Sheet Data – https://www.worldbank.org/en/research/commodity-markets

Output:

	Year	NGAS_US US ($/mmbtu)	SOYBEANS ($/mt)	CRUDE_PETRO ($/bbl)	MAIZE ($/mt)	SORGHUM ($/mt)	RICE_05 ($/mt)	WHEAT_US_SRW ($/mt)	WHEAT_US_HRW ($/mt)	UREA_EE_BULK ($/mt)
0	197901	1.02	284.00	17.45	105.41	97.06	278.52	140.36	137.79	143.00
1	197902	1.05	298.00	20.75	107.01	97.83	279.50	144.77	140.36	143.00
2	197903	1.10	310.00	22.02	109.52	99.21	293.00	144.04	139.99	131.88
3	197904	1.11	300.00	22.43	111.49	98.27	295.46	142.20	139.99	135.00
4	197905	1.15	300.00	33.50	112.57	100.49	296.78	141.43	143.30	141.75
...	...	...	...	...	...	...	...	...	...	...
501	202010	2.25	454.25	39.90	186.75	NaN	471.00	245.20	NaN	245.00
502	202011	2.59	499.98	42.30	190.38	NaN	489.00	247.95	NaN	245.00
503	202012	2.54	510.94	48.73	198.77	NaN	520.00	251.15	NaN	245.00
504	202101	2.67	576.30	53.60	234.47	NaN	545.00	276.45	NaN	265.00
505	202102	5.07	574.80	60.46	245.24	NaN	557.00	276.63	NaN	335.00

The Pink Sheet data is loaded from January 1979 to December 2020.

Step 2: Statistical properties of Data.

```
CMOHistoricalDataDF.describe()
```

Output:

	Year	NGAS_US US ($/mmbtu)	SOYBEANS ($/mt)	CRUDE_PETRO ($/bbl)	MAIZE ($/mt)	SORGHUM ($/mt)	RICE_05 ($/mt)	WHEAT_US_SRW ($/mt)	WHEAT_US_HRW ($/mt)	UREA_EE_BULK ($/mt)
count	506.000000	506.000000	506.000000	506.000000	506.000000	500.000000	506.000000	506.000000	498.000000	506.000000
mean	199964.978261	3.233636	322.219960	42.101403	139.801344	134.967520	337.408834	168.935119	180.688574	185.365059
std	1218.367423	2.014272	109.708207	28.549572	54.484496	51.105978	123.794435	57.446569	62.307556	109.216730
min	197901.000000	1.020000	183.000000	9.620000	65.310000	62.530000	163.750000	85.300000	101.780000	62.500000
25%	198907.250000	1.900000	241.000000	18.907500	103.667500	98.935000	251.265000	130.090000	140.005000	103.500000
50%	200001.500000	2.580000	283.500000	30.890000	118.835000	114.470000	303.750000	153.165000	164.240000	155.200000
75%	201007.750000	3.895000	383.560000	58.147500	163.300000	161.930000	418.940000	197.472500	199.865000	242.972500
max	202102.000000	13.520000	684.020000	132.830000	333.050000	302.530000	907.000000	419.610000	439.720000	785.000000

Count of records 506.

We might get a clearer idea of the distribution of production (millions of tons) values by visualizing the data. Common plot types for visualizing numeric data distributions are histograms and box plots.

```
import pandas as pd
import matplotlib.pyplot as plt

# This ensures plots are displayed inline in the Jupyter notebook
%matplotlib inline

# Get the label column
label = CMOHistoricalDataDF['UREA_EE_BULK ($/mt)']

# Create a figure for 2 subplots (2 rows, 1 column)
fig, ax = plt.subplots(2, 1, figsize = (9,12))

# Plot the histogram
ax[0].hist(label, bins=100)
ax[0].set_ylabel('Frequency')

# Add lines for the mean, median, and mode
ax[0].axvline(label.mean(), color='magenta', linestyle='dashed', linewidth=2)
ax[0].axvline(label.median(), color='cyan', linestyle='dashed', linewidth=2)

# Plot the boxplot
ax[1].boxplot(label, vert=False)
ax[1].set_xlabel('UREA_EE_BULK ($/mt)')

# Add a title to the Figure
fig.suptitle('UREA_EE_BULK ($/mt)')

# Show the figure
fig.show()
```

Output:

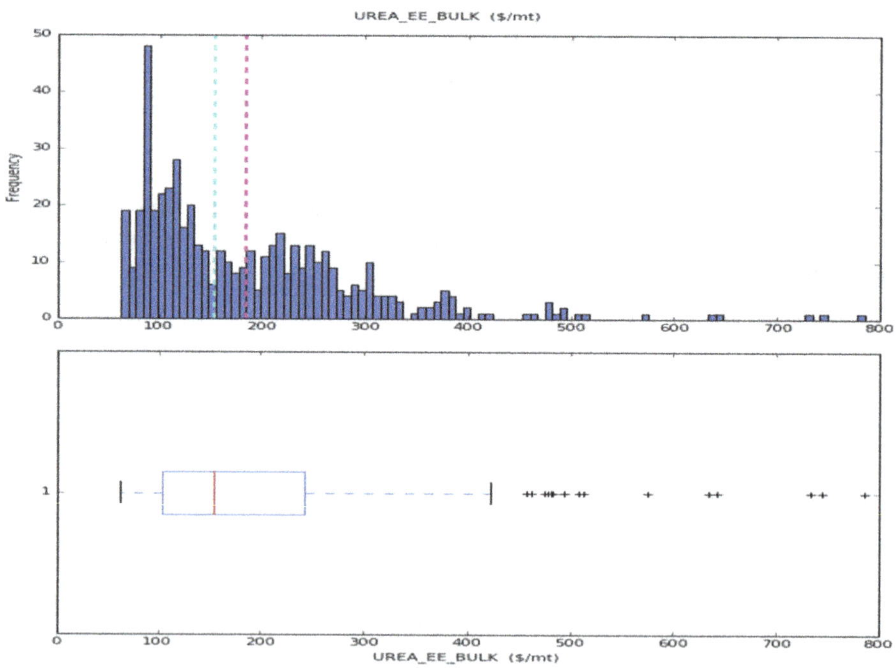

As you can see, the usage of urea is right skewed—evidently, peaks in production and price of urea have seen in the 2000s and 2010s.

Below are the price movements (please see Fig. 4.25) of UREA.[105]

And the following are the best prices for farmers—that is, cost of buying UREA is lowest[106]:

Month-Year	Price ($)
Jun-2016	142.63
May-2017	178.75
Jul-2016	181
Jul-2017	181
Aug-2016	186.25

Correlation analysis of production to all other features provide the following details:

```
plt.figure(figsize=(12, 9))
sns.heatmap(CMOHistoricalDataDF.corr())
```

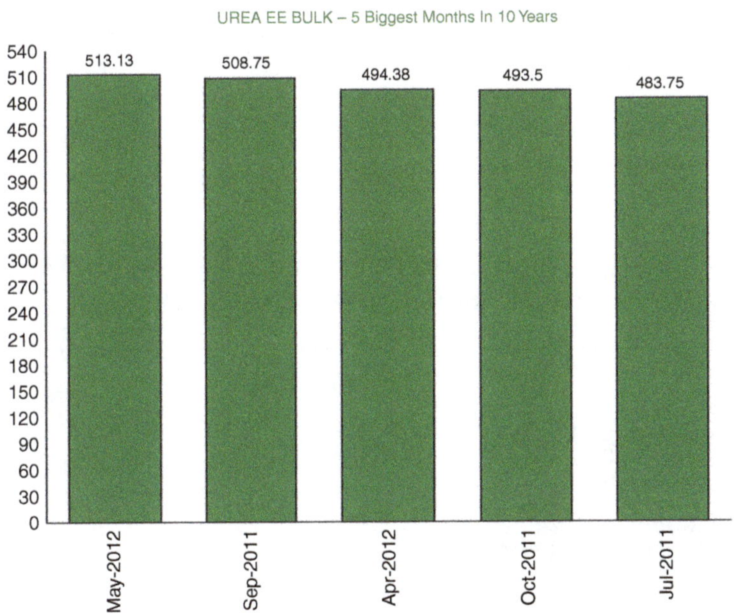

Fig. 4.25 Urea Prices

[105] UREA Price Movements – https://www.netcials.com/commodity-monthly-price-history/UREA-EE-BULK/

[106] Lower Prices – https://www.netcials.com/commodity-monthly-price-history/UREA-EE-BULK/

Output:

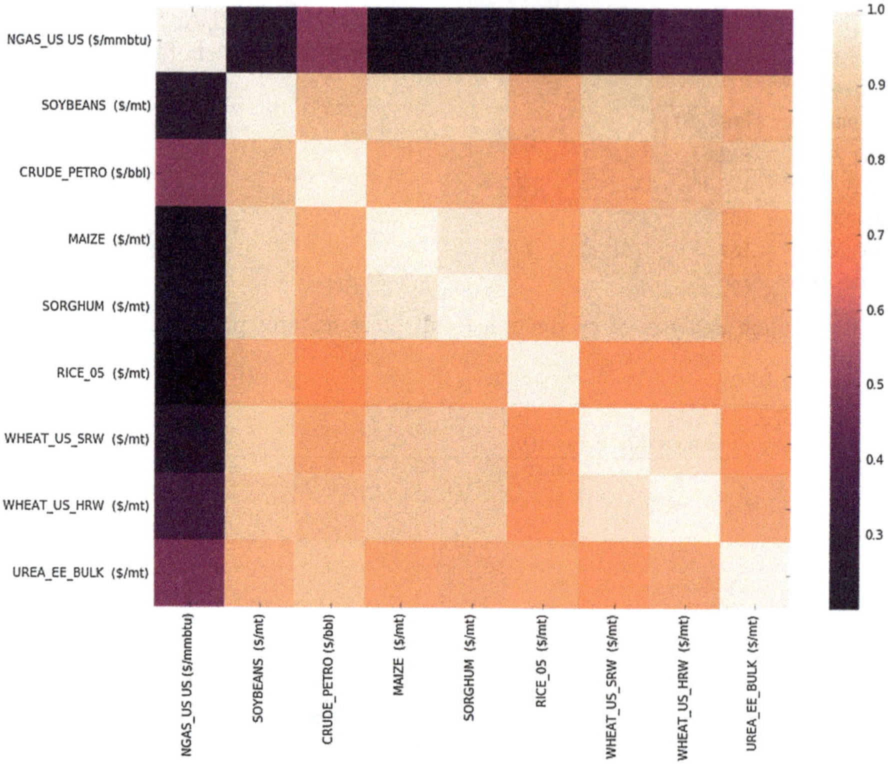

Highly correlated map between UREA and Commodity Prices.
Scatter plot provides the distribution of Feature and UREA usage:

```
fig = go.Figure()

# Add traces
fig.add_trace(go.Scatter(x=CMOHistoricalDataDF.index,
y=CMOHistoricalDataDF['UREA_EE_BULK ($/mt)'],
                mode='lines',
                name='UREA_EE_BULK ($/mt) '))
fig.update_layout(
   title="UREA_EE_BULK ($/mt)")

fig.show()
```

Output:

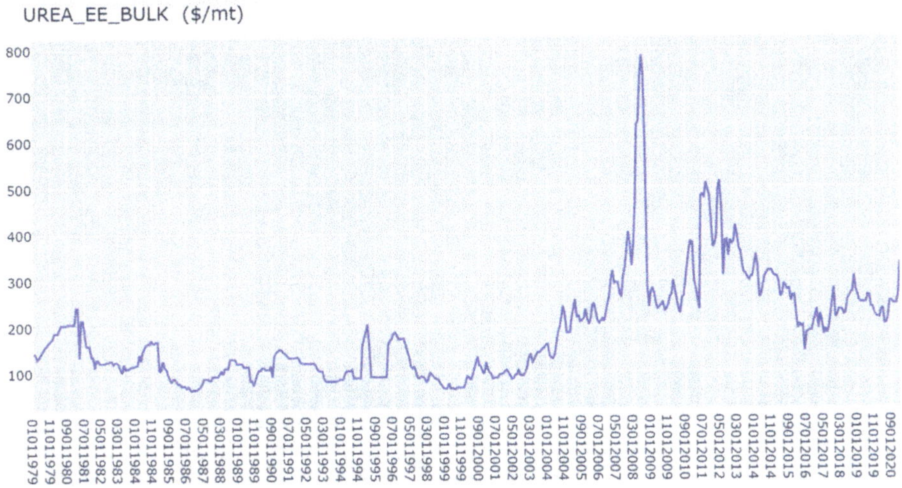

Urea peaked in 2008 to a $780 per metric tons.

```
fig = go.Figure()

# Add traces
fig.add_trace(go.Scatter(x=CMOHistoricalDataDF.index, y=CMOHistoricalDataDF['CRUDE_PETRO ($/bbl)'],
              mode='lines',
              name='CRUDE_PETRO ($/bbl)'))
fig.update_layout(
    title='CRUDE_PETRO ($/bbl)')

fig.show()
```

Crude Petrol had more fluctuations:

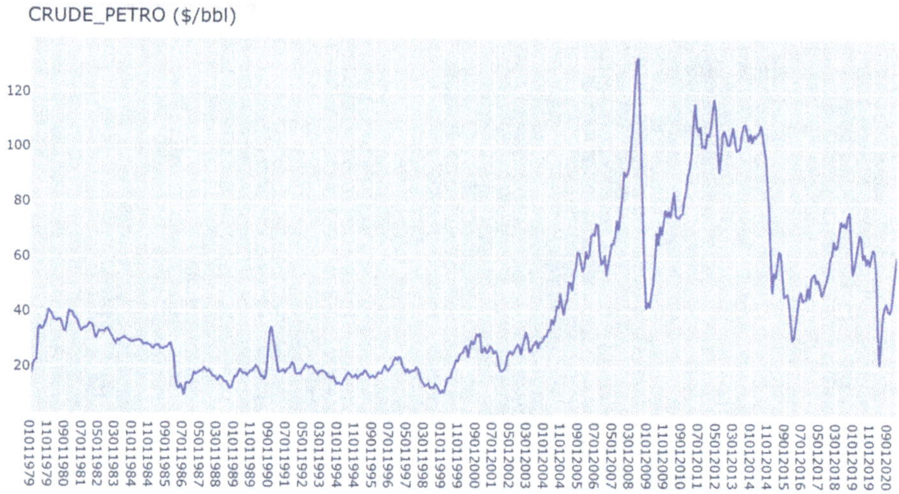


```
fig = go.Figure()

# Add traces
fig.add_trace(go.Scatter(x=CMOHistoricalDataDF.index, y=CMOHistoricalDataDF['MAIZE  ($/mt)'],
                mode='lines',
                name='MAIZE  ($/mt)'))
fig.update_layout(
   title="MAIZE  ($/mt)")

fig.show()
```

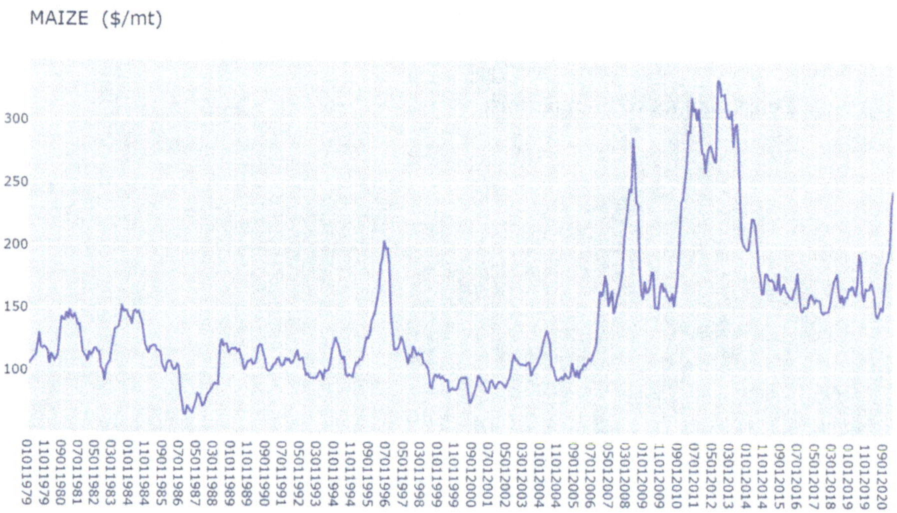

Perform linear regression:
Load Feature and Label Data Frames for performing regression task:

```
X, y = CMOHistoricalDataDF[['SOYBEANS  ($/mt)', 'CRUDE_PETRO ($/bbl)',
    'MAIZE  ($/mt)', 'SORGHUM  ($/mt)', 'RICE_05  ($/mt)',
    'WHEAT_US_SRW                  ($/mt)',        'WHEAT_US_HRW                  ($/mt)']].values,
CMOHistoricalDataDF['UREA_EE_BULK  ($/mt)'].values
print('Features:',X[:10], '\nLabels:', y[:10], sep='\n')
```

Output:

```
Features:
[[284.   17.45 105.41  97.06 278.52 140.36 137.79]
 [298.   20.75 107.01  97.83 279.5  144.77 140.36]
 [310.   22.02 109.52  99.21 293.   144.04 139.99]
 [300.   22.43 111.49  98.27 295.46 142.2  139.99]
 [300.   33.5  112.57 100.49 296.78 141.43 143.3 ]
 [322.   34.67 120.35 109.46 303.31 163.88 167.92]
 [322.   33.5  130.51 118.9  304.05 168.29 174.53]
 [302.   34.   118.85 111.38 328.43 162.77 169.02]
 [292.   35.5  117.96 112.6  338.17 164.98 174.16]
 [283.   37.25 119.1  116.2  340.38 163.1  178.21]]

Labels:
[143.   143.   131.88 135.   141.75 148.5  155.4  160.   163.75 167.5 ]
```

Perform Data Distribution Analysis.

```
# Create a function that we can re-use
def show_distribution(var_data,colname):
    from matplotlib import pyplot as plt

    # Get statistics
    min_val = var_data.min()
    max_val = var_data.max()
    mean_val = var_data.mean()
    med_val = var_data.median()
    mod_val = var_data.mode()[0]


print('Minimum:{:.2f}\nMean:{:.2f}\nMedian:{:.2f}\nMode:{:.2f}\nMaximum:{:.2f}\n'.format(min_v
al,
                                                                mean_val,
                                                                med_val,
                                                                mod_val,
                                                                max_val))

    # Create a figure for 2 subplots (2 rows, 1 column)
    fig, ax = plt.subplots(2, 1, figsize = (10,4))
    # Plot the histogram
    ax[0].hist(var_data)
    ax[0].set_ylabel('Frequency')
    # Add lines for the mean, median, and mode
    ax[0].axvline(x=min_val, color = 'gray', linestyle='dashed', linewidth = 2)
    ax[0].axvline(x=mean_val, color = 'cyan', linestyle='dashed', linewidth = 2)
    ax[0].axvline(x=med_val, color = 'red', linestyle='dashed', linewidth = 2)
    ax[0].axvline(x=mod_val, color = 'yellow', linestyle='dashed', linewidth = 2)
    ax[0].axvline(x=max_val, color = 'gray', linestyle='dashed', linewidth = 2)

    # Plot the boxplot
    ax[1].boxplot(var_data, vert=False)
    ax[1].set_xlabel('Value')

    # Add a title to the Figure
    fig.suptitle(colname + ' Data Distribution')

    # Show the figure
    fig.show()
# Get the variable to examine

for col in ['SOYBEANS  ($/mt)', 'CRUDE_PETRO ($/bbl)',
      'MAIZE  ($/mt)', 'SORGHUM  ($/mt)', 'RICE_05  ($/mt)',
      'WHEAT_US_SRW  ($/mt)', 'WHEAT_US_HRW  ($/mt)', 'UREA_EE_BULK  ($/mt)']:
    show_distribution(CMOHistoricalDataDF[col],col)

# Call the function
```

Output:

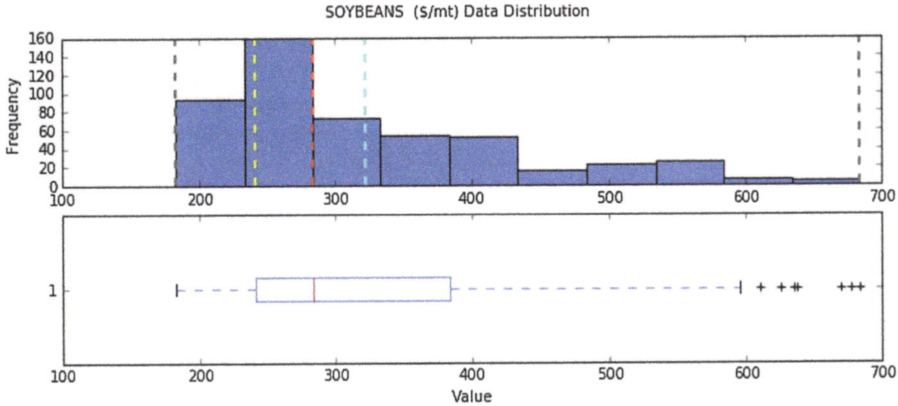

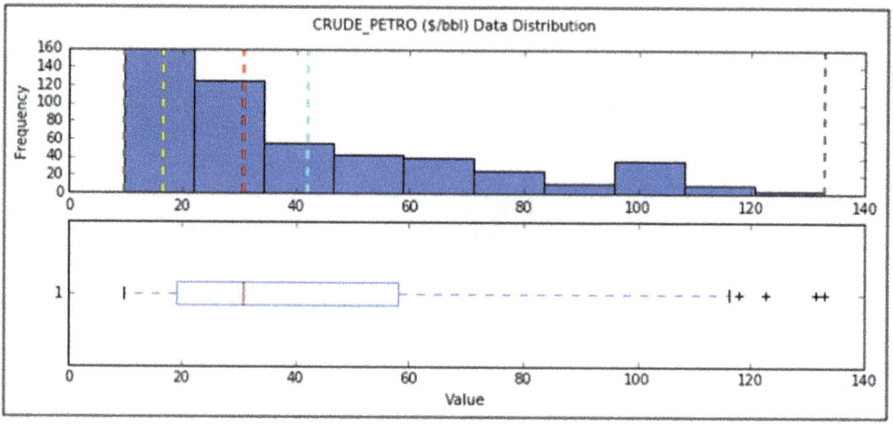
CRUDE_PETRO ($/bbl) Data Distribution
Frequency
Value

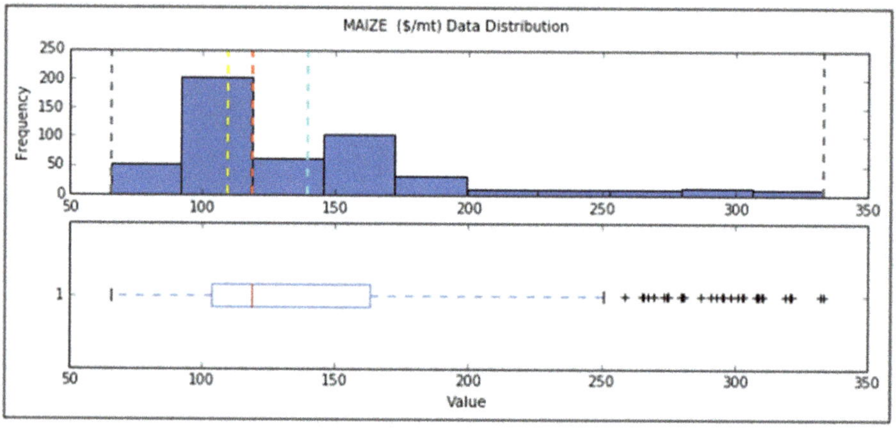
MAIZE ($/mt) Data Distribution
Frequency
Value

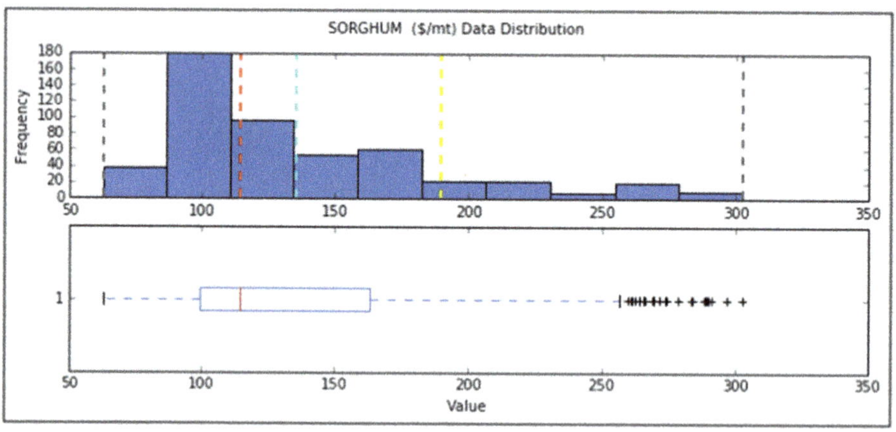
SORGHUM ($/mt) Data Distribution
Frequency
Value

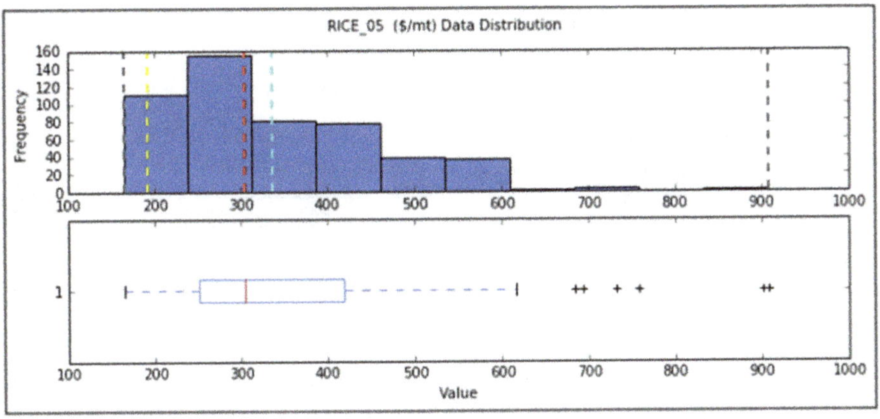
RICE_05 ($/mt) Data Distribution
Frequency
0 20 40 60 80 100 120 140 160
100 200 300 400 500 600 700 800 900 1000
1
Value

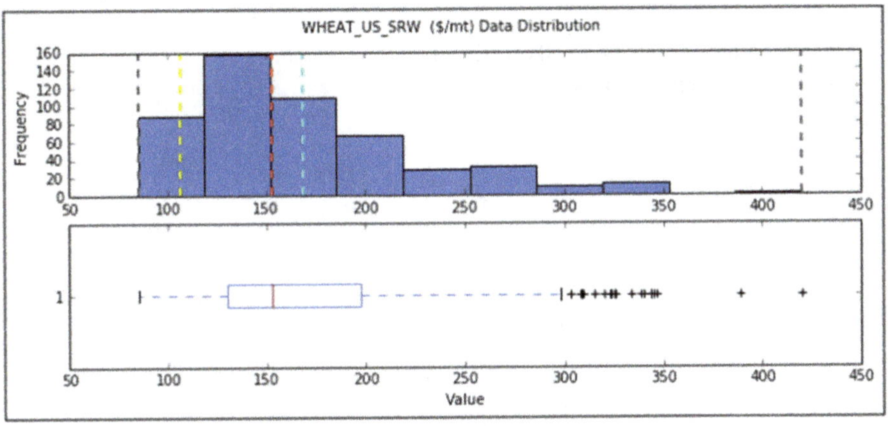
WHEAT_US_SRW ($/mt) Data Distribution
Frequency
0 20 40 60 80 100 120 140 160
50 100 150 200 250 300 350 400 450
1
Value

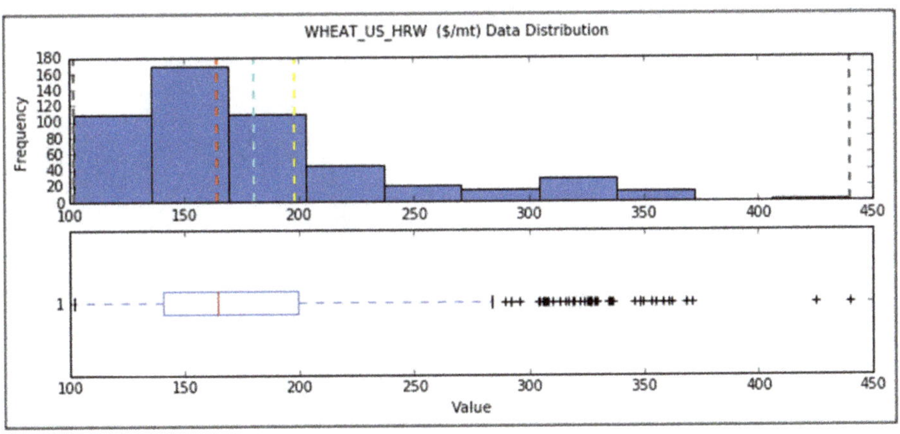
WHEAT_US_HRW ($/mt) Data Distribution
Frequency
0 20 40 60 80 100 120 140 160 180
100 150 200 250 300 350 400 450
1
Value

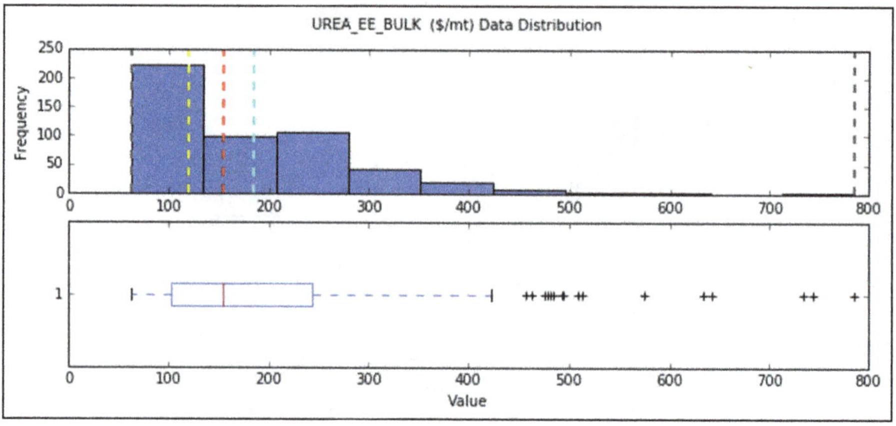

All the feature variables are skewed right of the graph.
Perform Distribution analysis:

```
def show_density(var_data,colname):
    from matplotlib import pyplot as plt

    fig = plt.figure(figsize=(10,4))

    # Plot density
    var_data.plot.density()

    # Add titles and labels
    plt.title(colname + '  Density')

    # Show the mean, median, and mode
    plt.axvline(x=var_data.mean(), color = 'cyan', linestyle='dashed', linewidth = 2)
    plt.axvline(x=var_data.median(), color = 'red', linestyle='dashed', linewidth = 2)
    plt.axvline(x=var_data.mode()[0], color = 'yellow', linestyle='dashed', linewidth = 2)

    # Show the figure
    plt.show()

# Get the density of Grade

for col in ['NGAS_US US ($/mmbtu)', 'SOYBEANS  ($/mt)', 'CRUDE_PETRO ($/bbl)',
        'MAIZE  ($/mt)', 'SORGHUM  ($/mt)', 'RICE_05  ($/mt)',
        'WHEAT_US_SRW  ($/mt)', 'WHEAT_US_HRW  ($/mt)', 'UREA_EE_BULK  ($/mt)']:
    show_density(CMOHistoricalDataDF[col],col)
```

Output:

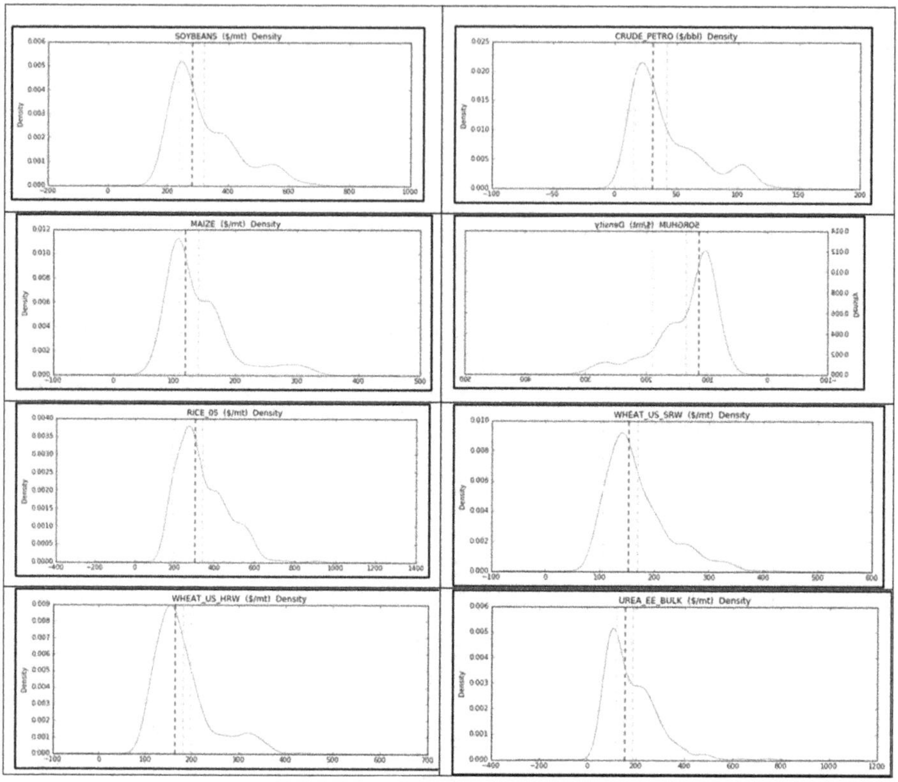

Prepare regression data by train and split:

```
X_train,X_test, y_train , y_test = train_test_split(X ,y, test_size = 0.2,random_state = 0)

regressor = LinearRegression()
regressor.fit(X_train,y_train)

# predicting the test
y_pred = regressor.predict(X_test)
y_pred
```

Plot the prediction:

```
import matplotlib.pyplot as plt

%matplotlib inline

plt.scatter(y_test, y_pred)
plt.xlabel('Actual Labels')
plt.ylabel('Predicted Labels')
plt.title('Milk Production (in Million Tonnes)')
# overlay the regression line
z = np.polyfit(y_test, y_pred, 1)
p = np.poly1d(z)
plt.plot(y_test,p(y_test), color='magenta')
plt.show()
```

Output:

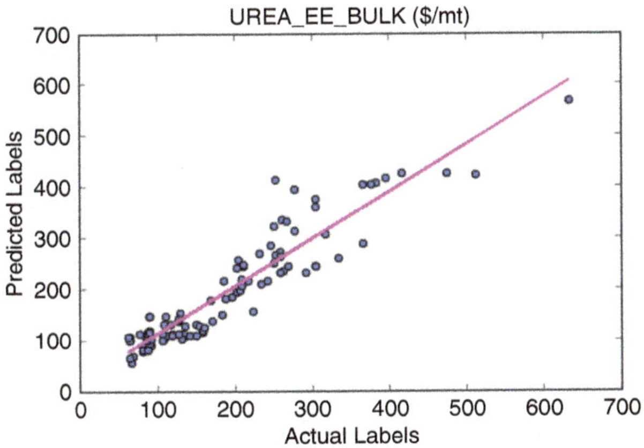

As you can see the actual and predicted values follow the linear prediction curve. Perform prediction to actual summary:

```
pred_y_df=pd.DataFrame({'Actual Value':y_test, 'Predicted value':y_pred,'Difference': y_test-y_pred})
pred_y_df
```

Output:

	Actual Value	Predicted value	Difference
0	210.00	245.005245	-35.005245
1	262.00	333.660915	-71.660915
2	135.00	126.877903	8.122097
3	382.50	408.368254	-25.868254
4	109.50	102.039228	7.460772
...	...	...	...
97	110.50	147.619785	-37.119785
98	217.75	216.860138	0.889862
99	116.50	130.170142	-13.670142
100	172.25	136.048094	36.201906
101	92.50	104.190909	-11.690909

102 rows × 3 columns

Evaluate Intercept Values:

```
'''
'SOYBEANS ($/mt)', 'CRUDE_PETRO ($/bbl)',
    'MAIZE ($/mt)', 'SORGHUM ($/mt)', 'RICE_05 ($/mt)',
    'WHEAT_US_SRW ($/mt)', 'WHEAT_US_HRW ($/mt)'
    '''
regressor.coef_
```

Output:

```
array([ 0.0280269 , 2.09727059, -0.11114963, 0.37699977, 0.29655391,
    -0.46832559, 0.34333731])
```

Please note: Except for MAIZE and WHEAT_US_SRW, and other coefficients are positive, implies any change in the positive demands will have increase in price for urea.

Check Model Score:

```
from sklearn.metrics import mean_squared_error, r2_score

mse = mean_squared_error(y_test, y_pred)
print("MSE:", mse)

rmse = np.sqrt(mse)
print("RMSE:", rmse)

r2 = r2_score(y_test, y_pred)
print("R2:", r2)
```

Output:

```
MSE: 1564.520028729579
RMSE: 39.554014065952636
R2: 0.8669524827821189
```

Perform explainability of the model (please see Figs. 4.26 and 4.27).

The following are major global and local influencers when performed explainability: CRUDE PETRO, RICE_05, WHEAT_US_SRW, and WHEAT_US_HRW. Rice fields have huge influence on UREA usage.[107]

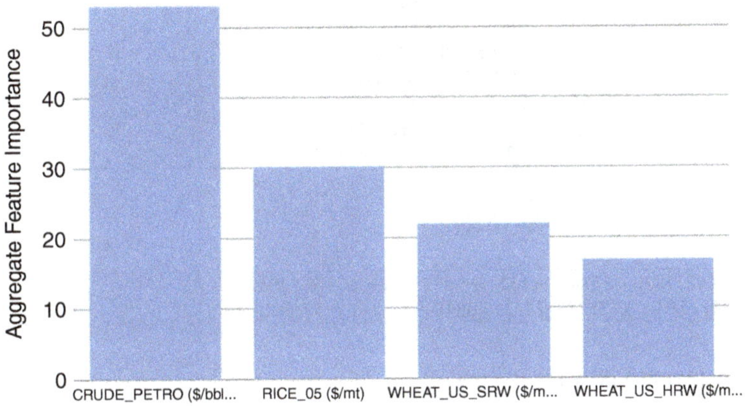

Fig. 4.26 Model explainability

Contribution?	Feature
+82.887	RICE_05 ($/mt)
+48.191	WHEAT_US_HRW ($/mt)
+43.518	CRUDE_PETRO ($/bbl)
+36.882	SORGHUM ($/mt)
+8.352	SOYBEANS ($/mt)
-11.894	MAIZE ($/mt)
-29.767	<BIAS>
-67.799	WHEAT_US_SRW ($/mt)

Fig. 4.27 Model explainability

[107] Urea fertilization and the N-cycle of rice-fields in the Camargue (S. France) – https://link.springer.com/article/10.1023/A:1003247327586#citeas

What if Analysis

Let's look how the following what if analysis would help small farmer.

Demand Spike for CRUDE Oil

What if demand for crude oil goes to $150.29 $/bbl in next 3 months: let input increase in demand for major commodities and predict the price of UREA:

```
WIF_X_test = [[267.  , 150.29, 109.4 , 111.88, 512.5 , 140.14, 195.16],
       [449.35,  74.88, 164.56, 166.35, 591.  , 206.5 , 206.25],
       [338.  ,  17.88, 124.4 , 118.  , 304.5 , 158.62, 183.49],
       [684.02, 105.27, 331.99, 273.34, 567.75, 333.73, 349.4 ],
       [223.  ,  26.38, 114.91,  99.01, 191.12, 117.77, 128.6 ],
       [252.  ,  36.58, 108.2 , 114.5 , 395.97, 155.8 , 156.53],
       [360.93,  57.67, 163.59, 147.93, 430.  , 197.52, 181.15],
       [334.  ,  65.16, 160.23, 150.01, 317.6 , 180.78, 195.72]]
```

Output: Urea cost goes to $476.29.

```
array([476.29221529, 333.66091462, 126.87790326, 408.36825409,
       102.03922783, 183.35776574, 236.09484681, 231.71579693])
```

The cost would go $476.

Demand Goes Down

What if demand for crude oil goes down in the next 3 months: the urea price goes to $ 224.61.

Output:

```
array([224.61974434, 333.66091462, 126.87790326, 408.36825409,
       102.03922783, 183.35776574, 236.09484681, 231.71579693])
```

The advantage of using ML to predict urea price is easy integration with the market and faster closed loop to small farmer.

Able to predict fertilizer price using machine learning and artificial intelligence will save lives of small farmers in the developing countries. Having ability to analyze the historical trends in price data of oil, staple crops, and major fertilizers would provide ML model all the avenue of predicting the price of fertilizer in a lead of one or two agricultural calendars. This is easy part! Disseminating these insights on a timely basis and enabling small farmers get these details and, importantly, having these insights embed into policy makers manual of governance is the most difficult part, the one require help from human counter parts.

Chapter Summary

After reading the chapter you should comfortably answer:

- Demand and supply framework.
- ML modeling of stocks to use ratio for predicting commodity items prices.
- Machine learning modeling of time series data and vector autoregression (VAR) model.
- Granger's causality.
- Durbin-Watson statistic.
- Fertilizers, crude oil, and agricultural commodity model.
- World Bank Commodity Market Data—Pink Sheet.
- Fertilizer price predict ML model.

References

1. Nigatu, G., Badau, F., Seeley, R., Hansen, J.: Factors Contributing to Changes in Agricultural Commodity Prices and Trade for the United States and the World. https://ageconsearch.umn.edu/record/301181/files/err-272.pdf (2020, January), Accessed Dec 2020
2. Tomek, W.G., Robinson, K.L.: Agricultural product prices. ASIN: 0801440939, Cornell University Press; 4th edition (2003, April 10)
3. van der Mensbrugghe, D., Osorio-Rodarte, I., Burns, A., Baffes, J.: Macroeconomic Environment, Commodity Markets: A Longer Term Outlook. http://www.fao.org/3/a-ak967e.pdf (2021). Accessed Mar 2021
4. Kapusuzoglu, A., Karacaer, U.M.: The interactions between agricultural commodity and oil prices: An empirical analysis. Agric. Econ. Czech. **61**, 410–421 (2015)
5. Timmer, C.: Did Speculation Affect World Rice Prices (2009)
6. Dezember, R., Maltais, K.: U.S. Crop Prices Are Rising, and China Is Buying. https://www.wsj.com/articles/u-s-crop-prices-are-rising-and-china-is-buying-11605522607 (2020, Nov. 16). Accessed Jan 2021
7. Li, J., Chavas, J.-P., Etienne, X.L., Li, C.: Commodity price bubbles and macroeconomics: Evidence from the Chinese agricultural markets. https://doi.org/10.1111/agec.12372 (2017, 19 July). Accessed 8 Jan 2021
8. Abbott, C.: Robust U.S. Economy, Higher Commodity Prices in 2021, Says USDA. https://www.agriculture.com/news/business/robust-us-economy-higher-commodity-prices-in-2021-says-usda (2020). Accessed 8 Jan 2021
9. Fajarnes, P.: An Overview Of Major Sources Of Data And Analyses Relating To Physical Fundamentals In International Commodity Markets. https://unctad.org/system/files/official-document/osgdp20112_en.pdf (2011, June). Accessed 18 Sept 2020

10. Barker, R.: The response of milk production to price: A regional analysis. https://core.ac.uk/download/pdf/38913888.pdf (1960). Accessed 22 Nov 2020
11. Irwin, S., Good, D.: The Relationship between Stocks-to-Use and Soybean Prices Revisited. farmdoc daily (5): 89, Department of Agricultural and Consumer Economics, University of Illinois at Urbana-Champaign. https://farmdocdaily.illinois.edu/2015/05/relationship-between-stock-to-use-and-soybean-prices.html (2015, May 14)
12. Bobenrieth, E., Brian Wright, Di Zeng, Stocks-to-use ratios and prices as indicators of vulnerability to spikes in global cereal markets, 28 June 2013, doi:https://doi.org/10.1111/agec.12049, Access Date: November 22, 2020
13. Columbus, L.: 10 Ways AI Has The Potential To Improve Agriculture In 2021. https://www.forbes.com/sites/louiscolumbus/2021/02/17/10-ways-ai-has-the-potential-to-improve-agriculture-in-2021/?sh=7f611e297f3b (2021, Feb 17). Accessed 10 Mar 2021
14. Elliot, G., Timmermann, A.: Handbook of Economic Forecasting, 1st edn. Elsevier Science and Technology Books, North Holland (2013) ISBN-13: 978-0444627322
15. Prabhakaran, S.: Vector Autoregression (VAR) – Comprehensive Guide with Examples in Python. https://www.machinelearningplus.com/time-series/vector-autoregression-examples-python/ (2019). Access Dec 2020
16. Introduction to Time Series Analysis, NIST/SEMATECH e-Handbook of Statistical Methods, http://www.itl.nist.gov/div898/handbook/ (2020). Accessed 23 Dec 2020
17. Dissanayake, B.: Introduction to Time Series Forecasting. https://towardsdatascience.com/introduction-to-time-series-forecasting-7e03c4bd83e0 (2020, July 14). Accessed 10 Oct 2020
18. Maitra, S.: Time Series Forecasting using Granger's Causality and Vector Auto-regressive Model. https://towardsdatascience.com/granger-causality-and-vector-auto-regressive-model-for-time-series-forecasting-3226a64889a6 (2019, Oct 7). Accessed 8 Nov 2020
19. Meredith, S.: Where's crude heading? Morgan Stanley says to watch the oil-gold ratio. https://www.cnbc.com/2020/07/14/oil-and-gold-prices-morgan-stanley-says-to-watch-the-oil-gold-ratio.html (2020, July 14). Accessed 15 Jan 2021
20. Ng, K.: Relationship Between Gold and Oil Prices: Why There's a Big Disconnection in 2020. https://ca.finance.yahoo.com/news/relationship-between-gold-oil-prices-153048345.html (2020, May 28). Accessed 12 Dec 2020
21. Hussan, A.: Project: Gold vs Silver Price. http://rstudio-pubs-static.s3.amazonaws.com/368841_224fe698bd7c45d8a5dd9a6b639988fc.html (2020). Accessed 20 Dec 2020
22. Adhikary, S.: Testing for Granger Causality Using Python. https://rishi-a.github.io/2020/05/25/granger-causality.html (2020, May 25). Accessed 14 Nov 2020
23. Vigna, P., Ostroff, C.: Stocks Cap Losing Week as Bond Yields Jump. https://www.wsj.com/articles/global-stock-markets-dow-update-02-26-2021-11614314418?page=1 (2021, Feb. 26). Accessed 1 Mar 2021
24. Quint, M.: Treasury bill rate at 14.54%. https://www.nytimes.com/1981/08/08/business/treasury-bill-rate-at-14.54.html (1981, Aug. 8). Accessed 30 Apr 2020
25. Leong, R.: TREASURIES-Weak jobs data knock U.S. yields lower. https://www.reuters.com/article/markets-usa-bonds-idUSL1E8H1A3Z20120601 (2012, June 1). Accessed 14 Dec 2020
26. Amadeo, K.: 10-Year Treasury Note and How It Works. https://www.thebalance.com/10-year-treasury-note-3305795 (2021, February 17). Accessed 1 Mar 2021
27. Verlaine, J.-A.: Gold's Climb Adds Momentum to Year-End Rally. https://www.wsj.com/articles/golds-climb-adds-momentum-to-year-end-rally-11609859184?page=1 (2021, Jan. 5). 1 Mar 2021
28. Batten, J.A., Lucey, B.M., Peat, M.: Gold and silver manipulation: what can be empirically verified? Econ. Model. **56**, 168–176 (2016). https://doi.org/10.1016/j.econmod.2016.03.005. ISSN 0264-9993
29. Brownlee, J.: How to Calculate Correlation Between Variables in Python. https://machinelearningmastery.com/how-to-use-correlation-to-understand-the-relationship-between-variables/ (2020, August 20). Accessed 14 Nov 2020

30. Brownlee, J.: A Gentle Introduction to Autocorrelation and Partial Autocorrelation. https://machinelearningmastery.com/gentle-introduction-autocorrelation-partial-autocorrelation/ (2020, August 14). Accessed 14 Nov 2020
31. Pinhanez, C.S.: C4AI: A New Center for Artificial Intelligence to accelerate science and innovation in Brazil. https://www.ibm.com/blogs/research/2020/10/c4ai-a-new-center-for-artificial-intelligence-to-accelerate-science-and-innovation-in-brazil/ (2020, October 15). 2 Jan 2021
32. Gowan, J.: "Free" Fertilizer is Saving Rural Farmers. https://www.permaculturenews.org/2017/10/02/free-fertilizer-saving-rural-farmers/ (2017, October 2). Accessed 25 Mar 2021
33. Komarek, A.M., Drogue, S., Chenoune, R., Hawkins, J., Msangi, S., Belhouchette, H., Flichman, G.: Agricultural household effects of fertilizer price changes for smallholder farmers in central Malawi. Agri. Syst. **154**, 168–178 (2017). https://doi.org/10.1016/j.agsy.2017.03.016. ISSN 0308-521X
34. Lbendhl, G.: Predicting Fertilizer Prices. https://www.agmanager.info/sites/default/files/pdf/2019.4.pdf (2019, April). Accessed 10 Mar 2021
35. Rezitis, A.N.: The relationship between agricultural commodity prices, crude oil prices and US dollar exchange rates: a panel VAR approach and causality analysis. https://www.tandfonline.com/doi/citedby/10.1080/02692171.2014.1001325?scroll=top&needAccess=true (2015, 22 Jan). Accessed 10 Dec 2021
36. Brunelle, T., Dumas, P., Souty, F., Dorin, B., Nadaud, F.: Evaluating the impact of rising fertilizer prices on crop yields. https://doi.org/10.1111/agec.12161 (2015, 31 March). Accessed 12 Mar 2021
37. Shah, S.: Consumer Price Index (CPI) vs. Producer Price Index (PPI): An Overview. https://www.investopedia.com/ask/answers/011915/what-difference-between-consumer-price-index-cpi-and-producer-price-index-ppi.asp (2020, Dec 15). 10 Mar 2021
38. Chen, P.-Y., Chang, C.-L., Chen, C.-C., McAleer, M.: Modeling the Effects of Oil Prices on Global Fertilizer Prices and Volatility, Working Papers in Economics 13/07, University of Canterbury, Department of Economics and Finance (2013)
39. Bhosale, J.: Cotton hurt by fall in crude, low demand. https://economictimes.indiatimes.com/markets/commodities/news/cotton-hurt-by-fall-in-crude-low-demand/articleshow/67431155.cms?from=mdr (2019, Jan 08). Accessed 1 Mar 2021
40. Ibendahl, G.: Predicting Fertilizer Prices in 2021. https://www.agmanager.info/sites/default/files/pdf/FertilizerProjection_2021-2022.pdf (2021, February 13). March 16 2021

Section III
Employment and Weather

Chapter 5
Weather Patterns and Machine Learning

This chapter covers:
- Climatic changes and modeling climate parameters as part of ML models
- Application NOAA Storm Events Database and deriving weather events
- USDA Quick Stats Database
- Rice crops in California and weather events
- Water use in rice production
- Weather events and model weather events
- Ensemble ML models
- USDA Crop Calendar
- Milk production and climate events modeling for Indian subcontinent

In this chapter, the effects of climate change and temperature variations on agricultural commodities are studied. Next, introduction to Machine Learning (ML) algorithms and ensemble modeling techniques that explain the influence of changing temperatures & climate on the agricultural yield is studied. Exploratory data analysis (EDA) and feature engineering of the following critical climate and agricultural attributes are studied: crop phenology, the time sensitivity & crop calendar, temperature variations & rice yield, precipitation, drought index (Palmer Drought Severity Index (PDSI), macrolevel economic indicators, weather events, water, and storm data. Finally, data from the National Oceanic and Atmospheric Administration (NOAA), National Centers for Environmental Information Strom Data, and the National Agricultural Statistics Services (NASS) Data are queried and analyzed.

Worldwide, agriculture is a $5-trillion industry and is fully ready for embracing a digital, data-fueled, smart intelligent transformation to overcome increasing demand and several disruptive forces and challenges [1]. Agriculture, the backbone of many economies and the largest employer in the world, is one of the sensitive industries which is undergoing seismic shift due to global warming and associated climate change. Modeling impacts of changes in key climatic elements on agricultural resources are important and top sustainable priority for formulating response strategies and

C. Vuppalapati, *Machine Learning and Artificial Intelligence for Agricultural Economics*, International Series in Operations Research & Management Science 314,
https://doi.org/10.1007/978-3-030-77485-1_5

addressing challenges faced by agricultural economies. Second, the demand for food is growing; at the same time, the supply side faces constraints in land [2], decrease in harvested land and farming inputs, constrained or limited water, and availability of human resources. ML and AI innovative technologies are critical and life-savers to address the topmost challenges faced by us as humanity in the face of ever-increasing intense climatic challenges that are witnessed by societies across the world—March 2021 Texas winter storm,[1] March 2021 Hawaii floods and landslides,[2] and March 2021 European[3] droughts and heat waves worst in 2000 years. The climatic challenges faced by humanity that need intervention from AI to overcome include:

- Constrained natural resources, i.e., by 2030, the water supply will fall, due to population, drought, and deforestation, and 40% short of meeting global water needs.
- Excessive chemical usage and degradation of arable land, i.e., about one-quarter of arable land is degraded (due to heavy usage of pesticides, which makes harvesting hard without applying more pesticides) and needs significant restoration before it can again sustain crops at scale.
- And then there are increasing environmental pressures, such as climate change and the economic impact of catastrophic weather events.

Understanding climate change, especially the temperature impacts, is critical if policy makers, agriculturalists, and crop breeders are to ensure global food security. Prognostic analytics that combine statistical models with data ensemble provide a guidance to small farmers and a better outlook to policy makers on concurring the impact of climate on the essential cereals, rice and wheat[4] [3]. As part of the chapter, ML modeling and AI are applied to develop models that assess the vulnerability of rice and milk production to develop predictive prognostic analytics for climate change.

> The core concept of sustainability[5] is that lasting success (and avoiding crises) requires an integrated approach to producing food and other products; farm profitability; quality of life for farmers, workers, and communities; and stewardship of natural resources. That is, sustainability requires recognizing and acting upon productivity, economic, social, and environmental goals as a simultaneous set of system attributes. The National Research Council (NRC) report "Toward Sustainable Agricultural Systems in the 21st Century" effectively summarized the current state of science and practice in the United States. It documented the considerable science-based progress in American agriculture: **producing more food and fiber on about the same acreage as a century with less labor, energy, and water per unit of output and considerably less soil erosion.**

[1] Texas weather: Deadly winter storm sweeps Texas and US southern states - https://www.bbc.com/news/world-us-canada-56076686

[2] Hawaii governor declares emergency after floods and landslides - https://www.theguardian.com/us-news/2021/mar/10/hawaii-landslides-emergency-declaration-weather-rain-floods

[3] Climate crisis: recent European droughts 'worst in 2,000 years' - https://www.theguardian.com/environment/2021/mar/15/climate-crisis-recent-european-droughts-worst-in-2000-years

[4] Temperature increase reduces global yields of major crops in four independent estimates - https://www.pnas.org/content/114/35/9326

[5] Sustainable Agricultural Systems Science White Paper - https://www.usda.gov/sites/default/files/documents/sustainable-agriculture-science-white-paper.pdf

As part of the chapter, rice production and modeling climate change and the impact on crop yield are developed.

Rice Production

California rice[8] industry is the most efficient and productive found anywhere in the world. The industry is fortunate to have a solid base of natural resources including

A Primer on Rice Production [4]
Rice production occurs over vast areas of the world, particularly in Asia where it is a staple crop and has significant cultural and historical roots. Largest rice producers from Asia[6] include China, India, Indonesia, Bangladesh, Vietnam, Thailand, Myanmar, Japan, and the Philippines.

Economically sound rice production requires agronomic requirements, such as a plentiful supply of water applied in a timely fashion (via rain or irrigation from groundwater or surface water sources), high average temperatures during the growing season, a smooth land surface to facilitate uniform flooding and drainage, and a subsoil hardpan that inhibits the percolation of water. Therefore, rice production in the United States is limited to certain areas.[7]

In the United States, four regions, which span portions of six states, meet the required characteristics and produce almost the entire US rice crop (USDA, ERS, 2017b):

- Arkansas Grand Prairie (Non-Delta)
- Mississippi Delta (parts of Arkansas, Mississippi, Missouri, and Louisiana)
- Gulf Coast (Texas and Southwest Louisiana)
- Sacramento Valley (California)

Different regions grow different types of rice—i.e., long, medium, or short grain—as determined by regional soil conditions:

- Long grain: Produced in the South and account for 80% of US production.
- Medium grain: The majority of US medium-grain rice is grown in California, while most of the Southern medium-grain rice production occurs in Arkansas. California has grown almost exclusively medium-grain rice, accounting for over 90% of production.
- Short grain: Short-grain rice accounts for only 1–2% of the total US rice production and is grown almost exclusively in California.

[6] Major Rice producing nations - http://nipunarice.com/rice-o-pedia/major-rice-producing-nations/#:~:text=China%20is%20the%20world's%20largest,of%20total%20world%20rice%20production

[7] U.S. Rice Production in the New Millennium: Changes in Structure, Practices, and Costs https://www.ers.usda.gov/webdocs/publications/90926/eib-202.pdf?v=9628

[8] CALIFORNIA RICE PRODUCTION: ECONOMIC AND ENVIRONMENTAL PARTNERSHIPS - http://worldfood.apionet.or.jp/web/3.htm

good soils, abundant water, manmade dams, and highly favorable climate. The industry is well positioned to produce high-quality Japonica rice for domestic and export markets for the foreseeable future [5].

The Sacramento Valley of Northern California is one of the ideal regions in the world to produce Japonica rice. Much of the rice production land is a fine-textured clay soil that drains slowly. It is well suited to rice culture, but not to other crops (corn or wheat). The soil in the counties north of Sacramento where rice growing is concentrated is a thick clay that holds water well. "It's basically a bathtub out there, and rice seems to thrive".[9] Historical planted rice acreage over the past 25 years fluctuates between 300,000 acres and 600,000 acres [6].

The Mediterranean climate of an extended summer dry season of warm days and cool nights allows rice to establish quickly, mature properly, and avoid many of the diseases that are common in other parts of the world. An extensive water storage and delivery system provides rice farmers with an abundant and reliable supply of surface water that is easily managed to assure top yields and high quality of each year's crop. The unique California water-seeded method of rice production results in yields on average over 8000 pounds of paddy rice per acre.

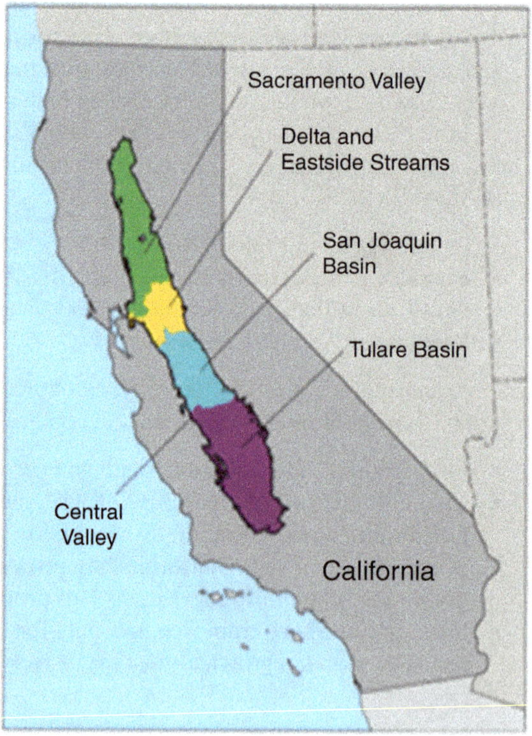

Fig. 5.1 The central valley [7]

[9]Amid drought, California fields to grow rice: Bloomberg View - https://www.oregonlive.com/opinion/2015/05/amid_drought_california_fields.html

Major Central Valley[10] (please see Fig. 5.1) counties [7] are as follows:

- Fresno
- Tulare
- Stanislaus
- San Joaquin
- Kings
- Merced
- Butte
- Colusa

The hot, dry summer weather also produces great rice, less likely to be hit by the diseases that plague wetter, more humid rice-growing states, such as Arkansas and Louisiana [6]. Another main resource is water to grow it. Rice's per-acre water needs, flooded fields, and all actually aren't off the charts relative to other crops, but they are higher.

Rice Crop Calendar

The US rice-growing season varies by region:

- Planting typically begins in early or mid-March in Texas and southwest Louisiana. Producers in the Delta States plant the bulk of their crop in April, and California's crop is planted from late April through May.
- Harvest begins by mid-July in Texas and southwest Louisiana. Peak harvest in the South is in September and early October when the Delta harvests the bulk of its crop. Some producers in Texas and southwest Louisiana can re-flood rice fields after harvest and achieve a partial second or "ratoon" crop from the stubble of the first. California typically begins harvest at the end of September and finishes by early or mid-November. Please refer to the USDA Crop Calendar[11] (please see Fig. 5.2).

Crop calendar is very important not only to understand the agricultural seasons and related agricultural activities but also to assess the global climate change historical analysis and to be able to develop a prognostic modeling
{Crop Calendar} + {NOAA Weather events} provide climate change effect!

[10] California's Central Valley - https://www.usgs.gov/centers/ca-water/science/californias-central-valley?qt-science_center_objects=0#qt-science_center_objects

[11] U.S. Crop Calendar - https://ipad.fas.usda.gov/rssiws/al/crop_calendar/us.aspx

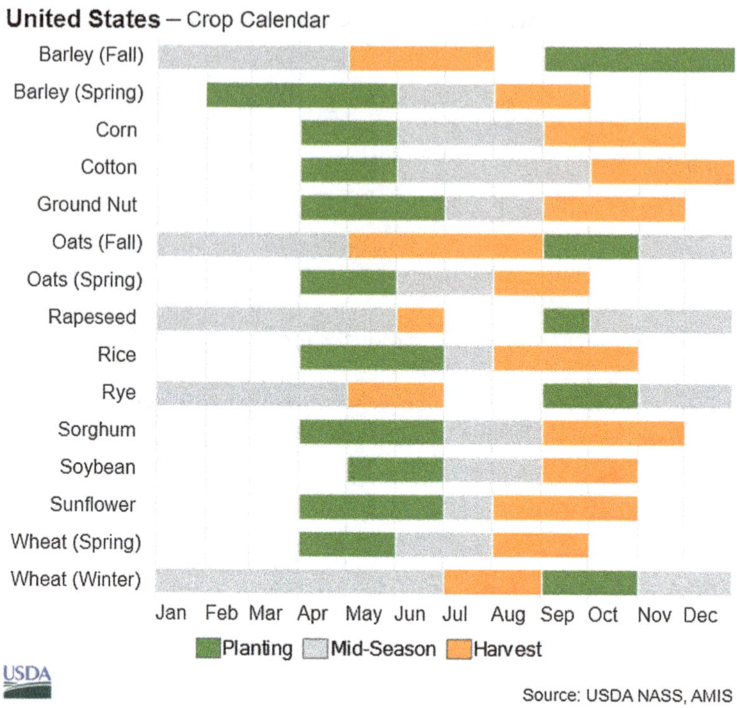

Fig. 5.2 USDA crop calendar

Rice Farm Consolidation

Like the trends seen in many agricultural and dairy farms, the number of US farms producing rice has declined continually over the past 20 years,[12] as rice farms have trended toward consolidation into larger operations.[13] In essence, many of the family-owned farms have disappeared (please see Fig. 5.3) [8]. There is still a large share of small farms—those with less than 250 acres—in the Southern rice-growing regions. These farms may be at a competitive disadvantage relative to larger farms due to economies of size. Many small to medium farms and small-scale farmers were hit by the brunt of the consolidation and were left helpless. Similar trend can be seen in Asia and other places [9].

The consolidation of small rice farms, ironically, can be seen in the developing countries—India and China. Surprisingly, the government regulation and policies could accelerate the issue, an unintended consequence, if you will. A good example

[12] U.S. Rice Production Changed Significantly in the New Millennium, but Remained Profitable - https://www.ers.usda.gov/amber-waves/2020/may/us-rice-production-changed-significantly-in-the-new-millennium-but-remained-profitable/#:~:text=Number%20of%20Rice%20Farms%20Fell%20From%201997%20to%202017,rice%2Dgrowing%20States%20reporting%20declines

[13] Working the Land With 10 Acres: Small Acreage Farming in the United States - https://www.ers.usda.gov/webdocs/publications/43847/46432_eib123.pdf?v=0

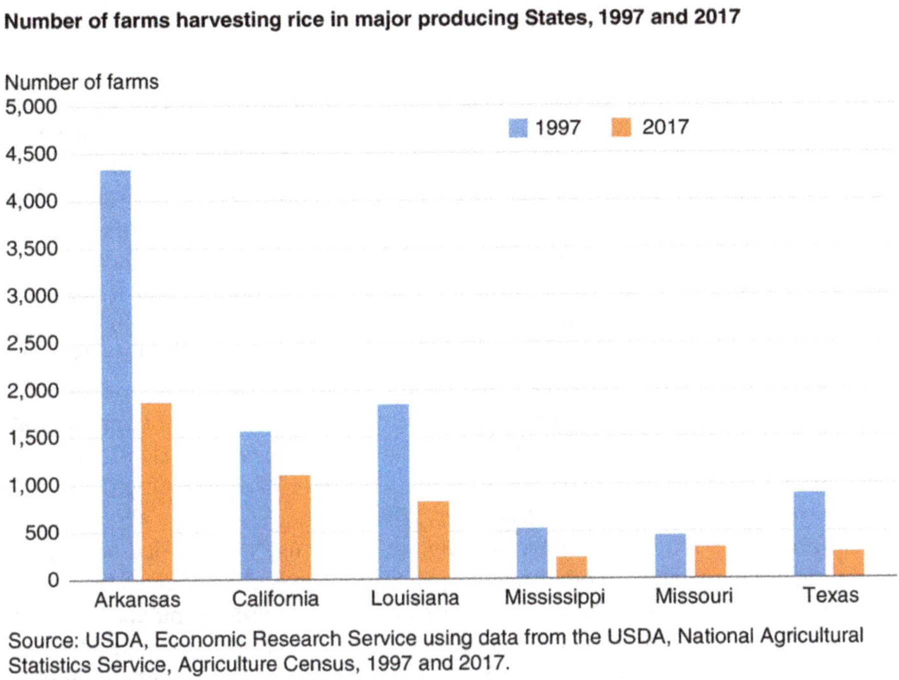

Fig. 5.3 Number of farms harvesting rice [8]

is the abolishment of Minimum Support Price (MSP) [10, 11] in India.[14] Assessing region-wide trends in East Asia and Pacific, the average farm size has been declining for many decades in most of these countries. The trend of large commercial farms taking over small farms is common phenomena one can observe in these countries. A similar trend has happened decades ago in industrialized countries.[15] The small-scale farmers in China face unprecedented challenges and farm consolidation due to modernization and globalization.[16] In Vietnam, nearly 90% of agricultural land falls under either agricultural households or farms, about 6% is under the enterprises, and the rest is held by other entities. The trend is reduction of small household farms[17] (which numbered 8.9 million in 2011, down from 10.1 million in 2001) [11, 12].

[14] India's farm laws are a global problem - https://www.brookings.edu/opinions/indias-farm-laws-are-a-global-problem/

[15] East Asia and Pacific - http://www.fao.org/3/Y1860E/y1860e08.htm

[16] Small Scale Farmers in China - https://pubs.iied.org/sites/default/files/pdfs/migrate/16515IIED.pdf

[17] Transforming Vietnamese Agriculture: Gaining More for Less - http://documents1.worldbank.org/curated/en/923211468310487558/pdf/VDR-report-word-version.pdf

Water Use

Rice on average requires 5.1 ft of water applied per growing season, edging out alfalfa as the most water-intensive of major California crops. When you multiply it by 550,000 acres planted in a normal year, you get 2.8 million acre-feet of water, or about 6% of the state's total consumption. The good news for rice growers is that most of them are clustered along the Sacramento River and its tributary, the Feather River, two of the state's most bounteous sources of irrigation water. Additionally, Shasta Dam on the Sacramento River and Oroville Dam on the Feather River have ensured good protection from spring floods and a more reliable summer water supply. In addition, agriculture in the Sacramento Valley (please see Fig. 5.4) depending on location relies on a variable combination of surface water and groundwater. Groundwater accounts for less than 30% of the annual supply used for agricultural and urban purposes in this area. The Sacramento Valley, generally rich in surface water, provides water for much of California's urban and agricultural needs[18] (*California Department of Water Resources, 2003*).

California rice producers[19] have the highest water use among rice-producing regions, about double that in other regions [4]. California is characterized by higher water use for rice production because most of the water used is purchased from irrigation districts and flows via gravity over rice fields until being drained for use on other down-slope farms (please see Fig. 5.5). Most water use in Southern regions is pumped from farm wells or surface sources and remains on rice fields through a system of levees.

Water temperature also plays an important role in rice yield. It is common knowledge that yields are low near a cold-water intake. Recent research has shown that the

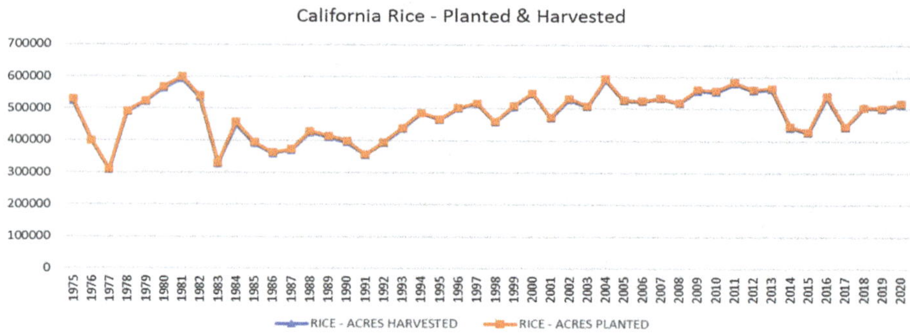

Fig. 5.4 California rice planted and harvested

[18] The Central Valley: Sacramento Valley - https://ca.water.usgs.gov/projects/central-valley/sacramento-valley.html

[19] U.S. Rice Production in the New Millennium: Changes in Structure, Practices, and Costs - https://www.ers.usda.gov/webdocs/publications/90926/eib-202.pdf?v=9628

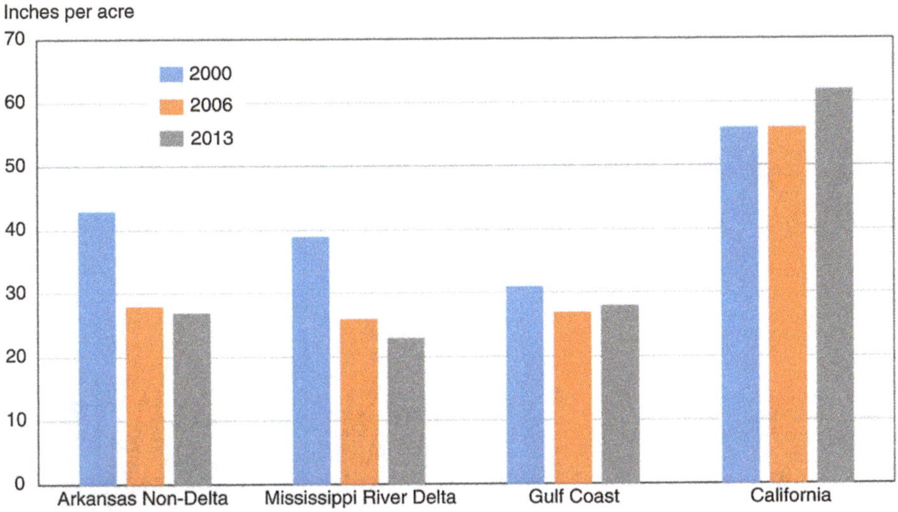

Source: USDA, National Agricultural Statistical Service and Economic Research Service, Agricultural Resource Management Survey.

Fig. 5.5 Water inches per acre [4]

cold water and the associated reduction in rice productivity extend well beyond the area where the effects are readily visible. The appropriate combination of temperature and water is required for a good rice yield. In other words, rice is sensitive to the abovementioned factors—temperature and water.[20]

> Water is essential to agriculture. Without it, farmers could not grow crops or produce livestock[21] [13]

The following figure contains phenological stages of rice. Under warm conditions, the number of days required to maturity from germinate stage is lesser (please see Fig. 5.6). And under cool conditions, the number of days required to maturity is higher[22] [14].

Weather Events

Climate change, i.e., increase in temperature, abnormal rains, or floods, impacts rice yields. Rice production is affected by increasing temperature, abnormal temperatures, high heat events, excessive rains, storms, and other climate change-related

[20] Rice Production Manual - http://rice.ucanr.edu/files/288581.pdf

[21] Rice Production Manual - http://rice.ucanr.edu/files/288581.pdf

[22] Arkansas Rice Production Handbook - MP192 - https://www.uaex.edu/publications/pdf/mp192/chapter-2.pdf

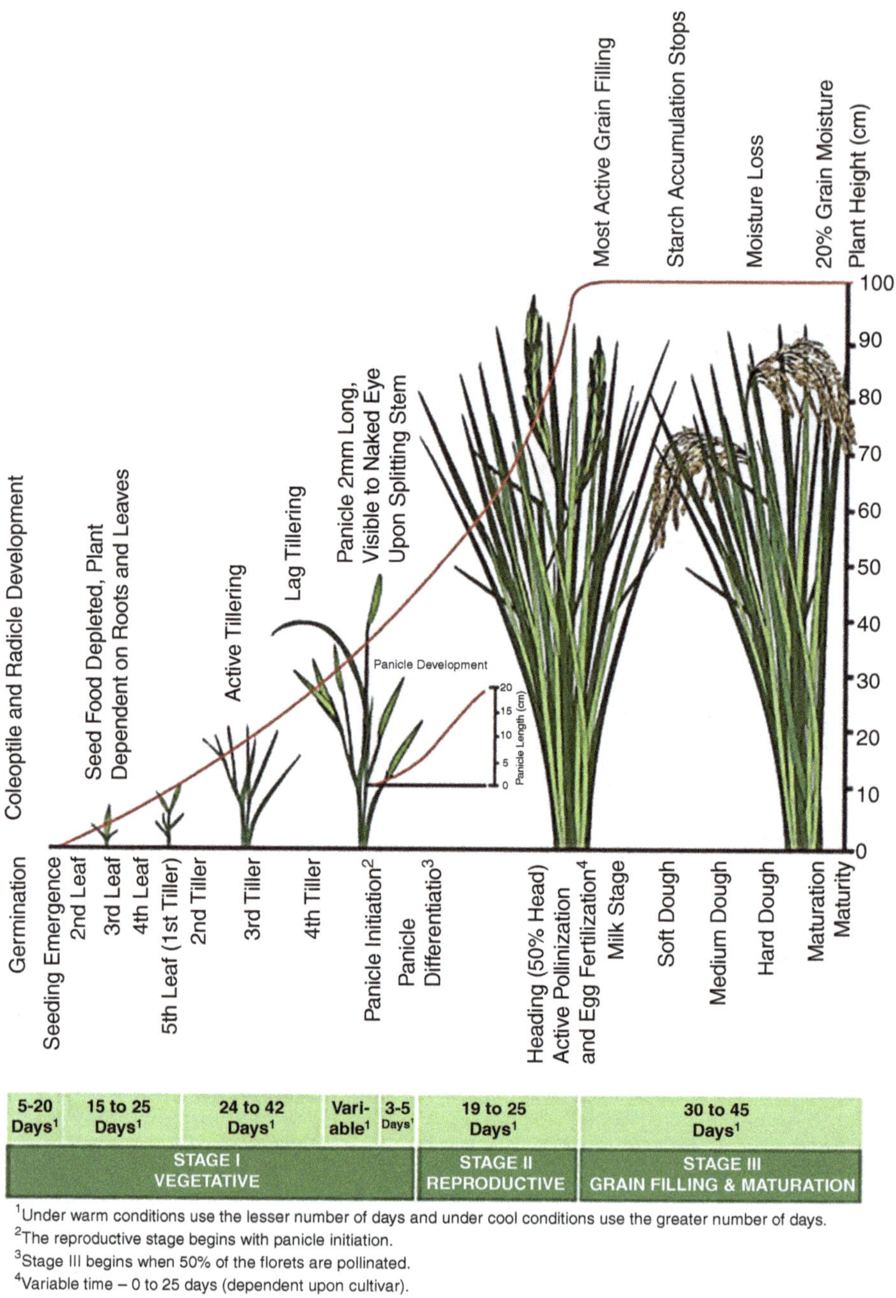

Fig. 5.6 Rice production [14]

events. The following is the summary of weather events that were recorded in the Sacramento Valley from March 07, 2010, to November 20, 2020.

In the Sacramento Valley, April would normally be a busy month for rice farmers who fire up their tractors to prepare fields before the busy planting season (please see Appendix to see US Agriculture Calendar). However, repeated rainstorms during this period are very detrimental to rice yields[23] [15].

The following is a summary of climate events:[24] 286 events were reported between March 07, 2010, and November 30, 2020 (3922 days).

Summary Info

Number of County/Zone areas affected:	31
Number of Days with Event:	86
Number of Days with Event and Death:	6
Number of Days with Event and Death or Injury:	8
Number of Days with Event and Property Damage:	4
Number of Days with Event and Crop Damage:	1
Number of Event Types reported:	3

Here are some of the events (not comprehensive):

Location	County/Zone	St.	Date	Time	T.Z.	Type	Mag	Dth	Inj	PrD	CrD
Totals:								0	0	75.00K	16.025M
FRESNO	FRESNO CO.	CA	05/18/2019	11:00	PST-8	Heavy Rain		0	0	0.00K	15.000M
CENTRAL SIERRA (ZONE)	CENTRAL SIERRA (ZONE)	CA	05/18/2019	11:00	PST-8	Heavy Snow		0	0	0.00K	0.00K
SEQUOIA KINGS (ZONE)	SEQUOIA KINGS (ZONE)	CA	05/18/2019	11:00	PST-8	Heavy Snow		0	0	0.00K	0.00K
SEQUOIA KINGS (ZONE)	SEQUOIA KINGS (ZONE)	CA	05/18/2019	11:00	PST-8	Heavy Snow		0	0	0.00K	0.00K
SEQUOIA KINGS (ZONE)	SEQUOIA KINGS (ZONE)	CA	05/18/2019	11:00	PST-8	Heavy Snow		0	0	0.00K	0.00K
CENTRAL SIERRA (ZONE)	CENTRAL SIERRA (ZONE)	CA	05/18/2019	11:00	PST-8	Heavy Snow		0	0	0.00K	0.00K
CENTRAL SIERRA (ZONE)	CENTRAL SIERRA (ZONE)	CA	05/18/2019	11:00	PST-8	Heavy Snow		0	0	0.00K	0.00K
SEQUOIA KINGS (ZONE)	SEQUOIA KINGS (ZONE)	CA	05/18/2019	11:00	PST-8	Heavy Snow		0	0	0.00K	0.00K

Column Definitions

Mag, magnitude; Dth, deaths; Inj, injuries; PrD, property damage; CrD, crop damage

[23] sacramento-valley-rice-crop-off-to-a-late-start-due-to-continued-rains - https://www.chicoer.com/2017/05/07/sacramento-valley-rice-crop-off-to-a-late-start-due-to-continued-rains/

[24] NOAA - https://www.ncdc.noaa.gov/stormevents/listevents.jsp?eventType=%28Z%29+Excessive+Heat&eventType=%28Z%29+Heat&eventType=%28C%29+Heavy+Rain&eventType=%28Z%29+Tropical+Storm&beginDate_mm=03&beginDate_dd=07&beginDate_yyyy=2010&endDate_mm=11&endDate_dd=30&endDate_yyyy=2020&county=BUTTE%3A7&county=COLUSA%3A11&county=FRESNO%3A19&county=KINGS%3A31&county=MERCED%3A47&county=SACRAMENTO%3A67&county=SAN%2BJOAQUIN%3A77&county=STANISLAUS%3A99&county=TULARE%3A107&hailfilter=0.00&tornfilter=0&windfilter=000&sort=DT&submitbutton=Search&statefips=6%2CCALIFORNIA

The most notable in the above table is Fresno heavy rain event.[25] Fresno rains had caused over 60% of the cherry crop, which was destroyed by the persistent rains that took place during the two storms that impacted the area between May 15 and May 19, 2019 [16]. There was also substantial damage to strawberries and carrots.[26] Rice farmers are the one who got affected the most due to May month as it's planting season. It took a while for them to get back out into the field and get rice planted. However, this rain has caused a few to fall behind. The farmers were 75% into planting season but could not start due to heavy rains. The total crop damage was $15 million.

Fresno Rain

A strong upper-level low pressure system moved into central California during the afternoon of May 18. A cold front associated with this system pushed across the area overnight, bringing periods of moderate to locally heavy precipitation to the area with much of the area picking up between 0.75 and 2 inches of liquid precipitation. As a result, many long-standing daily rainfall records were shattered on both May 18 and May 19. The precipitation mainly fell as snow above 6000 feet and brought an additional 8–14 inches of snow to much of the higher Sierra Nevada range from Yosemite Park southward.

The following table contains rain, heavy rain, winter storm, flood, and flash flood weather events details extracted from NOAA site for Sacramento Valley, California, from 1996 to 2020 (please see Fig. 5.7).

Similarly, temperature and drought events data are collected from NOAA site to see the influence of weather events on the rice yield.

Although torrential rains and heavy flooding cause huge economic loss to small farmers, the timing could not be better, especially late planting and early mid-season that could result into economic disaster. For instance, Jay County in eastern Indiana was hit with a deluge of rain during the week of June 22, 2015, and had destroyed as much as 5% of Indiana's corn and soybean crops and potentially had caused about $300 million in crop damage.[27]

There is general agreement among observers and agricultural experts that excess rainfall is likely to result in significantly lower-than expected yields. As is true for other grain crops, rice growth and development are also influenced by temperature. The seasonal temperatures within the expected range and normal monsoon would result in a condition for a great crop.[28] A recent study conducted on increase temperatures on rice yield has shown that the elevated temperature may cause a

[25] Northern California almond, walnut, rice farmers face problems with persistent rain - https://www.mercurynews.com/2019/05/18/almond-walnut-rice-farmers-face-problems-with-persistent-rain/

[26] Fresno rain - https://www.ncdc.noaa.gov/stormevents/eventdetails.jsp?id=815640

[27] About $300 million in Indiana crops' value lost to flooding so far - https://www.purdue.edu/newsroom/releases/2015/Q2/300-million-in-indiana-crops-value-lost-to-flooding-so-far.html

[28] Sep 5 2013 | Excess Monsoon Rainfall Raise Concerns for India's 2013/14 Soybean Productivity - https://ipad.fas.usda.gov/cropexplorer/pecad_stories.aspx?regionid=sasia&ftype=topstories

Year	Quarter	Month	Day	DAMAGE_CROPS_NUM	DAMAGE_PROPERTY_NUM	CZ_NAME_STR
2005	Qtr 2	May	8	21940000	0	FRESNO CO.
2006	Qtr 2	April	2	16000000	0	FRESNO CO.
1998	Qtr 1	February	2	7800000	4300000	SOUTHERN SACRAMENTO VALLEY (ZONE)
2003	Qtr 2	April	1	7800000	0	FRESNO CO.
1998	Qtr 2	May	1	7200000	0	FRESNO CO.
1998	Qtr 1	February	1	1800000	1600000	FRESNO CO.
2000	Qtr 2	May	15	300000	0	FRESNO CO.
2000	Qtr 3	September	1	250000	0	E CENTRAL SJ VALLEY (ZONE)
1998	Qtr 2	May	2	200000	0	FRESNO CO.
2000	Qtr 2	June	8	200000	0	FRESNO CO.
1996	Qtr 1	March	13	50000	25000	FRESNO CO.
2006	Qtr 2	April	2	50000	250000	SACRAMENTO CO.
1996	Qtr 1	February	20	0	0	SAN LUIS OBISPO CO.
1996	Qtr 1	February	25	0	0	SOUTHERN SACRAMENTO VALLEY (ZONE)
1996	Qtr 1	March	12	0	0	FRESNO CO.
1996	Qtr 2	April	16	0	30000	FRESNO CO.
1996	Qtr 4	October	29	0	0	FRESNO CO.
1996	Qtr 4	December	9	0	0	CARQUINEZ STRAIT AND DELTA (ZONE)
1996	Qtr 4	December	12	0	0	SOUTHERN SACRAMENTO VALLEY (ZONE)
1996	Qtr 4	December	29	0	0	FRESNO CO.
Total				**63590000**	**29783000**	

Fig. 5.7 Storm events

noteworthy reduction in the productions of these widely planted genotypes in central and southern regions of China. A study—Vulnerability of Rice and Wheat Yields in NW India to Future Changes in Climate, conducted in India—has shown acute water shortage conditions combined with thermal stress should adversely affect both the wheat and more severely the rice productivity in North West India even under the positive effects of elevated CO_2[29] [17]. To ensure the security of rice production in this region in an expected global warming environment, currently planted varieties will need to be replaced by heat-resistant varieties in the future[30] [18].

Data Sources

Data source	Description	
NOAA—Storm Events Database[31]	Storm data are geographically categorized by county or by NWS forecast zone. Smaller events (areal coverage) are collected by county (tornado, thunderstorm winds, flash floods, and hail), while larger-scale events are collected by forecast zone (heat, cold, drought, flood, tropical and winter weather)	
NOAA California Storm Events[32]	Storm Events Database[33]	

[29] Vulnerability of rice and wheat yields in NW India to future changes in climate - https://www.sciencedirect.com/science/article/abs/pii/S0168192397000646

[30] The Effect of Season-Long Temperature Increases on Rice Cultivars Grown in the Central and Southern Regions of China - https://www.ncbi.nlm.nih.gov/pmc/articles/PMC5682563/

[31] NOAA - https://www.ncdc.noaa.gov/

[32] NOAA Storm events - https://www.ncdc.noaa.gov/stormevents/choosedates.jsp?statefips=6%2CCALIFORNIA

[33] Storm Event Meta Data - https://adata.site.wesleyan.edu/files/2017/08/Storm_Event_Data_Codebook.pdf

Data source	Description	
Rice yield and harvested data —NASS Quick Stats[34]	The Quick Stats Database is the most comprehensive tool for accessing agricultural data published by National Agricultural Statistics Services (NASS).[35] It allows you to customize your query by commodity, location, or time period	
NOAA	The National Oceanic and Atmospheric Administration is an American scientific agency within the US Department of Commerce that focuses on the conditions of the oceans, major waterways, and the atmosphere.[36] The National Centers for Environmental Information[37]	

BEGIN_AZIMUTH[38]

The storm is in knots or nautical miles per hour (1 knot = 1.15 mph), with the direction expressed in terms of 16 compass points and to the nearest 5° (true) of azimuth (e.g., 360° = North, 90° = East, 180° = South, and 270° = West).

Strong US dollar makes it more expensive for buyers in other countries to purchase rice from the United States

"There's an old story," said Western Canal Water District Manager Ted Trimble, "after May 12, you lose a bag of rice a day," due to lower yield[39] [15]

Storm Rainfall Depth and Distribution[40]

The Clear Lake region of Lake County was hit hard by flooding. Clear Lake is a very slow-draining natural lake (on a day with no rainfall, the lake drains at about the rate of 1 inch per 24 hours when at these reported levels). Moreover, 500 residents in the town of Clearlake Oaks were evacuated and kept from their homes for weeks. This town alone suffered $5 million of the damage recorded above. The town of Lakeport

[34] Rice Yield Data - https://quickstats.nass.usda.gov/#2787C033-9661-3C63-9E82-679A76C4C650

[35] NASS - https://www.nass.usda.gov/

[36] The NOAA - https://www.noaa.gov/

[37] NCEI – NOAA - https://www.ncdc.noaa.gov/

[38] Begin Azimuth - https://www.nhc.noaa.gov/help/tcm.shtml?MOTION

[39] sacramento-valley-rice-crop-off-to-a-late-start-due-to-continued-rains - https://www.chicoer.com/2017/05/07/sacramento-valley-rice-crop-off-to-a-late-start-due-to-continued-rains/

[40] Storm rainfall depth and distribution - https://www.wcc.nrcs.usda.gov/ftpref/wntsc/H&H/NEHhydrology/ch4_Sept2015draft.pdf

had 60 homes flooded. Rainfall rates of 4 inches in 6 hours were observed during the second and third. Clear Lake reached all-time record levels on the 23rd, 2.5 feet above flood stage. The flooding continued into March and April. The main transportation artery through the county, State Route 20, was closed for weeks.

Rice Data Sources

Select commodity (please see Fig. 5.8)			
Program: Survey	Sectors: CROPS	Group: Field crops	Commodity: Rice
Category: Area Harvested and Area Planted[41]	Data item: MILK, FLUID GRADE—PRICE RECEIVED, MEASURED IN $ / CWT	Domain: Total	
Select location			
Geographical Level: State	State: California		
Select time			
Year: All available years			

California—especially the Sacramento Valley—is a heavily agricultural valley with the following offerings:

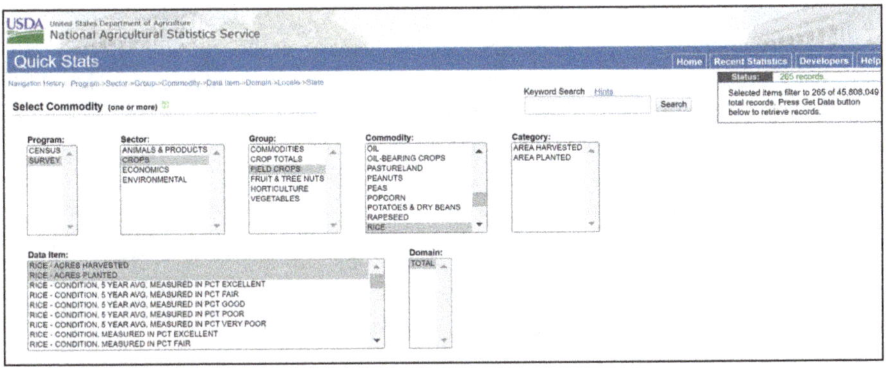

Fig. 5.8 USDA quick stats

[41] Rice Yield data - https://quickstats.nass.usda.gov/#2787C033-9661-3C63-9E82-679A76C4C650

Crop and livestock commodities in which California[42] leads the nation[43] [19]

Almonds Flowers, Bulbs Melons, Cantaloupe **Plums, Dried**
Apricots Flowers, Cut Melons, Honeydew Pluots
Artichokes Flowers, Potted Plants Milk and Cream **Pomegranates**
Asparagus Garlic Nectarines Raspberries
Avocados **Grapes, Raisins** Nursery, Bedding Plants **Rice, Sweet**
Beans, Dry Lima Grapes, Table Nursery Crops Safflower
Broccoli Grapes, Wine **Olives** Seed, Alfalfa
Brussels Sprouts Hay, Alfalfa Onions, Dry Seed, Bermuda Grass
Cabbage, F.M. Herbs Onions, Green **Seed, Ladino Clover**
Carrots, F.M. Jojoba Parsley Seed, Vegetable and Flower
Cauliflower Kale **Peaches, Clingstone** Spinach, F.M.
Celery **Kiwifruit** Peaches, Freestone Strawberries
Corn, Sweet Kumquats Peppers, Chili Tomatoes, Processing
Cotton, American Pima Lemons Peppers, Bell Triticale
Daikon Lettuce, Head Persimmons Vegetables, Greenhouse
Dates Lettuce, Leaf Pigeons and Squabs Vegetables, Oriental
Eggplant Lettuce, Romaine **Pistachios Walnuts**
Escarole/Endive Limes Plums Watercress **Figs** Mandarins and Mandarin Hybrids[2]

- *California is the sole producer (99% or more) of the commodities in bold*
- [2]*Includes tangelos, tangerines, and tangors*

Data Variables: Temperature, Precipitation, and PDSI

Monthly temperature, precipitation, and Palmer Drought Severity Index information were obtained from the National Oceanic and Atmospheric Administration's National Climatic Data Center via the nClimDiv FTP site[44] (please see Fig. 5.9). Climate weather data can be parsed using Python data frame.[45]

The files are retrieved from NOAA FTP site.

Data Files

Climate variable	File link
Minimum temperature	climdiv-tminst-v1.0.0-20210106[46]
Maximum temperature	climdiv-tmaxst-v1.0.0-20210106[47]
Palmer Modified Drought Index (PMDI)	climdiv-pmdist-v1.0.0-20210106[48]
Precipitation	climdiv-pcpnst-v1.0.0-20210106[49]

[42] California is the sole producer (99% or more) of the commodities in bold.

[43] California Agriculture - https://www.cdfa.ca.gov/statistics/pdfs/2016report.pdf

[44] Index of /pub/data/cirs/climdiv/ - ftp://ftp.ncdc.noaa.gov/pub/data/cirs/climdiv/

[45] Combining Data in Pandas With merge(), .join(), and concat() - https://realpython.com/pandas-merge-join-and-concat/

[46] USA Minimum Temperature - ftp://ftp.ncdc.noaa.gov/pub/data/cirs/climdiv/climdiv-tmaxst-v1.0.0-20210106

[47] US Maximum Temperature (1895 – Current) - ftp://ftp.ncdc.noaa.gov/pub/data/cirs/climdiv/climdiv-tmaxst-v1.0.0-20210106

[48] PMDI Data Source - ftp://ftp.ncdc.noaa.gov/pub/data/cirs/climdiv/climdiv-pmdist-v1.0.0-20210106

[49] Precipitation - ftp://ftp.ncdc.noaa.gov/pub/data/cirs/climdiv/climdiv-pcpnst-v1.0.0-20210106

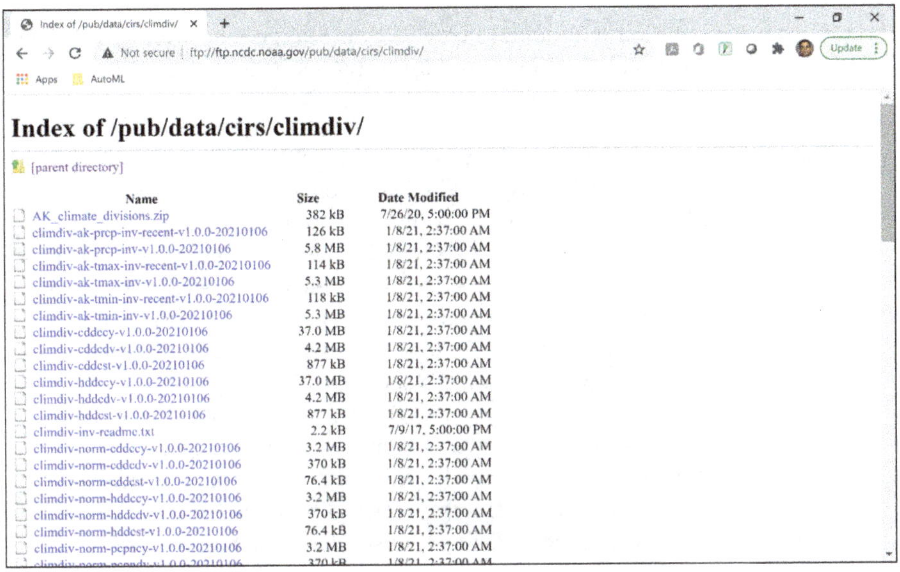

Index of /pub/data/cirs/climdiv/

[parent directory]

Name	Size	Date Modified
AK_climate_divisions.zip	382 kB	7/26/20, 5:00:00 PM
climdiv-ak-prcp-inv-recent-v1.0.0-20210106	126 kB	1/8/21, 2:37:00 AM
climdiv-ak-prcp-inv-v1.0.0-20210106	5.8 MB	1/8/21, 2:37:00 AM
climdiv-ak-tmax-inv-recent-v1.0.0-20210106	114 kB	1/8/21, 2:37:00 AM
climdiv-ak-tmax-inv-v1.0.0-20210106	5.3 MB	1/8/21, 2:37:00 AM
climdiv-ak-tmin-inv-recent-v1.0.0-20210106	118 kB	1/8/21, 2:37:00 AM
climdiv-ak-tmin-inv-v1.0.0-20210106	5.3 MB	1/8/21, 2:37:00 AM
climdiv-cddccy-v1.0.0-20210106	37.0 MB	1/8/21, 2:37:00 AM
climdiv-cddcdv-v1.0.0-20210106	4.2 MB	1/8/21, 2:37:00 AM
climdiv-cddcst-v1.0.0-20210106	877 kB	1/8/21, 2:37:00 AM
climdiv-hddccy-v1.0.0-20210106	37.0 MB	1/8/21, 2:37:00 AM
climdiv-hddcdv-v1.0.0-20210106	4.2 MB	1/8/21, 2:37:00 AM
climdiv-hddcst-v1.0.0-20210106	877 kB	1/8/21, 2:37:00 AM
climdiv-inv-readme.txt	2.2 kB	7/9/17, 5:00:00 PM
climdiv-norm-cddccy-v1.0.0-20210106	3.2 MB	1/8/21, 2:37:00 AM
climdiv-norm-cddcdv-v1.0.0-20210106	370 kB	1/8/21, 2:37:00 AM
climdiv-norm-cddcst-v1.0.0-20210106	76.4 kB	1/8/21, 2:37:00 AM
climdiv-norm-hddccy-v1.0.0-20210106	3.2 MB	1/8/21, 2:37:00 AM
climdiv-norm-hddcdv-v1.0.0-20210106	370 kB	1/8/21, 2:37:00 AM
climdiv-norm-hddcst-v1.0.0-20210106	76.4 kB	1/8/21, 2:37:00 AM
climdiv-norm-pcpncy-v1.0.0-20210106	3.2 MB	1/8/21, 2:37:00 AM

Fig. 5.9 CLIMDIV FTP site

All files are in the following format (please refer to divisional-readme.txt[50]):

nClimDiv

Divisional

Temperature-Precipitation-Drought

June 2014

The major parameters in this file are sequential climatic division monthly maximum, minimum, and average temperature (deg. F. to 10ths, national temperature to 100ths), precipitation (inches to 100ths), Standardized Precipitation Index (SPI), and Palmer Drought Indices (PDSI, PHDI, PMDI, and ZNDX). Period of record is 1895 through the latest month available, updated monthly.

Values from the most recent two calendar years will be updated on a monthly basis. Period of record updates will occur when the underlying dataset undergoes a version change.

Methodology

Divisional values in nClimDiv were derived from area-weighted averages of grid-point estimates interpolated from station data. A nominal grid resolution of 5 km was used to ensure that all divisions had sufficient spatial sampling (only four small divisions had less than 100 points) and because the impact of

[50] Divisional readme txt - ftp://ftp.ncdc.noaa.gov/pub/data/cirs/climdiv/divisional-readme.txt

elevation on precipitation is minimal below 5 km. Station data were gridded via climatologically aided interpolation to minimize biases from topographic and network variability.

The Global Historical Climatology Network (GHCN) Daily dataset is the source of station data for nClimDiv. GHCN-Daily contains several major observing networks in North America, five of which are used here. The primary network is the National Weather Service (NWS) Cooperative Observing (COOP) program, which consists of stations operated by volunteers as well as by agencies such as the Federal Aviation Administration. To improve coverage in western states and along international borders, nClimDiv also includes the National Interagency Fire Center (NIFC) Remote Automatic Weather Station (RAWS) network, the USDA Snow Telemetry (SNOTEL) network, the Environment Canada (EC) network (south of 52°N), and part of Mexicos Servicio Meteorologico Nacional (SMN) network (north of 24°N). Note that nClimDiv does not incorporate precipitation data from RAWS because that network's tipping-bucket gauges are unheated, leading to suspected cold-weather data.

Element Name	Record Position	Element Description
STATE-CODE	1–2	STATE-CODE as indicated in State Code Table as described in FILE 1. Range of values is 01–91.
DIVISION-NUMBER	3–4	DIVISION NUMBER—Assigned by NCDC. Range of values 01–10.
ELEMENT CODE	5–6 02 = Average Temperature 05 = PDSI 06 = PHDI 07 = ZNDX 08 = PMDI 25 = Heating Degree Days 26 = Cooling Degree Days 27 = Maximum Temperature 28 = Minimum Temperature 71 = 1-month Standardized Precipitation Index 72 = 2-month Standardized Precipitation Index 73 = 3-month Standardized Precipitation Index 74 = 6-month Standardized Precipitation Index 75 = 9-month Standardized Precipitation Index 76 = 12-month Standardized Precipitation Index 77 = 24-month Standardized Precipitation Index	01 = Precipitation
YEAR	7–10	This is the year of record. Range is 1895 to current year processed.
(All data values are right justified):		

JAN-VALUE	11–17	Palmer Drought Index format (f7.2) Range of values −20.00 to 20.00. Decimal point retains a position in seven-character field. Missing values in the latest year are indicated by −99.99. Monthly Divisional Temperature format (f7.2) Range of values −50.00–140.00 degrees Fahrenheit. Decimals retain a position in seven-character field. Missing values in the latest yea are indicated by −99.90. Monthly Divisional Precipitation format (f7.2) Range of values 00.00–99.99. Decimal point retains a position in seven-character field. Missing values in the latest year are indicated by −9.99. Monthly Divisional Degree Day format (f7.0) Range of values 0000.–9999. Decimal point retains a position in seven-character field. Missing values in the latest year are indicated by −9999.. Standardized Precipitation Index format (f7.2). Range of values −4.00 to 4.00. Decimal point retains a position in seven-character field. Missing values in the latest year are indicated by −99.99.

Header

Required to transform temperature and precipitation with the following fixed width: 0, 1, 2, 3, 4, 6, 10, 17, 24, 31, 38, 45, 52, 59, 66, 73, 80, 87, and 94 (please see Fig. 5.10).

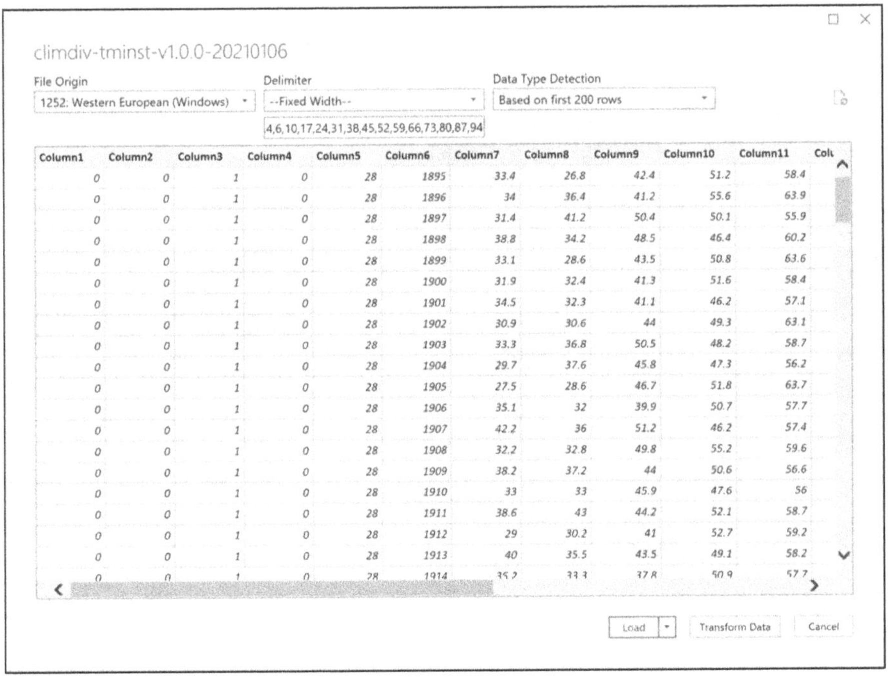

Column1	Column2	Column3	Column4	Column5	Column6	Column7	Column8	Column9	Column10	Column11
0	0	1	0	28	1895	33.4	26.8	42.4	51.2	58.4
0	0	1	0	28	1896	34	36.4	41.2	55.6	63.9
0	0	1	0	28	1897	31.4	41.2	50.4	50.1	55.9
0	0	1	0	28	1898	38.8	34.2	48.5	46.4	60.2
0	0	1	0	28	1899	33.1	28.6	43.5	50.8	63.6
0	0	1	0	28	1900	31.9	32.4	41.3	51.6	58.4
0	0	1	0	28	1901	34.5	32.3	41.1	46.2	57.1
0	0	1	0	28	1902	30.9	30.6	44	49.3	63.1
0	0	1	0	28	1903	33.3	36.8	50.5	48.2	58.7
0	0	1	0	28	1904	29.7	37.6	45.8	47.3	56.2
0	0	1	0	28	1905	27.5	28.6	46.7	51.8	63.7
0	0	1	0	28	1906	35.1	32	39.9	50.7	57.7
0	0	1	0	28	1907	42.2	36	51.2	46.2	57.4
0	0	1	0	28	1908	32.2	32.8	49.8	55.2	59.6
0	0	1	0	28	1909	38.2	37.2	44	50.6	56.6
0	0	1	0	28	1910	33	33	45.9	47.6	56
0	0	1	0	28	1911	38.6	43	44.2	52.1	58.7
0	0	1	0	28	1912	29	30.2	41	52.7	59.2
0	0	1	0	28	1913	40	35.5	43.5	49.1	58.2
0	0	1	0	28	1914	35.2	33.3	37.8	50.9	57.7

Fig. 5.10 Excel transform data

NOAA: National Centers for Environmental Information—Storm Data

Storm data collected for the Sacramento Valley[51] are geographically categorized by county or by NWS forecast zone. Smaller events (areal coverage) are collected by county (tornado, thunderstorm winds, flash floods, and hail), while larger-scale events are collected by forecast zone (heat, cold, drought, flood, tropical and winter weather).

Storm Events Database contains weather events data from January 1950 to November 2020. The following variables are selected to get California Sacramento Valley storm data (please see Fig. 5.11):

Heavy Rains

State/area: California	Begin date: 01/01/1991	End date: 11/30/2020
County: • Fresno • Sacramento • San Luis Obispo	Event type (s): • Flood • Flash flood • Heavy rains • Winter storm	

Some of the weather events include:

- Vineyards and farmland along the Russian River are inundated by widespread flooding following days of torrential rain as viewed on January 11, 2017, near Healdsburg, California[52] [20]:

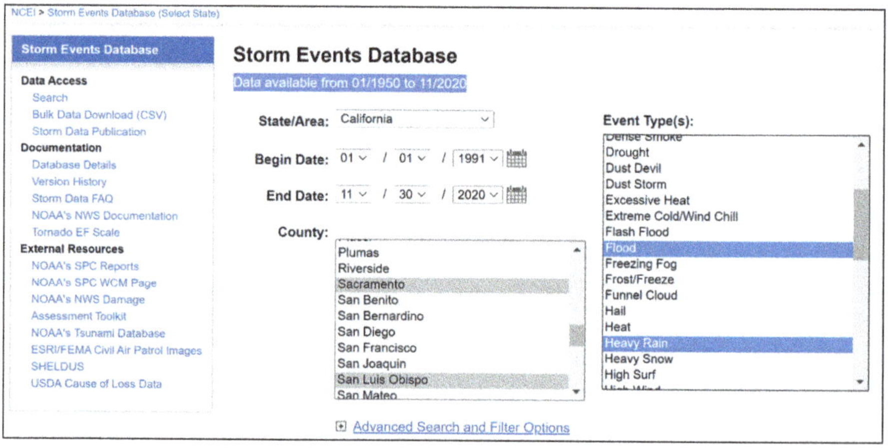

Fig. 5.11 Storm events data

[51] NOAA Storm Data - https://www.ncdc.noaa.gov/stormevents/choosedates.jsp?statefips=6%2CCALIFORNIA#

[52] California farmers can't win as monster storms threaten crops - https://www.cnbc.com/2017/01/25/california-farmers-cant-win-as-monster-storms-threaten-crops.html

- Agricultural lands became muddy and wet—making it difficult for many of our farmers to get in to plant or leaving crops in the ground that are hard to harvest. Loss of revenue due to muddy conditions.
- The USDA and the state food and agriculture agency didn't have damage estimates (please see Fig. 5.12).

- Floods in northern California left peach[53] and walnut orchards with disease, which could lead to tree death [21].

Heat and Excessive Heat

State/area: California	Begin date: 01/01/1991	End date: 11/30/2020
County: • Fresno • Sacramento • San Luis Obispo	Event type (s): • Heat • Excessive heat	

Heat affects rice yields[54] [22]. Some of the important events include:

- Scorching heat can affect rice yield[55] [23]:

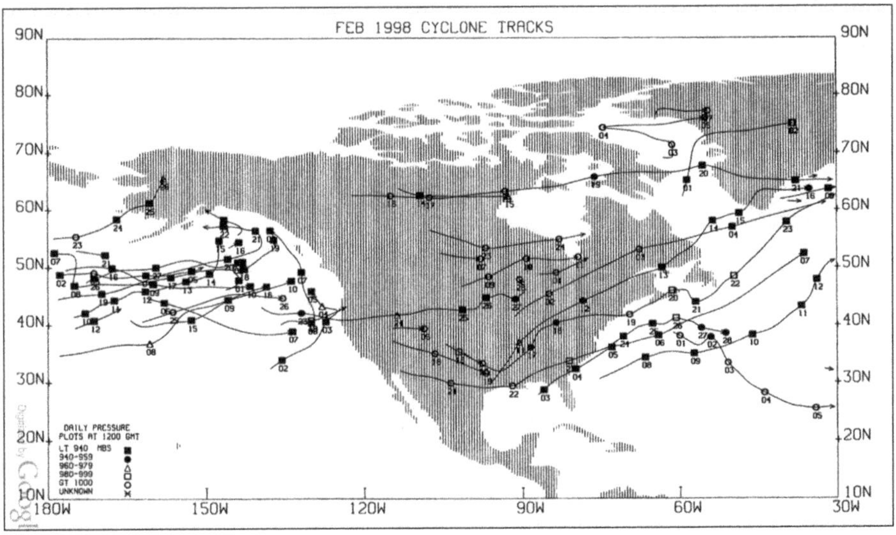

Fig. 5.12 Storm data

[53] Some walnut, peach orchards show "collapse" symptoms after NorCal winter floods - https://www.farmprogress.com/orchard-crops/some-walnut-peach-orchards-show-collapse-symptoms-after-norcal-winter-floods

[54] Golden harvest at risk: Climate change puts heat on rice yields - https://www.koreatimes.co.kr/www/nation/2020/12/371_296304.html

[55] Scorchingheat-https://www.thestar.com.my/news/nation/2016/03/19/scorching-heat-can-affect-rice-yield-mardi-extreme-hot-weather-could-damage-harvest-quality-and-caus/

  - Rice is highly susceptible to heat stress, particularly during the reproductive and ripening stages.
  - Extremely high temperatures—even for a few hours—during flowering can cause complete sterility, while during ripening, this can lead to reduced grain filling and poor milling quality, resulting in more broken grains.
  - A 2°C rise in temperature could lead to a 13% reduction in padi yield, while drought, or a 15% decrease in seasonal rainfall, could lead to a drop in yield of up to 80%.
  - Extremely high temperatures can irreversibly damage rice yield and grain quality, causing a shortage in production.

- Rice is subjected to yield loss due to heat stress[56] [24]:

  - Crop production will likely face enormous challenges against the occurrences of extreme climatic events projected under future climate change. Heat waves that occur at critical stages of the reproductive phase have detrimental impacts on the grain yield formation of rice!

Climate change threatens India's crop production; rice yields may be hit hard[57] [25]:

Storm Data February 1998 Volume 40 Number 2[58]

Model Development

Software code for this model can be found on GitHub Link: Severe_weather_impact_Sacremanto_California_Rice. Ipynb (Jupyter Notebook)

[56] Current rice models underestimate yield losses from short-term heat stresses - https://onlinelibrary.wiley.com/doi/abs/10.1111/gcb.15393

[57] Climate Change - https://www.livemint.com/news/india/climate-change-threatens-india-s-crop-production-rice-yields-may-be-hard-hit-1561435191944.html

[58] Storm Data - https://www.google.com/books/edition/Storm_Data/TavgYM2vpREC?hl=en&gbpv=1&bsq=deciduous

Step 1: *Load Required Libraries*

```
import numpy as np
import pandas as pd
import functools
from datetime import date
import plotly.graph_objects as go
import seaborn as sns; sns.set()
import statsmodels.api as sm
from statsmodels.tsa.arima_process import ArmaProcess
from scipy import stats
from scipy.stats import pearsonr
import matplotlib.pyplot as plt
from sklearn.preprocessing import MinMaxScaler
from sklearn.model_selection import train_test_split
from sklearn.linear_model import LinearRegression

import logging
import pandas as pd
import numpy as np
from numpy import random
import nltk
from sklearn.model_selection import train_test_split
from sklearn.feature_extraction.text import CountVectorizer, TfidfVectorizer
from sklearn.metrics import accuracy_score, confusion_matrix
import matplotlib.pyplot as plt
from nltk.corpus import stopwords
import re
from bs4 import BeautifulSoup
import pandas as pd
import matplotlib.pyplot as plt
import matplotlib as mpl
import numpy as np
import scipy.stats as spstats
import seaborn as sns

%matplotlib inline
mpl.style.reload_library()
mpl.style.use('classic')
mpl.rcParams['figure.facecolor'] = (1, 1, 1, 0)
mpl.rcParams['figure.figsize'] = [6.0, 4.0]
mpl.rcParams['figure.dpi'] = 600
```

Required libraries include:

- scikit-learn[59]—scikit-learn is a free software machine learning library for the Python programming language.
- pandas[60]—pandas is a software library written for the Python programming language for data manipulation and analysis.

---

[59] SCIKIT Machine Learning Library - https://scikit-learn.org/stable/

[60] Pandas Data Analysis Library - https://pandas.pydata.org/

- Statistical models[61]
- SciPy[62]—open-source Python library used for scientific computing and technical computing.
- Graphic libraries.

Step 2: *Load Rice Harvested, Rice Yields, Rice Price, PDSI Values, PCP, and Rice Consumer Prices*

```
Rice_Harvest_df = pd.read_csv("Rice_Acres_Harvested.csv")
Rice_Production_df = pd.read_csv("sacramentoValleyRiceProduction.csv")
Rice_Yield_df = pd.read_csv("Rice_yields.csv")
Rice_Price_df = pd.read_csv("Rice_price.csv")
Pdsi_df = pd.read_csv("PDSIvalues.csv")
pcp_df = pd.read_csv("pcpdv.csv")
Rice_consumer = pd.read_csv("Rice_consumer_price.csv")
```

Step 3: *Feature Engineer and Validate Each Data Frames*

Load PDSI values for California state and year after 1990:

```
Pdsidf_cal = Pdsi_df[(Pdsi_df["State"] == 4) & (Pdsi_df["Year"]> 1990) &
(Pdsi_df["DivisionNo"]==2)]
Pdsidf_cal
```

Output

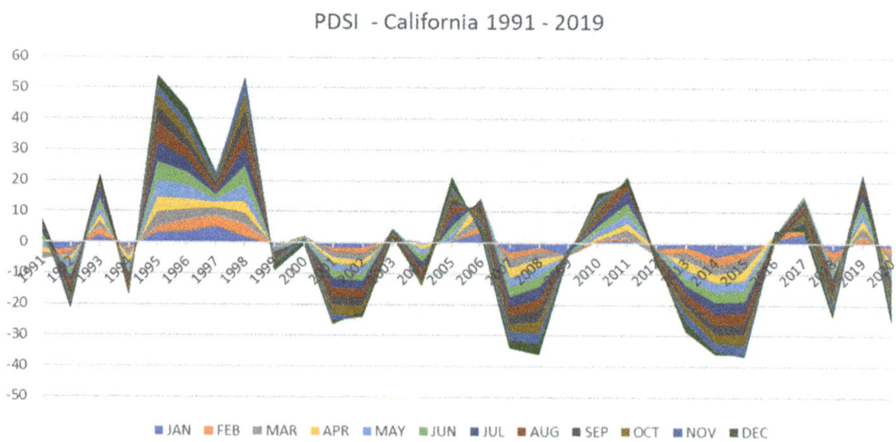

[61] Statsmodels - https://www.statsmodels.org/stable/index.html

[62] SCIPY - https://www.scipy.org/

The Palmer Drought Severity Index (PDSI)[63] uses readily available temperature and precipitation data to estimate relative dryness. It is a standardized index that generally spans from −10 (dry) to +10 (wet). Maps of operational agencies like NOAA typically show a range from −4 to +4, but more extreme values are possible. The PDSI has been reasonably successful at quantifying long-term drought.

The Palmer Drought Severity Index (please see Fig. 5.13) data is for the state of California from 1991 to 2020. The data are collected from NOAA site monthly. Since 2000, the longest duration of drought (D1, moderate drought; D4, exceptional drought[64]) in California lasted 376 weeks beginning on December 27, 2011, and ending on March 5, 2019[65,66] [26].

Drought has economic, social, cultural, and agricultural impact. Drought also has negative impact on hydropower generation in California. Finally, negative economic impacts on California's agricultural sector as a whole from the recent drought

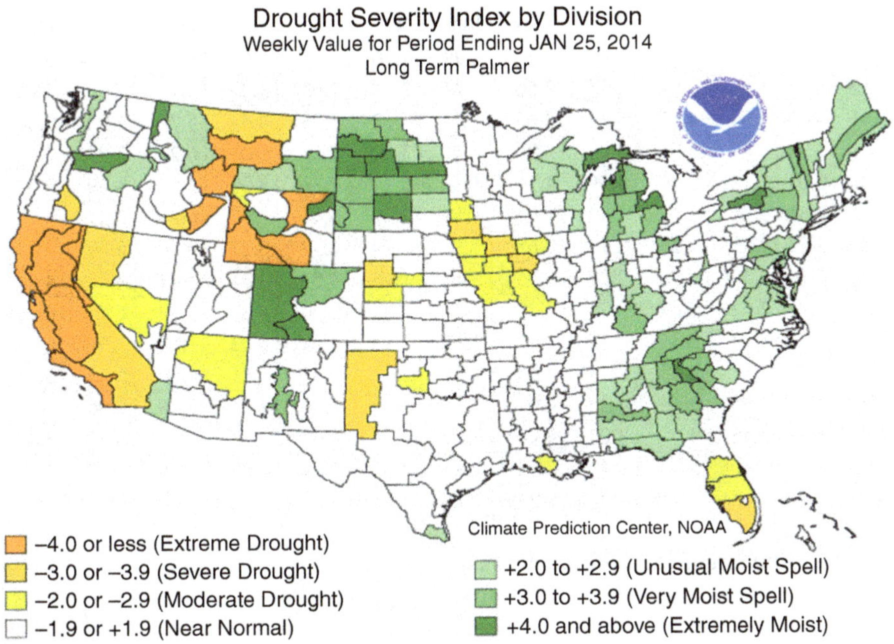

Fig. 5.13 PDSI

---

[63] PDSI - https://climatedataguide.ucar.edu/climate-data/palmer-drought-severity-index-pdsi

[64] Current U.S. Drought Monitor Conditions for California - https://www.drought.gov/states/california

[65] Impacts of the California drought from 2007 to 2009 - https://pacinst.org/wp-content/uploads/sites/21/2013/02/ca_drought_impacts_full_report3.pdf

[66] Lessons from California's 2012–2016 Drought - https://ascelibrary.org/doi/full/10.1061/%28ASCE%29WR.1943-5452.0000984

were significant. Impacts included abandoned orchards and vineyards, fallowed land (more than 500,000 acres, or 6% of irrigated acreage, were fallowed in 2015), and lost jobs (DWR, 2015[67]; PPIC, 2016[68]). The livelihoods of many farmworkers disappeared[69] [27].

Local PCP Values

Monthly precipitation values for California are loaded. Monthly divisional precipitation format (f7.2) has a range of values from 00.00 to 99.99. Decimal retains a position in seven-character field. Missing values in the latest year are indicated by −9.99:

```
pcp_df = pd.read_csv("pcpdv.csv")
pcp_df.head()
```

Output (please note only sample 2007–2014 precipitation data)

Year	Jan	Feb	Mar	Apr	May	Jun	Jul	Aug	Sep	Oct	Nov	Dec
2007	7.21	4.39	1.77	2.02	2.6	2.98	1.76	3.92	6.65	6.5	10.84	5.78
2008	5.1	1.98	3.99	2.78	3.56	1.74	3	2.03	3.03	6.17	3.52	6.45
2009	3.66	5.03	4.07	5.03	3.35	2.9	6.53	3.98	4.64	5.26	4.98	7.87
2010	1.38	1.19	2.74	3.04	3.26	2.61	3.52	5.28	5.06	6.23	3.94	2.67
2011	6.03	2.76	2.36	4.59	4.17	4.48	5.22	4.31	8.07	4.8	4.69	7.51
2012	3.61	5.98	6.41	3.42	1.97	1.58	3.29	5.41	7.97	6.56	3.37	4.16
2013	3.57	2.9	4.25	2.13	2.68	3.32	3.79	6.38	6.27	9.79	5.09	6.44
2014	5.78	3.39	2.18	2.02	0.95	2.49	1.95	4.5	7.34	4.47	4.7	4.07
2015	2.24	2.62	3.84	2.78	5.21	1.21	4.03	2.06	4.67	5.99	6.44	6.26
2016	4.9	3.07	3.37	4.27	2.47	2.22	2.18	4.64	8.02	10.54	5.5	6.43
2017	5.32	2.3	0.13	2.14	3.33	2.61	2.64	5.08	4.81	6.89	5.75	5.01
2018	3.24	3.86	3.46	4.36	4.09	3.2	5.46	7.48	4.21	9.25	5.08	6.7
2019	5.91	5.46	4.9	3.07	4.44	3.35	2.04	4.06	6.13	10.14	6.58	5.4
2020	3.21	4.6	2.03	1.4	2.04	1.94	2.76	3.24	6.04	4.18	6.79	2.79
2021	7.52	−9.99	−9.99	−9.99	−9.99	−9.99	−9.99	−9.99	−9.99			

Please find graphical representation.

[67] Delivery Capability Report and Studies 2015 - https://water.ca.gov/Library/Modeling-and-Analysis/Central-Valley-models-and-tools/CalSim-2/DCR2015

[68] PPIC Statewide Survey: Californians and the Environment - https://www.ppic.org/publication/ppic-statewide-survey-californians-and-the-environment-july-2016/

[69] California Drought - https://oehha.ca.gov/media/epic/downloads/cc_drought2018.pdf

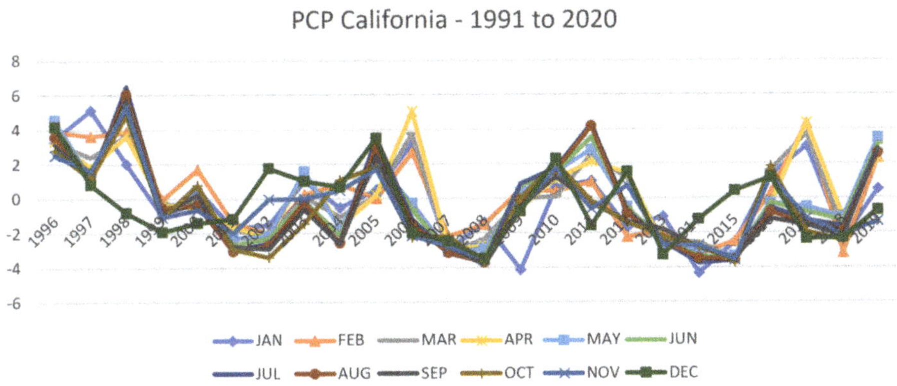

Palmer Drought Severity Index (PDSI)[70]

Step 4a: *Load Rice Agricultural Data*

Rice data is downloaded from the US Department of Agriculture (USDA) Quick Stats Database.[71] Data process the rice dataset to load only survey data, load only rice harvest data after 1990, and set year as index of the dataset:

```
Rice_Yield_df1 =Rice_Yield_df[['Value',"Year"]]
Rice_Yield_df1.rename(columns = {"Value" : "YIELD(LB/ACRE)"}, inplace = True)
Rice_Yield_df1["YIELD(LB/ACRE)"] = Rice_Yield_df1["YIELD(LB/ACRE)"].apply(lambda x :
x.replace(',',''))
Rice_Yield_df1["YIELD(LB/ACRE)"] = pd.to_numeric(Rice_Yield_df1["YIELD(LB/ACRE)"])
Rice_Yield_df1.set_index("Year", inplace = True)
Rice_Yield_df2= Rice_Yield_df1.groupby(['Year'])["YIELD(LB/ACRE)"].mean()
Rice_Yield_df3 = pd.DataFrame(Rice_Yield_df2)
rice_yield = Rice_Yield_df3[Rice_Yield_df3.index>1990]
rice_yield = np.round(rice_yield,2)
```

Output:

[70] PDSI - https://climatedataguide.ucar.edu/climate-data/palmer-drought-severity-index-pdsi

[71] Quick Stats Database - https://quickstats.nass.usda.gov/

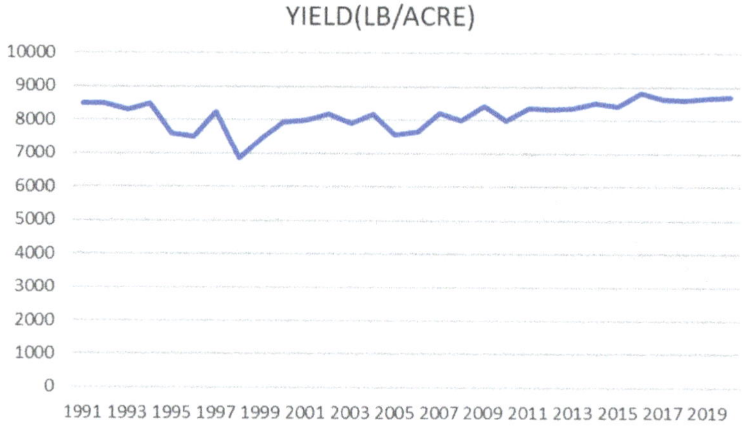

As it can be seen from the above graph, the rice yield per acre is between 8000 and 8704 LBS per acre, the only exception in 1998 when rice yield dropped to 6850 LBs per acre. The overall US national rice planted,[72] harvested, and yield () exhibits the following pattern with increasing yield since 1959, surprisingly dropping harvested and planted acres. One can explain this trend in disappearance of family small farms and increase in productivity (economy of scale) of the large farms.

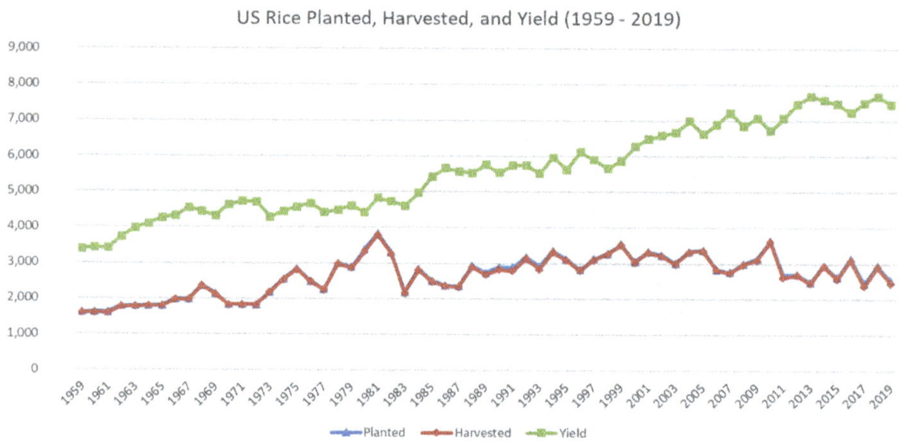

The correlation between planted, harvested, and yield provides with the following details: there is 100% correlation between planted and harvested columns and 55% and 56% correlation between planted, harvested, and yield, respectively.

[72] US Rice Acreage, Production and Yield – USDA ERS - https://www.ers.usda.gov/webdocs/DataFiles/53146/U_S_and_State_acreage_yield_and_production_all_rice_and_by_class_marketing_year.xls?v=43042

Step 4b*: Max and Min Temperature Data*

The temperature data is collected from NOAA site:[73]

```
Tmax_df = pd.read_csv("weather_fft_climdiv-tmaxdv.csv" )
Tmax_d.head(10)
```

Output:

Year	Jan	Feb	Mar	Apr	May	Jun	Jul	Aug	Sep	Oct	Nov	Dec
2010	48.1	52.1	56.5	57.4	64.4	79	89.4	85.8	82.5	68.8	55	47.5
2011	52	50.7	51.2	59.6	65.3	75.4	85.3	87.6	86.3	69.6	54.4	53.2
2012	53.5	53.8	52.9	62.9	74.3	79.4	87.9	91.5	87	71.7	57.6	46.1
2013	49.6	55	61.7	67.8	73.6	83	91.7	86.5	78.8	70	60.9	52.7
2014	59.7	54.3	59.7	66.9	75.7	84.4	91.8	87.1	83.2	73.4	58.5	49.7
2015	57.1	60.5	66.2	67	71.6	87.8	87.6	88	83.1	75.4	54	45.9
2016	47.4	59.5	57.4	67.4	72.2	83.9	88.9	89.8	82.7	65.1	57.2	47.9
2017	45.7	50.2	58.9	60.5	75.5	83.5	92.4	91.6	81.4	72.7	55.1	55
2018	52.4	54.9	54.5	64.1	73.8	83.6	92.6	88.2	83	71.6	60.8	49.2
2019	51.8	44	53.7	66.1	68.5	83.3	87.4	89	78.4	69.3	63	48.2
2020	49.5	59.2	55.1	66.6	73.5	82.2	90.4	91.1	86.5	78.7	58.1	52.5

The graphical representation of the data using excel charts:

[73] NOAA Site - ftp://ftp.ncdc.noaa.gov/pub/data/cirs/climdiv/

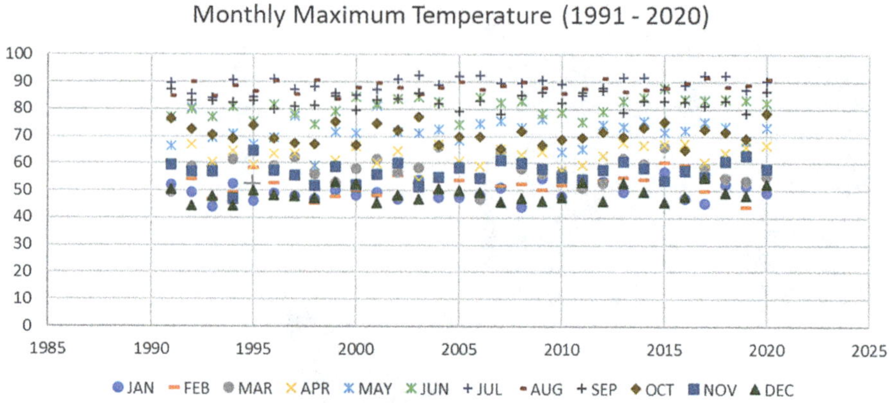

NOAA Divisional values in nClimDiv[74] were derived from area-weighted averages of grid-point estimates interpolated from station data. A nominal grid resolution of 5 km was used to ensure that all divisions had sufficient spatial sampling (only four small divisions had less than 100 points) and because the impact of elevation on precipitation is minimal below 5 km. Station data were gridded via climatologically aided interpolation to minimize biases from topographic and network variability.

The Global Historical Climatology Network (GHCN) Daily dataset is the source of station data for nClimDiv. GHCN-Daily contains several major observing networks in North America, five of which are used here. The primary network is the National Weather Service (NWS) Cooperative Observing (COOP) program, which consists of stations operated by volunteers as well as by agencies such as the Federal Aviation Administration. To improve coverage in western states and along international borders, nClimDiv also includes the National Interagency Fire Center (NIFC) Remote Automatic Weather Station (RAWS) network, the USDA Snow Telemetry (SNOTEL) network, the Environment Canada (EC) network (south of 52°N), and part of Mexicos Servicio Meteorologico Nacional (SMN) network (north of 24°N). The temperature is increasing in the Sacramento Valley[75] and in general [28, 29].

Load minimum temperatures:

```
Tmin_df = pd.read_csv("weather_fft_climdiv-tmindv.csv" )
Tmin_df.head(10)
```

[74] Improved Historical Temperature and Precipitation Time Series for U.S. Climate Divisions - https://journals.ametsoc.org/view/journals/apme/53/5/jamc-d-13-0248.1.xml

[75] Another day of record high temperatures for the Sacramento Valley - https://www.abc10.com/article/weather/record-high-april-temps/103-6dcdf546-deee-437b-bbc0-903fec3f0555

Output:

Year	Jan	Feb	Mar	Apr	May	Jun	Jul	Aug	Sep	Oct	Nov	Dec
1996	33	36	35.6	39	44.4	51.6	58.9	56.4	48.7	42.1	37.7	35.2
1997	32.4	32.8	36.8	39.4	48.8	51	55.6	54.5	51.2	41.6	39.1	30.7
1998	34.9	33.7	35.4	37.3	40.7	49.5	56.8	56.2	53.1	40.9	35.8	27.3
1999	31.4	31.2	32.4	36.7	42.5	49.2	52.7	53.5	51.6	43.9	38.4	30.4
2000	33.1	35.9	35.5	40.5	45.4	53.2	53	54.7	49.8	42.3	31.9	32.6
2001	29.2	30.4	37.4	35.4	49.7	51.5	55.4	55.3	51.5	45.9	38	32.7
2002	28.8	32.9	32.6	39.5	43.6	51.9	57.7	54.2	50.6	41.1	36.8	33.3
2003	37.6	31.6	36.4	35.2	44.4	52.9	59	54.8	52.8	45.6	33.8	33.2
2004	31	32.6	39.6	40.2	44.6	51.7	56.8	55.5	49.7	42.6	35	32.8
2005	31.7	35.2	36.7	37.2	45.6	47.5	59.1	55.9	47.4	42.5	36.9	34.1
2006	33.2	31.8	31	39.1	46.6	54.1	59.3	53.6	49.8	41.1	36.6	30.9
2007	26.7	33.5	37.4	38.7	45.4	50.5	56.6	55.1	48.2	41.4	37.1	29.1
2008	29.1	31.1	33.3	35.1	45.7	50.6	56.7	56.7	51	43.5	39.2	28.4
2009	31.8	32.8	33.6	37.2	47.8	50.9	56.7	54.9	52.8	41.9	35	28.4
2010	35	34.7	33.5	35.1	39.4	51.2	56.5	53.2	50.2	45.6	34.2	34.6
2011	31.3	29.3	34.7	36.4	40.6	49.4	55.3	54.3	53.4	44.4	34.1	28.7
2012	31	31.9	33.9	39.2	44.7	49.3	54.9	56.8	52.2	44.4	38.3	30.4
2013	27.1	30	36.5	40.9	46.1	53	58.6	54.5	50.4	40.5	36.8	27.7
2014	33.9	35.5	37.9	40.1	46.3	52.6	60.2	56.8	53	46.1	38.4	37.2
2015	35.4	38.4	39.8	39	45.7	56.1	58.2	56.1	52.1	48.1	32.6	30.6
2016	33.8	36.3	37.8	42.2	47	53.1	56.2	56.2	50.7	44	38.2	29.2
2017	29.8	35.1	37.6	38.8	46.5	54.2	58.8	59.8	52.8	42.1	38.9	31.6
2018	35.8	30.4	33.2	39	47	51.8	60.2	55.4	49.7	44.1	36.8	32.2
2019	34.2	28.8	34.8	42.8	44.2	53.4	56.6	57.3	50.1	39.8	37.3	34.2

The average temperature of the Sacramento Valley from 1996 through 2019 is depicted in the following graph. The warmest month is July with average temperatures in 90° Fahrenheit, and the coldest month is December with a temperature around 30° F.

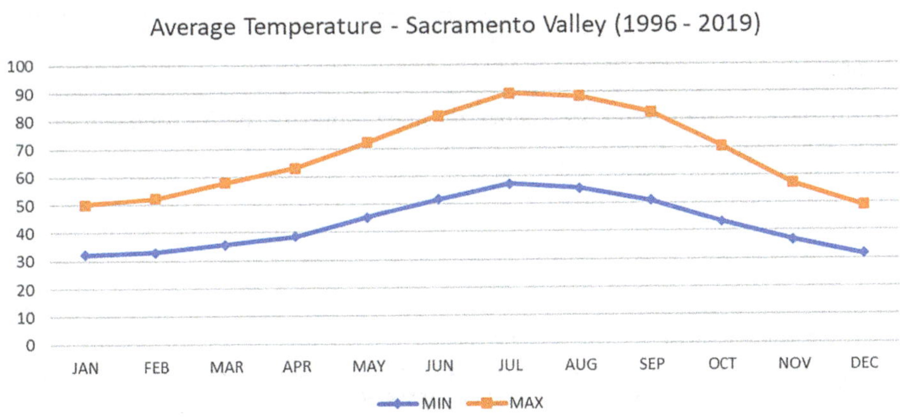

Combine weather and temperature datasets into a single data frame:

```
Weather_df = pd.concat([pcpdf_California,Pdsifinaldf_cal,Tmax_df2,Tmin_df2],axis=1)
```

Average rainfall and precipitation in Sacramento from 1996 to 2020[76] is as follows (please refer to 1971–2000[77]):

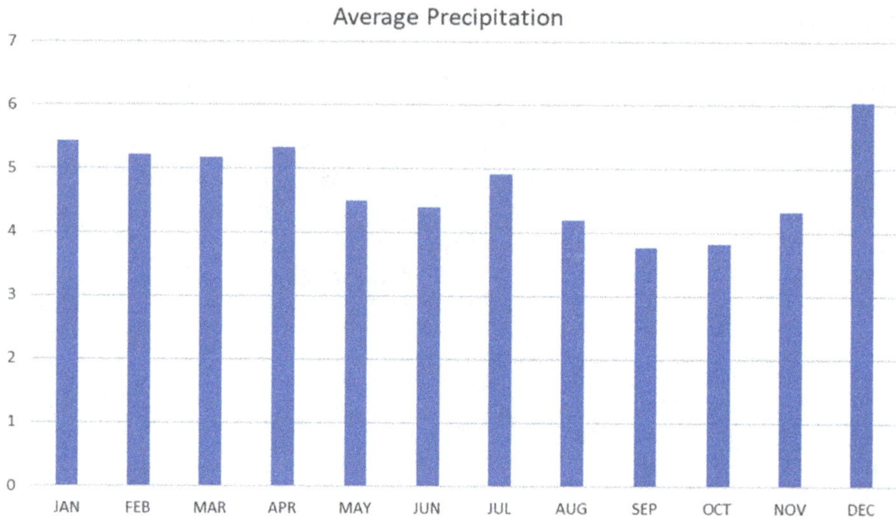

Step 5: *Load Weather Events Data*

Load weather events data—rain and heavy rain weather events data:

```
werainheavyrain_floodDF = pd.read_csv("updated_noaa_storm_we_rain_heavyrain_flood.csv")
werainheavyrain_floodDF .head()
```

Output:

The following table summarizes heavy weather events in the Sacramento Valley from 1996 through 2017:

[76] PCP - ftp://ftp.ncdc.noaa.gov/pub/data/cirs/climdiv/

[77] Precipitation - http://www.rssweather.com/climate/California/Sacramento/#:~:text=Sacramento's%20coldest%20month%20is%20December,rises%20to%2092.4%C2%B0F

year	Jan	Feb	Mar	Apr	May	Jun	July	Aug	Sep	Oct	Nov	Dec
1996	0	2	2	1	0	0	0	0	0	1	0	3
1997	7	0	0	0	0	0	0	0	0	0	3	2
1998	4	12	1	0	6	1	0	0	0	0	1	0
1999	1	2	0	0	0	0	0	0	0	0	1	0
2000	14	16	2	2	2	1	0	0	2	3	0	0
2001	8	9	12	3	0	0	0	0	1	1	4	7
2002	3	0	2	1	2	0	0	0	0	0	4	7
2003	0	3	1	5	1	0	0	1	1	0	3	6
2004	1	3	0	0	0	0	1	0	0	5	1	6
2005	9	1	8	2	6	2	0	1	2	0	0	6
2006	6	1	6	8	1	0	4	0	0	0	1	2
2007	1	5	1	0	0	0	0	0	2	3	0	8
2008	4	4	0	0	0	0	0	0	0	1	1	3
2009	0	4	2	0	1	0	0	0	0	6	0	4
2010	4	5	4	4	2	0	0	0	0	1	7	5
2011	4	5	9	0	3	9	2	0	0	1	0	0
2012	2	1	5	2	0	0	0	0	0	1	6	3
2013	0	2	0	0	1	0	0	0	0	0	0	0
2014	1	4	0	1	2	0	0	0	0	0	0	6
2015	0	0	0	1	1	1	0	0	0	0	0	0
2016	15	0	0	0	19	0	0	0	0	1	1	12
2017	30	7	0	0	0	0	0	0	0	0	0	0

Heavy rain events are grayed in the above table. Please kindly note that the temperatures need to be sufficiently warm to allow the rice seed to germinate and the seedling to grow. Optimal temperatures[78] for seedling emergence and growth are above 77° F in the Sacramento Valley [30]. Such temperatures typically occur late April in most parts of the Sacramento Valley. Any rains and cool weather during April will have detrimental effect on the rice yield as it delays growth and harvest and could encroach late October and early November; by then, cool weather will increase drying costs due to having to harvest grain at higher moisture contents. In essence, the temperatures during early planting and lower temperatures during late harvest months will have huge impact on the rice yield. The below radar view provides excessive weather events in Sacramento from 1996 to 2020. The following depicts the radar view of heavy rain events (please see Fig. 5.14).

As can be seen from the radar view, January 2017 had witnessed highest weather events—30—followed by May 2016 (19 events).

Frost/Freeze Event:[79] Event ID 287791[80]

Begin Date	2011-02-26 21:00 PST-8
End Date	2011-02-27 09:00 PST-8
Crop Damage	100.00K

[78] When is the optimal time to plant rice in the Sacramento Valley? - http://rice.ucanr.edu/files/232444.pdf

[79] Hard Freeze: California Wine Country - https://wattsupwiththat.com/2011/02/26/hard-freeze-california-wine-country/

[80] Freeze Event - https://www.ncdc.noaa.gov/stormevents/eventdetails.jsp?id=287791

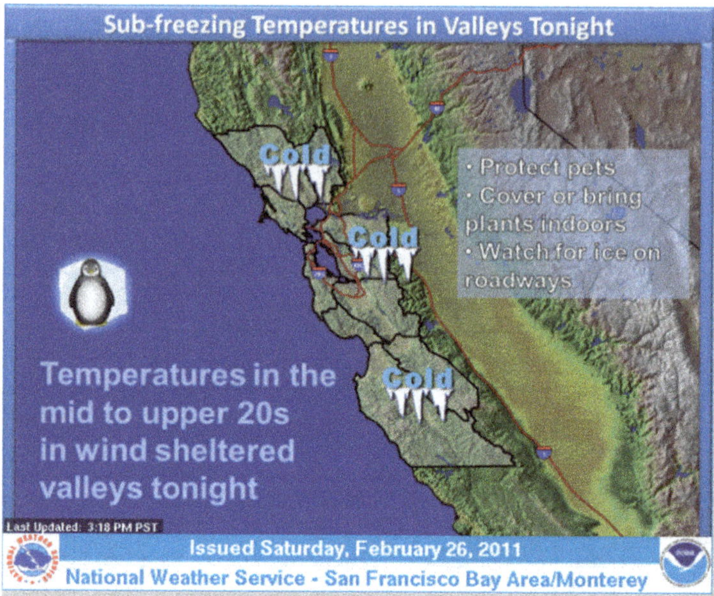

Weak ridging returned [32] to the area by the 20th, with dry, but relatively cool, conditions throughout the central California interior during the next few days. However, by February 24th, the next storm was approaching the California coast and weakening the ridge. Precipitation associated with this storm reached the region late on the 25th, with rain continuing into the 26th. Heavy rain caused some flooding in the south end of the San Joaquin Valley, with road flooding reported as far north as the Lamont area. This was a cold storm system, bringing measurable snowfall to the foothills along the Southern Sierra Nevada and the mountains along the west side of the San Joaquin Valley. A half inch of snow fell at Granite Station in the foothills of the Kern County Sierra Nevada, at an elevation of 2600 feet. The higher elevations of the Southern Sierra Nevada received up to 3 feet with this storm. During the daytime of the 26th, trace amounts of snow were even observed in some areas of the San Joaquin Valley, including Delano, Taft, Coalinga, Arvin, and some locations near Bakersfield. The end of the month remained cooler than average, with low temperatures falling to freezing or below due to the residual cold air behind the previous system. On the 27th, some locations, mainly north of Kern County, had minimum temperatures as low as the mid-20s to upper 20s. Lows the next morning were a couple of degrees warmer but were 32° or colder in several locations. In summary, February began dry and cool, followed by periods of mild temperatures throughout the central California interior. Until the middle of the month, there was no rain reported in Fresno and Bakersfield, and significant rain and snow did not occur throughout the region until the latter half of the month. Temperatures were cooler than average during this part of the month due to repeated influxes of cold air. Frosty to freezing conditions occurred at times throughout the central and southern San Joaquin Valley, especially during the last 2 days of the month.

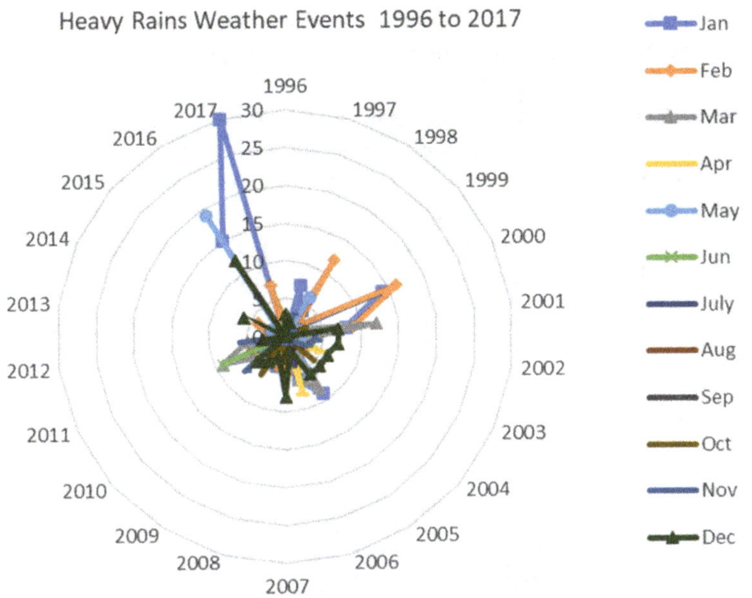

Fig. 5.14 Heavy rains

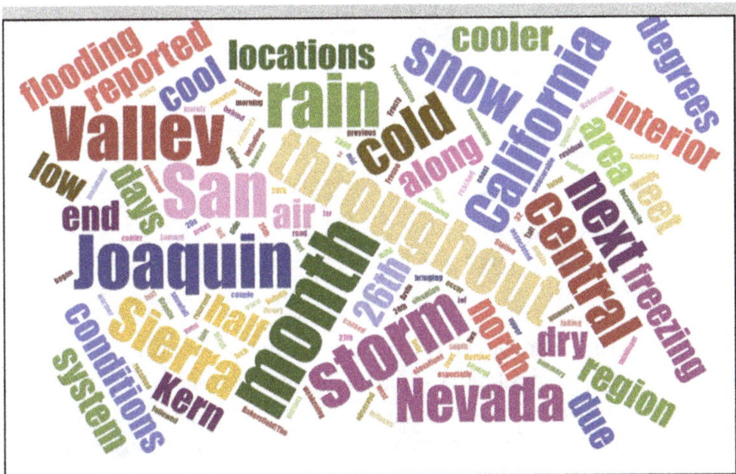

Frost/Freeze Event: Event ID 358907[81]

Begin Date	2011-12-08 00:00 PST-8
End Date	2011-12-08 09:00 PST-8
Crop Damage	25.00M

[81] Freeze Event - https://www.ncdc.noaa.gov/stormevents/eventdetails.jsp?id=358907

Widespread freezing temperatures continued into December 8th across the San Joaquin Valley. Seven consecutive days of freezing temperatures took its toll on the citrus crop in the San Joaquin Valley.

According to Citrus Growers' Association California Citrus Mutual, about 35% of the mandarin orange crop was lost to frost and freeze damage. The news was a little better for the navel crop; about 15% were not expected to make it. Initial estimates were not expected to be severe; however, after several weeks of inspecting the citrus crop, it was determined that damages were much worse than first expected.

87 million dollars was spent to protect crops from the extended frost and freeze.

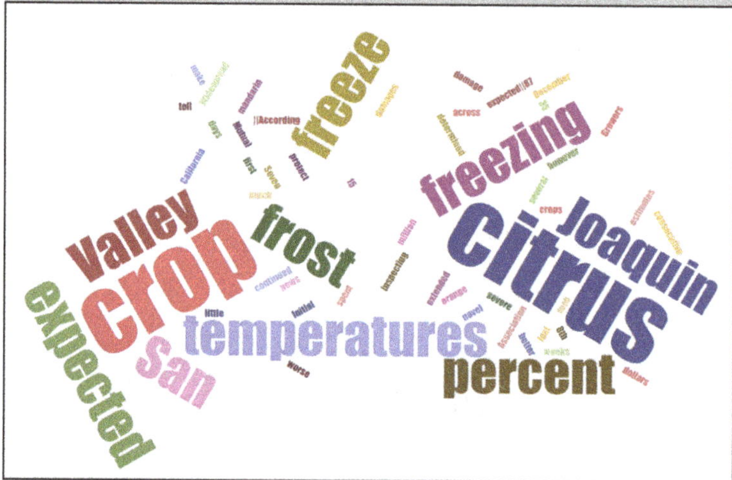

Frost/Freeze[82]

Begin Date 2013-12-05 00:00 PST-8
End Date 2013-12-10 00:00 PST-8
Crop Damage 110.00M

December began with an upper-level ridge of high pressure over the central California interior. This ridge brought clear skies and above-normal temperatures to the region for the first 2 days of the month. On December 2nd, some dense fog was also observed in the San Joaquin Valley, especially from Selma to Hanford. By the afternoon of the 2nd, the ridge of high pressure began to break down, and winds began to increase over the crest of the southern Sierra Nevada. This was a signal of an approaching storm system.

A cold front brought colder weather to the region on December 3rd, followed by a stronger system on December 6th and 7th. The first storm brought gusty winds to the mountains and desert, as well as light rain and drizzle over the San Joaquin Valley. Some light snow fell in Yosemite and other parts of the Sierra Nevada. Daytime high temperatures cooled to below normal in quite a few locations, especially north of Kern County, due to increased cloud cover and light precipitation.

[82] Freeze Event - https://www.ncdc.noaa.gov/stormevents/eventdetails.jsp?id=510057

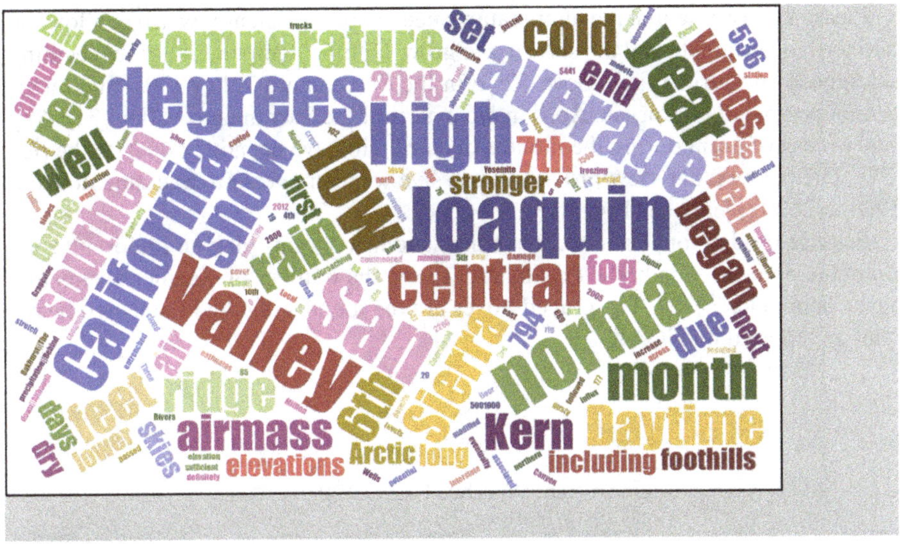

Excessive Heat

Excessive heat and excessive heat weather events play an important role in affecting the rice production. The following code loads NOAA Storm Events[83] data for heat and excessive heat:

Please load excessive heat-related weather events:

```
weheat_ExcessiveHeat_floodDF =
pd.read_csv("updated_noaa_storm_we_heat_ExcessiveHeat_flood.csv")
```

Output:

The following table summarizes excessive heat weather events in the Sacramento Valley from 1996 to 2017:

year	Jan	Feb	Mar	Apr	May	Jun	July	Aug	Sep	Oct	Nov	Dec
1999	0	0	0	0	0	0	2	0	0	0	0	1
2000	0	0	0	0	4	4	1	3	2	0	0	0
2001	0	0	0	0	1	0	2	0	0	0	0	0
2002	0	0	0	0	0	0	2	0	0	0	0	0
2003	0	0	0	0	1	4	2	0	0	2	0	0
2004	0	0	2	2	0	0	0	2	2	0	0	0
2005	0	0	0	0	0	0	3	3	0	0	0	0
2006	1	0	0	0	2	2	7	0	0	0	0	0
2007	0	1	2	0	0	0	5	0	4	0	0	0
2008	0	0	0	1	3	2	4	1	0	0	3	0
2009	0	0	0	0	2	0	3	0	0	0	0	0
2011	0	0	0	0	0	0	2	0	0	0	0	0
2012	0	0	0	0	0	0	2	5	0	0	0	0
2013	0	0	0	0	0	3	2	0	0	0	0	0
2017	0	0	0	0	0	7	2	4	3	0	0	0
2018	0	0	0	0	0	3	3	0	0	0	0	0
2019	0	0	0	0	0	7	4	10	0	0	0	0
2020	0	0	0	0	4	8	5	8	5	0	0	0

[83] Storm Events Database - https://www.ncdc.noaa.gov/stormevents/

Please kindly note that the temperatures need to be sufficiently warm to allow the rice seed to germinate and the seedling to grow. Optimal temperatures[84] for seedling emergence and growth are above 77° F in the Sacramento Valley [30]. Such temperatures typically occur late April in most parts of the Sacramento Valley [31]. Any heat excessive event during the months of April will have negative effect on the rice [32]. The below radar view provides excessive weather events in Sacramento from 1996 to 2020 (please see Fig. 5.15).

For rice production, both excessive heat and heavy rains play an important role. Combine both the weather events dataset to see the number of weather events that have taken place[85] [33]. The grayed boxes are the months that had witnessed heavy rains as reported by NOAA.

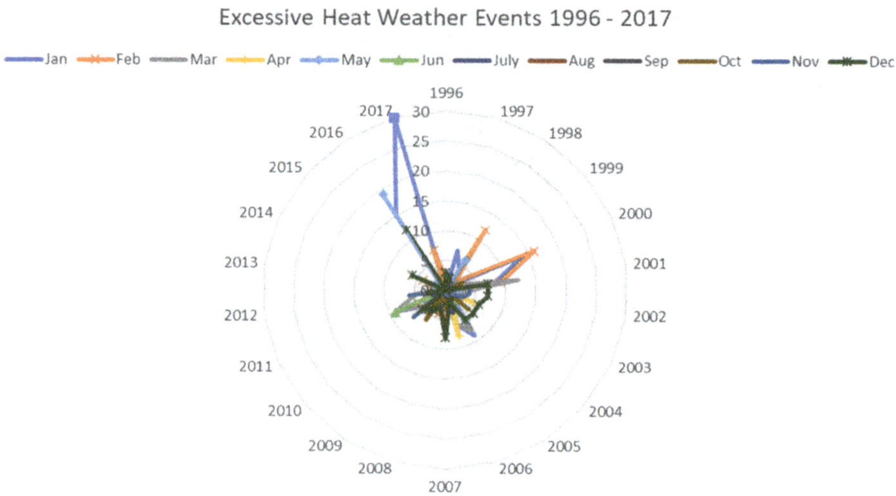

Fig. 5.15 Excessive heat

[84] When is the optimal time to plant rice in the Sacramento Valley? - http://rice.ucanr.edu/files/232444.pdf

[85] Rice yields decline with higher night temperature from global warming - https://www.pnas.org/content/101/27/9971

Year	JAN	FEB	MAR	APR	MAY	JUN	JUL	AUG	SEP	OCT	NOV	DEC
1996	0	2	2	1	0	0	2	0	0	1	0	4
1997	7	0	0	0	0	0	2	0	0	0	3	3
1998	4	12	1	0	6	1	2	0	0	0	1	1
1999	1	2	0	0	0	0	2	0	0	0	1	1
2000	14	16	2	2	6	5	1	3	4	3	0	0
2001	8	9	12	3	1	0	2	0	1	1	4	7
2002	3	0	2	1	2	0	2	0	0	0	4	7
2003	0	3	1	5	2	4	2	1	1	2	3	6
2004	1	3	2	2	0	0	1	2	2	5	1	6
2005	9	1	8	2	6	2	3	4	2	0	0	6
2006	7	1	6	8	3	2	11	0	0	0	1	2
2007	1	6	3	0	0	0	5	0	6	3	0	8
2008	4	4	0	1	3	2	4	1	0	1	4	3
2009	0	4	2	0	3	0	3	0	0	6	0	4
2010	4	5	4	4	2	0	0	0	0	1	7	5
2011	4	5	9	0	3	9	4	0	0	1	0	0
2012	2	1	5	2	0	0	2	5	0	1	6	3
2013	0	2	0	0	1	3	2	0	0	0	0	0
2014	1	4	0	1	2	0	0	0	0	0	0	6
2015	0	0	0	1	1	1	0	0	0	0	0	0
2016	15	0	0	0	19	0	0	0	0	1	1	12
2017	30	7	0	0	0	7	2	4	3	0	0	0
2018	0	0	0	0	0	3	3	0	0	0	0	0
2019	0	0	0	0	0	7	4	10	0	0	0	0

Combination of heavy rains and excessive heat are critical factors in determining the weather events on rice production (please see Fig. 5.16).

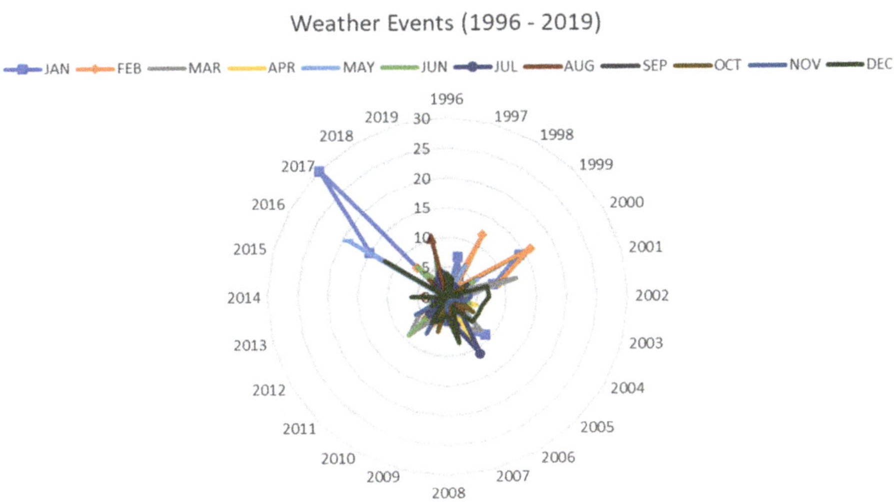

Fig. 5.16 Weather events

Step 6: *Construct Model*

Combine rice yield, minimum temperature, maximum temperature, PDSI, precipitation (PCP), and weather events (heavy rains and excessive temperature) data frames into final data frame to run machine learning models:

```
finalcalfdf = pd.read_csv('californiaRiceAllDatesetFinalwithWeatherEvents.csv')
finalcalfdf.columns
```

Output: List all the columns of data frame:

```
Index(['Year', 'YIELD(LB/ACRE)_rice', 'ACRES HARVESTED_rice',
    'PRODUCTION VALUE $_rice', 'JAN_mintemp', 'FEB_mintemp', 'MAR_mintemp',
    'APR_mintemp', 'MAY_mintemp', 'JUN_mintemp', 'JUL_mintemp',
    'AUG_mintemp', 'SEP_mintemp', 'OCT_mintemp', 'NOV_mintemp',
    'DEC_mintemp', 'JAN_maxtemp', 'FEB_maxtemp', 'MAR_maxtemp',
    'APR_maxtemp', 'MAY_maxtemp', 'JUN_maxtemp', 'JUL_maxtemp',
    'AUG_maxtemp', 'SEP_maxtemp', 'OCT_maxtemp', 'NOV_maxtemp',
    'DEC_maxtemp', 'JAN_pcp', 'FEB_pcp', 'MAR_pcp', 'APR_pcp', 'MAY_pcp',
    'JUN_pcp', 'JUL_pcp', 'AUG_pcp', 'SEP_pcp', 'OCT_pcp', 'NOV_pcp',
    'DEC_pcp', 'JAN_pdsi', 'FEB_pdsi', 'MAR_pdsi', 'APR_pdsi', 'MAY_pdsi',
    'JUN_pdsi', 'JUL_pdsi', 'AUG_pdsi', 'SEP_pdsi', 'OCT_pdsi', 'NOV_pdsi',
    'DEC_pdsi', 'JAN_weatherevents', 'FEB_weatherevents',
    'MAR_weatherevents', 'APR_weatherevents', 'MAY_weatherevents',
    'JUN_weatherevents', 'JUL_weatherevents', 'AUG_weatherevents',
    'SEP_weatherevents', 'OCT_weatherevents', 'NOV_weatherevents',
    'DEC_weatherevents', 'Year.1', 'year'],
   dtype='object')
```

Please kindly note all the columns are combined with Year as index. Plot correlation map among rice dataset columns:

```
sns.heatmap(finalricewithallmonthsdf.corr(method='pearson'))
```

Pearson's correlation statistic, or Pearson's correlation coefficient, is also known in statistical models as the r value. For any two variables, it returns a value that indicates the strength of the correlation. In rice dataset, we have analyzed Pearson's correlation coefficient between 'YIELD(LB/ACRE)_rice' and all other feature variables. Pearson coefficient is computed by taking the covariance of two variables and dividing the product of their standard deviations. The coefficient is not affected by changes of scale in the two variables.

Output: The correlation map provides insights into columns FEB_PCP, MAY_PCP, AUG_PCP, FEB_PDSI, and NOV_PDSI, which have higher correlation.

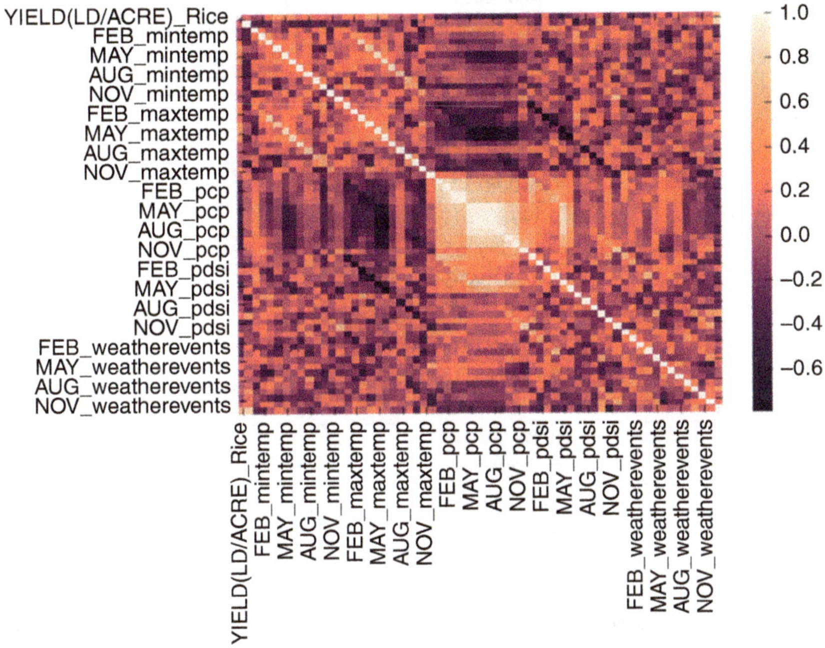

	YIELD(LB/ACRE)_rice		
YIELD(LB/ACRE)_rice	1	JAN_pcp	**−0.315468745**
ACRES HARVESTED_rice	0.101745314	FEB_pcp	**−0.357489939**
PRODUCTION VALUE $_rice	**0.614537943**	MAR_pcp	**−0.249580837**
JAN_mintemp	-0.095272335	APR_pcp	**−0.323798202**
FEB_mintemp	-0.057488123	MAY_pcp	**−0.458860643**
MAR_mintemp	0.274875293	JUN_pcp	**−0.443619981**
APR_mintemp	**0.467697382**	JUL_pcp	**−0.457098738**
MAY_mintemp	**0.398833481**	AUG_pcp	**−0.468094655**
JUN_mintemp	**0.450511599**	SEP_pcp	**−0.518252067**
JUL_mintemp	0.138560993	OCT_pcp	**−0.458464047**
AUG_mintemp	**0.307772046**	NOV_pcp	**−0.506010122**
SEP_mintemp	0.140268828	DEC_pcp	-0.200030374
OCT_mintemp	0.163180036	JAN_pdsi	-0.181988819
NOV_mintemp	0.099777518	FEB_pdsi	-0.285016822
DEC_mintemp	0.040271585	MAR_pdsi	0.174088871
JAN_maxtemp	**0.342983226**	APR_pdsi	-0.240646773
FEB_maxtemp	**0.343437482**	MAY_pdsi	**−0.464401351**
MAR_maxtemp	0.170495724	JUN_pdsi	**−0.210706989**
APR_maxtemp	0.50240163	JUL_pdsi	-0.12702632

	YIELD(LB/ACRE)_rice		
MAY_maxtemp	**0.461358226**	AUG_pdsi	0.005936639
JUN_maxtemp	**0.466046518**	SEP_pdsi	0.014826052
JUL_maxtemp	0.065587307	OCT_pdsi	0.180998429
AUG_maxtemp	0.044978245	NOV_pdsi	**−0.316718726**
SEP_maxtemp	0.080528104	DEC_pdsi	-0.121638942
OCT_maxtemp	-0.026440265		
NOV_maxtemp	0.37669215		
DEC_maxtemp	0.04008137		
JAN_weatherevents	0.147208617		
FEB_weatherevents	-0.35560265		
MAR_weatherevents	**−0.223067044**		
APR_weatherevents	**−0.309511868**		
MAY_weatherevents	0.045692022		
JUN_weatherevents	0.238715718		
JUL_weatherevents	**−0.197450344**		
AUG_weatherevents	**0.221506944**		
SEP_weatherevents	0.00295264		
OCT_weatherevents	0.090366981		
NOV_weatherevents	-0.093335316		
DEC_weatherevents	0.084551146		

As you can see, the bolded variables exhibit positive/negative correlation. For instance, YIELD(LB/ACRE)_rice is highly correlated with the amount of rice produced[86] [33]. Makes sense as more rice produced[87] [34], the yield (LB/acre) would also be tended to go up. On the other hand, if less is produced, the yield (lb per acre) would also lower. Minimum temperatures during April, May, June, and August positively correlated with YIELD(LB/ACRE)_rice. January, February, May, and June maximum temperatures positively correlated with the YIELD(LB/ACRE)_rice. January through November precipitation values would negatively affect the rice yield. May, June, and November PDSI values negatively affect rice yield. Finally, climate change[88,89] [35] and weather events in March, April, and July would have negative correlation with yield, meaning increased number of weather events, excessive temperature, or heavy rains would reduce rice yield. It is very clear that the temperature, precipitation, rains, weather events, and drought influence the rice yield[90] [36]. The same applies to other commodities such as wheat.[91]

[86] Rice yields decline with higher night temperature from global warming - https://www.pnas.org/content/101/27/9971

[87] The response of milk production to price: a regional analysis - https://core.ac.uk/download/pdf/38913888.pdf

[88] Indicator of Climate Change in California - https://oehha.ca.gov/media/epic/downloads/cc_drought2018.pdf

[89] Climate change, the monsoon, and rice yield in India - http://hpccc.gov.in/PDF/Agriculture/Climate%20Change,%20the%20Monsoon,%20and%20Rice%20Yield%20in%20India.pdf

[90] Economies of Size in U.S. Field Crop Farming - https://naldc.nal.usda.gov/download/CAT87202149/PDF

[91] Wheat Data - https://www.ers.usda.gov/data-products/wheat-data/

Next, get X and Y values:

```
X = finalricewithallmonthsdf.iloc[:,0:62].values
y = finalricewithallmonthsdf.iloc[:,62].values
```

Split data into train and test datasets:

```
X_train,X_test,y_train,y_test = train_test_split(X,y, test_size = 0.2 )
```

Model: Linear Regression

Develop linear regression model:

```
# Train the model
from sklearn.linear_model import LinearRegression

# Fit a linear regression model on the training set
model = LinearRegression(normalize=False).fit(X1_train, y1_train)
print (model)
```

Output:
LinearRegression()

```
import numpy as np

predictions = model.predict(X1_test)
np.set_printoptions(suppress=True)
print('Predicted labels: ', np.round(predictions)[:10])
print('Actual labels   : ' ,y1_test[:10])
```

Output
The predicted values for the class variable YIELD(LB/ACRE)_rice:

```
Predicted labels: [44509708. 35703587. 40380595. 41597394. 45913399.]
Actual labels   : [41959000. 38631000. 41990000. 42346000. 46076000.]
```

Plot the accuracy of the model:

```
import matplotlib.pyplot as plt

%matplotlib inline

plt.scatter(y1_test, predictions)
plt.xlabel('Actual Labels')
plt.ylabel('Predicted Labels')
plt.title('Rice Yield Predictions')
# overlay the regression line
z = np.polyfit(y1_test, predictions, 1)
p = np.poly1d(z)
plt.plot(y1_test,p(y1_test), color='magenta')
plt.show()
```

Output:

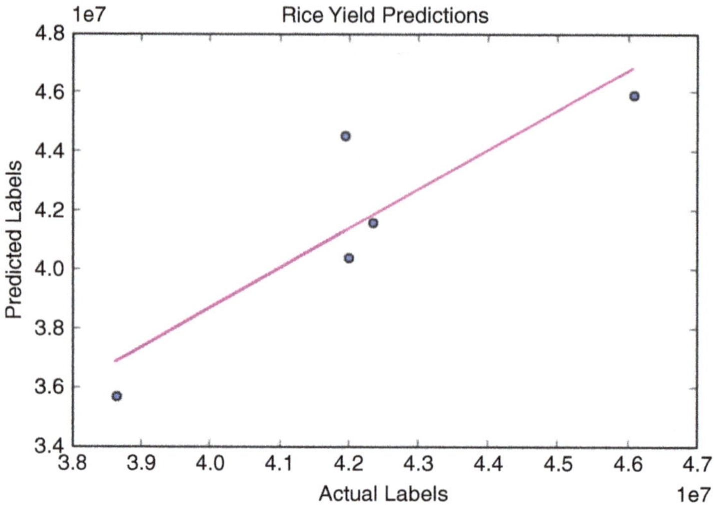

Calculate the accuracy of the model to determine feature variables relationship to class variable—YIELD(LB/ACRE)_rice. The R^2 score value can be derived from:

```
from sklearn.metrics import mean_squared_error, r2_score

mse = mean_squared_error(y1_test, predictions)
print("MSE:", mse)

rmse = np.sqrt(mse)
print("RMSE:", rmse)

r2 = r2_score(y1_test, predictions)
print("R2:", r2)
```

Output:

```
MSE: 3650578517681.7656
RMSE: 1910648.716452547
R2: 0.3454139253870899
```

As you can see, the model R^2 value can be improved by optimizing features.

Step 7: *Construct Model with Optimized Features Based on Rice Crop Calendar*

The number of features included in the dataset consists of monthly minimum temperatures, monthly maximum temperatures, monthly precipitation, monthly PDSI, weather events pertaining excessive temperature, and heavy floods. The pragmatic approach would include considering months that are rice agricultural crop months and remodel the dataset to see any increase in model accuracy.

Crop Calendar[92] [4]

The US rice-growing season varies by region. Here are very important months regarding the rice crop calendar in California (please see USDA US crop calendar[93]):

- Planting
  - California: from late April through May.
  - Texas and southwest Louisiana typically begins in early or mid-March.
  - Delta States: April.
- Harvest
  - California: end of September and finishes by early or mid-November
  - Texas and southwest Louisiana typically mid-July
  - Delta States: September and early October

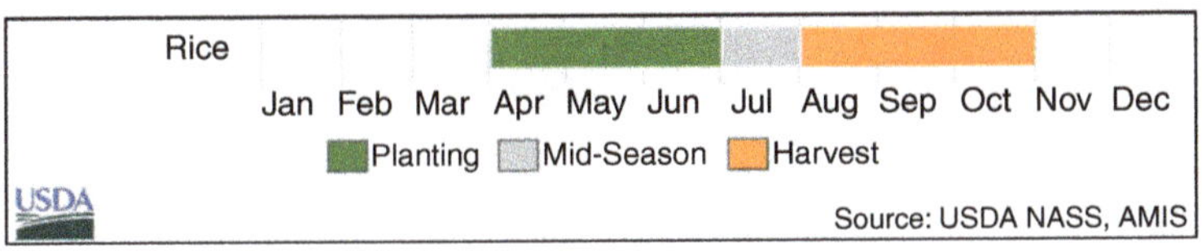

> A major difference between California and the southern US rice-growing region is nighttime temperatures. In California, even when the daytime high temperature is well above 100, night temperatures will usually drop into the 60s. In the south, during a period of very high temperatures, often the nighttime temperatures will only drop to the high 70s or low 80s. These high nighttime temperatures can cause significant issues with both rice yield and grain quality.

Re-modify the features of rice data to include months as per the crop calendar:

- Planting months (heat temperature stress is one of the most environmental factors that influence rice growth, development, and yield.[94] Higher temperatures, importantly, have negative impact[95] on the rice crop but are more damaging when plants are first heading, and pollination is occurring!) [37]

---

[92] U.S. Rice Production in the New Millennium: Changes in Structure, Practices, and Costs https://www.ers.usda.gov/webdocs/publications/90926/eib-202.pdf?v=9628

[93] U.S. Crop Calendar - https://ipad.fas.usda.gov/rssiws/al/crop_calendar/us.aspx

[94] Effect of High Temperature at Different Growth Stages on Rice Yield and Grain Quality Traits - https://citeseerx.ist.psu.edu/viewdoc/download?doi=10.1.1.467.5427&rep=rep1&type=pdf

[95] The Heat is On for California Crop - https://www.usarice.com/news-and-events/publications/usa-rice-daily/article/usa-rice-daily/2020/06/24/the-heat-is-on-for-california-crop

- April, May, June, July
  - Key features: minimum temperature, maximum temperature
- Harvesting months
  - August, September, October

```
finalcalfdf.columns
```

Output:

```
'YIELD(LB/ACRE)_rice', 'ACRES HARVESTED_rice',
    'PRODUCTION VALUE $_rice', 'JAN_mintemp', 'FEB_mintemp', 'MAR_mintemp',
    'APR_mintemp', 'MAY_mintemp', 'JUN_mintemp', 'JUL_mintemp',
    'AUG_mintemp', 'SEP_mintemp', 'OCT_mintemp', 'NOV_mintemp',
    'DEC_mintemp', 'JAN_maxtemp', 'FEB_maxtemp', 'MAR_maxtemp',
    'APR_maxtemp', 'MAY_maxtemp', 'JUN_maxtemp', 'JUL_maxtemp',
    'AUG_maxtemp', 'SEP_maxtemp', 'OCT_maxtemp', 'NOV_maxtemp',
    'DEC_maxtemp', 'JAN_pcp', 'FEB_pcp', 'MAR_pcp', 'APR_pcp', 'MAY_pcp',
    'JUN_pcp', 'JUL_pcp', 'AUG_pcp', 'SEP_pcp', 'OCT_pcp', 'NOV_pcp',
    'DEC_pcp', 'JAN_pdsi', 'FEB_pdsi', 'MAR_pdsi', 'APR_pdsi', 'MAY_pdsi',
    'JUN_pdsi', 'JUL_pdsi', 'AUG_pdsi', 'SEP_pdsi', 'OCT_pdsi', 'NOV_pdsi',
    'DEC_pdsi', 'JAN_weatherevents', 'FEB_weatherevents',
    'MAR_weatherevents', 'APR_weatherevents', 'MAY_weatherevents',
    'JUN_weatherevents', 'JUL_weatherevents', 'AUG_weatherevents',
    'SEP_weatherevents', 'OCT_weatherevents', 'NOV_weatherevents',
    'DEC_weatherevents'
```

Keep only columns that are relevant from rice crop point of view:

```
# Only load harvest months

finalcalfdf.drop(['JAN_mintemp',  'FEB_mintemp',  'MAR_mintemp','AUG_mintemp','SEP_mintemp',
'OCT_mintemp', 'NOV_mintemp',
    'DEC_mintemp', 'JAN_maxtemp', 'FEB_maxtemp', 'MAR_maxtemp','AUG_maxtemp',
    'SEP_maxtemp', 'OCT_maxtemp', 'NOV_maxtemp',
    'DEC_maxtemp', 'JAN_pcp', 'FEB_pcp', 'MAR_pcp', 'AUG_pcp','SEP_pcp','OCT_pcp', 'NOV_pcp',
    'DEC_pcp', 'JAN_pdsi', 'FEB_pdsi', 'MAR_pdsi', 'AUG_pdsi','SEP_pdsi', 'OCT_pdsi', 'NOV_pdsi',

'DEC_pdsi','JAN_weatherevents','FEB_weatherevents','AUG_weatherevents','SEP_weatherevents','OC
T_weatherevents',
    'NOV_weatherevents','DEC_weatherevents'],axis=1 , inplace = True)
finalcalfdf.columns
```

Output:

```
Index(['YIELD(LB/ACRE)_rice', 'ACRES HARVESTED_rice',
       'PRODUCTION VALUE $_rice', 'APR_mintemp', 'MAY_mintemp', 'JUN_mintemp',
       'JUL_mintemp', 'APR_maxtemp', 'MAY_maxtemp', 'JUN_maxtemp',
       'JUL_maxtemp', 'APR_pcp', 'MAY_pcp', 'JUN_pcp', 'JUL_pcp', 'APR_pdsi',
       'MAY_pdsi', 'JUN_pdsi', 'JUL_pdsi', 'MAR_weatherevents',
       'APR_weatherevents', 'MAY_weatherevents', 'JUN_weatherevents',
       'JUL_weatherevents'],
      dtype='object')
```

The correlation matrix is:

```
finalricewithplantharvestmonthsdf = finalcalfdf
sns.heatmap(finalricewithplantharvestmonthsdf.corr(method='pearson'))
```

Output:

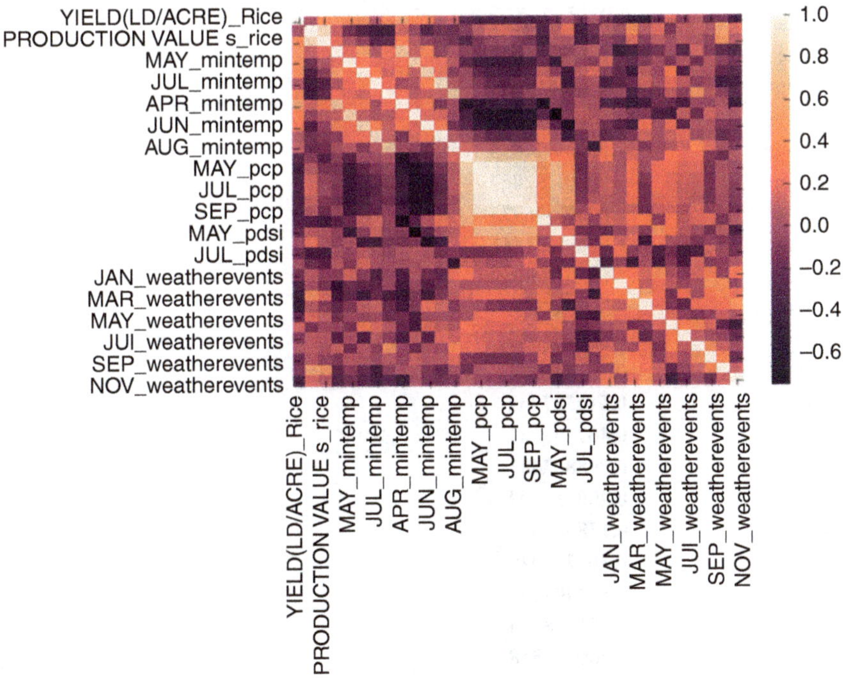

Except for few columns, all most all features have a correlation with the YIELD(LB/ACRE)_rice.

	YIELD(LB/ ACRE)_rice		
YIELD(LB/ACRE)_rice	1	JAN_weatherevents	0.147208617
PRODUCTION VALUE $_rice	0.101745314	FEB_weatherevents	**−0.35560265**
APR_mintemp	**0.614537943**	MAR_weatherevents	**−0.223067044**
MAY_mintemp	**0.467697382**	APR_weatherevents	**−0.309511868**
JUN_mintemp	**0.398833481**	MAY_weatherevents	0.045692022
JUL_mintemp	**0.450511599**	JUN_weatherevents	**0.238715718**
AUG_mintemp	0.138560993	JUL_weatherevents	**−0.197450344**
APR_maxtemp	**0.307772046**	AUG_weatherevents	**0.221506944**
MAY_maxtemp	**0.50240163**	SEP_weatherevents	0.00295264
JUN_maxtemp	**0.461358226**	OCT_weatherevents	0.090366981
JUL_maxtemp	**0.466046518**	NOV_weatherevents	−0.093335316
AUG_maxtemp	0.065587307	DEC_weatherevents	0.084551146
PRODUCTION VALUE $_rice	0.044978245		
APR_mintemp	**−0.323798202**		
MAY_mintemp	**−0.458860643**		
JUN_mintemp	**−0.443619981**		
JUL_mintemp	**−0.457098738**		
AUG_mintemp	**−0.468094655**		
APR_maxtemp	**−0.518252067**		
MAY_maxtemp	**−0.240646773**		
JUN_maxtemp	**−0.464401351**		
JUL_maxtemp	**−0.210706989**		
AUG_maxtemp	−0.12702632		
APR_pcp	0.005936639		
MAY_pcp	0.101745314		
JUN_pcp	**0.614537943**		
JUL_pcp	**0.467697382**		
AUG_pcp	**0.398833481**		
SEP_pcp	**0.450511599**		
APR_pdsi	**0.138560993**		
MAY_pdsi	**0.307772046**		
JUN_pdsi	**0.50240163**		
JUL_pdsi	**0.461358226**		
AUG_pdsi	**0.466046518**		

As it can be seen from the correlation matrix, the following signals can be inferenced:

- April, May, June, and July minimum temperature have positively correlated[96] with the Yield (LB per hectare):

[96] The Heat is On for California Crop - https://www.usarice.com/news-and-events/publications/usa-rice-daily/article/usa-rice-daily/2020/06/24/the-heat-is-on-for-california-crop

```
finalricewithplantharvestmonthsdf = finalcalfdf
finalricewithplantharvestmonthsdf['yPRODUCTION            VALUE            $_rice']            =
finalricewithplantharvestmonthsdf['PRODUCTION VALUE $_rice']
finalricewithplantharvestmonthsdf.drop(["PRODUCTION VALUE $_rice"],axis=1 , inplace = True)
finalricewithplantharvestmonthsdf.columns
```

Output:

```
Index(['YIELD(LB/ACRE)_rice', 'ACRES HARVESTED_rice', 'APR_mintemp',
    'MAY_mintemp', 'JUN_mintemp', 'JUL_mintemp', 'APR_maxtemp',
    'MAY_maxtemp', 'JUN_maxtemp', 'JUL_maxtemp', 'APR_pcp', 'MAY_pcp',
    'JUN_pcp', 'JUL_pcp', 'APR_pdsi', 'MAY_pdsi', 'JUN_pdsi', 'JUL_pdsi',
    'MAR_weatherevents', 'APR_weatherevents','MAY_weatherevents',
    'JUN_weatherevents', 'JUL_weatherevents', 'yPRODUCTION VALUE $_rice'],
   dtype='object')
```

Get X and y values of the dataset:

```
X = finalricewithplantharvestmonthsdf.iloc[:,0:23].values
y = finalricewithplantharvestmonthsdf.iloc[:,23].values
```

Perform train and test split. The purpose is to split the data, 80% training and 20% testing:

```
X_train,X_test,y_train,y_test = train_test_split(X,y, test_size = 0.2 )
```

Perform ML modeling with regression models. But ensemble models are performing well in explaining model outcomes.

Before applying linear regression model, use StandardScaler to convert the values into normal form.

Column transformer is used to transform the features and test values of the dataset into a normal StandardScaler.

```
# Train the model
from sklearn.compose import ColumnTransformer
from sklearn.pipeline import Pipeline
from sklearn.impute import SimpleImputer
from sklearn.preprocessing import StandardScaler, OneHotEncoder
from sklearn.linear_model import LinearRegression
import numpy as np

# Define preprocessing for numeric columns (scale them)
numeric_features = [0,1,2,3,4,5,6,7,8,9,10,11,12,13,14,15,16,17,18,19,20
,21,22]
numeric_transformer = Pipeline(steps=[
   ('scaler', StandardScaler())])
```

Construct the model by creating pipeline and performing fit on the pipeline.

Perform ordinary least squares linear regression. The linear regression fits a linear model with coefficients w = (w1, …, wp) to minimize the residual sum of squares between the observed targets in the dataset and the targets predicted by the linear approximation:[97]

```
# Combine preprocessing steps
preprocessor = ColumnTransformer(
  transformers=[
    ('num', numeric_transformer, numeric_features)])

# Create preprocessing and training pipeline
pipeline = Pipeline(steps=[('preprocessor', preprocessor),
                  ('regressor', LinearRegression(normalize=True))])
```

Normalize parameter is set to True to perform standard scaling on input dataset X. If True, the regressors X will be normalized before regression by subtracting the mean and dividing the l2-norm. If you wish to standardize, please use StandardScaler before calling fit on an estimator with normalize=False:

Build the model:

```
# fit the pipeline to train a linear regression model on the training set
model = pipeline.fit(X_train, (y_train))
print (model)
```

Output—Based on the number of features of the dataset, prepare StandardScaler and create pipeline:

```
Pipeline(steps=[('preprocessor',
         ColumnTransformer(transformers=[('num',
                          Pipeline(steps=[('scaler',
                                  StandardScaler())]),
                          [0, 1, 2, 3, 4, 5, 6, 7, 8, 9,
                           10, 11, 12, 13, 14, 15, 16,
                           17, 18, 19, 20, 21, 22])])),
        ('regressor', LinearRegression(normalize=True))])
```

Predict the model:

```
predictions = model.predict(X_test)
np.set_printoptions(suppress=True)
print('Predicted labels: ', np.round(predictions)[:10])
print('Actual labels: ' ,y_test[:10])
```

Output:

---

[97] Linear Regression - https://scikit-learn.org/stable/modules/generated/sklearn.linear_model.LinearRegression.html

Predicted labels: [45303900. 43829441. 34086771. 41373030. 37870405.]
Actual labels: [42346000. 43773000. 36881000. 41959000. 36767000.]

Generate R^2 matrix value:

```
mse = mean_squared_error(y_test, predictions)
print("MSE:", mse)
rmse = np.sqrt(mse)
print("RMSE:", rmse)
r2 = r2_score(y_test, predictions)
print("R2:", r2)
```

Output:

MSE: 3624187102692.093
RMSE: 1903729.7872051308
R^2: 0.5801603283665204

The model has 58% accuracy!
The plot of predicted to test values shows a linear curve:

```
plt.scatter(y_test, predictions)
plt.xlabel('Actual Labels')
plt.ylabel('Predicted Labels')
plt.title('Rice Predictions - Preprocessed')
z = np.polyfit(y_test, predictions, 1)
p = np.poly1d(z)
plt.plot(y_test,p(y_test), color='magenta')
plt.show()
```

Output:

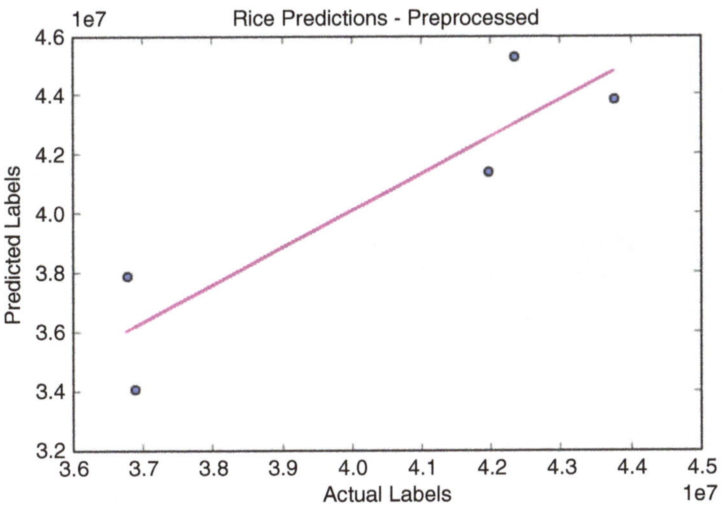

Ensemble Model: Extra Trees Regressor

Like random forest, the Extra Trees algorithm will randomly sample the features at each split point of a decision tree. Unlike random forest, which uses a greedy algorithm to select an optimal split point, the Extra Trees algorithm selects a split point at random.[98]

Set up extra tree regressor:

```
from sklearn.ensemble import ExtraTreesRegressor

pipeline = Pipeline(steps=[('preprocessor', preprocessor),
                ('regressor', ExtraTreesRegressor())])


# train a logistic regression model on the training set
model = pipeline.fit(X_train, (y_train))
print (model)
```

Output:

```
Pipeline(steps=[('preprocessor',
         ColumnTransformer(transformers=[('num',
                                 Pipeline(steps=[('scaler',
                                              StandardScaler())]),
                                 [0, 1, 2, 3, 4, 5, 6, 7, 8, 9,
                                  10, 11, 12, 13, 14, 15, 16,
                                  17, 18, 19, 20, 21, 22])])),
        ('regressor', ExtraTreesRegressor())])
```

Train the model by executing the pipeline and predict the values:

```
predictions = model.predict(X_test)

np.set_printoptions(suppress=True)
print('Predicted labels: ', np.round(predictions)[:10])
print('Actual labels: ' ,y_test[:10])
```

Output:

```
Predicted labels: [41782490. 43934980. 37992920. 41703630. 39505980.]
Actual labels: [42346000. 43773000. 36881000. 41959000. 36767000.]
```

Print output metrics:

---

[98] How to Develop an Extra Trees Ensemble with Python - https://machinelearningmastery.com/extra-trees-ensemble-with-python/

```
mse = mean_squared_error(y_test, predictions)
print("MSE:", mse)
rmse = np.sqrt(mse)
print("RMSE:", rmse)
r2 = r2_score(y_test, predictions)
print("R2:", r2)
```

Output:

MSE: 1829474480840.0
RMSE: 1352580.6744294404
R^2: 0.7880666909478399

The R^2 value is: 78.8%
The predicted test values:

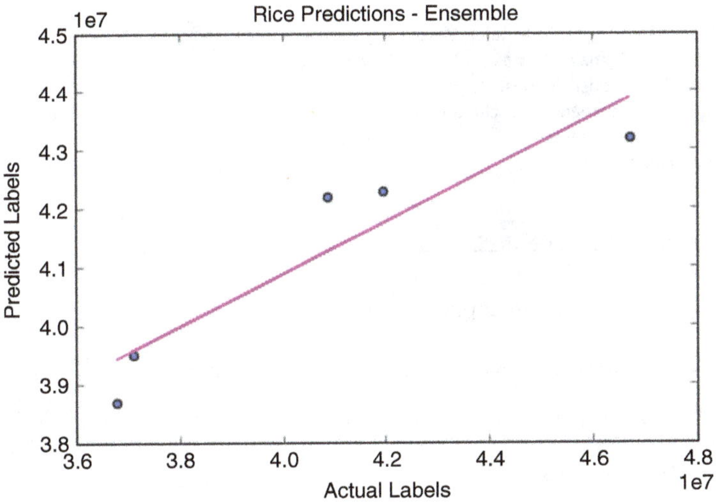

Step 8: *Model Interpretability*

Model interpretability is one of the important features of machine learning. Model interpretability enables to understand the model prediction process as well as local and global aggregated feature importance.

Set up explainer model:

```
#from interpret.ext.blackbox import MimicExplainer
# You can use one of the following four interpretable models as a global surrogate to the black box
model
from interpret.ext.glassbox import LGBMExplainableModel
from interpret.ext.glassbox import LinearExplainableModel
from interpret.ext.glassbox import SGDExplainableModel
from interpret.ext.glassbox import DecisionTreeExplainableModel

# OR

# 3. PFI Explainer
#from interpret.ext.blackbox import PFIExplainer
```

Create a table explainer. This process requires setting up of feature columns of the dataset:

```
feature_names= ['YIELD(LB/ACRE)_rice', 'ACRES HARVESTED_rice', '4_min', '5_min',
     '6_min', '7_min', '4_max', '5_max', '6_max', '7_max', '4_pcp', '5_pcp',
     '6_pcp', '7_pcp', '4_pdsi', '5_pdsi', '6_pdsi', '7_pdsi',
     '3_climatechange', '4_climatechange', '5_climatechange',
     '6_climatechange', '7_climatechange']
# 1. Using SHAP TabularExplainer
explainer = TabularExplainer(model,
                    X1_train,
                    features = feature_names)
```

Create global explanation by calling explain global:

```
# Passing in test dataset for evaluation examples - note it must be a representative sample of the
original data
# x_train can be passed as well, but with more examples explanations will take longer although they
may be more accurate
global_explanation = explainer.explain_global(X1_test)
```

Calculate global rank values:

```
# Sorted SHAP values
print('ranked global importance values: {}'.format(global_explanation.get_ranked_global_values()))
# Corresponding feature names
print('ranked global importance names: {}'.format(global_explanation.get_ranked_global_names()))
# Feature ranks (based on original order of features)
print('global importance rank: {}'.format(global_explanation.global_importance_rank))

# Print out a dictionary that holds the sorted feature importance names and values
print('global importance rank: {}'.format(global_explanation.get_feature_importance_dict()))
```

Output:

```
ranked global importance values: [1896222.273320876, 617335.914198136, 274546.13615135656, 188980.72409290433, 85801.95819643616, 85576.9786045961, 78904.03553408835, 60849.26726792266, 34618.51523726943, 27485.53135214704, 22741.231478163314, 21262.56799512282, 15968.93089076177, 5713.26996038683, 3419.3004313693846, 1418.8825128339522, 0.0, 0.0, 0.0, 0.0, 0.0, 0.0, 0.0]
ranked global importance names: ['ACRES HARVESTED_rice', 'YIELD(LB/ACRE)_rice', 'MAY_maxtemp', 'APR_pdsi', 'JUL_mintemp', 'MAY_mintemp', 'JUL_pdsi', 'JUN_mintemp', 'MAY_pdsi', 'JUN_pcp', 'JUN_pdsi', 'APR_weatherevents', 'APR_mintemp', 'MAR_weatherevents', 'JUL_pcp', 'MAY_weatherevents', 'JUL_weatherevents', 'APR_maxtemp', 'JUN_maxtemp', 'JUL_maxtemp', 'APR_pcp', 'JUN_weatherevents', 'MAY_pcp']
global importance rank: [1, 0, 7, 14, 5, 3, 17, 4, 15, 12, 16, 19, 2, 18, 13, 20, 22, 6, 8, 9, 10, 21, 11]
global importance rank: {'ACRES HARVESTED_rice': 1896222.273320876, 'YIELD(LB/ACRE)_rice': 617335.914198136, 'MAY_maxtemp': 274546.13615135656, 'APR_pdsi': 188980.72409290433, 'JUL_mintemp': 85801.95819643616, 'MAY_mintemp': 85576.9786045961, 'JUL_pdsi': 78904.03553408835, 'JUN_mintemp': 60849.26726792266, 'MAY_pdsi': 34618.51523726943, 'JUN_pcp': 27485.53135214704, 'JUN_pdsi': 22741.231478163314, 'APR_weatherevents': 21262.56799512282, 'APR_mintemp': 15968.93089076177, 'MAR_weatherevents': 5713.26996038683, 'JUL_pcp': 3419.3004313693846, 'MAY_weatherevents': 1418.8825128339522, 'JUL_weatherevents': 0.0, 'APR_maxtemp': 0.0, 'JUN_maxtemp': 0.0, 'JUL_maxtemp': 0.0, 'APR_pcp': 0.0, 'JUN_weatherevents': 0.0, 'MAY_pcp': 0.0}
```

Check local values:

```
print('local importance values: {}'.format(global_explanation.local_importance_values))
```

Output:

```
local importance values: [[600650.8043038812, 294542.7913893126, -79844.65445380885, 32579.65807881899, 47148.12546274508, 51259.96281523962, 0.0, 199370.61637286807, 0.0, 0.0, 0.0, 0.0, 12589.357256063493, 17096.502156846924, 206064.4470245731, 13597.221411405248, 0.0, 50960.95765574463, 0.0, 0.0, 0.0, 0.0, 0.0], [-727022.2344351062, -1011991.1706259972, 0.0, 63482.9171242899, -108435.50243318907, 0.0, 0.0, -463478.63077535015, 0.0, 0.0, 0.0, 0.0, -124838.29950467171, 0.0, 173700.23188445577, 0.0, 0.0, 0.0, 0.0, -27481.52176074125, 0.0, 0.0, 0.0], [758862.8776090331, -1609621.9894016956, 0.0, 141643.34739213722, 87711.18391146144, 263440.28600678925, 0.0, 257079.0968315135, 0.0, 0.0, 0.0, 0.0, 0.0, 0.0, 155149.87433836184, 8540.627841687208, 58905.06019824583, 42924.66238005241, 28566.34980193415, 0.0, 7094.412564169761, 0.0, 0.0], [0.0, 3189147.511228302, 0.0, -138961.7331754867, 34465.4644278224, 58318.05581476021, 0.0, -289695.4417640769, 0.0, 0.0, 0.0, 0.0, 0.0, 0.0, 132778.49597948862, 131305.25979458477, -54801.09719257074, -237282.04385404533, 0.0, 78831.31821487285, 0.0, 0.0, 0.0], [1000143.6546426592, 3375807.903959072, 0.0, 51217.237252285704, 26486.060104395263, 55991.48634539172, 0.0, 163106.89501297427, 0.0, 0.0, 0.0, 0.0, 0.0, 0.0, 277210.5712376423, 19649.46713866992, 0.0, 63352.51378059946, 0.0, 0.0, 0.0, 0.0, 0.0]]
```

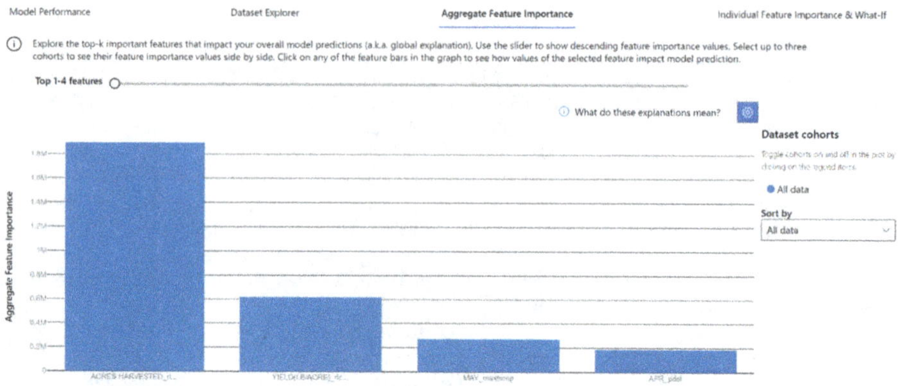

Individual features:

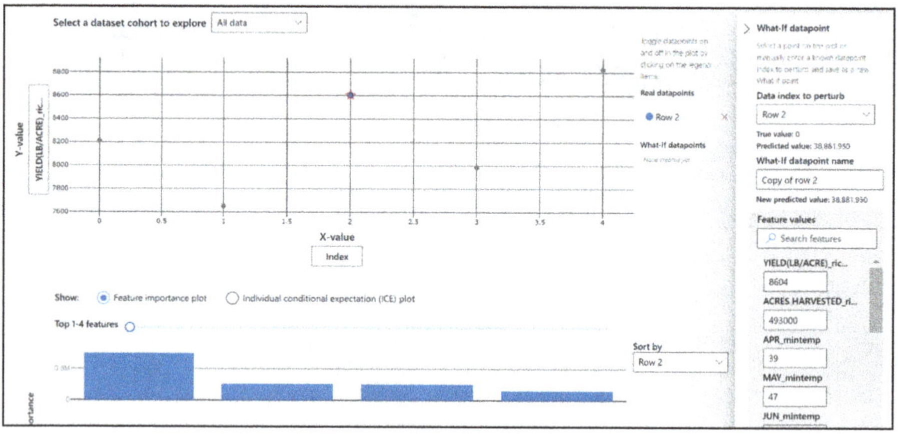

Milk Production

Farmers all over the world, in dealing with costs, returns, and risks, are calculating economic agents and citizen data scientists. Within their small individual, allocative domain, they are fine-tuning entrepreneurs, tuning so subtly that many experts fail to see how efficient they are.[99] Small farmers around the world face economic challenges and stresses of maintaining farm afloat—running farm and maintaining their individual family and preserving their inheritance, farm. Farmers rely outside farm revenue to sustain their lives. In this regard, milk plays an important role. Some farmers consider dairy cattle as part of their family. Milk in essence is a cash cow for small farmers.

[99] Relationship between milk production and price variations in the EC - http://aei.pitt.edu/36837/1/A67.pdf

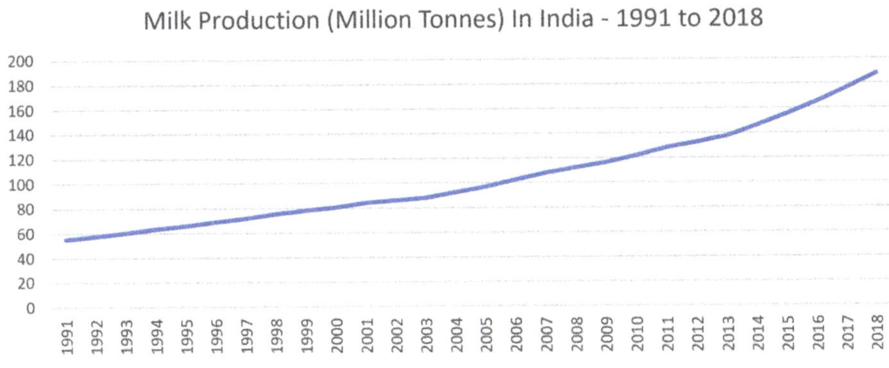

Fig. 5.17 Milk production in India

Generally, agricultural crops need at least 3–6 months to get to fruition from seeding to marketing and selling of the crops. Milk, on the other hand, offer immediate cash to farmers and help them to sustain long selling cycles of crops.

Milk production in India (please see Fig. 5.17) has increased dramatically at the rate of 4.44% CAGR with 55.6 million tons in 1991–1992 and reached 187.7 million in 2018–2019.[100] Despite the increase, India still faces daunting task of making all citizens have the availability of milk as daily consumption staple. While the Indian Council of Medical Research (ICMR) recommends an average daily intake of 300 g per day of milk to deliver the requisite macro-micronutrients, the national average of per capita consumption is about 185 and 146 g a day, respectively, in urban and rural India. Ironically, despite the fact that the National Dairy Development Board (NDDB) statistics provide that per capita availability per day reached 394 gm, still there are many citizens in India that don't have minimum consumption per day.[101]

Milk production has increased 0.4% in 2019 to 218 billion pounds.[102] The rate per cow, at 23,391 pounds, was 241 pounds above 2018. The annual average number of milk cows on farms was 9.34 million head, down to 62,000 head from 2018 (please see Fig. 5.18).

Above normal rains, flooding, and abnormal rainfall not only lower crop yield production but also cause stress in dairy cows to the point of limited milk production[103]

California produced the most milk in the United States in 2019, followed by Wisconsin and Idaho. Seven (7) states produced over ten billion pounds of milk in

[100] NDDB - https://www.nddb.coop/information/stats/milkprodindia

[101] Interventions in food systems can help push up India's nutritional figures to higher levels - https://economictimes.indiatimes.com/blogs/et-commentary/interventions-in-food-systems-can-help-push-up-indias-nutritional-figures-to-higher-levels/

[102] Ranking of States That Produce The Most Milk - https://beef2live.com/story-milk-production-ranking-state-0-111564#:~:text=California%20produced%20the%20most%20milk,the%20United%20States%20in%202019

[103] Storm Data - https://www.google.com/books/edition/Storm_Data/TavgYM2vpREC?hl=en&gbpv=1&bsq=deciduous

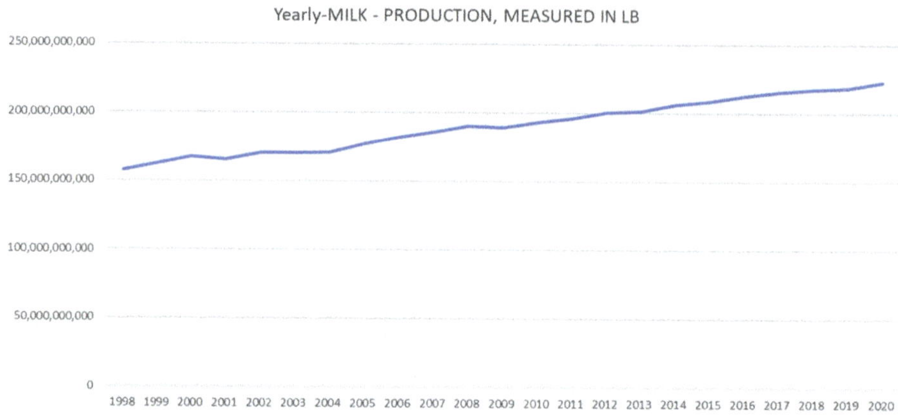

Fig. 5.18 Per capita

2019: California, Wisconsin, Idaho, New York, Texas, Michigan, and Pennsylvania. California accounted for nearly 19% of the milk produced in the United States in 2019.

Relationship Between Milk Production and Price Variations

One of the main difficulties to precisely quantify the relationship between milk production and price variations for milk is established by the fact that these price changes are only one factor among others influencing the behaviors of producers.[104]

Price Elasticity of Supply

The first step is analyzing the relationship between milk production and producer price variation for milk in the form of a supply function, where the milk production[105] is regarded as a dependent variable and changes of which are caused by changes of an independent variable, milk price. In other words, milk production would be expressed as a function of milk prices.

The notion of price elasticity of supply:

- Milk prices: In the below graph (please see Fig. 5.19), we can see the following trend: A 1% change in the price farmers receive for milk changes national milk production about half of 1% in the same direction over four periods.

[104] Relationship between milk production and price variations in the EC - http://aei.pitt.edu/36837/1/A67.pdf

[105] The response of milk production to price - https://core.ac.uk/download/pdf/38913888.pdf

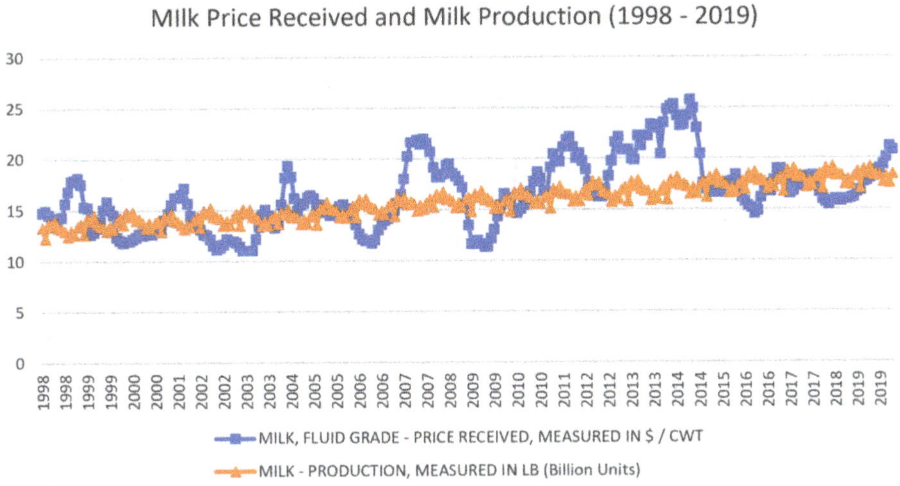

Fig. 5.19 Milk prices received

- Input costs: Input costs of running a dairy farm are represented by feed (hay and corn) prices. A 1% movement in corn price per bushel changes production by 0.075%. Corn prices significantly affect milk supply in majority of north states.[106]
- Profits in alternative farm enterprise: This factor is measured by feed and beef prices. A 1% change in beef price changes national milk production—0.056%.
- General economic conditions: These conditions are measured by the unemployment rate. A 1% change in unemployment affects national milk production by 0.085%.
- The difference between milk production's response to milk price increases, while its response decreases (a trend we can see in the below figure).

Simply put, the price elasticity of milk price to milk produced follows demand and supply curve.

$P_m, Q_m; Y$	Demand for milk
$P_m, Q_{m,} P_{ff}; P_l$	Supply of milk
$P_m, P_{ff}, Q_{ff,}; P_l$	Demand for feed

Where:

P_m = Price of milk
Q_m = Quantity of milk produced
Y = Per capital consumer income
P_{ff} = Price of feed fed to dairy cows
P_l =Price received for competing livestock (hogs and beef)

[106] Factors affecting U.S. Milk Production - https://naldc.nal.usda.gov/download/CAT87202245/PDF

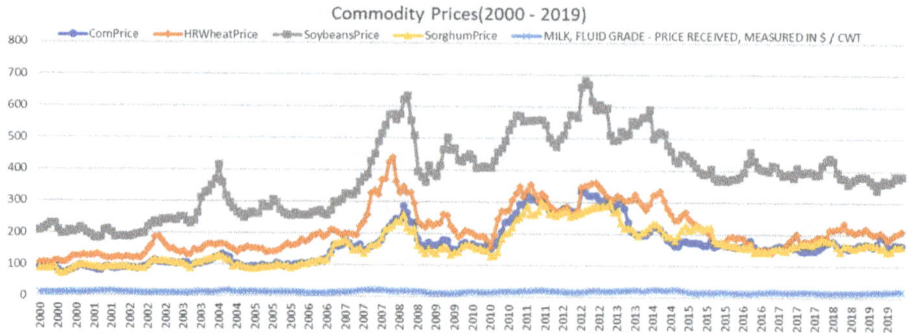

Fig. 5.20 Commodity prices

Select commodity			
Program: Survey	Sectors: Animals & products	Group: Dairy	Commodity: Milk
Category: Price received	Data item: MILK—PRODUCTION, MEASURED IN LB	Domain: Total	
Select location			
Geographical level: National	State: US total		
Select time			
Year: 2012–2020			

The price swings of basic commodities (please see Fig. 5.20), namely, corn, soybeans, wheat, and sorghum, have huge implication on dairy industry. With tiny margins on the milk, any spike in the price of dairy feed will throw the dairy farmers on the verge of losing their dairy farms. According to American Dairy Products Institute (ADPI), the major swings in corn prices during 2011 had a detrimental effect on the dairy industry in California.[107] As per ADPI website, the nation's drought and high corn prices are devastating California's $8-billion dairy industry to the point where farmers can't afford to feed their cows—and their professional trade organization has been regularly referring despondent dairymen to suicide hotlines.

The prognosis of price increases in basic commodities is very dire. California dairy farmers, unlike those in Midwest, rely much more on purchasing feed rather than growing grain for feed. Due to this dependency, many dairy farmers in California had gone through enduring tough time.[108] Many have lost their family-owned dire, and some reduce their farm size due to lack of capacity to feed. Small dairy farmers hit very hard when price of basic commodities, especially animal

[107] California Expected to Lose 100 Dairy Farms By Year's End - https://www.adpi.org/Publications/PublicationCatalog/TabId/128/ArtMID/1671/ArticleID/1780/California-Expected-to-Lose-100-Dairy-Farms-By-Years-End.aspx

[108] High cost of feed hits livestock farmers - https://www.startribune.com/high-cost-of-feed-hits-livestock-farmers/182597701/

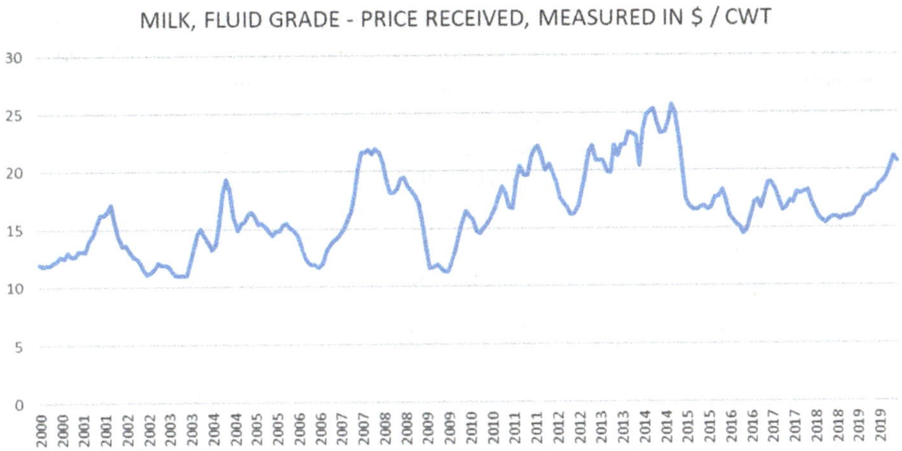

Fig. 5.21 Milk fluid grade

feed, increases or spikes.[109] One dire effect is many small farmers lose their dairy farms to large farms, a trend that can be witnessed from the last 20 years the US dairy industry, which has been transitioning to larger dairies of 1,000 head or more from smaller family farm dairies that might milk 50–200 cows. This would result in social imbalance and more causalities to small farmers, socially and financially (please see Fig. 5.21).

Agricultural Prices[110]

Dairy Feed:

- Corn—Corn stalks can be cost-effectively included in lactation cow rations. Producers need to use caution and not add corn stalks at excessive inclusion rates.[111]
- Wheat—Traditionally, the milling of wheat produces flour for human consumption and animal byproduct. Wheat becomes an attractive feed ingredient for the cattle, swine, and poultry industries when price gap between corn and wheat shrinks.[112] Both hard winter wheat (generally in high protein) and soft winter wheat (tends to be lower in protein[113]) are good dairy feed. As you can see from the above graph, the price shrink could be observed after 2015.
- Soybeans—Incorporating soybeans and their byproducts in the rations for dairy cattle is a fairly common practice.[114] They are an excellent source of essential

[109] Analysis: Small dairy farms worst hit by historic drought - https://www.reuters.com/article/us-usa-drought-dairy/analysis-small-dairy-farms-worst-hit-by-historic-drought-idUSBRE8790XN20120810

[110] Agricultural Prices - https://downloads.usda.library.cornell.edu/usda-esmis/files/c821gj76b/z316q273n/gt54kp41h/AgriPric-02-27-2018_correction.pdf

[111] Feeding corn stalks on dairy farms: Lactating cows - https://www.canr.msu.edu/news/feeding_corn_stalks_on_dairy_farms_lactating_cows

[112] U.S. Wheat Prices Competitive with Corn for Domestic Feed Use - https://www.uswheat.org/wheatletter/u-s-wheat-prices-competitive-with-corn-for-domestic-feed-use/

[113] Wheat for Animal Feed - http://www.omafra.gov.on.ca/english/livestock/beef/facts/wheat.htm

[114] Soybeans and Soybean Byproducts for Dairy Cattle - https://extension.psu.edu/soybeans-and-soybean-byproducts-for-dairy-cattle

amino acids, and they fit into any type of forage-based ration. If soybeans have been properly heat-treated, then they can provide additional rumen undegradable protein (RUP) and fat. Soybeans that have not been heated provide a source of degradable and soluble protein.

- Sorghum—Sorghum contains more crude protein than corn.[115] Sorghum grain can be utilized in the rations of dairy cattle as a replacement for corn. Although research has shown sorghum grain to be comparable to corn in lactating dairy cow diets, the market often values sorghum less than corn. Sorghum and corn can be interchanged in the diet of lactating dairy cows.

Ranking of States that Produce the Most Milk[116]

FeedVal[117] is a decision support tool to evaluate the actual value of feed ingredients and hence to help dairy farm producers, dairy farm nutritionists, and dairy farm consultants make economical optimal decisions for purchasing and using feed ingredients for dairy farm feed rations.

Analyze | Download Results | Convert all to kg | Refresh

Remove nutrients with negative predicted unit costs.

			Nutrients				As Fed Basis			Calculated	
		Ingredient	RUP %	RDP %	NEl3x Mcal/lb	peNDF %	DM %	Unit	Price* $/Unit	Predicted Value $/Unit	Actual Price as % of Predicted Value
1	☑	Shelled Corn	4.5	4.5	0.91	0	86	ton	142.86		
2	☑	Soybean Meal 48%	21	33	1	0	89	ton	388		
3	☑	Soybean Meal 44%	17.5	32.5	0.97	0	89	ton	376		
4	☐	Soybean Meal, expeller	30	16	1.09	0	92	ton			
5	☑	Soybeans, raw	12	28	1.25	0	87	ton	325.67		
6	☐	Soybeans, heated	22	21	1.24	0	92	ton			
7	☑	Good Quality Hay	6	14	0.6	35	87	ton	227		
8	☑	Poor Quality Hay	4.8	11.2	0.5	50	87	ton	156		
9	☑	Corn Silage	2.8	4.2	0.67	30	35	ton	40		
10	☐	Earlage/Snaplage	3.6	5.4	0.82	0	60	ton			
11	☑	Distillers Dried Grains	15	15	0.9	0	89	ton	178		
12	☐	High Moisture Corn	3.6	5.4	0.95	0	70	ton			
13	☑	Tallow	0	0	2.06	0	99	cwt	29		
14	☑	Blood Meal	76	19	1.06	0	94	ton	1100		
15	☑	Urea	0	287	0	0	99	ton	472		

Milk Prices Received

Two data sources were used to collect the milk prices received by farmers at the national US level. Prices received per month (dollar per cwt) from the USDA[118] are shown in Fig. 5.22.

The second source of milk prices received is NASS Quick Stats Database. Please kindly note that the Quick Stats Database contains till 2010 prices received for milk.

[115] Sorghum in Dairy Cattle Production - https://amarillo.tamu.edu/files/2010/11/dairy_handbook-FINAL.pdf

[116] Ranking of States That Produce The Most Milk - https://beef2live.com/story-milk-production-ranking-state-0-111564#:~:text=California%20produced%20the%20most%20milk,the%20United%20States%20in%202019.

[117] FeedVal - https://dairymgt.info/tools/feedval_12_v2/

[118] Prices Received per month - https://www.nass.usda.gov/Charts_and_Maps/Agricultural_Prices/pricemk.php

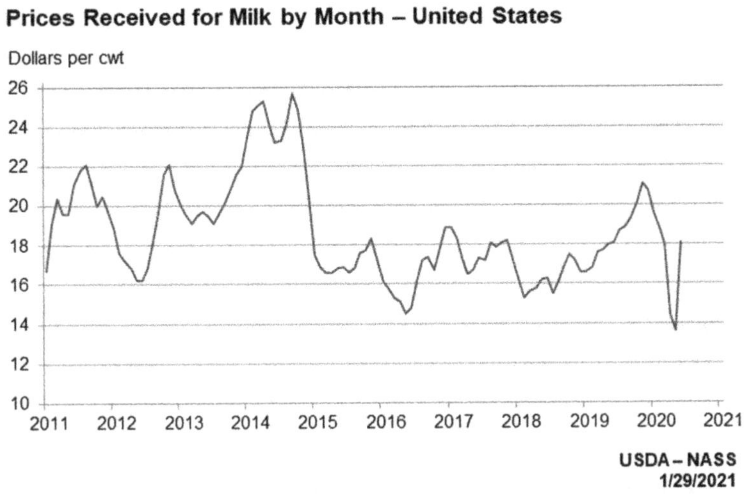

Fig. 5.22 Prices received for milk by month

Merged dataset from both sources to represent the complete data for milk received is from 1991 to 2020.

NASS Quick Stats Settings:

Select commodity (please see Fig. 5.23)			
Program: Survey	Sectors: Animals & products	Group: Dairy	Commodity: Milk
Category: Price received	Data item: MILK, FLUID GRADE—PRICE RECEIVED, MEASURED IN $ / CWT	Domain: Total	
Select location			
Geographical level: National	State: US total		
Select time			
Year: 2012–2020			

Basic Commodity Prices and Milk Production

Let's see if any relationship exists between milk production and commodity prices (please see Fig. 5.24). More precisely, the price fluctuation of 2008 and 2011has any influence on milk production across the United States. Temperature increase in summer will impact the milk prices in the world. In this analysis, we would like to see price changes have resulted in milk production decrease.

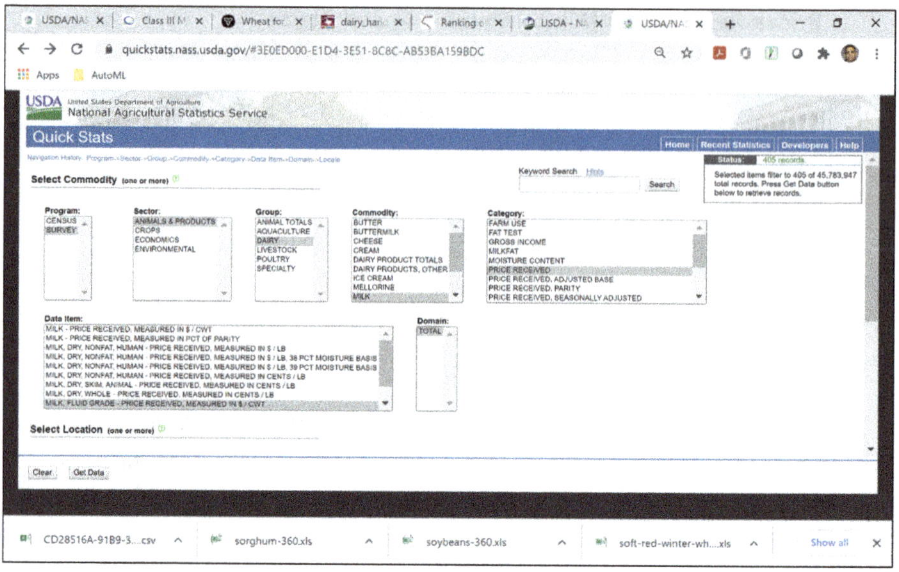

Fig. 5.23 USDA quick stats

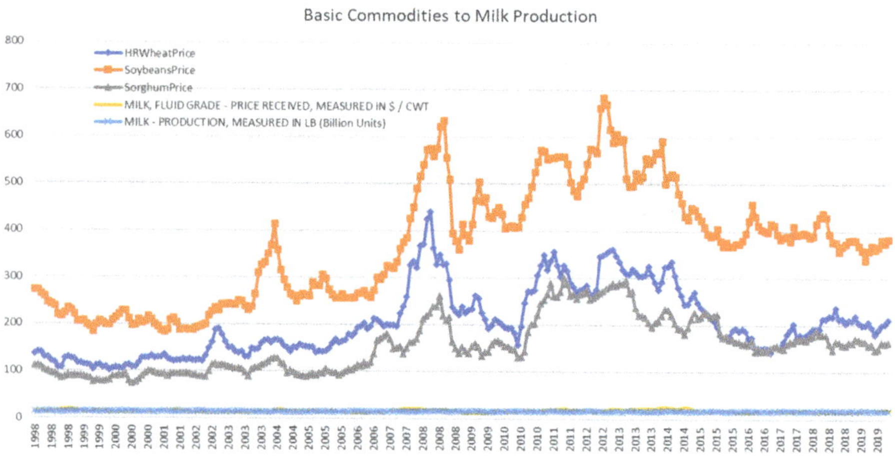

Fig. 5.24 Basic commodities to milk

	The software code for this model can be found on GitHub Link: FeatureEngineeringSilverGoldAgCommodities. Ipynb (Jupyter Notebook)

Step 1: *Load Dataset*

Load commodity prices (HR wheat, corn, sorghum, and soybeans):

```
import pandas as pd
DATA_DIR = "./data"
milkCommoditiesProductionRelation_df = pd.read_csv(os.path.join(DATA_DIR, 'wheat-corn-sorghum-milkproceswithmilkproduction1998-2019.csv'))
print(milkCommoditiesProductionRelation_df.shape)
milkCommoditiesProductionRelation_df.head()
```

Output:

Month	CornPrice	Change	MonthDate	month_number	month_name	add	year	HRWheatPrice	SoybeansPrice	SorghumPrice	MILK, FLUID GRADE - PRICE RECEIVED, MEASURED IN $ / CWT	MILK - PRODUCTION, MEASURED IN LB (Billion Units)
Jan-98	114.71	1.11%	1/1/1998	1	JAN	3	1998	137.10	273.0	111.44	14.7	13.28
Feb-98	113.38	-1.16%	2/1/1998	2	FEB	6	1998	140.05	273.0	111.73	14.9	12.19
Mar-98	114.21	0.73%	3/1/1998	3	MAR	9	1998	138.92	265.0	110.47	14.5	13.69
Apr-98	107.12	-6.21%	4/1/1998	4	APR	12	1998	130.15	259.0	103.90	14.1	13.49
May-98	106.10	-0.95%	5/1/1998	5	MAY	15	1998	128.92	247.0	100.81	13.4	13.99

The data frame consists of month, year, corn price, HR wheat price, soybean price, sorghum price, milk full-grade price received, and milk production measured in LB.

Data frame info provides all columns' type and number of rows present in the dataset:

```
milkCommoditiesProductionRelation_df.info()
```

Output:

```
<class 'pandas.core.frame.DataFrame'>
RangeIndex: 264 entries, 0 to 263
Data columns (total 14 columns):
 #   Column                                                  Non-Null Count  Dtype
---  ------                                                  --------------  -----
 0   Month                                                   264 non-null    object
 1   CornPrice                                               264 non-null    float64
 2   Change                                                  264 non-null    object
 3   MonthDate                                               264 non-null    object
 4   month_number                                            264 non-null    int64
 5   month_name                                              264 non-null    object
 6   add                                                     264 non-null    int64
 7   year                                                    264 non-null    int64
 8   HRWheatPrice                                            264 non-null    float64
 9   SoybeansPrice                                           264 non-null    float64
 10  SorghumPrice                                            264 non-null    float64
 11  MILK, FLUID GRADE - PRICE RECEIVED, MEASURED IN $ / CWT  264 non-null    float64
 12  MILK - PRODUCTION, MEASURED IN LB (Billion Units)       264 non-null    float64
 13  MILK - PRODUCTION, MEASURED IN LB                       264 non-null    object
dtypes: float64(6), int64(3), object(5)
memory usage: 29.0+ KB
```

Step 2: *Missing Values*

As part of exploration of data analysis, check any missing or null values in the data. This step will provide feedback into filling null values into missing columns' values:

```
milkCommoditiesProductionRelation_df.isnull().sum()
```

Output:

```
Month                                                   0
CornPrice                                               0
Change                                                  0
MonthDate                                               0
month_number                                            0
month_name                                              0
add                                                     0
year                                                    0
HRWheatPrice                                            0
SoybeansPrice                                           0
SorghumPrice                                            0
MILK, FLUID GRADE - PRICE RECEIVED, MEASURED IN $ / CWT 0
MILK - PRODUCTION, MEASURED IN LB (Billion Units)       0
MILK - PRODUCTION, MEASURED IN LB                       0
dtype: int64
```

It is clear from the above dataset there were no missing commodity prices and milk produced values since 1998–2019.

Step 3: *Compute Basic Statistics of Data Frame*

Explore data in the Data Frame Now that we've identified with no missing values. Let's start by comparing the mean milk produced and commodity prices:

```
numeric_features = ['CornPrice', 'HRWheatPrice', 'SoybeansPrice', 'SorghumPrice']

milkCommoditiesProductionRelation_df[numeric_features + ['MILK - PRODUCTION, MEASURED IN LB (Billion Units)']].describe()
```

Output:

	CornPrice	HRWheatPrice	SoybeansPrice	SorghumPrice	MILK - PRODUCTION, MEASURED IN LB (Billion Units)
count	264.000000	264.000000	264.000000	264.000000	264.000000
mean	159.356970	203.954129	367.693939	156.115833	15.746553
std	64.913058	74.173206	126.205292	59.340504	1.678630
min	75.270000	102.160000	183.000000	73.720000	12.190000
25%	102.522500	147.000000	256.687500	100.592500	14.310000
50%	158.425000	191.075000	376.360000	152.855000	15.765000
75%	178.732500	250.452500	450.000000	188.382500	17.122500
max	333.050000	439.720000	684.000000	302.530000	19.120000

Corn prices reached above $300 in 2011 and 2012. Soybeans reached $684 in 2012.

The statistics reveal some information about the distribution of the data in each of the numeric fields, including the number of observations (there are 731 records), the mean standard deviation, minimum and maximum values, and the quartile values (the threshold values for 25%, 50% which is also the median, and 75% of the data). From this, we can see that the mean of MILK - PRODUCTION, MEASURED IN LB (Billion Units) is around 15.74; but there's a comparatively small standard deviation, indicating a smaller variance in MILK—PRODUCTION, MEASURED IN LB (Billion Units).

US farmers are planting more row crops than ever

We might get a clearer idea of the distribution of MILK—PRODUCTION, MEASURED IN LB (Billion Units) values by visualizing the data. Common plot types for visualizing numeric data distributions are histograms and box plots, so let's use Python's Matplotlib library to create one of each of these for the MILK—PRODUCTION, MEASURED IN LB (Billion Units) column.

Step 4: *Statistical Plots*

```
import pandas as pd
import matplotlib.pyplot as plt

# This ensures plots are displayed inline in the Jupyter notebook
%matplotlib inline

# Get the label column
label = milkCommoditiesProductionRelation_df['MILK - PRODUCTION, MEASURED IN LB
(Billion Units)']


# Create a figure for 2 subplots (2 rows, 1 column)
fig, ax = plt.subplots(2, 1, figsize = (9,12))

# Plot the histogram
ax[0].hist(label, bins=100)
ax[0].set_ylabel('Frequency')

# Add lines for the mean, median, and mode
ax[0].axvline(label.mean(), color='magenta', linestyle='dashed', linewidth=2)
ax[0].axvline(label.median(), color='cyan', linestyle='dashed', linewidth=2)

# Plot the boxplot
ax[1].boxplot(label, vert=False)
ax[1].set_xlabel('MILK - PRODUCTION, MEASURED IN LB (Billion Units)')

# Add a title to the Figure
fig.suptitle('MILK - PRODUCTION, MEASURED IN LB (Billion Units) Distribution')

# Show the figure
fig.show()
```

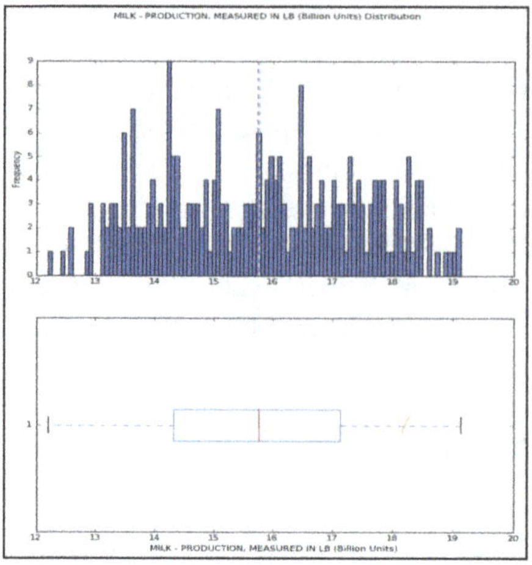

Step 5: *Plot Histogram of Each Numerical Feature—Corn Price, Soybean Price, HW Wheat Price, and Sorghum Price*

```
# Plot a histogram for each numeric feature
for col in numeric_features:
    fig = plt.figure(figsize=(9, 6))
    ax = fig.gca()
    feature = milkCommoditiesProductionRelation_df[col]
    feature.hist(bins=100, ax = ax)
    ax.axvline(feature.mean(), color='magenta', linestyle='dashed', linewidth=2)
    ax.axvline(feature.median(), color='cyan', linestyle='dashed', linewidth=2)
    ax.set_title(col)
plt.show()
```

Output:

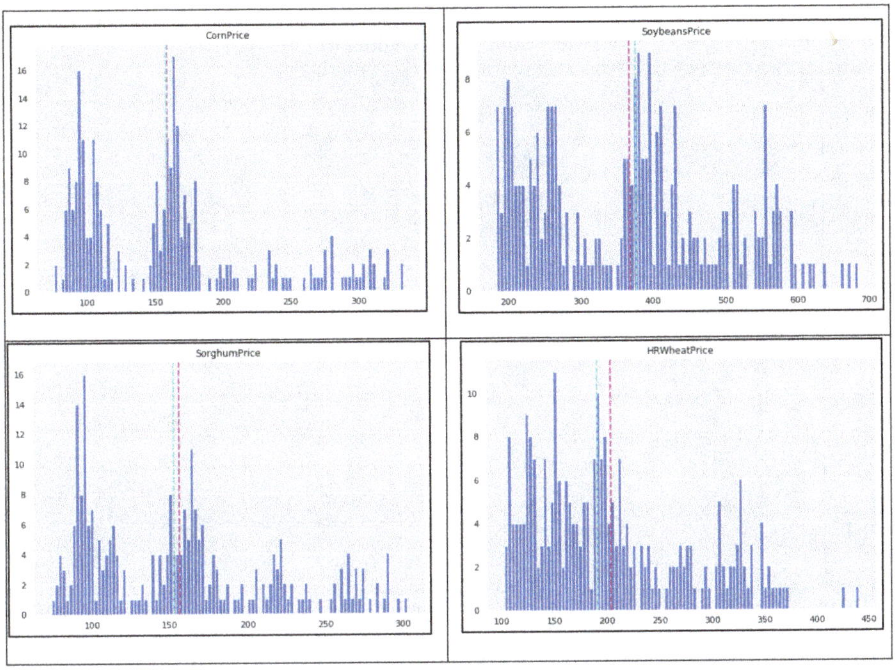

Step 6: *Scatter Plot*

For the numeric features, we can create scatter plots that show the intersection of feature and label values. We can also calculate the correlation statistic to quantify the apparent relationship:

```
for col in numeric_features:
    fig = plt.figure(figsize=(9, 6))
    ax = fig.gca()
    feature = milkCommoditiesProductionRelation_df[col]
    label = milkCommoditiesProductionRelation_df['MILK - PRODUCTION, MEASURED IN LB
(Billion Units)']
    correlation = feature.corr(label)
    plt.scatter(x=feature, y=label)
    plt.xlabel(col)
    plt.ylabel('MILK - PRODUCTION, MEASURED IN LB (Billion Units)')
    ax.set_title('MILK - PRODUCTION, MEASURED IN LB (Billion Units) vs ' + col + '- correlation: '
+ str(correlation))
plt.show()
```

Output:

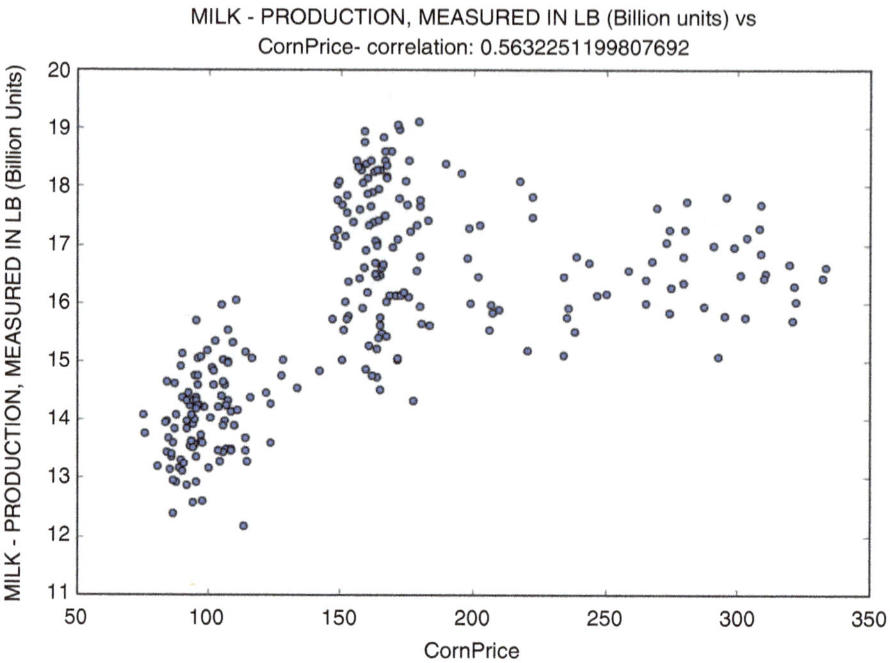

From the above graph, the following trends can be identified: (1) correlation between milk production measured and corn price is 0.56, positively correlated. The spread of corn prices to milk produced is spread in the central of distribution with corn prices between $150 and $200 US dollars per metric ton.[119]

> During 2006–2008, price runup of commodities (corn, wheat, sorghum, and soybeans) sharply increased to historic highs of more than 60% above the previous year's levels; the major contributors include global demand for biofuels, adverse weather conditions in 2006 and 2007 in some major grain- and oilseed-producing areas, global food commodity price inflation, declining US dollar, rising energy prices, and increasing agricultural costs of production[120]

Another reason grain price is increasing is due to stock-to-use ratio (please see Fig. 5.25). Corn prices are higher due to higher consumption than yield. With wheat, the harvest (762 million tons) does exceed consumption (750 million tons). As a result, this season ends with a record stock of wheat. This puts pressure on the wheat price.[121]

[119] Corn Prices - https://www.indexmundi.com/commodities/?commodity=corn&months=360

[120] Global Agricultural Supply and Demand: Factors Contributing to the Recent Increase in Food Commodity Prices - https://www.hsdl.org/?view&did=485652

[121] IGC: Grain records for production and consumption - https://www.allaboutfeed.net/animal-feed/raw-materials/igc-grain-records-for-production-and-consumption/

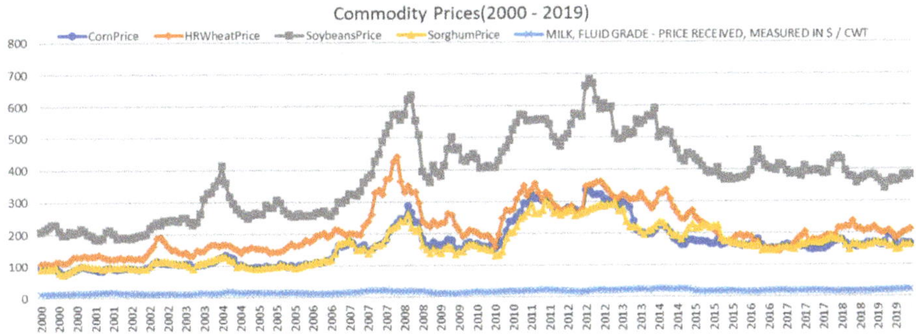

Fig. 5.25 Commodities prices

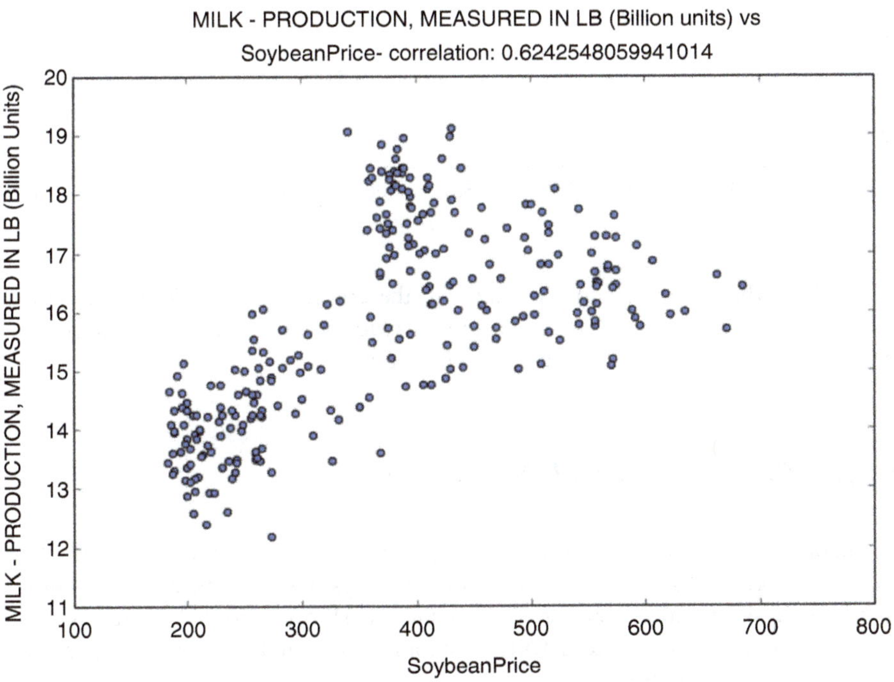

The correlation between soybeans price and milk produced is 0.62. The spread is more evenly as compared to corn price.

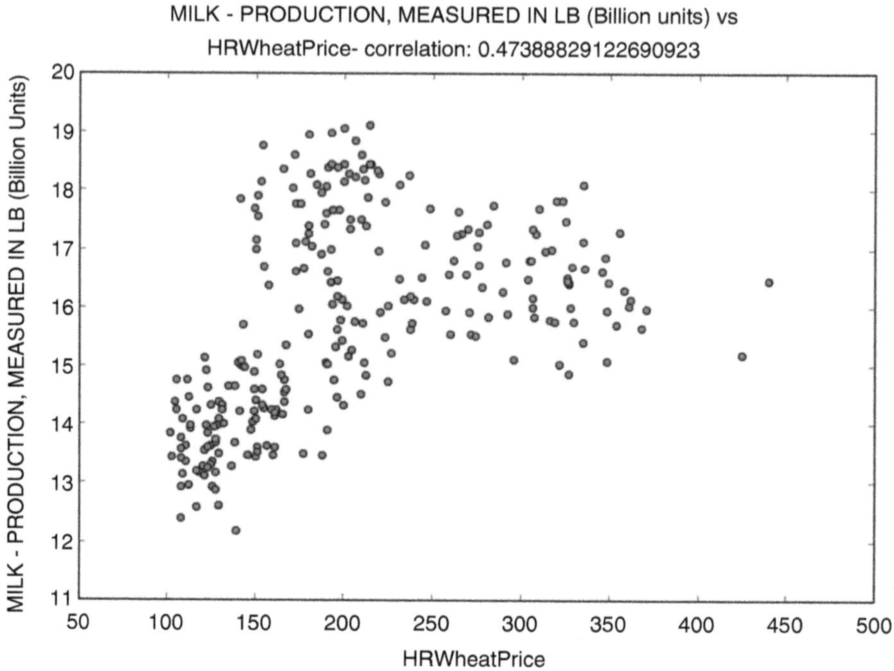

For HR wheat price to milk production, the correlation is 0.47 and the graph trend is left aligned. The increase in milk production observed for wheat price is between $150 and $220 US dollars per metric ton.[122]

Step 7: *Check for Normal Distribution*

To get maximum information out of features and class variable, the data distribution should be normal. To check for that, normality test is based on the null and alternative hypothesis intuition. scipy.stats.normaltest performs normality test based on D'Agostino and Pearson's test, which combines skew and kurtosis to produce an omnibus test of normality:[123]

[122] HR Wheat Price - https://www.indexmundi.com/commodities/?commodity=soft-red-winter-wheat

[123] D'Agostino, R., & Pearson, E. (1973). Tests for Departure from Normality. Empirical Results for the Distributions of b2 and $\sqrt{b1}$. Biometrika, 60(3), 613–622. doi:10.2307/2335012

```
from scipy import stats

milkProduction=milkCommoditiesProductionRelation_df['MILK - PRODUCTION, MEASURED IN LB (Billion Units)'].values


stat,p = stats.normaltest(milkProduction)

print("milkProduction Statistics = %.3f, p=%.3f" % (stat,p))
alpha = 0.05
if p> alpha:
   print('Data looks Gaussian (fail to reject null hypothesis)')
else:
   print('Data looks non-Gaussian (reject null hypothesis)')
```

Output: the *p* value of 0 confirms the distribution is indeed normal!

```
milkProduction Statistics = 61.542, p=0.000
Data looks non-Gaussian (reject null hypothesis)
```

Similar check for feature variables, i.e., corn, wheat, soybeans, and sorghum prices:

```
from scipy import stats

CornPrices=milkCommoditiesProductionRelation_df['CornPrice'].values


stat,p = stats.normaltest(CornPrices)

print("CornPrices Statistics = %.3f, p=%.3f" % (stat,p))
alpha = 0.05
if p> alpha:
   print('Data looks Gaussian (fail to reject null hypothesis)')
else:
   print('Data looks non-Gaussian (reject null hypothesis)')
```

Output:

```
CornPrices Statistics = 31.037, p=0.000
Data looks non-Gaussian (reject null hypothesis)
```

Wheat prices:

```
from scipy import stats

HRWheatPrice=milkCommoditiesProductionRelation_df['HRWheatPrice'].values


stat,p = stats.normaltest(HRWheatPrice)

print("HRWheatPrice Statistics = %.3f, p=%.3f" % (stat,p))
alpha = 0.05
if p> alpha:
   print('Data looks Gaussian (fail to reject null hypothesis)')
else:
   print('Data looks non-Gaussian (reject null hypothesis)')
```

Output:

```
HRWheatPrice Statistics = 22.376, p=0.000
Data looks non-Gaussian (reject null hypothesis)
```

Soybeans price:

```
from scipy import stats
SoybeansPrice=milkCommoditiesProductionRelation_df['SoybeansPrice'].values
stat,p = stats.normaltest(SoybeansPrice)

print("SoybeansPrice Statistics = %.3f, p=%.3f" % (stat,p))
alpha = 0.05
if p> alpha:
   print('Data looks Gaussian (fail to reject null hypothesis)')
else:
   print('Data looks non-Gaussian (reject null hypothesis)')
```

Output:

```
SoybeansPrice Statistics = 34.000, p=0.000
Data looks non-Gaussian (reject null hypothesis)
```

Sorghum price:

```
from scipy import stats

SorghumPrice=milkCommoditiesProductionRelation_df['SorghumPrice'].values


stat,p = stats.normaltest(SorghumPrice)

print("SorghumPrice Statistics = %.3f, p=%.3f" % (stat,p))
alpha = 0.05
if p> alpha:
   print('Data looks Gaussian (fail to reject null hypothesis)')
else:
   print('Data looks non-Gaussian (reject null hypothesis)')
```

Output:

```
SorghumPrice Statistics = 22.078, p=0.000
Data looks non-Gaussian (reject null hypothesis)
```

Step 8a: *Kurtosis and Skewness*

A normal distributin is a key assumption to many models we will later be using. The graph of the normal distribution depends on two factors—the mean and the standard deviation. The mean of the distribution determines the location of the center of the graph, and the standard deviation determines the height of the graph. When the standard deviation is large, the curve is short and wide; when the standard deviation is small, the curve is tall and narrow. All normal distributions look like a symmetric, bell-shaped curve.[124]

Skewness: In probability theory and statistics, skewness is a measure of the asymmetry of the probability distribution of a real-valued random variable about its mean. The skewness value can be positive or negative, or even undefined.

Skewness is defined as:[125]

$$a^3 = \sum \frac{(x_i - X)^3}{ns^3}$$

where n is the sample size, x_i is the ith value of x, X is the average, and S is the standard deviation. Note the exponent in the summation. It is "3". The skewness is referred to as the "third standardized central moment for the probability model".

The rule of thumb for skewness:

- If the skewness is between −0.5 and 0.5, the data are fairly symmetrical.
- If the skewness is between −1 and −0.5 or between 0.5 and 1, the data are moderately skewed.
- If the skewness is less than −1 or greater than 1, the data are highly skewed.

Kurtosis: Kurtosis is a measure of whether the data are peaked or flat relative to a normal distribution. That is, datasets with high kurtosis tend to have a distinct peak near the mean, decline rather rapidly, and have heavy tails:

$$a^4 = \sum \frac{(x_i - X)^4}{ns^4}$$

where n is the sample size, x_i is the ith value of x, X is the average, and S is the standard deviation. Note the exponent in the summation. It is "4". The kurtosis is referred to as the "fourth standardized central moment for the probability model".

Most often, kurtosis is measured against the normal distribution. If the kurtosis is close to 0, then a normal distribution is often assumed. These are called mesokurtic distributions. If the kurtosis is less than 0, then the distribution is light tails and is called a platykurtic distribution. If the kurtosis is greater than 0, then the distribution has heavier tails and is called a leptokurtic distribution.

Though skewness and kurtosis provide statistical properties of the data distributions of commodity prices, the real value these data signals provide to small farmers is the price movements of commodities over the decade and root causing the price movements. Negative values for the skewness indicate that observations are skewed leftward, or that the left tail is heavier than the right tail. The indication here is that there were more episodes of downward spikes than upward spikes in commodity

[124] Exploring distributions - https://sphinx-dat.readthedocs.io/en/latest/02-01-exploring-distributions.html

[125] Skewness and Kurtosis - https://www.spcforexcel.com/knowledge/basic-statistics/are-skewness-and-kurtosis-useful-statistics

prices for periods with heavier left tails. These price spikes could be due to macroeconomic policies, productivity gains or loss as a matter of fact, or evolving climate/weather changes. Whatever the causation factors, the signals could be witnessed through skewness and kurtosis. Positive skewness, on the other hand, indicates more upward spikes (episodes of rising prices) than negative ones.

For example, nominal wheat prices display a negative skewness[126] during the 1980s–1990s and a positive skewness in decades thereafter (1990–2000, 2000–2010, and 2010–2020). Hence, nominal wheat prices exhibit a higher frequency of rising prices in the decades 1990–2000, 2000–2010, and 2010–2019.

Maize has substantial kurtosis observed during the 1990s–2000s, 2000–2010, and later. Maize exhibited negative kurtosis till the 1990s, a clear indication that not larger price swings till the 1990s. Nonetheless, maize exhibited large price movements during the 1990s–2000s and thereafter relative to the other earlier decades (1970s–1990s).

Milk Production Skewness and Kurtosis

```
import matplotlib.pyplot as plt
import scipy.stats as stats
from scipy.stats import kurtosis

milkCommoditiesProductionRelation_df['MILK - PRODUCTION, MEASURED IN LB (Billion
Units)'].plot(kind = 'density')

print('This distribution has skew', milkCommoditiesProductionRelation_df['MILK - PRODUCTION,
MEASURED IN LB (Billion Units)'].skew())
print('This distribution has kurtosis', milkCommoditiesProductionRelation_df['MILK -
PRODUCTION, MEASURED IN LB (Billion Units)'].kurt())
```

Output:

The milk production distribution has skew of 0.043698484581727856 and has kurtosis of −1.0499339270136985.

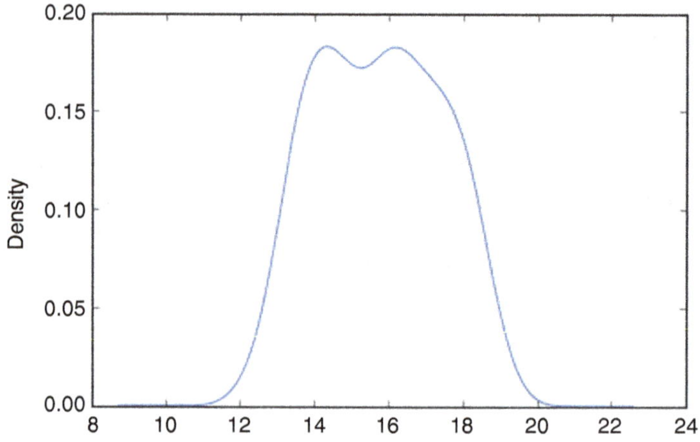

[126]The statistical distributions of commodity prices in both real and nominal terms - http://www.fao.org/3/y4344e/y4344e0i.htm

Skewness of 0.043 is a fairly distributed milk production. This implies milk production was maintained fairly symmetrical—not varied! Kurtosis less than 0 implies the distribution has light tailed with respect to normal distribution.

Corn Price Skewness and Kurtosis

```
import matplotlib.pyplot as plt
import scipy.stats as stats
from scipy.stats import kurtosis

milkCommoditiesProductionRelation_df['CornPrice'].plot(kind = 'density')

print('This distribution has skew', milkCommoditiesProductionRelation_df['CornPrice'].skew())
print('This distribution has kurtosis', milkCommoditiesProductionRelation_df['CornPrice'].kurt())
```

Output:

The corn price has distribution with skewness of 0.9481 and has kurtosis of 0.1466.

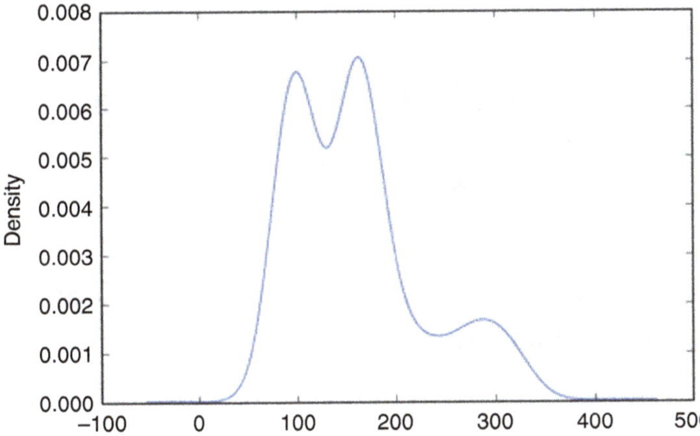

Statistically, skewness of 0.9481 is a moderately skewed distribution, and it reflects in corn prices. Kurtosis greater than 0 (0.1466) signifies heavier tails in the higher corn prices.

Investment Strategy: systematically *buying* commodities with *low skewness* and *shorting* commodities with *high skewness* is a profitable strategy. The performance is not merely a compensation for the inherent backwardation and contango risks present in commodity futures markets, for the transaction costs incurred in implementing the trades, or for the relative lack of liquidity of the assets traded. The long-short skewness portfolio is also found to act as a good risk diversifier but not as an inflation hedge[127]

[127] Skewness Strategies in Commodity Futures Markets - https://risk.edhec.edu/sites/risk/files/edhec-working-paper-skewness-strategies-in-commodity-futures-markets-f_1460018026076.pdf

These two, skewness and kurtosis, distributions give us some intuition about the distribution of data. A value close to 0 for kurtosis indicates a normal distribution, where asymmetrical nature is signified by a value between −0.5 and +0.5 for skewness. The tails are heavier for kurtosis greater than 0 and vice versa. Moderate skewness refers to the value between −1 and −0.5 or 0.5 and 1.

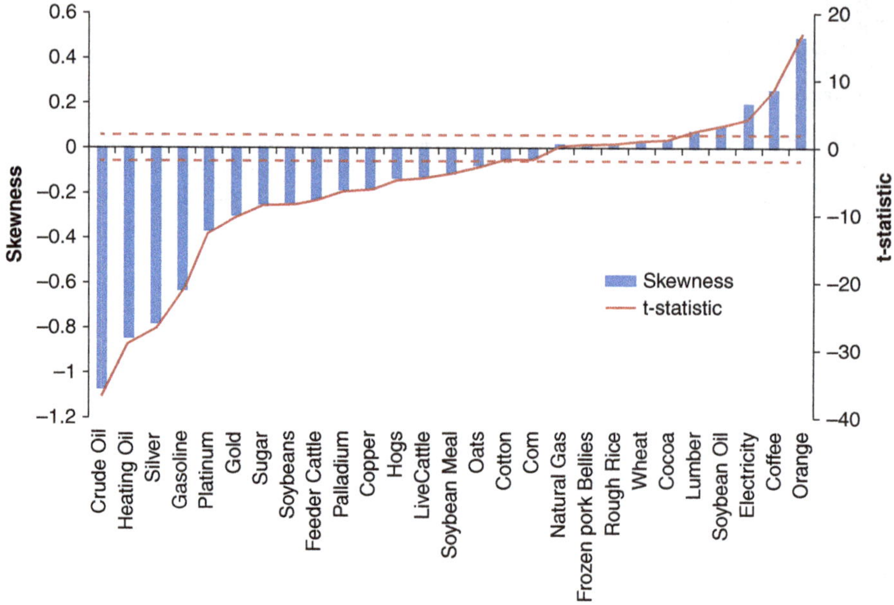

Commodity skewness: Sample period—January 5, 1987, to November 28, 2014[128]

HR Wheat Price Skewness and Kurtosis

```
import matplotlib.pyplot as plt
import scipy.stats as stats
from scipy.stats import kurtosis

milkCommoditiesProductionRelation_df['HRWheatPrice'].plot(kind = 'density')

print('This distribution has skew',
milkCommoditiesProductionRelation_df['HRWheatPrice'].skew())
print('This distribution has kurtosis',
milkCommoditiesProductionRelation_df['HRWheatPrice'].kurt())
```

Output:
HR wheat distribution has skewness of 0.756 and kurtosis of −0.285.

[128] Skewness Strategies in Commodity Futures Markets - https://risk.edhec.edu/sites/risk/files/edhec-working-paper-skewness-strategies-in-commodity-futures-markets-f_1460018026076.pdf

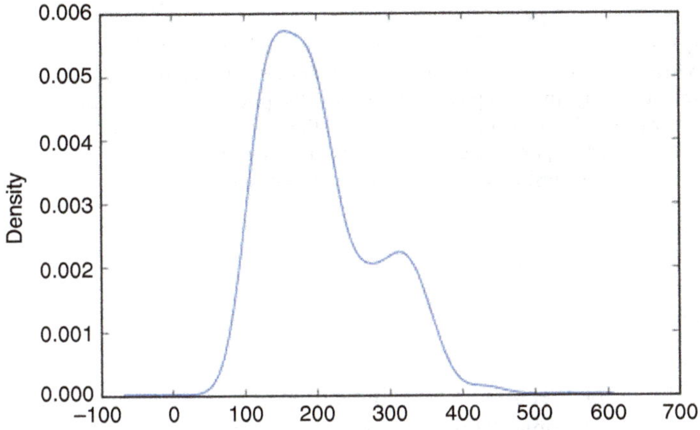

The skewness of 0.756 suggests the HR wheat prices are moderately skewed, and kurtosis less than 0.5 suggests lightly tailed distribution. What this implies is during the observed period from 1998 to 2019, the HR wheat exhibits higher prices.[129]

Sorghum Skewness and Kurtosis

```
import matplotlib.pyplot as plt
import scipy.stats as stats
from scipy.stats import kurtosis

milkCommoditiesProductionRelation_df['SorghumPrice'].plot(kind = 'density')

print('This distribution has skew', milkCommoditiesProductionRelation_df['SorghumPrice'].skew())
print('This distribution has kurtosis', milkCommoditiesProductionRelation_df['SorghumPrice'].kurt())
```

Output:

This distribution has skewness of 0.65 and kurtosis of −0.53.

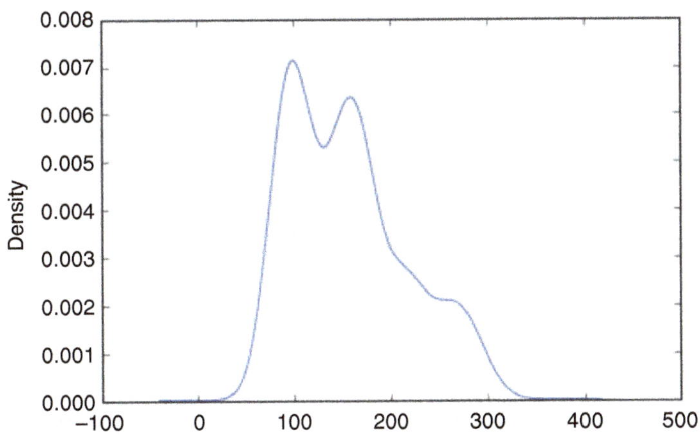

[129] The statistical distributions of commodity prices in both real and nominal terms - http://www.fao.org/3/y4344e/y4344e0i.htm

Market Participation: Commodities[130]

Farmers usually sell large parts of their crop in the futures markets well before they have brought it in. When a disastrous event occurs (drought, flooding, tsunami, etc.), this becomes a double whammy: their crop is destroyed, and the price shoots up in the market. The farmer is unable to meet its delivery obligation and is forced to buy back his shorts at highly inflated prices, exacerbating the market imbalance. In a bizarre consequence, the farmer in question faces bankruptcy in a period of very high commodity prices

Step 8b: *Boxplots*

Boxplots are very useful graph that gives a good indication of how the values of data are spread out. Please note that some datasets need more information about the variability or dispersion of data than the centrality of data (bell curve, mean, standard deviation).[131] Boxplots are the best candidates!

Boxplots are a standardized way of displaying the distribution of data based on a five-number summary (“minimum”, first quartile (Q1), median, third quartile (Q3), and “maximum”):

```
# plot a boxplot for the label by each categorical feature
for col in categorical_features:
    fig = plt.figure(figsize=(9, 6))
    ax = fig.gca()
    milkCommoditiesProductionRelation_df.boxplot(column = 'MILK - PRODUCTION, MEASURED
IN LB (Billion Units)', by = col, ax = ax)
    ax.set_title('Label by ' + col)
    ax.set_ylabel("MILK - PRODUCTION, MEASURED IN LB (Billion Units)")
plt.show()
```

Output:

[130] Skewness and Kurtosis: What Commodities Investors Need to Know - https://seekingalpha.com/article/269053-skewness-and-kurtosis-what-commodities-investors-need-to-know

[131] Understanding Box Plots - https://towardsdatascience.com/understanding-boxplots-5e2df7bcbd51

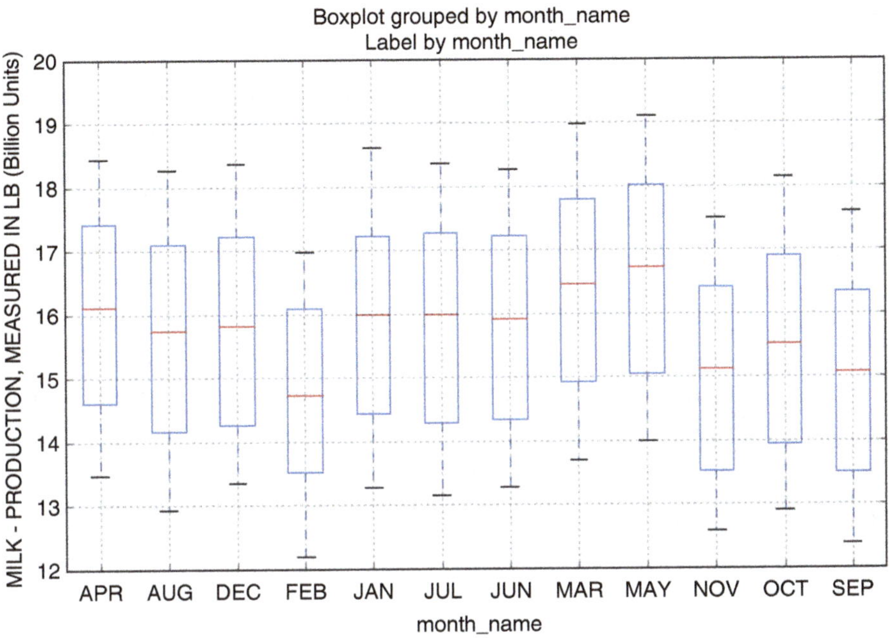

As indicated in the above boxplot, aggregated milk production (for years 1998–2019) is lowest in the month of February and peaks in May. This is true across US national. The same can be witnesses over milk production during 2018, 2019, and 2020.[132]

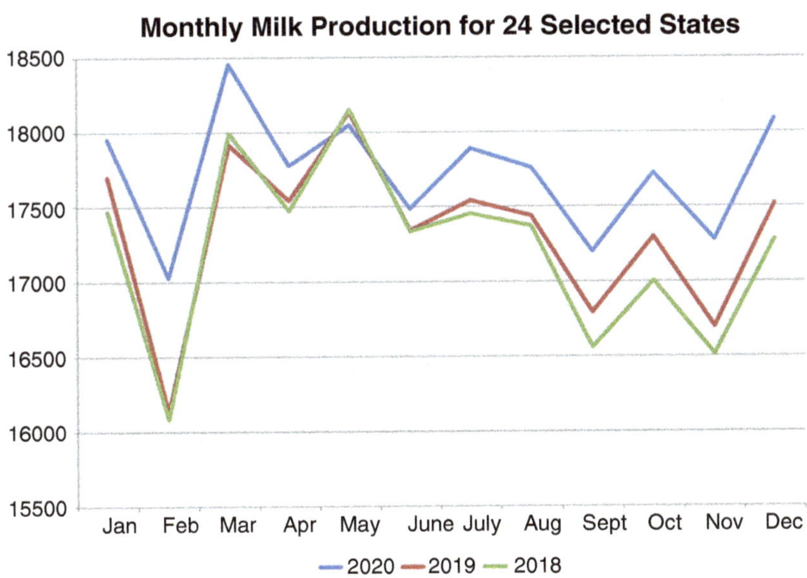

[132] Monthly Milk production - https://hoards.com/article-21-monthly-milk-production.html

Step 9: *Train a Regression Model*

Now that we've explored the data, it's time to use it to train a regression model that uses the features we've identified as potentially predictive to predict the MILK—PRODUCTION, MEASURED IN LB (Billion Units) label. The first thing we need to do is to separate the features we want to use to train the model from the label we want it to predict.

Feature variables include CornPrice, HRWheatPrice, SoybeansPrice, and SorghumPrice:

```
# Separate features and labels
X, y = milkCommoditiesProductionRelation_df[['CornPrice', 'HRWheatPrice', 'SoybeansPrice',
'SorghumPrice']].values, milkCommoditiesProductionRelation_df['MILK - PRODUCTION, MEASURED IN
LB (Billion Units)'].values
print('Features:',X[:10], '\nLabels:', y[:10], sep='\n')


from sklearn.model_selection import train_test_split

# Split data 70%-30% into training set and test set
X_train, X_test, y_train, y_test = train_test_split(X, y, test_size=0.30, random_state=0)

print ('Training Set: %d, rows\nTest Set: %d rows' % (X_train.size, X_test.size))
```

Now we have the following four datasets:

- X_train: The feature values that'll use to train the model.
- y_train: The corresponding labels that'll use to train the model.
- X_test: The feature values that'll use to validate the model.
- y_test: The corresponding labels that'll use to validate the model.

Now we're ready to train a model by fitting a suitable regression algorithm to the training data.

Linear Regression

Use the scikit-learn implementation of the linear regression algorithm, a common starting point for regression that works by trying to find a linear relationship between the X values and the y label. The resulting model is a function that conceptually defines a line where every possible X and y value combination intersects:

```
# Train the model
from sklearn.linear_model import LinearRegression

# Fit a linear regression model on the training set
model = LinearRegression(normalize=False).fit(X_train, y_train)
print (model)
```

Output:

```
Features:
[[114.71 137.1  273.   111.44]
 [113.38 140.05 273.   111.73]
 [114.21 138.92 265.   110.47]
 [107.12 130.15 259.   103.9 ]
 [106.1  128.92 247.   100.81]
 [104.23 120.64 242.    96.51]
 [100.51 118.9  238.    95.44]
 [ 87.86 107.95 219.    91.  ]
 [ 86.31 107.93 216.    85.04]
 [ 95.53 126.54 223.    87.7 ]]

Labels:
[13.28 12.19 13.69 13.49 13.99 13.28 13.16 12.93 12.4  12.92]
Training Set: 736, rows
Test Set: 320 rows
LinearRegression()
```

Evaluate Trained Model

Let's use the trained model to predict MILK—PRODUCTION, MEASURED IN LB (Billion Units) for the features that held back in validation dataset. The goal was to compare these predictions to the actual label values to evaluate how well (or not!) the model is working:

```
import numpy as np

predictions = model.predict(X_test)
np.set_printoptions(suppress=True)
print('Predicted labels: ', np.round(predictions)[:10])
print('Actual labels   : ',y_test[:10])
```

Output:

```
Predicted labels: [16. 14. 17. 15. 18. 17. 14. 17. 16. 17.]
Actual labels   : [16.79 15.33 18.43 12.4  15.76 16.8  14.15 18.43 17.28 16.4 ]
```

Comparing each prediction with its corresponding "ground truth" actual value could be useful for smaller dataset but is not a very efficient way to determine how well the model is in predicting for larger datasets.

Visualization could help by creating a scatter plot that compares the predictions to the actual labels. Overlay a trend line to get a general sense for how well the predicted labels align with the true labels:

```
import matplotlib.pyplot as plt

%matplotlib inline

plt.scatter(y_test, predictions)
plt.xlabel('Actual Labels')
plt.ylabel('Predicted Labels')
plt.title('MILK - PRODUCTION, MEASURED IN LB (Billion Units) Predictions')
# overlay the regression line
z = np.polyfit(y_test, predictions, 1)
p = np.poly1d(z)
plt.plot(y_test,p(y_test), color='magenta')
plt.show()
```

There's a definite diagonal trend, and the intersections of the predicted and actual values are generally following the path of the trend line; but there's a fair amount of difference between the ideal function represented by the line and the results. This variance represents the residuals of the model—in other words, the difference between the label predicted when the model applies the coefficients it learned during training to the validation data and the actual value of the validation label. These residuals when evaluated from the validation data indicate the expected level of error when the model is used with new data for which the label is unknown.

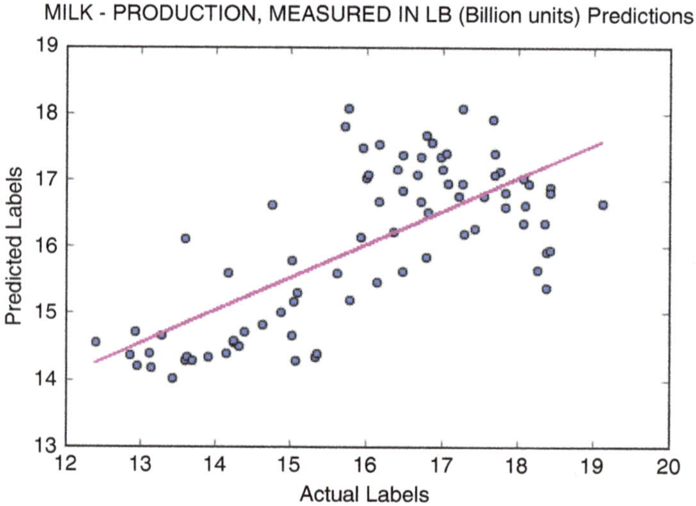

The following can quantify the residuals by calculating a number of commonly used evaluation metrics:

- *Mean Square Error (MSE)*: The mean difference between predicted and actual squared. This yields a relative metric in which the smaller the value, the better the fit of the model.
- *Root Mean Square Error (RMSE)*: The square root of the MSE. This yields an absolute metric in the same unit as the label (in this case, MILK—PRODUCTION,

MEASURED IN LB (Billion Units)). The smaller the value, the better the model (in a simplistic sense, it represents the average number of MILK - PRODUCTION, MEASURED IN LB (Billion Units) by which the predictions are wrong!).

- *Coefficient of Determination (usually known as R-squared or R^2)*: A relative metric in which the higher the value, the better the fit of the model. In essence, this metric represents how much of the variance between predicted and actual label values the model is able to explain.

Let's use scikit-learn to calculate these metrics for our model, based on the predictions it generated for the validation data:

```
from sklearn.metrics import mean_squared_error, r2_score

mse = mean_squared_error(y_test, predictions)
print("MSE:", mse)

rmse = np.sqrt(mse)
print("RMSE:", rmse)

r2 = r2_score(y_test, predictions)
print("R2:", r2)
```

Output:

```
MSE: 1.4328486474105033
RMSE: 1.197016561042704 8
R²: 0.5077677578995581
```

So now the above code quantified the ability of our model to predict the MILK—PRODUCTION, MEASURED IN LB (Billion Units). It definitely has some predictive power, but we can probably do better!

Step 10: *Train Model Pipeline*

The above code trained a model with data that was loaded straight from a source file, with only moderately successful results.

In practice, it's common to perform some preprocessing of the data to make it easier for the algorithm to fit a model to it. There's a huge range of preprocessing transformations that can be performed to get the data ready for modeling (not exhaustive):

- Scaling numeric features so they're on the same scale. This prevents features with large values from producing coefficients that disproportionately affect the predictions.

- Encoding categorical variables. For example, by using a one-hot encoding technique, you can create individual binary (true/false) features for each possible category value.

To apply these preprocessing transformations, use of a scikit-learn feature named *pipelines*.[133] These enable us to define a set of preprocessing steps that end with an algorithm. One can then fit the entire pipeline to the data, so that the model encapsulates all of the preprocessing steps as well as the regression algorithm. This is useful because when data scientists want to use the model to predict values from new data, they need to apply the same transformations (based on the same statistical distributions and category encodings used with the training data):

```
from sklearn import linear_model

pipeline = Pipeline(steps=[('preprocessor', preprocessor),
                ('regressor', linear_model.Ridge(alpha=.5))])


# train a logistic regression model on the training set
model = pipeline.fit(X_train, (y_train))
print (model)
```

Output:

```
Pipeline(steps=[('preprocessor',
          ColumnTransformer(transformers=[('num',
                                Pipeline(steps=[('scaler',
                                          StandardScaler())]),
                                [0, 1, 2, 3])])),
          ('regressor', Ridge(alpha=0.5))])
```

The model is trained, including the preprocessing steps. Let's see how it performs with the validation data:

```
predictions = model.predict(X_test)


np.set_printoptions(suppress=True)
print('Predicted labels: ', np.round(predictions)[:10])
print('Actual labels: ' ,y_test[:10])


mse = mean_squared_error(y_test, predictions)
print("MSE:", mse)
rmse = np.sqrt(mse)
print("RMSE:", rmse)
r2 = r2_score(y_test, predictions)
print("R2:", r2)
```

[133] sklearn.pipeline.Pipeline - https://scikit-learn.org/stable/modules/generated/sklearn.pipeline.Pipeline.html

Output:

```
Predicted labels: [16. 14. 17. 15. 18. 17. 14. 17. 16. 17.]
Actual labels: [16.79 15.33 18.43 12.4  15.76 16.8  14.15 18.43 17.28 16.4 ]
MSE: 1.438046503747142
RMSE: 1.1991857669882269
R2: 0.5059821174668973
```

There's a definite diagonal trend, and the intersections of the predicted and actual values are generally following the path of the trend line; but there's a fair amount of difference between the ideal function represented by the line and the results. This variance represents the residuals of the model—in other words, the difference between the label predicted when the model applies the coefficients it learned during training to the validation data, and the actual value of the validation label.

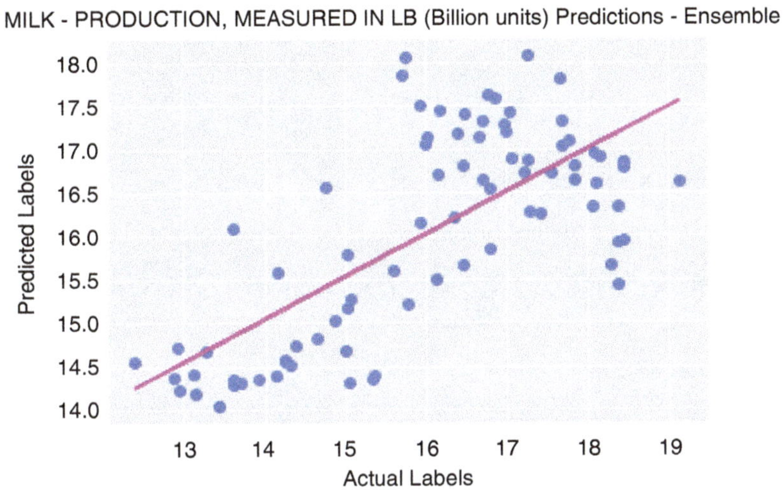

Linear Model: Lasso

Linear model is trained with L1 prior as regularizer (aka the Lasso[134]). The optimization objective for Lasso is:

```
(1 / (2 * n_samples)) * ||y - Xw||^2_2 + alpha * ||w||_1
```

Technically, the Lasso model is optimizing the same objective function as the elastic net with l1_ratio = 1.0 (no L2 penalty):

[134] Lasso - https://scikit-learn.org/stable/modules/generated/sklearn.linear_model.Lasso.html

```
from sklearn import linear_model

pipeline = Pipeline(steps=[('preprocessor', preprocessor),
                    ('regressor', linear_model.Lasso(alpha=0.1))])


# train a logistic regression model on the training set
model = pipeline.fit(X_train, (y_train))
print (model)
```

Output:

```
Pipeline(steps=[('preprocessor',
         ColumnTransformer(transformers=[('num',
                                          Pipeline(steps=[('scaler',
                                                           StandardScaler())]),
                                          [0, 1, 2, 3])])),
        ('regressor', Lasso(alpha=0.1))])
```

Prediction:

```
predictions = model.predict(X_test)
np.set_printoptions(suppress=True)
print('Predicted labels: ', np.round(predictions)[:10])
print('Actual labels: ' ,y_test[:10])
mse = mean_squared_error(y_test, predictions)
print("MSE:", mse)
rmse = np.sqrt(mse)
print("RMSE:", rmse)
r2 = r2_score(y_test, predictions)
print("R2:", r2)
```

Output:

```
Predicted labels:  [16. 15. 16. 15. 18. 17. 15. 16. 17. 17.]
Actual labels:  [16.79 15.33 18.43 12.4  15.76 16.8  14.15 18.43 17.28 16.4 ]
MSE: 1.8100377032604391
RMSE: 1.3453764169407902
R2: 0.37819048887515516
```

As you can see, the linear lasso R^2 value is suboptimal than linear regressor. Let's visualize:

```
plt.scatter(y_test, predictions)
plt.xlabel('Actual Labels')
plt.ylabel('Predicted Labels')
plt.title('MILK - PRODUCTION, MEASURED IN LB (Billion Units) Predictions - Ensemble')
z = np.polyfit(y_test, predictions, 1)
p = np.poly1d(z)
plt.plot(y_test,p(y_test), color='magenta')
plt.show()
```

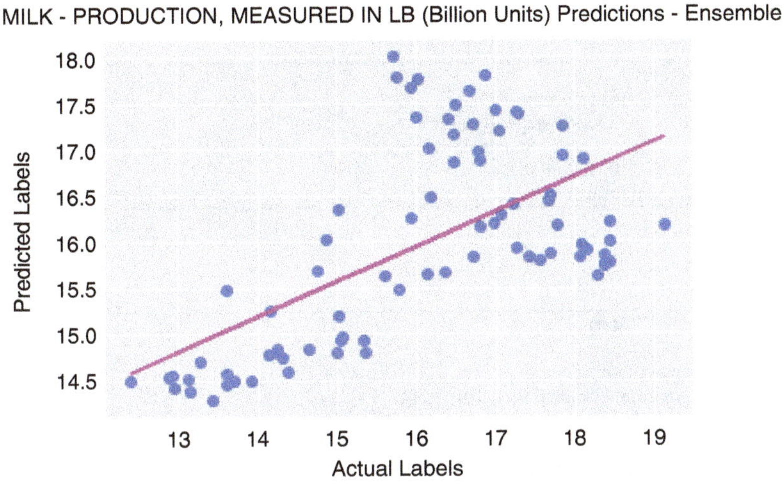

Random Forrest Regressor

A random forest is an ensemble technique capable of performing both regression and classification tasks with the use of multiple decision trees and a technique called bootstrap aggregation, commonly known as bagging. Bagging, in the random forest[135] method, involves training each decision tree on a different data sample where sampling is done with replacement. A random forest[136] is a meta-estimator that fits a number of classifying decision trees on various subsamples of the dataset and uses averaging to improve the predictive accuracy and control overfitting. The subsample size is controlled with the max_samples parameter if bootstrap=True (default); otherwise the whole dataset is used to build each tree:

```
from sklearn.ensemble import RandomForestRegressor

pipeline = Pipeline(steps=[('preprocessor', preprocessor),
                  ('regressor', RandomForestRegressor())])


# train a logistic regression model on the training set
model = pipeline.fit(X_train, (y_train))
print (model)
```

[135] A Beginners Guide to Random Forest Regression - https://medium.com/datadriveninvestor/random-forest-regression-9871bc9a25eb

[136] Random Forrest Regressor - https://scikit-learn.org/stable/modules/generated/sklearn.ensemble.RandomForestRegressor.html

Output:

```
Pipeline(steps=[('preprocessor',
        ColumnTransformer(transformers=[('num',
                        Pipeline(steps=[('scaler',
                                StandardScaler())]),
                        [0, 1, 2, 3])])),
        ('regressor', RandomForestRegressor())])
```

Predict the classifier:

```
predictions = model.predict(X_test)

np.set_printoptions(suppress=True)
print('Predicted labels: ', np.round(predictions)[:10])
print('Actual labels: ' ,y_test[:10])

mse = mean_squared_error(y_test, predictions)
print("MSE:", mse)
rmse = np.sqrt(mse)
print("RMSE:", rmse)
r2 = r2_score(y_test, predictions)
print("R2:", r2)
```

Output:

```
Predicted labels: [16. 16. 18. 14. 17. 16. 14. 16. 16. 16.]
Actual labels: [16.79 15.33 18.43 12.4  15.76 16.8  14.15 18.43 17.28 16.4 ]
MSE: 0.5300754653750039
RMSE: 0.7280628169155489
R2: 0.8179010495801368
```

Plot the model outcome:

```
plt.scatter(y_test, predictions)
plt.xlabel('Actual Labels')
plt.ylabel('Predicted Labels')
plt.title('MILK - PRODUCTION, MEASURED IN LB (Billion Units) Predictions - Ensemble')
z = np.polyfit(y_test, predictions, 1)
p = np.poly1d(z)
plt.plot(y_test,p(y_test), color='magenta')
plt.show()
```

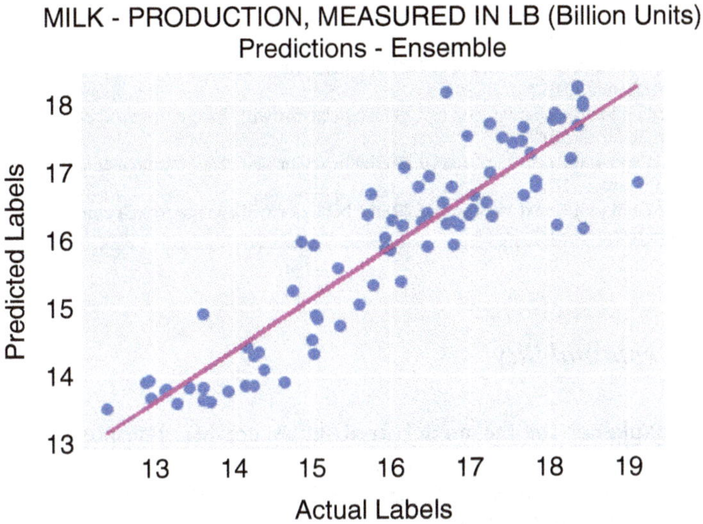

Explainers:

```
# Explainers:
# 1. SHAP Tabular Explainer
from interpret.ext.blackbox import TabularExplainer

# OR

# 2. Mimic Explainer
#from interpret.ext.blackbox import MimicExplainer
# You can use one of the following four interpretable models as a global surrogate to the black box
model
from interpret.ext.glassbox import LGBMExplainableModel
from interpret.ext.glassbox import LinearExplainableModel
from interpret.ext.glassbox import SGDExplainableModel
from interpret.ext.glassbox import DecisionTreeExplainableModel
```

```
feature_names= ['CornPrice', 'HRWheatPrice', 'SoybeansPrice', 'SorghumPrice']
# 1. Using SHAP TabularExplainer
explainer = TabularExplainer(model,
                             X_train,
                             features = feature_names)
```

```
# Passing in test dataset for evaluation examples - note it must be a representative sample of the
original data
# x_train can be passed as well, but with more examples explanations will take longer although they
may be more accurate
global_explanation = explainer.explain_global(X_test)
```

Output:

```
l1_reg="auto" is deprecated and in the next version (v0.29) the behavior will change from a conditional use of AIC to simply "num_features(10)"!
l1_reg="auto" is deprecated and in the next version (v0.29) the behavior will change from a conditional use of AIC to simply "num_features(10)"!
l1_reg="auto" is deprecated and in the next version (v0.29) the behavior will change from a conditional use of AIC to simply "num_features(10)"!
l1_reg="auto" is deprecated and in the next version (v0.29) the behavior will change from a conditional use of AIC to simply "num_features(10)"!
l1_reg="auto" is deprecated and in the next version (v0.29) the behavior will change from a conditional use of AIC to simply "num_features(10)"!
```

Step 11: *Model Explainability*

Let's get a suitable explainer for the model from the Azure ML interpretability library you installed earlier. There are many kinds of explainer. In this example, you'll use a Tabular Explainer, which is a "black box" explainer that can be used to explain many kinds of model by invoking an appropriate SHAP model explainer:

```
# Explainers:
# 1. SHAP Tabular Explainer
from interpret.ext.blackbox import TabularExplainer

# OR

# 2. Mimic Explainer
#from interpret.ext.blackbox import MimicExplainer
# You can use one of the following four interpretable models as a global surrogate to the black box model
from interpret.ext.glassbox import LGBMExplainableModel
from interpret.ext.glassbox import LinearExplainableModel
from interpret.ext.glassbox import SGDExplainableModel
from interpret.ext.glassbox import DecisionTreeExplainableModel
```

```
feature_names= ['CornPrice', 'HRWheatPrice', 'SoybeansPrice', 'SorghumPrice']
# 1. Using SHAP TabularExplainer
explainer = TabularExplainer(model,
                    X_train,
                    features = feature_names)
```

Output:

```
Setting feature_perturbation = "tree_path_dependent" because no background data was given.
The option feature_dependence has been renamed to feature_perturbation!
The option feature_perturbation="independent" is has been renamed to
feature_perturbation="interventional"!
```

```
# Passing in test dataset for evaluation examples - note it must be a representative sample of the original data
# x_train can be passed as well, but with more examples explanations will take longer although they may be more accurate
global_explanation = explainer.explain_global(X_test)
```

Output:

```
l1_reg="auto" is deprecated and in the next version (v0.29) the behavior will change from a conditional use of AIC to simply "num_features(10)"!
l1_reg="auto" is deprecated and in the next version (v0.29) the behavior will change from a conditional use of AIC to simply "num_features(10)"!
l1_reg="auto" is deprecated and in the next version (v0.29) the behavior will change from a conditional use of AIC to simply "num_features(10)"!
```

```
# Sorted SHAP values
print('ranked global importance values: {}'.format(global_explanation.get_ranked_global_values()))
# Corresponding feature names
print('ranked global importance names: {}'.format(global_explanation.get_ranked_global_names()))
# Feature ranks (based on original order of features)
print('global importance rank: {}'.format(global_explanation.global_importance_rank))

# Print out a dictionary that holds the sorted feature importance names and values
print('global importance rank: {}'.format(global_explanation.get_feature_importance_dict()))
```

Output:

```
ranked global importance values: [1.5958548054970099, 1.3749127799757606, 1.0705944072953855, 0.9083334522985625]
ranked global importance names: ['SorghumPrice', 'SoybeansPrice', 'CornPrice', 'HRWheatPrice']
global importance rank: [3, 2, 0, 1]
global importance rank: {'SorghumPrice': 1.5958548054970099, 'SoybeansPrice': 1.3749127799757606, 'CornPrice': 1.0705944072953855, 'HRWheatPrice': 0.9083334522985625}
```

Explain overall model predictions as a collection of local (instance-level) explanations:

```
# Sorted local feature importance information; reflects the original feature order
sorted_local_importance_names = local_explanation.get_ranked_local_names()
sorted_local_importance_values = local_explanation.get_ranked_local_values()

print('sorted local importance names: {}'.format(sorted_local_importance_names))
print('sorted local importance values: {}'.format(sorted_local_importance_values))
```

```
sorted local importance names: ['SoybeansPrice', 'SorghumPrice', 'CornPrice', 'HRWheatPrice']
sorted local importance values: [1.2983374789150002, 0.28364662319934775, -0.5178476726426763, -0.8546466046895271]
```

```
from interpret_community.widget import ExplanationDashboard
ExplanationDashboard(global_explanation, model, datasetX=X_test, true_y=y_test)
```

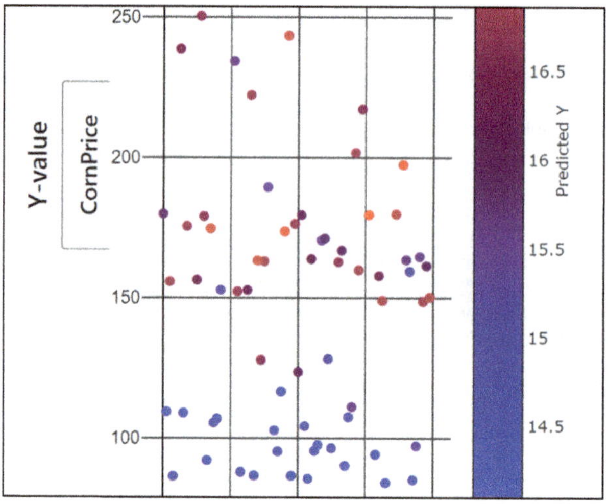

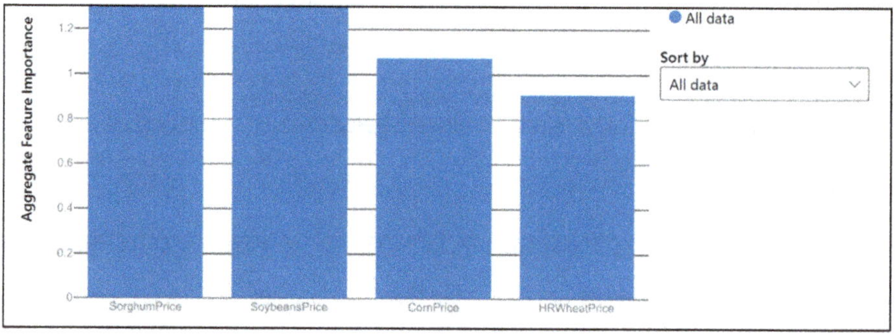

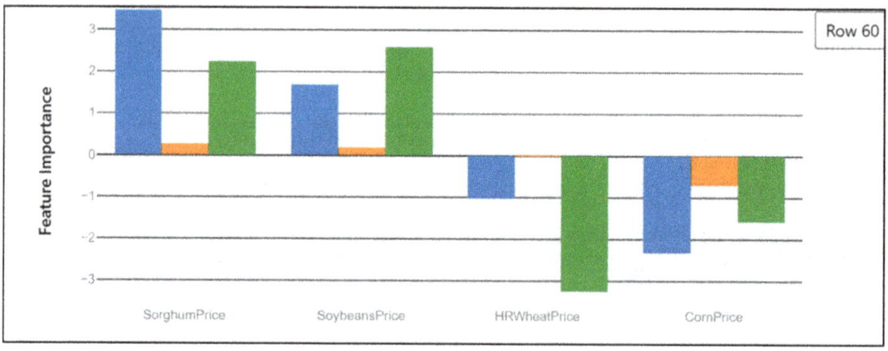

Random Forest Regressor

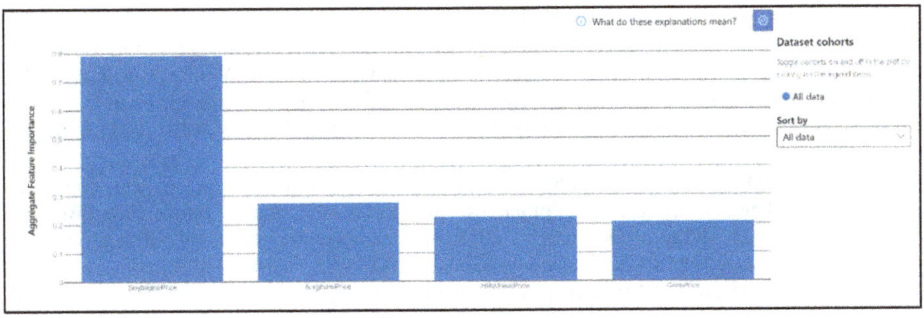

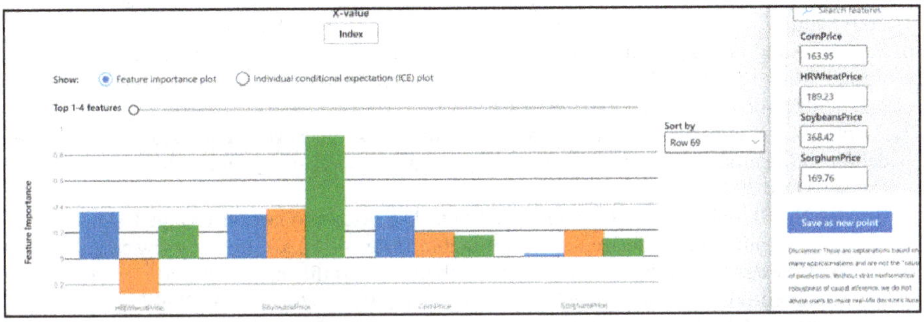

Chapter Summary
After reading the chapter, you should comfortably answer:

- NOAA Weather Events Database
- Machine learning modeling of temperature, rain, precipitation, and Palmer Drought Severity Index (PDSI)
- Weather Events modeling as part of ML
- Rice and milk production modeling and step-by-step development of the models
- Usage of crop calendar and overlapping crop calendar and weather events to develop ML models

References

1. Goedde, L., Katz, J., Ménard, A., Revellat, J.: Agriculture's connected future: How technology can yield new growth (October 9, 2020). https://www.mckinsey.com/industries/agriculture/our-insights/agricultures-connected-future-how-technology-can-yield-new-growth. Access Date: 18 Sept 2020
2. Walch, K.: How AI is transforming agriculture (July 5, 2019). https://www.forbes.com/sites/cognitiveworld/2019/07/05/how-ai-is-transforming-agriculture/?sh=7ce84c004ad1. Access Date: 8 Jan 2021

3. Zhao, C., Liu, B., Piao, S., Wang, X., Lobell, D.B., Huang, Y., Huang, M., Yao, Y., Bassu, S., Ciais, P., Durand, J.-L., Elliott, J., Ewert, F., Janssens, I.A., Li, T., Lin, E., Liu, Q., Martre, P., Müller, C., Peng, S., Peñuelas, J., Ruane, A.C., Wallach, D., Wang, T., Wu, D., Liu, Z., Zhu, Y., Zhu, Z., Asseng, S.: Temperature increase reduces global yields of major crops in four independent estimates. **114**, 9326–9331 (2017). https://doi.org/10.1073/pnas.1701762114. https://www.pnas.org/content/114/35/9326
4. Mcbride, W.D., Skorbiansky, S.R., Childs, N.: U.S. rice production in the new millennium: Changes in structure, practices, and costs. Economic Information Bulletin No. (EIB-202) 62pp (December 2018). https://www.ers.usda.gov/publications/pub-details/?pubid=90925#:~:text=December%202018-,U.S.%20Rice%20Production%20in%20the%20New%20Millennium,in%20Structure%2C%20Practices%2C%20and-%20Costs&text=The%20adoption%20of%20new%20technologies,increased%2C%20offsetting%20the%20higher%20costs. Access Date: 18 Sept 2019
5. Shaffer, S.: California rice production: Economic and environmental partnerships (March 2001). http://worldfood.apionet.or.jp/web/3.htm. Access Date: 18 Sept 2019
6. Fox, J.: Amid drought, California fields to grow rice: Bloomberg View (January 9, 2019); Posted May 17, 2015. https://www.oregonlive.com/opinion/2015/05/amid_drought_california_fields.html. Access Date: 10 Dec 2020
7. Faunt, C.: California's Central Valley. https://www.usgs.gov/centers/ca-water/science/californias-central-valley?qt-science_center_objects=0#qt-science_center_objects. Access Date: 10 Dec 2020
8. Childs, N., Skorbiansky, S.R., McBride, W.D.: U.S. rice production changed significantly in the new millennium, but remained profitable (May 4, 2020). https://www.ers.usda.gov/amber-waves/2020/may/us-rice-production-changed-significantly-in-the-new-millennium-but-remained-profitable/#:~:text=Number%20of%20Rice%20Farms%20Fell%20From%201997%20to%202017,rice%2Dgrowing%20States%20reporting%20declines. Access Date: 10 Dec 2020
9. Newton, D.J.: Working the land with 10 acres: Small acreage farming in the United States (April 2014). https://www.ers.usda.gov/webdocs/publications/43847/46432_eib123.pdf?v=0. Access Date: 10 Dec 2020
10. Basu, K.: India's farm laws are a global problem, Friday (February 19, 2021). https://www.brookings.edu/opinions/indias-farm-laws-are-a-global-problem/. Access Date: 20 Mar 2021
11. Huang, J., Wang, X., Qiu, H.: Small-scale farmers in China in the face of modernisation and globalisation. ISBN: 9781843698692. https://pubs.iied.org/16515iied. Access Date: 20 Mar 2021
12. World Bank Group: Transforming Vietnamese Agriculture: Gaining More for Less. Vietnam Development Report. World Bank, Washington, DC (2016). © World Bank. https://openknowledge.worldbank.org/handle/10986/24375. License: CC BY 3.0 IGO
13. Linquist, B., Espino, L., Leinfelder-Miles, M., Brim-Deforest, W., Grettenberger, I., Al-Khatib, K.: Rice Production Manual (2018). http://rice.ucanr.edu/files/288581.pdf. Access Date: 10 Dec 2020
14. Hardke, J.T.: Arkansas Rice Production Handbook – MP192 (2018). https://www.uaex.edu/publications/mp-192.aspx. Access Date: 10 Dec 2020
15. Hackin, H.: Sacramento Valley rice crop off to a late start due to continued rains (April 20, 2018). https://www.chicoer.com/2017/05/07/sacramento-valley-rice-crop-off-to-a-late-start-due-to-continued-rains/. 10 Dec 2020
16. Fernandez, B.: Northern California almond, walnut, rice farmers face problems with persistent rain (May 18, 2019). https://www.mercurynews.com/2019/05/18/almond-walnut-rice-farmers-face-problems-with-persistent-rain/. 11 Dec 2020
17. Lal, M., Singh, K.K., Rathore, L.S., Srinivasan, G., Saseendran, S.A.: Vulnerability of rice and wheat yields in NW India to future changes in climate. Agric. For. Meteorol. **89**(2), 101–114 (1998). ISSN 0168-1923, https://doi.org/10.1016/S0168-1923(97)00064-6. 11 Dec 2020

18. Yang, Z., Zhang, Z., Zhang, T., et al.: The effect of season-long temperature increases on rice cultivars grown in the central and southern regions of China. Front Plant Sci. **8**:1908, 2017. Published 2017 Nov 6. https://doi.org/10.3389/fpls.2017.01908
19. Ross, K.: California Agricultural Statistics Review 2015–2016. https://www.cdfa.ca.gov/statistics/pdfs/2016report.pdf. Access Date: 10 Dec 2020
20. Daniels, J.: California farmers can't win as monster storms threaten crops. Published January 25, 2017 Est updated January 27, 2017. https://www.cnbc.com/2017/01/25/california-farmers-cant-win-as-monster-storms-threaten-crops.html. Access Date: 8 Jan 2020
21. Fitchette, T.: Some walnut, peach orchards show "collapse" symptoms after NorCal winter floods (June 18, 2017). https://www.farmprogress.com/orchard-crops/some-walnut-peach-orchards-show-collapse-symptoms-after-norcal-winter-floods. 10 Sept 2019
22. Dong-Hwan, K.: Golden harvest at risk: Climate change puts heat on rice yields. Posted: September 20, 2020. Updated: September 24, 2020. https://www.koreatimes.co.kr/www/nation/2020/12/371_296304.html. 8 Jan 2021
23. Hamdan, N.: Scorching heat can affect rice yield, Saturday (March 19, 2016). https://www.thestar.com.my/news/nation/2016,/03/19/scorching-heat-can-affect-rice-yield-mardi-extreme-hot-weather-could-damage-harvest-quality-and-caus/. 11 Dec 2020.
24. Sun, T., Hasegawa, T., Liu, B., Tang, L., Liu, L., Cao, W., Zhu, Y.: Current rice models underestimate yield losses from short-term heat stresses (October 16, 2020), https://doi.org/10.1111/gcb.15393. 8 Jan 2021
25. Choudhary, S.: Climate change threaten's India's crop production, rice yields may be hit hard, Updated: June 25, 2019. https://www.livemint.com/news/india/climate-change-threatens-india-s-crop-production-rice-yields-may-be-hard-hit-1561435191944.html. 11 Dec 2020
26. Lund, J., Medellin-Azuara, J., Durand, J., Stone, K.: Lessons from California's 2012–2016 drought. J. Water Resour. Plann. Manag. **144**(10) (October 2018). https://ascelibrary.org/doi/full/10.1061/%28ASCE%29WR.1943-5452.0000984. 22 Nov 2020
27. Baldassare, M., Bonner, D., Kordus, D., Lopes, L.: PPIC Statewide Survey: Californians and the Environment (July 2016). https://www.ppic.org/publication/ppic-statewide-survey-californians-and-the-environment-july-2016/. 18 Sept 2020
28. Vose, R.S., Applequist, S., Squires, M., Durre, I., Menne, M.J., Williams, C.N., Jr., Fenimore, C., Gleason, K., Arndt, D.: Improved historical temperature and precipitation time series for U.S. climate divisions. Print Publication: May 1, 2014, Page(s): 1232–1251. https://journals.ametsoc.org/view/journals/apme/53/5/jamc-d-13-0248.1.xml. 22 Nov 2020
29. Gomez, C., Humphrey, T.: Another day of record high temperatures for the Sacramento Valley. https://www.abc10.com/article/weather/record-high-april-temps/103-6dcdf546-deee-437b-bbc0-903fec3f0555. Published: April 25, 2020 Updated: April 25, 2020. Access Date: 18 Sept 2020
30. Linquist, B., Espe, M.: When is the optimal time to plant rice in the Sacramento Valley? http://rice.ucanr.edu/files/232444.pdf. Access Date: 8 Jan 2021
31. Watts, A., Freeze, H.: California Wine Country (2011). https://wattsupwiththat.com/2011/02/26/hard-freeze-california-wine-country/. 14 Nov 2018
32. Peng, S., Huang, J., Sheehy, J.E., Laza, R.C., Visperas, R.M., Zhong, X., Centeno, G.S., Khush, G.S., Cassman, K.G.: Rice yields decline with higher night temperature from global warming. Proc. Natl Acad. Sci. **101**(27):9971–9975, Jul 2004. https://doi.org/10.1073/pnas.0403720101. 14 Nov 2018
33. Barker, R.: The response of milk production to price: A regional analysis (1960). https://core.ac.uk/download/pdf/38913888.pdf. Access Date: 22 Nov 2020
34. Auffhammer, M., Ramanathan, V., Vincent, J.R.: Climate change, the monsoon, and rice yield in India. Clim. Change. **111**, 411–424 (2012). https://doi.org/10.1007/s10584-011-0208-4
35. Miller, T.A., Rodewald, G.E., McElroy, R.G.: Economies of Size in U.S. Field Crop Farming (1981). https://naldc.nal.usda.gov/download/CAT87202149/PDF. Access Date: 22 Nov 2020

36. Shrivastava, P., Saxena, R.R., Xalxo, M.S., Verulkar, S.B.: Effect of high temperature at different growth stages on rice yield and grain quality traits. Adv. Agron. **111**:87–206, December 2011. https://doi.org/10.1016/B978-0-12-387689-8.00004-7. Access Date: 22 Nov 2020
37. D'Agostino, R., Pearson, E.: Tests for departure from normality. Empirical results for the distributions of b2 and $\sqrt{}$ b1. Biometrika. **60**(3), 613–622 (1973). https://doi.org/10.2307/2335012

Chapter 6
Agriculture Employment and the Role of AI in Improving Productivity

This chapter covers:

- Factors influencing agriculture's contribution to GDP
- The role of rural sector and employment in agriculture on the overall GDP of economies
- Metrics to evaluate classification and regression models
- Agricultural machine learning use cases:
  - ML enabled adaptive edge device coefficients
  - Wine multi-class classification
- The role of AI and ML in agriculture
- AI economics
- The call for policy makers

Explanations on the importance of the agricultural sector in the economy as economic growth progresses have benefitted greatly from the dual sector theory of Arthur Lewis.[1] In this theory, the modern service or industrial sector utilizes the surplus labor in the agricultural or primary sector as its source of growth, along with the capital generated by the investment of savings, to expand its production and thus the gross output of the economy [1]. As the services or industrial (modern) sector expands in importance, there is a concomitant reduction in the percentage contribution to gross output by the agricultural sector (please see Table 6.1). This growth process thus generally requires the movement of labor from rural areas to urban areas with a decline of the rural population as a percentage of the national population. Paradoxically, the rural population and percentage of agriculture employment to total employment play an important role in the growth of agriculture. In the following figure, you can observe employment change trend in the agriculture (decline) and service industries (increase).

[1] W.A Lewis Economic Development with Unlimited Supplies of Labour (1954) - https://la.utexas.edu/users/hcleaver/368/368lewistable.pdf

C. Vuppalapati, *Machine Learning and Artificial Intelligence for Agricultural Economics*, International Series in Operations Research & Management Science 314,
https://doi.org/10.1007/978-3-030-77485-1_6

Table 6.1 Employment in agriculture vs. service sector

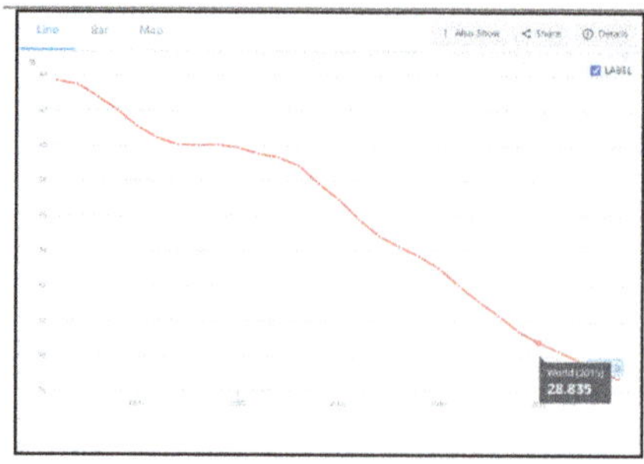	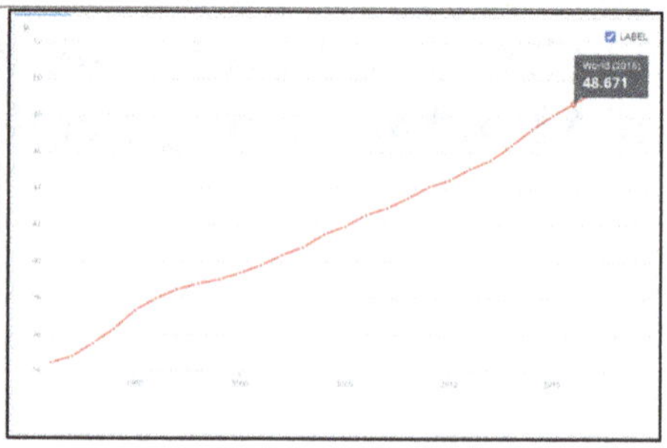
Employment in agriculture (% of total employment) (modeled by ILO)—World	Employment in services (% of total employment) (modeled by ILO)—World

Employment in agriculture—https://data.worldbank.org/indicator/SL.AGR.EMPL.ZS?end=2019&locations=1W&start=1991&view=chart
Employment in Services—https://data.worldbank.org/indicator/SL.SRV.EMPL.ZS?end=2019&locations=1W&start=1991&view=chart

It is general knowledge that agriculture does contribute not just food and fiber to the economy but also labor, capital, and foreign exchange, which all go toward economic development. Let's look at the following Haiti model:

> If agriculture fails to develop at a suitable pace, this could prove to be a critical constraint to the growth of the industrial sector as well as other sectors of the economy[2] [2].

Factors Influencing to Agriculture's Contribution to GDP

The purpose of the model is to go through some of the key macrolevel indicators and study the role they play in agriculture's contribution to GDP. Some of the important factors that influence agriculture's contribution to GDP include:

Population Growth

Population growth is derived from the total population. Population sources are as follows: (1) UN Population Division. World Population Prospects: 2019 Revision, (2) census reports and other statistical publications from national statistical offices, (3) Eurostat: Demographic Statistics, (4) UN Statistical Division. Population and

[2] Factors Influencing Agriculture's Contribution to GDP - https://ageconsearch.umn.edu/record/187331/files/Factors%20Influencing%20Agriculture_s%20Contribution%20to%20GDP%20Latin%20America%20and%20the%20Caribbean.pdf

Vital Statistics Report (various years), (5) US Census Bureau: International Database, and (6) Secretariat of the Pacific Community: Statistics and Demography Program.

Definition: Population growth is the average annual percent change in the population, resulting from a surplus (or deficit) of births over deaths and the balance of migrants entering and leaving a country.[3] The rate may be positive or negative. The growth rate is a factor that determines how great a burden would be imposed on a country by the changing needs of its people for infrastructure (e.g., schools, hospitals, housing, roads), resources (e.g., food, water, electricity), and jobs. Rapid population growth can be seen as a threat by neighboring countries.

With respect to development and macroeconomic factors, population growth could be attributed to be positive or negative, depending upon the economic philosophies. If a country experiences high population growth and therefore has a larger population base, then it can transfer labor to the expanding modern sectors, without reducing the agricultural labor supply. In fact, a rapidly expanding population could allow a country to expand its labor surplus and thus allow a prolonged period of economic growth by expansion of both the agricultural and modern sector, particularly if there are technological innovations taking place in the agricultural sector, such as improved genetic planting material. Artificial intelligence plays a huge role in upgrading as well as improving productivity. As noted by McKinsey report,[4] the highly rural population countries with huge agricultural base experience a lower penetration of AI. This is a game-changer opportunity to tap into and create products that enable agricultural productivity [3].

Population growth could therefore allow the rural sector to play a role in fostering economic growth (Pemberton 2002). Under these circumstances, population growth could be expected to have a positive influence on the percentage contribution of agriculture to GDP.

Data Source: The World Bank—Population growth (annual %)[5]

---

[3] Population Growth - https://www.indexmundi.com/world/population_growth_rate.html#:~:text=Population%20growth%20rate%3A,1.03%25%20(2020%20est.)&text=Definition%3A%20The%20average%20annual%20percent,entering%20and%20leaving%20a%20country

[4] McKinsey Global Institute (MGI)—Notes from the AI frontier Modeling the impact of AI on the world economy September-2018 - https://www.mckinsey.com/~/media/McKinsey/Featured%20Insights/Artificial%20Intelligence/Notes%20from%20the%20frontier%20Modeling%20the%20impact%20of%20AI%20on%20the%20world%20economy/MGI-Notes-from-the-AI-frontier-Modeling-the-impact-of-AI-on-the-world-economy-September-2018.ashx

[5] Population growth (annual %) - https://data.worldbank.org/indicator/SP.POP.GROW?end=2017&locations=1W&start=1981&view=chart

Rural Population (% of the Total Population)

Definition[6]: Rural population refers to people living in rural areas as defined by national statistical offices. It is calculated as the difference between total population and urban population.

Description: The map below shows how rural population (% of the total population) varies by country. The shade of the country corresponds to the magnitude of the indicator. The darker the shade, the higher the value (please see Fig. 6.1). The country with the highest value in the world is Burundi, with a value of 86.97. The country with the lowest value in the world is Nauru, with a value of 0.00.

In general, if a country has been able to maintain a high percentage of its population in the rural sector, then it could be expected that its agricultural sector would constitute a larger percentage of its gross output. In accordance with the Lewis model,[7] this will imply a large proportion of the country's labor has not transferred to the modern sectors and being resident in the rural sector is still engaged in the agricultural sector, which could be expected to be making a greater percentage contribution to GDP [1].

Data Source: The World Bank—Rural population (% of the total population), World[8]

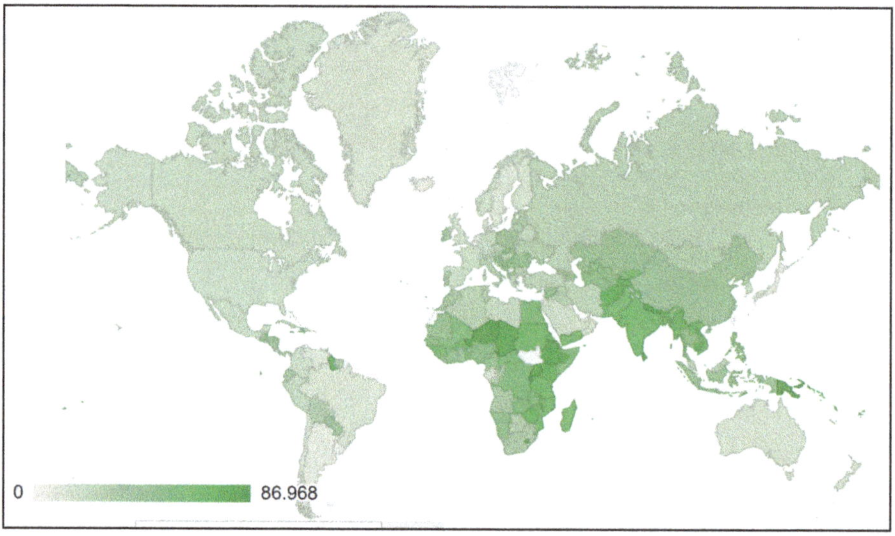

Fig. 6.1 Worldwide rural population

---

[6] Rural Population - https://www.indexmundi.com/facts/indicators/SP.RUR.TOTL.ZS

[7] W.A Lewis Economic Development with Unlimited Supplies of Labour (1954) - https://la.utexas.edu/users/hcleaver/368/368lewistable.pdf

[8] The World Bank - Rural population (% of total population) - https://data.worldbank.org/indicator/SP.RUR.TOTL.ZS?end=2017&locations=1W&start=1981&view=chart

Life Expectancy at Birth (LEB): Total (Years)

Life expectancy has a strong and positive effect on capital accumulation [4]. Increase in life expectancy is often associated with higher economic growth.[9] A 1998 World Bank study showed that life expectancy displays a strong tendency to improve with per capita income, ranging from as low as 37 years in Sierra Leone to as high as 77 in Costa Rica, more than 12 times richer. As the process of economic growth proceeds, it would be expected to be accompanied by a process of economic development, as the increasing gross output of the country is utilized by its citizens to improve the standard of living of the citizens. One important aspect of improved standard of living is the health status of the population and in particular the levels of maternal and childcare. As these aspects of the population improve, they will have a direct impact on the life span of citizens, improving the life expectancy at birth of the population.

Thus, it could be expected that the higher levels of life expectancy would be associated with countries at a higher state of economic development[10] (coefficient of regression tends to be positive). Importantly, agriculture is a major employer of many nations; higher agriculture contribution of agriculture to the GDP would result in higher economic status and would lead into higher life expectancy at birth. Conversely, countries with a lower percentage of agriculture sector in the gross output of the country would have lower LEB, and hence, the coefficient of LEB is expected to have a negative sign.

Calculation: (1) UN Population Division. World Population Prospects: 2019 Revision or derived from male and female life expectancy at birth from sources such as (2) census reports and other statistical publications from national statistical offices, (3) Eurostat: Demographic Statistics, (4) UN Statistical Division. Population and Vital Statistics Report (various years), (5) US Census Bureau: International Database, and (6) Secretariat of the Pacific Community: Statistics and Demography Program

Data Source: Life expectancy at birth, total (years)—World[11]

External Debt Stocks (% of GNI)

External debt[12] is that part of the total debt in a country that is owed to creditors outside the country. The debtors can be the government, corporations, or private households. The debt includes money owed to private commercial banks, other

---

[9] A Closer Look at the relationship between life expectancy and economic growth - https://www.gla.ac.uk/media/Media_43616_smxx.pdf

[10] Human Development - http://hdr.undp.org/en/content/latest-human-development-index-ranking

[11] Life expectancy at birth, total (years)—World - https://data.worldbank.org/indicator/SP.DYN.LE00.IN?end=2017&locations=1W&start=1981&view=chart

[12] External Debt - https://www.indexmundi.com/facts/indicators/DT.DOD.DECT.GN.ZS

governments, or international financial institutions. *External indebtedness affects a country's creditworthiness and investor perceptions.* Countries with fast-growing economies and exports are likely to be able to sustain higher debt levels. Various indicators determine a sustainable level of external debt, including (a) debt-to-GDP ratio, (b) foreign debt-to-exports ratio, (c) government debt-to-current fiscal revenue ratio, (d) share of foreign debt, (e) short-term debt, and (f) concessional debt in the total debt stock.

From agriculture point of view, higher external debt of a country implies the country has low foreign reserves and hence may not have resources to import food. This makes the country to produce locally, and hence the agriculture industry in the country expands. It is expected that as the external debt of a country increases, perhaps caused by the international un-competitiveness of its domestic production limiting exporting, the country may not be able to import as much agricultural and food products, due to lower availability of foreign currency. Thus, the country would be forced to produce more of its own food, resulting in increased agricultural production and an increased contribution of agriculture to GDP. Therefore, the higher the external debt of the country, the greater the contribution of agriculture is expected to be, and therefore the coefficient of external debt is hypothesized to be positive.

Source: External debt stocks (% of GNI)—World

Exports of goods and services (% of GDP)

https://data.worldbank.org/indicator/NE.EXP.GNFS.ZS?end=2017&locations=1W&start=1981&view=chart

Foreign Direct Investment and Net (BoP, Current US$)

Definition: Foreign direct investments are the net inflows of investment to acquire a lasting management interest[13] (10% or more of voting stock) in an enterprise operating in an economy other than that of the investor. It is the sum of equity capital, reinvestment of earnings, other long-term capital, and short-term capital as shown in the balance of payments. This series shows the total net FDI. In BPM6, financial account balances are calculated as the change in assets minus the change in liabilities. Net FDI outflows are assets and net FDI inflows are liabilities. Data are in current US dollars. The map below shows how foreign direct investment and net (BoP, current US$) varies by country (please see Fig. 6.2).

The darker the shade, the higher the value. The FDI into an economy will insert dollars and will have impact on GDP. The net effect would be more economic growth and potential impact to agricultural values[14] [5].

Data Source: Foreign direct investment, net (BoP, current US$)[15]

---

[13] Foreign Direct Investment - https://www.indexmundi.com/facts/indicators/BN.KLT.DINV.CD

[14] Analysis of Foreign Direct Investment, Agricultural Sector and Economic Growth in Tanzania - https://www.scirp.org/journal/paperinformation.aspx?paperid=73719

[15] FDI - https://data.worldbank.org/indicator/BN.KLT.DINV.CD?end=2017&locations=1W&start=1981&view=chart

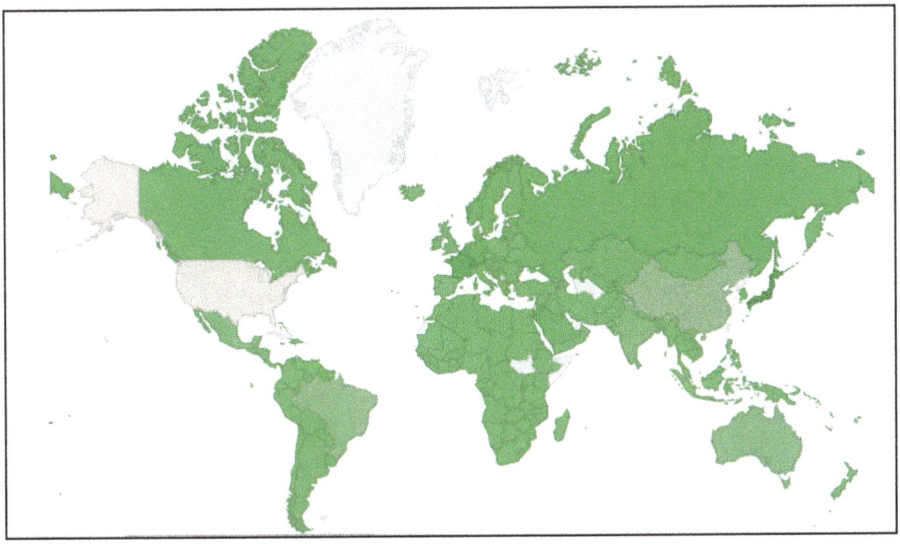

Fig. 6.2 FDI

Inflation, Consumer Prices (Annual %) (ICP)

The Consumer Price Index (CPI)[16] for food is a component of the all-items CPI. The CPI measures the average change over time in the prices paid by urban consumers for a representative market basket of consumer goods and services. While the all-items CPI measures the price changes for all consumer goods and services, including food, the CPI for food measures the changes in the retail prices of food items only.

Definition[17]: Inflation as measured by the consumer price index reflects the annual percentage change in the cost to the average consumer of acquiring a basket of goods and services that may be fixed or changed at specified intervals, such as yearly. The Laspeyres formula is generally used. The country with the highest value in the world is Venezuela, with a value of 254.95. The country with the lowest value in the world is Comoros, with a value of −4.29.

[16] Food Price Outlook - https://www.ers.usda.gov/data-products/food-price-outlook/

[17] Inflation - https://www.indexmundi.com/facts/indicators/FP.CPI.TOTL.ZG

Laspeyres Formula[18]

Laspeyres suggested this index formula in 1871. In case of calculating the price index, assuming that for individual item i, the price at the base period is $\mathbf{p}_{i0}$ and at the observation period is $\mathbf{p}_{i\,t}$, and quantity at the base period is $\mathbf{q}_{i\,0}$, the following equation is called the "Laspeyres formula":

$$\frac{\sum_i pit\, qi0}{\sum_i pi0\, qi0} \qquad \text{Equation (1)}$$

where the denominator and numerator are total expenditures for all items, at the base and the observation period, respectively, if consumers purchase the same number of commodities both at the base period and the observation period.

Equation 1 can be rewritten as:

$$\frac{\sum_i \frac{pit}{pi0} wi0}{\sum_i wi0} \text{ where } wi0 = pi0\, qi0 \qquad \text{Equation (1 weighted)}$$

This is the weighted average of price ratios of each item, weighted by expenditures at the base period.

Economic theory states that there are two sources of inflation: cost-push and demand-pull inflation. When in a country there is demand-pull inflation, due to increasing demand for food, producers are expected to invest more in the agricultural sector, resulting in increased production, and therefore increasing demand pull inflation should lead to increasing percentage contribution of agriculture to GDP. ICP's coefficient in this case is hypothesized to be positive. However, where cost-push inflation results, because of a decrease in aggregate agricultural supply, which may be caused by either an increase in wages or an increase in the prices of raw materials, the greater costs of agricultural production result in the amount of agricultural production falling, thus reducing the percentage contribution of agriculture to GDP. The coefficient of ICP, in this case, is expected to be negative. Thus, the coefficient of ICP is hypothesized to have either a negative or a positive sign, depending on the source of inflation in the countries (please see Fig. 6.3).

Data Source: Inflation, Consumer Prices[19]

The CPI in the United States reached[20] greater than 5% in 1989, 1990, and 2008. The world had double digit (10.22%) inflation in 1982.

[18] Q&A about the Consumer Price Index - https://www.stat.go.jp/english/data/cpi/1587.html

[19] Inflation - https://data.worldbank.org/indicator/FP.CPI.TOTL.ZG?end=2017&locations=1W&start=1981&view=chart

[20] Food Price Outlook - https://www.ers.usda.gov/data-products/food-price-outlook/food-price-outlook/#Consumer%20Price%20Index

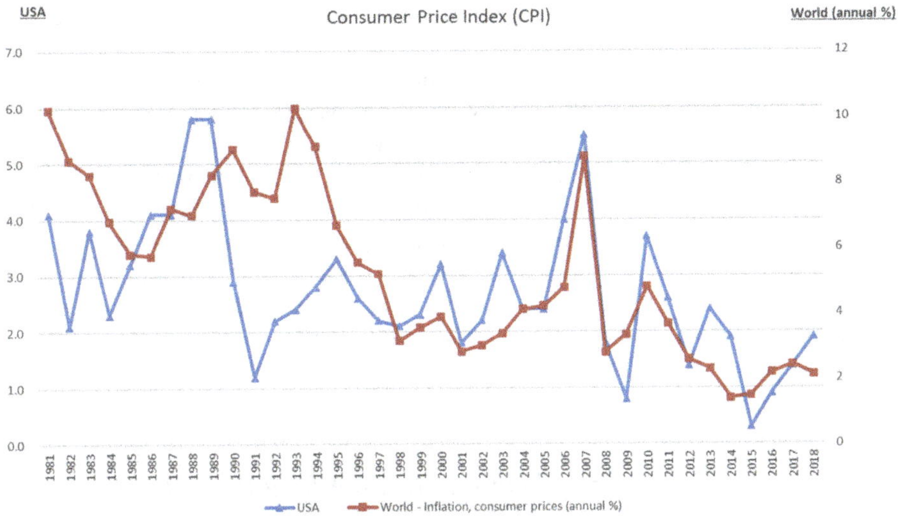

Fig. 6.3 CPI—the United States and world

Ratio of Export to Import Agricultural Products (EXIM)

The ratio of agricultural trade is the ratio of exports and imports of total agricultural merchandise for country:

$$\text{EXIM} = \frac{\text{Exports}}{\text{Imports}}$$

Data Source: Agricultural raw materials exports (% of merchandise exports)[21]
Data Source: Agricultural raw materials imports (% of merchandise imports)[22]

Employment in Agriculture (% of Total Employment) (Modeled ILO Estimate)

Employment is defined as persons of working age who were engaged in any activity to produce goods or provide services for pay or profit, whether at work during the reference period or not at work due to temporary absence from a job, or to working-time arrangement. The agriculture[23] sector consists of activities in agriculture,

[21] Agricultural raw materials exports (% of merchandise exports) - https://data.worldbank.org/indicator/TX.VAL.AGRI.ZS.UN?end=2017&locations=1W&start=1981&view=chart

[22] Agricultural raw materials imports (% of merchandise imports) - https://data.worldbank.org/indicator/TM.VAL.AGRI.ZS.UN?end=2017&locations=1W&start=1981&view=chart

[23] Employment in agriculture - https://www.indexmundi.com/facts/indicators/SL.AGR.EMPL.ZS

hunting, forestry, and fishing, in accordance with division 1 (ISIC 2) or categories A–B (ISIC 3) or category A (ISIC 4).

Sectoral information is particularly useful in identifying broad shifts in employment and stages of development. In the textbook case of economic development, labor flows from agriculture and other labor-intensive primary activities to industry and finally to the services sector; in the process, workers migrate from rural to urban areas. The breakdown of the indicator by gender allows for analysis of gender segregation of employment by specific sector. Women may be drawn into lower-paying service activities that allow for more flexible work schedules, thus making it easier to balance family responsibilities with work life. The country with the highest value in the world is Burundi, with a value of 92.04. The country with the lowest value in the world is Argentina, with a value of 0.06.

Source: Agriculture, forestry, and fishing, value added (annual % growth)[24]

Agriculture, Forestry, and Fishing: Value Added (% of GDP)

Agriculture corresponds to ISIC divisions[25] 1–5 and includes forestry, hunting, and fishing, as well as cultivation of crops and livestock production. Value added is the net output of a sector after adding up all outputs and subtracting intermediate inputs. It is calculated without making deductions for depreciation of fabricated assets or depletion and degradation of natural resources. The country with the highest value in the world is Somalia, with a value of 62.74. The country with the lowest value in the world is Singapore, with a value of 0.02.

Data Source: Agriculture, forestry, and fishing, value added (% of GDP)—World[26]

Brazil's Machine Learning Model: Agriculture's Contribution to Economy and GDP and the Role of AI Readiness

In the following section, we will prepare the machine learning model that prepares agriculture contribution to GDP using the abovementioned key macrolevel indictors and the exogenous role of AI to enhance the productivity.

[24] Employment in Agriculture - https://data.worldbank.org/indicator/SL.AGR.EMPL.ZS?end=2017&locations=1W&start=1981&view=chart

[25] Agriculture, value added (% of GDP) - https://www.indexmundi.com/facts/indicators/NV.AGR.TOTL.ZS

[26] Agriculture, forestry, and fishing, value added (% of GDP)—World - https://data.worldbank.org/indicator/NV.AGR.TOTL.KD.ZG?end=2017&locations=1W&start=1981&view=chart

AI readiness: Group 4—economies that need to strengthen foundations

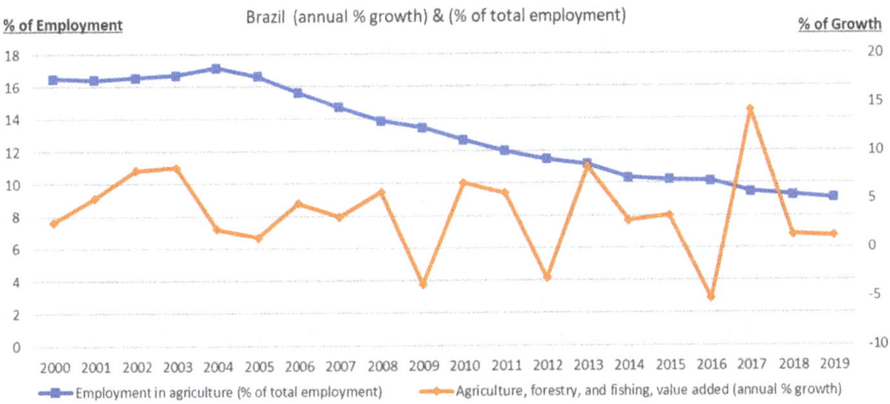

Fig. 6.4 Brazil—employment in agriculture and GDP

Brazil had 16.49% (2000) employment[27] in agriculture (% of the total employment). That is, out of 100 total employment, 16 people belong to agriculture in 2000. Two decades later, that number is 9.07% (only 9 out of 100 people employed in agriculture) (2019). Agriculture productivity growth[28] in recent decades in Brazil has been mainly driven by investments in agriculture innovation, facilitation of sector financing, and trade liberalization. Nonetheless, low productivity growth is impacting Brazil's economic growth.[29] Agriculture productivity has room to grow further (please see Fig. 6.4), improving productivity of lagging midsize farmers and regions and reforming agriculture policies toward agriculture financing [6].

Data Source: The World Bank[30]

Artificial intelligence can foster such productivity and growth. Given its value and importance, Brazilian government recently initiated AI policy[31] and created national

[27] Data Source—Employment in agriculture (% of total employment) (modeled ILO estimate) - https://data.worldbank.org/indicator/SL.AGR.EMPL.ZS?end=2017&locations=1W&start=1981&view=chart

[28] API_NV.AGR.TOTL.KD.ZG_DS2_en_excel_v2_2063913.XLS - https://api.worldbank.org/v2/en/indicator/NV.AGR.TOTL.KD.ZG?downloadformat=excel

[29] Agriculture Productivity Growth in Brazil—Recent trends and future prospects - http://documents1.worldbank.org/curated/en/268351520343354377/pdf/123948-WP-6-3-2018-8-39-22-AriasetalAgriculturalgrowthinBrazil.pdf

[30] API_SL.AGR.EMPL.ZS_DS2_en_excel_v2_2059645.XLS - https://api.worldbank.org/v2/en/indicator/SL.AGR.EMPL.ZS?downloadformat=excel

[31] AI In Brazil - https://www.oecd.ai/dashboards/countries/Brazil

AI innovation network[32]—focused on artificial intelligence (AI) with the aim of increasing the production capacity and competitiveness of local companies [7].

Brazil AI Policies[33]

- Applied research centers in AI
  - Brazil has announced its aims to establish eight applied research centers hosted by established academic and research institutions.
- The Brazilian artificial intelligence strategy
  - The Brazilian artificial intelligence strategy will cover relevant topics on AI and its impact on economy, ethics, development, education, and jobs, as well as coordinate specific public policies addressing such issues.
- Brazil's digital transformation strategy
  - Brazil's digital transformation strategy (E-Digital), published on March 2018, harmonizes and coordinates different governmental initiatives on digital issues to advance the Sustainable Development Goals in Brazil.
- General Personal Data Protection Law
  - The Brazilian data protection law includes information on the protection of personal data as well as a section on database ownership and database rights.
- National Internet of Things Plan: Degree 9854
  - Institutes the National Internet of Things Plan and provides for the management and monitoring chamber for the development of communication systems machine to machine and Internet of things.

First, we develop ML model with key macrolevel parameters.

Please find Brazil data (from 1990 to 2019) that includes population (annual growth %), inflation, consumer prices (annual growth %), rural population (% of total population), export of goods and services (% of GDP), export and import ration, FDI, external debt stocks (% of GNI), life expectancy at birth, employment in agriculture (% of total employment), and agriculture, forestry, and fishing, value added (% of GDP).

---

[32] Brazil creates national AI innovation network - https://www.zdnet.com/article/brazil-creates-national-ai-innovation-network/

[33] Brazil AI Policies - https://www.oecd.ai/dashboards/policy-initiatives?conceptUris=http:%2F%2Fkim.oecd.org%2FTaxonomy%2FGeographicalAreas%23Brazil

The software code for this model can be found on GitHub Link: BrazilAgriculturalGDPModelWithFinal.Ipynb (Jupyter Notebook)

Data

Year	Populatior	Inflation, c	Rural popu	Exports of good:	EXIM	FDI	External d	Life expec	Employme	Agriculture, forestry, and fishing, value added (% of GDP)
1990	1.811518	2947.733	26.078	8.199990015	0.746516	-8000000	26.75549	66.343	19.4	6.873300859
1991	1.759418	432.7867	25.31	8.676973095	0.773204	-1.2E+07	20.58823	66.742	19.64	6.788806688
1992	1.707335	951.9621	24.556	10.86824925	0.825162	2200000	32.95725	67.141	19.75	6.779952159
1993	1.664859	1927.381	23.819	10.50327024	1.412276	2800000	33.85293	67.539	19.31	6.563569585
1994	1.633793	2075.888	23.097	9.665051387	0.958919	0	28.46723	67.932	18.69	8.536411053
1995	1.609715	66.00703	22.39	7.528875038	0.523181	-7400000	21.23934	68.318	18.14	5.011666465
1996	1.588838	15.75767	21.698	6.730209683	0.808029	-4100000	21.96856	68.695	17.1	4.768411211
1997	1.563725	6.926713	20.952	6.983649575	0.717432	-4000000	23.27202	69.061	16.85	4.685047029
1998	1.528964	3.195076	20.219	7.030500339	0.558841	-1.1E+07	29.31409	69.419	16.2	4.750044784
1999	1.481112	4.858447	19.504	9.564898261	0.44451	-3E+07	43.14807	69.769	16.69	4.64598154
2000	1.424167	7.044141	18.808	10.188048	0.408439	-1.3E+07	38.09808	70.116	16.49	4.750459976
2001	1.366957	6.840359	18.447	12.37171067	0.324697	-4400000	42.48989	70.462	16.42	4.801043016
2002	1.312744	8.450164	18.12	14.23059027	0.375657	-5700000	47.24673	70.813	16.53	5.475274893
2003	1.257383	14.71492	17.797	15.18078371	0.384776	-1.4E+07	43.63802	71.17	16.68	6.167184386
2004	1.201294	6.597185	17.479	16.54576151	0.433634	-5900000	33.97677	71.531	17.14	5.665311576
2005	1.145881	6.869537	17.166	15.24382927	0.394671	-2.6E+07	21.75472	71.896	16.62	4.651168823
2006	1.090197	4.183568	16.857	14.3743163	0.404768	-1.6E+08	17.9923	72.26	15.62	4.370043778
2007	1.037862	3.641273	16.552	13.3276751	0.370104	-7.5E+07	17.43768	72.618	14.68	4.416915436

Model Development

The following are the steps to build the model:

Step 1: Load into Python Data Frame. Load ML-specific libraries:

```
import pandas as pd
import numpy as np
import plotly.graph_objects as go
import seaborn as sns; sns.set()
import matplotlib.pyplot as plt
from scipy import stats
from scipy.stats import pearsonr
import matplotlib.pyplot as plt
from sklearn.preprocessing import MinMaxScaler
from sklearn.model_selection import train_test_split
from sklearn.linear_model import LinearRegression
from sklearn.ensemble import RandomForestClassifier
from sklearn.metrics import r2_score
from sklearn import metrics

import eli5
import matplotlib.pyplot as plt
import matplotlib as mpl
import numpy as np
import scipy.stats as spstats
import seaborn as sns

%matplotlib inline
mpl.style.reload_library()
mpl.style.use('classic')
mpl.rcParams['figure.facecolor'] = (1, 1, 1, 0)
mpl.rcParams['figure.figsize'] = [6.0, 4.0]
mpl.rcParams['figure.dpi'] = 100
```

Step 2: Load Brazilian data sources:

```
# Importing data

dataBrazil_df = pd.read_csv("BrazilRegressionFinalwGDPPG.csv")
dataBrazil_df
```

Output:

Year	Population growth (annual %)	Inflation, consumer prices (annual %)	Rural population (% of total population)	Exports of goods and services (% of GDP)	EXIM	FDI	External debt stocks (% of GNI)	Life expectancy at birth, total (years)	Employment in agriculture (% of total employment) (modeled ILO estimate)	Agriculture, forestry, and fishing, value added (% of GDP)
1990	1.811518	2947.732772	26.078	8.199990	0.746516	-8000000.0	26.755490	66.343	19.400000	6.873301
1991	1.759418	432.786662	25.310	8.676973	0.773204	-11800000.0	20.588226	66.742	19.639999	6.788807
1992	1.707335	951.962053	24.556	10.868249	0.825162	2200000.0	32.957248	67.141	19.750000	6.779952
1993	1.664859	1927.380790	23.819	10.503270	1.412276	2800000.0	33.852930	67.539	19.309999	6.563570
1994	1.633793	2075.888398	23.097	9.665051	0.958919	0.0	28.467227	67.932	18.690001	8.536411
1995	1.609715	66.007034	22.390	7.528875	0.523181	-7400000.0	21.239336	68.318	18.139999	5.011666
1996	1.588838	15.757666	21.698	6.730210	0.808029	-4100000.0	21.968558	68.695	17.100000	4.768411
1997	1.563725	6.926713	20.952	6.983650	0.717432	-4000000.0	23.272024	69.061	16.850000	4.685047

Step 3: DataFrame Describe—To find statistical properties of the data:

```
dataBrazil_df.describe()
```

Output:

	Population growth (annual %)	Inflation, consumer prices (annual %)	Rural population (% of total population)	Exports of goods and services (% of GDP)	EXIM	FDI	External debt stocks (% of GNI)	Life expectancy at birth, total (years)	Employment in agriculture (% of total employment) (modeled ILO estimate)	Agriculture, forestry, and fishing, value added (% of GDP)
count	30.000000	30.000000	30.000000	30.000000	30.000000	3.000000e+01	30.000000	30.000000	30.000000	30.000000
mean	1.220800	285.320955	18.165067	11.526055	0.476303	-6.566112e+07	27.388445	71.529733	14.833333	5.095717
std	0.346284	728.014678	3.799873	2.688455	0.270416	8.291088e+07	9.129862	2.941477	3.456146	1.054377
min	0.751548	3.195076	13.176000	6.730210	0.199399	-3.748550e+08	15.851329	66.343000	9.080000	4.115756
25%	0.899139	4.925708	15.150000	9.795801	0.296799	-1.050000e+08	20.026365	69.150500	11.600000	4.423039
50%	1.173588	6.738404	17.322500	11.662373	0.389724	-2.790000e+07	26.790104	71.713500	16.310000	4.668108
75%	1.555034	13.293665	20.768750	13.482419	0.549926	-6.275000e+06	32.666126	74.137000	17.037500	5.359373
max	1.811518	2947.732772	26.078000	16.545762	1.412276	2.800000e+06	47.246726	75.672000	19.750000	8.536411

The statistics reveal some information about the distribution of the data in each of the numeric fields, including the number of observations (there are 30 records), the mean and standard deviation, minimum and maximum values, and the quartile values (the threshold values for 25%, 50% which is also the median, and 75% of the data). From this, we can see that the mean number of population growth (annual %) is around 1.22; but there's a comparatively small standard deviation, indicating a not much variance in the population growth (annual %).

The record count is 30. The population growth (annual %) was 1.8% max in 1990; the growth percentage is declining thereafter.

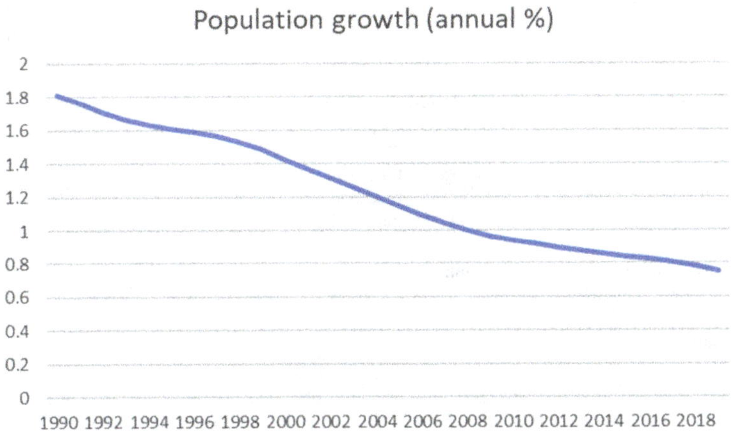

Step 4: Perform distribution function.

Distribution function provides more clues into the data:

```
# Create a function that we can re-use
def show_distribution(var_data,colname):
    from matplotlib import pyplot as plt

    # Get statistics
    min_val = var_data.min()
    max_val = var_data.max()
    mean_val = var_data.mean()
    med_val = var_data.median()
    mod_val = var_data.mode()[0]


print('Minimum:{:.2f}\nMean:{:.2f}\nMedian:{:.2f}\nMode:{:.2f}\nMaximum:{:.2f}\n'.format(min_v
al,
                                                    mean_val,
                                                    med_val,
                                                    mod_val,
                                                    max_val))

    # Create a figure for 2 subplots (2 rows, 1 column)
    fig, ax = plt.subplots(2, 1, figsize = (10,4))

    # Plot the histogram
    ax[0].hist(var_data)
    ax[0].set_ylabel('Frequency')

    # Add lines for the mean, median, and mode
    ax[0].axvline(x=min_val, color = 'gray', linestyle='dashed', linewidth = 2)
    ax[0].axvline(x=mean_val, color = 'cyan', linestyle='dashed', linewidth = 2)
    ax[0].axvline(x=med_val, color = 'red', linestyle='dashed', linewidth = 2)
    ax[0].axvline(x=mod_val, color = 'yellow', linestyle='dashed', linewidth = 2)
    ax[0].axvline(x=max_val, color = 'gray', linestyle='dashed', linewidth = 2)

    # Plot the boxplot
    ax[1].boxplot(var_data, vert=False)
    ax[1].set_xlabel('Value')

    # Add a title to the Figure
    fig.suptitle(colname + ' Data Distribution')

    # Show the figure
    fig.show()


# Get the variable to examine

for col in ['Population growth (annual %)', 'Inflation, consumer prices (annual %)',
       'Rural population (% of total population)',
       'Exports of goods and services (% of GDP)', 'EXIM', 'FDI',
       'External debt stocks (% of GNI)',
       'Life expectancy at birth, total (years)',
       'Employment in agriculture (% of total employment) (modeled ILO estimate)']:
    show_distribution(dataHaiti_df[col],col)

# Call the function
```

Output:

The population growth (annual %) feature has the following distribution properties: minimum, 0.75; mean, 1.22; median, 1.17; mode, 0.75; and maximum, 1.81. What it tells is that the population growth (annual %) is declining in Brazil.

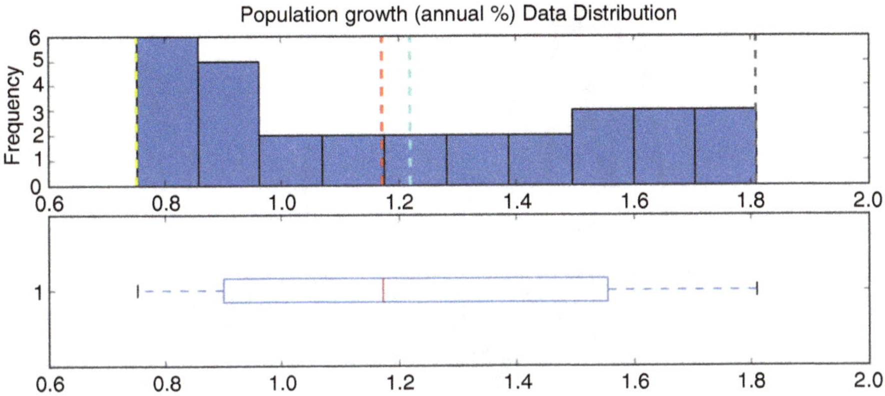

Inflation, consumer prices (annual %) have the following properties: minimum, 3.20; mean, 285.32; median, 6.74; mode, 3.20; and maximum, 2947.73.

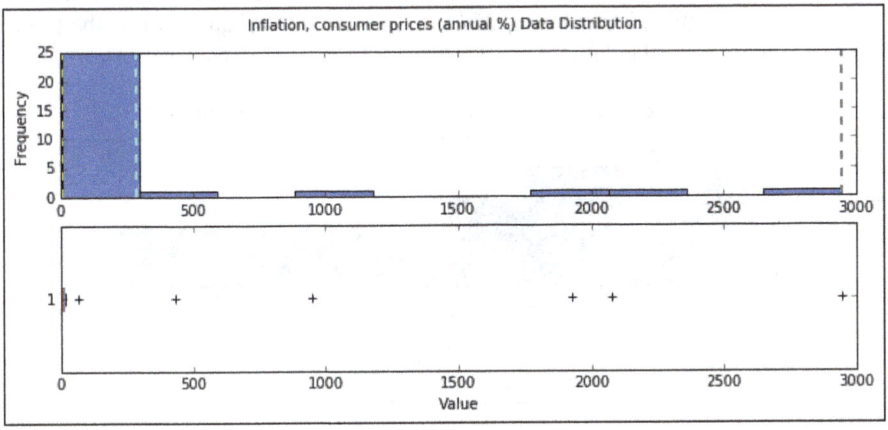

Brazil has huge inflation[34] during the 1990s (2947% in 1990). The detailed inflation changes are as described in the following figure (hyperinflation in Brazil occurred between the first 3 months of 1990): During the hyperinflation times, shop keepers used color codes to update price tags as the changes were so frequent. The banks that kept money more time were biggest sufferers.

[34] Colour-code clue to Brazil's inflation past - https://www.bbc.com/news/business-11477974

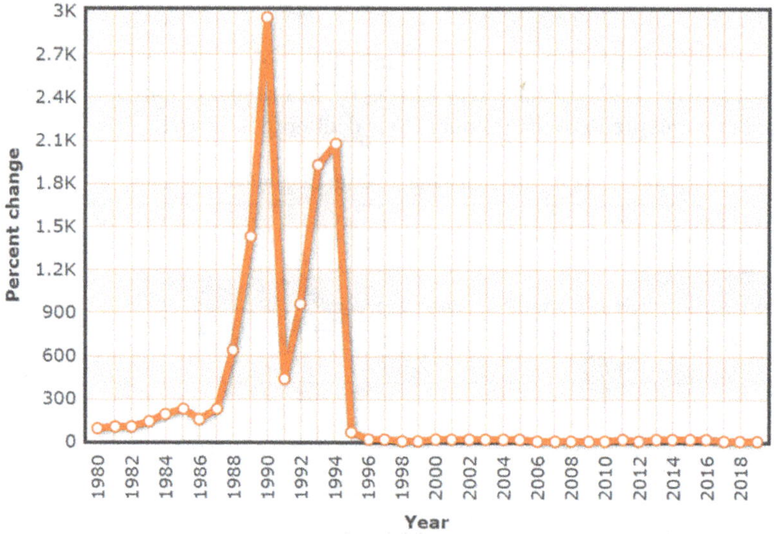

Rural population (% of the total population): the trend is clear—declining rural population. Brazil had peak rural population (% of the total population) in the 1990s, and now the trend is declining rural population as percentage of the total population. Some statistics are as follows: minimum, 13.18; mean, 18.17; median, 17.32; mode, 13.18; and maximum, 26.08. It is clear that urbanization is the trend one can witness from the data.

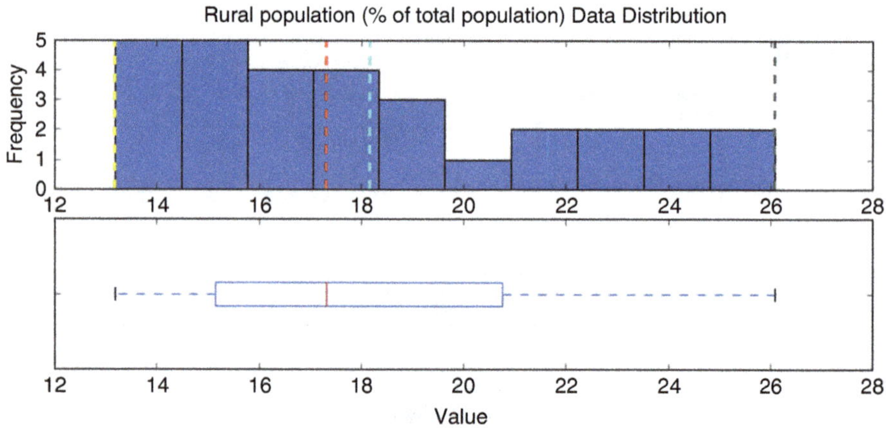

Exports of goods and services (% of GDP) are improving in Brazil. One can witness double-digit exports after 2006. The statistics include minimum, 6.73; mean, 11.53; median, 11.66; mode, 6.73; and maximum, 16.55.

In 2019, Brazil[35] was the number 9 economy in the world in terms of GDP (current US$), the number 25 in total exports, the number 27 in total imports, the number 79 economy in terms of GDP per capita (current US$), and the number 49 most complex economy according to the Economic Complexity Index (ECI) exports: the top exports of Brazil are soybeans ($26.1B), crude petroleum ($24.3B), iron ore ($23B), corn ($7.39B), and sulfate chemical woodpulp ($7.35B), exporting mostly to China ($63.5B), the United States ($30.5B), Argentina ($9.85B), the Netherlands ($9.13B), and Japan ($5.58B).

Employment in agriculture (% of total employment) (modeled ILO estimate): Employment in agriculture (% of total employment) is trending toward lower percentages. In 1990, 19.4% were employed in agriculture, and in 2019, that percentage is reduced to 9.08%. Here are the additional statistics: minimum, 9.08; mean, 14.83; median, 16.31; mode, 9.08; and maximum, 19.75.

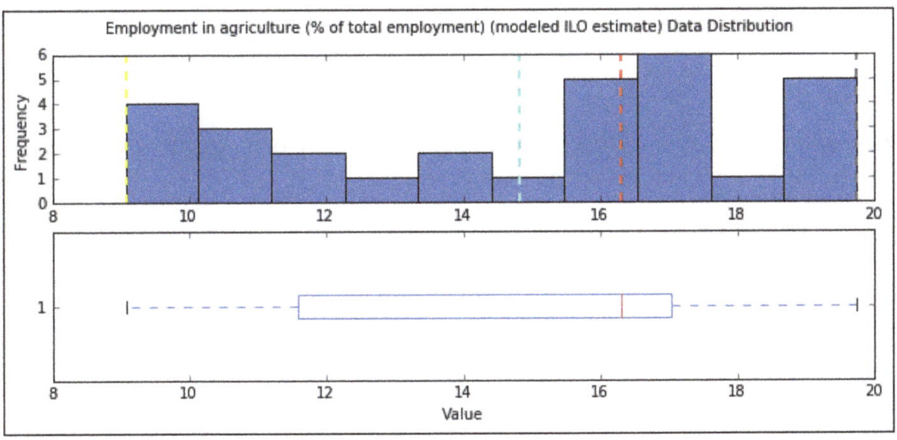

[35] Brazil - https://oec.world/en/profile/country/bra#:~:text=Exports%3A%20The%20top%20exports%20of,and%20Japan%20(%245.58B)

Prior the 1990s, small farms and rural cities migrated to urban centers. Nonetheless, in the 1990s and afterward, the trend has become different in rural Brazil. Most of the rural parts have expanded with urbanization, and nonfarm income has become more prevalent.[36] This provides an opportunity to sustain the rural agriculture.

Step 5: Density map—the density map would provide skewness analysis.

```
def show_density(var_data,colname):
    from matplotlib import pyplot as plt

    fig = plt.figure(figsize=(10,4))

    # Plot density
    var_data.plot.density()

    # Add titles and labels
    plt.title(colname + ' Density')

    # Show the mean, median, and mode
    plt.axvline(x=var_data.mean(), color = 'cyan', linestyle='dashed', linewidth = 2)
    plt.axvline(x=var_data.median(), color = 'red', linestyle='dashed', linewidth = 2)
    plt.axvline(x=var_data.mode()[0], color = 'yellow', linestyle='dashed', linewidth = 2)

    # Show the figure
    plt.show()

# Get the density of Grade

for col in ['Population growth (annual %)', 'Inflation, consumer prices (annual %)',
      'Rural population (% of total population)',
      'Exports of goods and services (% of GDP)', 'EXIM', 'FDI',
      'External debt stocks (% of GNI)',
      'Life expectancy at birth, total (years)',
      'Employment in agriculture (% of total employment) (modeled ILO estimate)']:
    show_density(dataHaiti_df[col],col)
```

Output:

Population growth (annual %): as you can see, the population growth has peaks during the early 1990s but has falling peaks thereafter.

---

[36] Rural Nonfarm Employment and Incomes in Brazil - http://www.oda-alc.org/documentos/1341945443.pdf

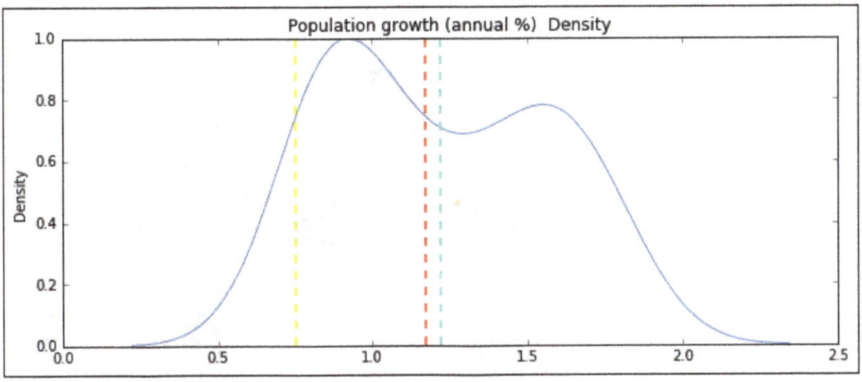

Inflation, consumer prices (annual %) are heavy to the left of the mean value. What is shown is that during earlier years (1990–2000), the inflation rate was higher compared to recent years.

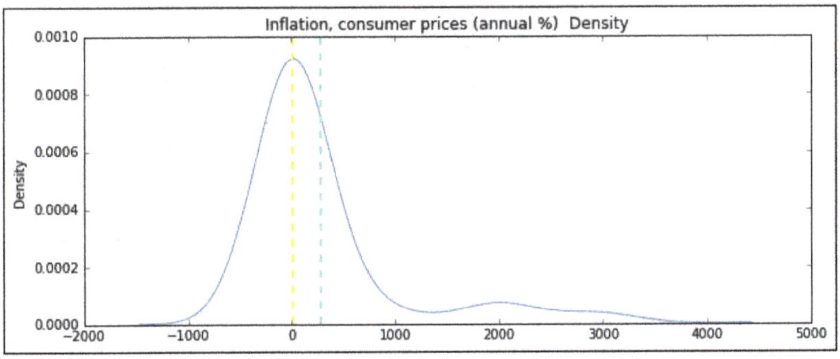

Step 6: Correlation analysis.

Develop correlation analysis for the Brazil feature engineer parameters:

```
plt.figure(figsize=(12, 9))

sns.heatmap(dataBrazil_df.corr())
```

Output:

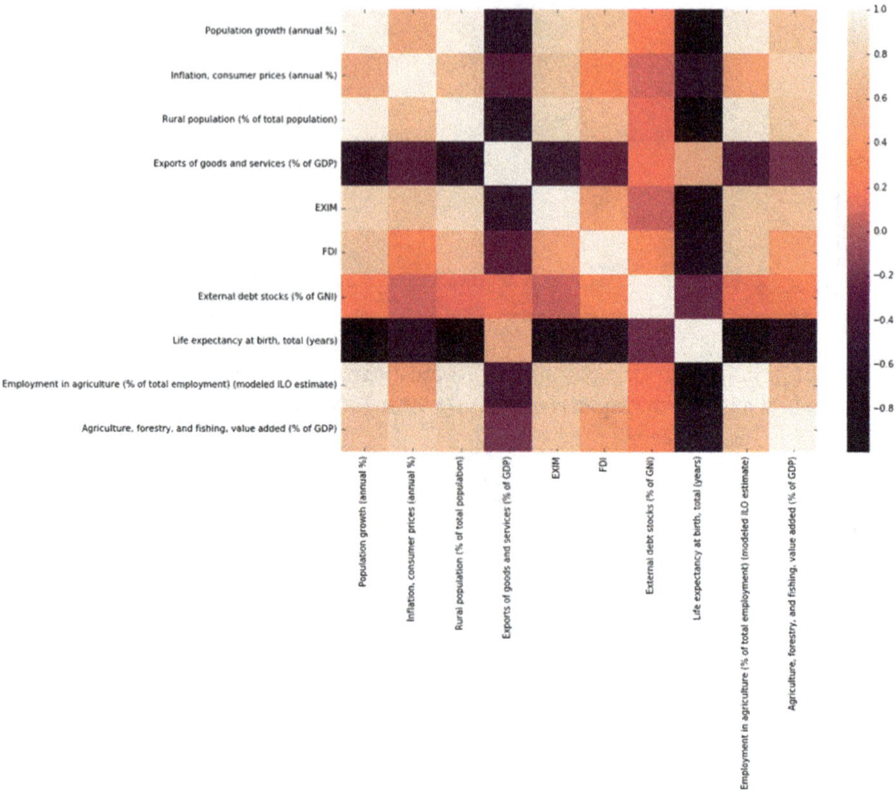

The correlation values:

	Population growth (annual %)	Inflation, consumer prices (annual %)	Rural population (% of total population)	Exports of goods and services (% of GDP)	EXIM	FDI	External debt stocks (% of GNI)	Life expectancy at birth, total (years)	Employment in agriculture (% of total employment) (modeled ILO estimate)	Agriculture, forestry, and fishing, value added (% of GDP)
Population growth (annual %)	1.000000	0.569671	0.977073	-0.624680	0.801096	0.708630	0.263615	-0.994746	0.938864	0.689581
Inflation, consumer prices (annual %)	0.569671	1.000000	0.669924	-0.311679	0.710225	0.302779	0.077750	-0.592640	0.506759	0.785943
Rural population (% of total population)	0.977073	0.669924	1.000000	-0.622985	0.845730	0.659076	0.136371	-0.985671	0.919991	0.747307
Exports of goods and services (% of GDP)	-0.624680	-0.311679	-0.622985	1.000000	-0.522336	-0.271369	0.190357	0.592037	-0.395792	-0.199953
EXIM	0.801096	0.710225	0.845730	-0.522336	1.000000	0.503315	0.059053	-0.813265	0.749973	0.719193
FDI	0.708630	0.302779	0.659076	-0.271369	0.503315	1.000000	0.341913	-0.716049	0.741195	0.480429
External debt stocks (% of GNI)	0.263615	0.077750	0.136371	0.190357	0.059053	0.341913	1.000000	-0.216579	0.227994	0.296129
Life expectancy at birth, total (years)	-0.994746	-0.592640	-0.985671	0.592037	-0.813265	-0.716049	-0.216579	1.000000	-0.959378	-0.708217
Employment in agriculture (% of total employment) (modeled ILO estimate)	0.938864	0.506759	0.919991	-0.395792	0.749973	0.741195	0.227994	-0.959378	1.000000	0.681605
Agriculture, forestry, and fishing, value added (% of GDP)	0.689581	0.785943	0.747307	-0.199953	0.719193	0.480429	0.296129	-0.708217	0.681605	1.000000

The variables that are highly and positively correlated with "**Agriculture, forestry, and fishing, value added (% of GDP)**" are:

- Population growth (annual %)—68.95 %
- Inflation, consumer prices (annual %)—78.59%
- Rural population (% of total population)—74.73%
- EXIM—71.91%
- Employment in agriculture (% of total employment) (modeled ILO estimate)—68.16%

The variables are highly negatively correlated with:

- Life expectancy at birth, total (years)—70.82%

The correlation makes sense as the value of agriculture added to the economy (%GDP) is depended upon the inflation rate—healthy inflation rate drives up the food production needs, and as a result, the agricultural economy would see high positive net effect. The rural population is the backbone of the economy as the contribution to the GDP depends upon the stronger base roots of a country and agriculture is such gross root infrastructure. The population growth and employment in agriculture both have a positive net effect on the overall economy. Brazil witnessed a miracle[37] (1965–1980) and lost a decade (1980–1990) in its economic expansion in the 1960s–1990s, with remarkable recovery on the world market. The economic growth should have positive influence on life expectancy at birth; nonetheless, with such boom-and-bust extreme cycles, a negative coefficient could be justified [8].

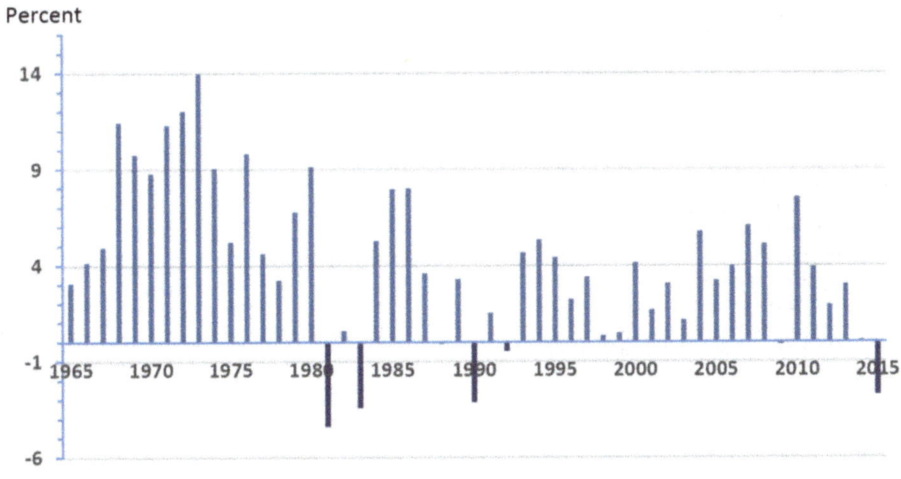

Source: USDA, Economic Research Service using data from World Bank, World Development Indicators

[37] Brazil's Changing Macroeconomic Conditions: Impacts on Agriculture - https://ageconsearch.umn.edu/record/236207/files/Brazil%20Macro%20and%20Agriculture.pdf

Step 7: Regression.

Let's train and test split the dataset:

```
X, y = dataBrazil_df[['Population growth (annual %)', 'Inflation, consumer prices (annual %)',
    'Rural population (% of total population)',
    'Exports of goods and services (% of GDP)', 'EXIM', 'FDI',
    'External debt stocks (% of GNI)',
    'Life expectancy at birth, total (years)',
    'Employment in agriculture (% of total employment) (modeled ILO estimate)']].values,
dataBrazil_df['Agriculture, forestry, and fishing, value added (% of GDP)'].values

X_train,X_test, y_train , y_test = train_test_split(X ,y, test_size = 0.2,random_state = 0)
```

Train size is 80% and test size is 20% of the dataset:

```
regressor = LinearRegression()
regressor.fit(X_train,y_train)
```

Output:

```
array([6.77843491, 4.81521233, 5.30994602, 4.673084 , 4.5871731 ,
    4.26800343])
```

Plot the regression line:

```
import matplotlib.pyplot as plt

%matplotlib inline

plt.scatter(y_test, y_pred)
plt.xlabel('Actual Labels')
plt.ylabel('Predicted Labels')
plt.title('Agriculture, forestry, and fishing, value added (annual % growth)')
# overlay the regression line
z = np.polyfit(y_test, y_pred, 1)
p = np.poly1d(z)
plt.plot(y_test,p(y_test), color='magenta')
plt.show()
```

Output:

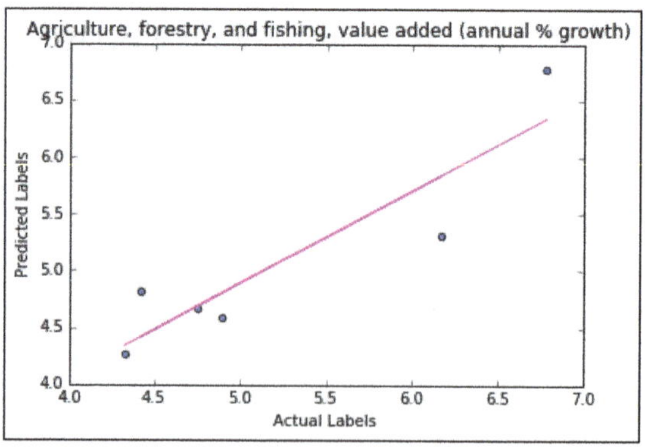

Check the actual and predicted values:

```
pred_y_df = pd.DataFrame({'Actual Value':y_test, 'Predicted value':y_pred,'Difference': y_test-
y_pred})
pred_y_df
```

Output:

	Actual Value	Predicted value	Difference
0	6.779952	6.778435	0.001517
1	4.418538	4.815212	-0.396675
2	6.167184	5.309946	0.857238
3	4.750460	4.673084	0.077376
4	4.891354	4.587173	0.304180
5	4.325611	4.268003	0.057607

The regression coefficients play an extreme role in analyzing the impact:

```
regressor.coef_
```

Output:

```
array([2.58484953, 0.00049126, 0.51888517, 0.12940318, 0.45223404,
    0.        , 0.01348183, 0.8248196 , 0.050797  ])
```

Feature engineer variable	Coefficient	Macro-influence
Population growth (annual %)	2.58484953	Population growth is a positive coefficient—implies Brazil can transfer valuable agricultural labor to the modern sectors without affecting the agricultural growth. If there are technological innovations taking place in the agricultural sector, such as artificial intelligence augmented analytics and climate defense models enabling the agricultural productivity, Brazil will witness golden and agricultural miracle, despite currency wars levied[38] against it [8]
Inflation, consumer prices (annual %)	0.00049126	Good news is inflation is not negative (cost-push[39] is not an exogenous influencer!), and demand-pull seems to be playing an important role [9]

[38] Brazil's changing Macroeconomic Conditions - https://ageconsearch.umn.edu/record/236207/files/Brazil%20Macro%20and%20Agriculture.pdf

[39] Economics by Lipsey, Richard, Chrystal, Alec 13th Revised edition (2015) Paperback Paper back – January 1, 1601 – Oxford University Press; 13th Revised edition (January 1, 1601).

Feature engineer variable	Coefficient	Macro-influence
Rural population (% of total population)	0.51888517	Brazil has been able to maintain a high percentage of its population in the rural sector, and it could be expected that its agricultural sector would constitute a larger percentage of its gross output—AI would have a very positive effect if it reached to the farmers at the gross root levels and will have a net positive effect on the overall economy
Exports of goods and services (% of GDP)	0.12940318	Brazil welcomes positive competition from international market and prepares agricultural market more efficient and competitive.[40] Artificial intelligence will enable agricultural sector to be more productive [10]
EXIM	0.45223404	Positive coefficient implies the ratio of Brazil exports to imports is positive—in simple terms, more agricultural exports as compared to imports
FDIs	0.01348183	More competitive business markets and hence a possible thriving economy
External Debt	0.01348183	A positive external debt coefficient would force Brazil to produce more of its own food, resulting in increased agricultural production and an increased contribution of agriculture to GDP. Therefore, the higher the external debt of the country, the greater the contribution of agriculture is expected to be
Life expectancy at birth, total (years)	0.8248196	Correlation analysis showed negative life expectance at birth—positive regression coefficient implies capital accumulation tend to be higher, making market more competitive. Artificial intelligence can reinforce the economic growth
Employment in agriculture (% of total employment) (modeled ILO estimate)	0.050797	Positive employment in agriculture (% of total employment) coefficient implies the contribution of agriculture to the GDP is highly positive and tends to be bearish. Artificial intelligence plays an important role in increasing the productivity

Regression metrics—the following table summarizes regression metrics:

```
from sklearn.metrics import mean_squared_error, r2_score

mse = mean_squared_error(y_test, y_pred)
print("MSE:", mse)

rmse = np.sqrt(mse)
print("RMSE:", rmse)

r2 = r2_score(y_test, y_pred)
print("R2:", r2)
```

[40] Robert Carbaugh, International Economics 15th Edition, Cengage Learning; 15th edition (September 8, 2014), ISBN-13: 978-1285854359.

Output:

```
MSE: 0.16567367914813866
RMSE: 0.40703031723464855
R²: 0.8051406453529305
```

Step 8: Durbin-Watson Test.

One of the assumptions of linear regression is that there is no correlation between the residuals. In other words, the residuals are assumed to be independent. One way to determine if this assumption is met is to perform a Durbin-Watson test, which is used to detect the presence of autocorrelation in the residuals of a regression:

```
from statsmodels.stats.stattools import durbin_watson

residual = (y_test - y_pred)
#perform Durbin-Watson test
durbin_watson(residual)
```

Output:

2.465970729113519

- A test statistic of 2 indicates no serial correlation.
- The closer the test statistics is to 0, the more evidence of positive serial correlation.
- The closer the test statistics is to 4, the more evidence of negative serial correlation.

As a rule of thumb, test statistic values between the range of 1.5 and 2.5 are considered normal. However, values outside of this range could indicate that autocorrelation is a problem.

Explainability of Model

Set up explainer model:

```
# Explainers:
# 1. SHAP Tabular Explainer
from interpret.ext.blackbox import TabularExplainer

# OR

# 2. Mimic Explainer
#from interpret.ext.blackbox import MimicExplainer
# You can use one of the following four interpretable models as a global surrogate to the black box
model
from interpret.ext.glassbox import LGBMExplainableModel
from interpret.ext.glassbox import LinearExplainableModel
from interpret.ext.glassbox import SGDExplainableModel
from interpret.ext.glassbox import DecisionTreeExplainableModel

# OR

# 3. PFI Explainer
#from interpret.ext.blackbox import PFIExplainer
```

Create table explainer:

```
feature_names= ['Population growth (annual %)',
        'Inflation, consumer prices (annual %)','Rural population (% of total population)',
        'Life expectancy at birth, total (years)','Exports of goods and services (% of GDP)','EXIM','FDI'
        ,'External debt stocks (% of GNI)','Employment in agriculture (% of total employment)
(modeled ILO estimate)']
# 1. Using SHAP TabularExplainer
explainer = TabularExplainer(regressor,
                    X_train,
                    features = feature_names)
```

```
# Passing in test dataset for evaluation examples - note it must be a representative sample of the
original data
# x_train can be passed as well, but with more examples explanations will take longer although they
may be more accurate
global_explanation = explainer.explain_global(X_test)
```

Generate local values:

```
# Sorted SHAP values
print('ranked global importance values: {}'.format(global_explanation.get_ranked_global_values()))
# Corresponding feature names
print('ranked global importance names: {}'.format(global_explanation.get_ranked_global_names()))
# Feature ranks (based on original order of features)
print('global importance rank: {}'.format(global_explanation.global_importance_rank))

# Print out a dictionary that holds the sorted feature importance names and values
print('global importance rank: {}'.format(global_explanation.get_feature_importance_dict()))
```

Output:

```
ranked global importance values: [2.3672322404935393, 1.7723820195820865, 0.8295202775600748,
0.2257853057729148, 0.19624572083247074, 0.1778905585927376, 0.11052147492270692,
0.09725317833436832, 0.023666685361365448]
ranked global importance names: ['External debt stocks (% of GNI)', 'Rural population (% of total
population)', 'Population growth (annual %)', 'Life expectancy at birth, total (years)', 'Employment in
agriculture (% of total employment) (modeled ILO estimate)', 'Inflation, consumer prices (annual %)',
'FDI', 'Exports of goods and services (% of GDP)', 'EXIM']
global importance rank: [7, 2, 0, 3, 8, 1, 6, 4, 5]
global importance rank: {'External debt stocks (% of GNI)': 2.3672322404935393, 'Rural population
(% of total population)': 1.7723820195820865, 'Population growth (annual %)': 0.8295202775600748,
'Life expectancy at birth, total (years)': 0.2257853057729148, 'Employment in agriculture (% of total
employment) (modeled ILO estimate)': 0.19624572083247074, 'Inflation, consumer prices (annual %)':
0.1778905585927376, 'FDI': 0.11052147492270692, 'Exports of goods and services (% of GDP)':
0.09725317833436832, 'EXIM': 0.023666685361365448}
```

Ranked global importance generated feature names and values.
Generate local importance values:

```
# Note: Do not run this cell if using PFIExplainer, it does not support local explanations
# feature shap values for all features and all data points in the training data
print('local importance values: {}'.format(global_explanation.local_importance_values))
```

Output values:

```
local importance values: [[1.2067276181951 39, 0.3127670032509725, 3.187900763465544, -
0.05574185572852 7565, 0.14819484158199248, 0.03517890903796566, 0.09463736476736911, -
3.4516638043224015, 0.23633301670690157], [-1.1803491824118741, -0.15309330377634223, -
2.5846967532879384, 0.4646466310513975, -0.12605041716182577, -0.018411698335490615,
0.06072128152575 8335, 3.584872169074939, -0.2965274552539237], [0.04366795159035493, -
0.14766484872570876, -0.319244100870346, 0.5023138019745548, -0.0509624229506872,
0.027180310922524428, 0.23863366675379055, -0.12846565207555036, 0.0803862560353291],
[0.474781294946879, -0.15143319402034883, 0.20534880604540837, -0.14376205810836165, -
0.0402616023298782, 0.02745526273274272, 0.16394514171308555, -0.9978255062498346,
0.07073479944549163], [-1.0772119981888175, -0.1506005074259497, -2.3112442686740873,
0.1511000364405266, -0.1061081971200649, -0.018361707097269107, 0.06702286974211787,
3.22030190764702, -0.2518260935084851], [-0.9943836200273842, -0.1517844943571037, -
2.0258574251491965, -0.03714745133412097, -0.11194158886176142, -0.015412224042200154, -
0.038168525034120035, 2.8202644035914917, -0.24166670404469337]]
```

```
# E.g., Explain the first data point in the test set
local_explanation = explainer.explain_local(X_test[0,:])
# Sorted local feature importance information; reflects the original feature order
sorted_local_importance_names = local_explanation.get_ranked_local_names()
sorted_local_importance_values = local_explanation.get_ranked_local_values()

print('sorted local importance names: {}'.format(sorted_local_importance_names))
print('sorted local importance values: {}'.format(sorted_local_importance_values))
```

Output:

```
Sorted local importance names: ['Rural population (% of total population)', 'Population growth (annual
%)', 'Inflation, consu mer prices (annual %)', 'Employment in agriculture (% of total employment)
(modeled ILO estimate)', 'Exports of goods and services (% of GDP)', 'FDI', 'EXIM', 'Life expectancy
at birth, total (years)', 'External debt stocks (% of GNI)']
sorted local import ance values: [3.187900763465544, 1.206727618195139, 0.3127670032509725,
0.23633301670690157, 0.14819484158199248, 0.09463736476736911, 0.03517890903796566,           -
0.055741855728527565,  -3.4516638043224015]
```

Aggregate feature importance: external debt stocks (% of GNP), rural population (% of total population), population growth, and life expectancy at birth.

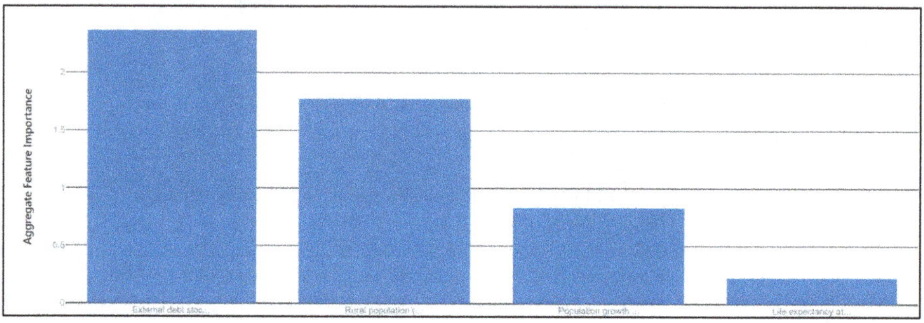

Individual feature importance:

Corn: A Classic Classification Problem
One key effort, Sanchez says, are AI systems "that allow me to tell whether grain being harvested is good quality or low quality and to make automatic adjustment systems for the harvester". The company is already selling an early version of this image analysis technology. But the many differences between grain types, and grains grown under different conditions, make this task a tough one for machine learning[41] [11]

India's Machine Learning Model: Agriculture's Contribution to Economy and GDP and the Role of AI Readiness

The software code for this model can be found on GitHub Link: IndiaAgriculturalGDPModelWithFinal.Ipynb (Jupyter Notebook)

In the following section, we will prepare the machine learning model that prepares agriculture contribution to GDP using the abovementioned key macrolevel indictors and the exogenous role of AI to enhance the productivity:

[41] Want a Really Hard Machine Learning Problem? Try Agriculture, Says John Deere Labs - https://spectrum.ieee.org/view-from-the-valley/robotics/artificial-intelligence/want-a-really-hard-machine-learning-problem-try-agriculture-say-john-deere-labs-leaders

India agriculture employs more than 50% of the total workforce in India and contributes around 17–18% to the country's GDP[42] [12]. Agriculture, with its allied sectors, is the largest source of livelihoods in India. 70% of its rural households still depend primarily on agriculture for their livelihood, with 82% of farmers being small and marginal. In 2017–2018, the total food grain production was estimated at 275 million tons (MT). India is the largest producer[43] (25% of global production), consumer (27% of world consumption), and importer (14%) of pulses in the world. India's annual milk production was 165 MT (2017–2018), making India the largest producer of milk, jute, and pulses, and was the world's second-largest cattle population with 190 million in 2012. It is the second-largest producer of rice, wheat, sugarcane, cotton, and groundnuts, as well as the second-largest fruit and vegetable producer, accounting for 10.9% and 8.6% of the world fruit and vegetable production, respectively. Agricultural success is vital to India's economy and the sustenance of its citizens. Indian farmers, however, are *facing incessant debt cycles* due to running bank loans that they often cannot pay off. Expensive farming chemicals,[44] low yielding crops, and drought are the reasons for these mounting debts, subsequently leading to a high yearly rate of farmer suicides.[45] Fertilizer price prediction and enabling lower cost of fertilizers save huge number of small farms in India and could help the humanity [13].

AI readiness: Group 3—economies with moderate foundations	

India had 59.65% (2000) employment[46] in agriculture (% of total employment). That is, out of 100 TOTAL employment, 57 people belong to agriculture in 2000. Two decades later, that number is 42% (only 42 out of 100 people employed in

[42] India economic survey 2018: Farmers gain as agriculture mechanisation speeds up, but more R&D needed - https://www.financialexpress.com/budget/india-economic-survey-2018-for-farmers-agriculture-gdp-msp/1034266/

[43] FAO In India - http://www.fao.org/india/fao-in-india/india-at-a-glance/en/#:~:text=Agriculture%2C%20with%20its%20allied%20sectors,275%20million%20tonnes%20(MT)

[44] "Free" Fertilizer is Saving Rural Farmers - https://www.permaculturenews.org/2017/10/02/free-fertilizer-saving-rural-farmers/

[45] India may become farmer suicide capital: Agri Adviser - https://www.deccanherald.com/national/india-may-become-farmer-suicide-capital-agri-adviser-773068.html

[46] Data Source—Employment in agriculture (% of total employment) (modeled ILO estimate) - https://data.worldbank.org/indicator/SL.AGR.EMPL.ZS?end=2017&locations=1W&start=1981&view=chart

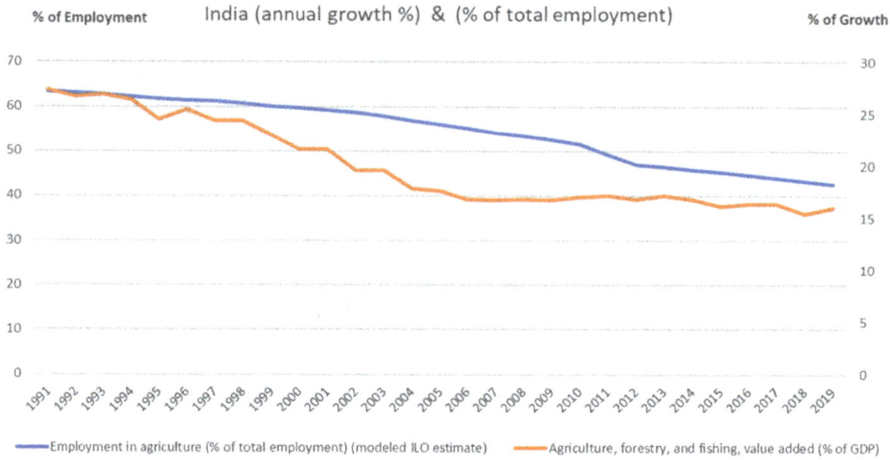

Fig. 6.5 India—employment in agriculture and GDP

agriculture) (2019). Agriculture productivity growth[47] in recent decades in India has been mainly driven by investments in agriculture innovation, facilitation of sector financing, and trade liberalization (please see Fig. 6.5). Nonetheless, low productivity growth is impacting India's economic growth.[48] Agriculture productivity has room to grow further, improving productivity of lagging midsize farmers and regions and reforming agriculture policies toward agriculture financing [6].

Data

Year	EXIM	Rural population (% of t	Life expec	Inflation, c	Populatior	Exports of	FDI	External d	Employme	Agriculture
1991	0.643579	74.222	58.353	13.87024	2.03973	8.494241	-73.5376	31.89993	63.32	27.33322
1992	0.432833	74.016	58.851	11.78782	2.003178	8.842927	-276.512	30.9287	62.99	26.66977
1993	0.583447	73.809	59.349	6.326893	1.970635	9.834217	-550.019	33.15743	62.66	26.8818
1994	0.274661	73.601	59.84	10.24793	1.943245	9.888085	-890.688	30.29897	62.18	26.38862
1995	0.326554	73.393	60.32	10.22488	1.918941	10.84397	-2026.44	26.33217	61.76	24.45686
1996	0.809801	73.183	60.783	8.977147	1.895219	10.38517	-2186.73	24.14264	61.35	25.41703
1997	0.597925	72.972	61.233	7.164247	1.869173	10.69072	-3464.41	22.81254	61.12	24.34536
1998	0.56378	72.76	61.669	13.23085	1.839658	11.01847	-2587.06	23.53646	60.72	24.37882
1999	0.449508	72.547	62.093	4.669815	1.80556	11.45206	-2089.23	21.91706	60.09	22.98896
2000	0.389858	72.333	62.505	4.009439	1.768125	12.99724	-3074.68	21.82291	59.65	21.60896
2001	0.312569	72.082	62.907	3.77929	1.728768	12.55838	-4073.96	20.67606	59.1	21.6205
2002	0.342195	71.756	63.304	4.297158	1.689563	14.26438	-3947.9	20.67325	58.6	19.5367
2003	0.354674	71.428	63.699	3.805869	1.651491	14.94791	-2444.14	19.70922	57.76	19.58061
2004	0.466028	71.097	64.095	3.767237	1.615308	17.85912	-3592.19	17.55893	56.78	17.81469
2005	0.642919	70.765	64.5	4.246351	1.57971	19.60525	-4628.65	14.88005	56	17.62021
2006	1.035082	70.431	64.918	5.796518	1.545696	21.26794	-5992.29	17.09968	55.08	16.80944
2007	1.150636	70.094	65.35	6.372881	1.509222	20.7997	-8201.63	16.84142	54.11	16.75012
2008	1.18333	69.754	65.794	8.349267	1.46489	24.09736	-24149.7	19.0572	53.51	16.79094
2009	0.862072	69.413	66.244	10.88235	1.410582	20.40052	-19485.8	19.21558	52.5	16.74427

47 API_NV.AGR.TOTL.KD.ZG_DS2_en_excel_v2_2063913.XLS - https://api.worldbank.org/v2/en/indicator/NV.AGR.TOTL.KD.ZG?downloadformat=excel

48 Agriculture Productivity Growth in Brazil - Recent trends and future prospects - http://documents1.worldbank.org/curated/en/268351520343354377/pdf/123948-WP-6-3-2018-8-39-22-AriasetalAgriculturalgrowthinBrazil.pdf

Model Development

The following are the steps to build the model.

Step 1: Load into Python Data Frame. Load ML-specific libraries:

```
import pandas as pd
import numpy as np
import plotly.graph_objects as go
import seaborn as sns; sns.set()
import matplotlib.pyplot as plt
from scipy import stats
from scipy.stats import pearsonr
import matplotlib.pyplot as plt
from sklearn.preprocessing import MinMaxScaler
from sklearn.model_selection import train_test_split
from sklearn.linear_model import LinearRegression
from sklearn.ensemble import RandomForestClassifier
from sklearn.metrics import r2_score
from sklearn import metrics

import eli5
import matplotlib.pyplot as plt
import matplotlib as mpl
import numpy as np
import scipy.stats as spstats
import seaborn as sns

%matplotlib inline
mpl.style.reload_library()
mpl.style.use('classic')
mpl.rcParams['figure.facecolor'] = (1, 1, 1, 0)
mpl.rcParams['figure.figsize'] = [6.0, 4.0]
mpl.rcParams['figure.dpi'] = 100
```

Step 2: Load Brazilian data sources:

```
# Importing data

dataIndia_df = pd.read_csv("IndiaRegressionFinalwGDPPG.csv")
dataIndia_df.set_index("Year",inplace = True)
dataIndia_df
```

Output:

Year	EXIM	Rural population (% of total population)	Life expectancy at birth, total (years)	Inflation, consumer prices (annual %)	Population growth (annual %)	Exports of goods and services (% of GDP)	FDI	External debt stocks (% of GNI)	Employment in agriculture (% of total employment) (modeled ILO estimate)	Agriculture, forestry, and fishing, value added (% of GDP)
1991	0.643579	74.222	58.353	13.870240	2.039730	8.494241	-73.537638	31.899926	63.320000	27.333218
1992	0.432833	74.016	58.851	11.787821	2.003178	8.842927	-276.512439	30.928699	62.990002	26.669772
1993	0.583447	73.809	59.349	6.326893	1.970635	9.834217	-550.019384	33.157433	62.660000	26.881801
1994	0.274661	73.601	59.840	10.247934	1.943245	9.888085	-890.688166	30.298974	62.180000	26.388619
1995	0.326554	73.393	60.320	10.224885	1.918941	10.843968	-2026.439031	26.332172	61.759998	24.456855
1996	0.809801	73.183	60.783	8.977147	1.895219	10.385169	-2186.732315	24.142642	61.349998	25.417031
1997	0.597925	72.972	61.233	7.164247	1.869173	10.690717	-3464.411052	22.812537	61.119999	24.345357
1998	0.563780	72.760	61.669	13.230849	1.839658	11.018469	-2587.058630	23.536455	60.720001	24.378821

Step 3: DataFrame Describe—To find the statistical properties of the data:

```
dataIndia_df.describe()
```

Output:

	EXIM	Rural population (% of total population)	Life expectancy at birth, total (years)	Inflation, consumer prices (annual %)	Population growth (annual %)	Exports of goods and services (% of GDP)	FDI	External debt stocks (% of GNI)	Employment in agriculture (% of total employment) (modeled ILO estimate)	Agriculture, forestry, and fishing, value added (% of GDP)
count	29.000000	29.000000	29.000000	29.000000	29.000000	29.000000	29.000000	29.000000	29.000000	29.000000
mean	0.825068	70.460966	64.444586	7.415585	1.543773	16.763568	-12645.785789	22.008889	54.452758	19.967552
std	0.505102	2.666139	3.491797	3.210119	0.333737	5.516694	13050.944582	4.630871	6.994408	4.068554
min	0.274661	65.528000	58.353000	2.490887	1.015106	8.494241	-39411.278940	14.880053	42.599998	15.406726
25%	0.449508	68.366000	61.669000	4.669815	1.231623	11.018469	-23890.659990	19.215581	47.000000	16.790942
50%	0.643579	70.765000	64.500000	6.372881	1.579710	18.414457	-4628.652265	20.676061	56.000000	17.620207
75%	1.035082	72.760000	67.545000	10.224885	1.839658	20.799700	-2444.138426	23.300868	60.720001	24.345357
max	2.505034	74.222000	69.416000	13.870240	2.039730	25.430861	-73.537638	33.157433	63.320000	27.333218

The statistics reveal some information about the distribution of the data in each of the numeric fields, including the number of observations (there are 29 records), the mean and standard deviation, minimum and maximum values, and the quartile values (the threshold values for 25%, 50% which is also the median, and 75% of the data). From this, we can see that the mean number of population growth (annual %) is around 1.22; but there's a comparatively small standard deviation, indicating a not much variance in the population growth (annual %).

The record count is 29. The population growth (annual %) was 2% max in 1990; the growth percentage is declining thereafter.

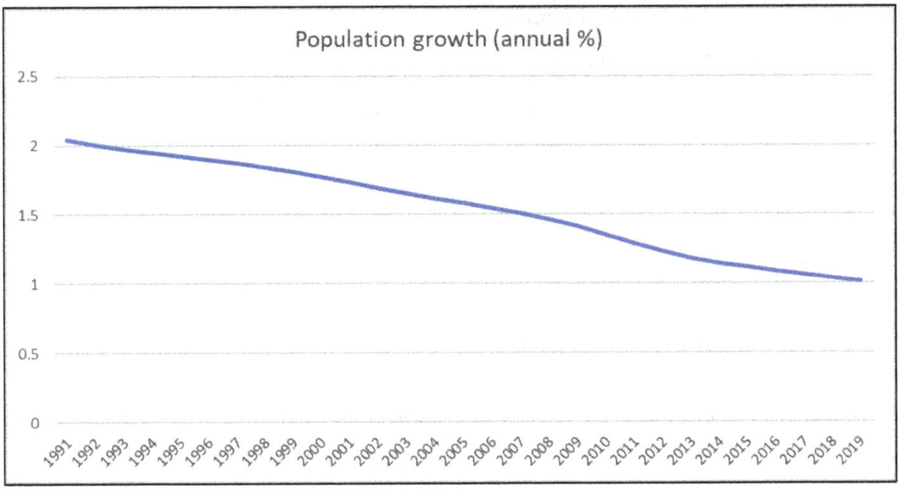

Population growth (annual %)
2.5
2
1.5
1
0.5
0
1991
1992
1993
1994
1995
1996
1997
1998
1999
2000
2001
2002
2003
2004
2005
2006
2007
2008
2009
2010
2011
2012
2013
2014
2015
2016
2017
2018
2019

Step 4: Perform distribution function.

Distribution function provides more clues into the data:

```
# Create a function that we can re-use
def show_distribution(var_data,colname):
    from matplotlib import pyplot as plt

    # Get statistics
    min_val = var_data.min()
    max_val = var_data.max()
    mean_val = var_data.mean()
    med_val = var_data.median()
    mod_val = var_data.mode()[0]


print('Minimum:{:.2f}\nMean:{:.2f}\nMedian:{:.2f}\nMode:{:.2f}\nMaximum:{:.2f}\n'.format(min_v
al,
                                                    mean_val,
                                                    med_val,
                                                    mod_val,
                                                    max_val))

    # Create a figure for 2 subplots (2 rows, 1 column)
    fig, ax = plt.subplots(2, 1, figsize = (10,4))

    # Plot the histogram
    ax[0].hist(var_data)
    ax[0].set_ylabel('Frequency')

    # Add lines for the mean, median, and mode
    ax[0].axvline(x=min_val, color = 'gray', linestyle='dashed', linewidth = 2)
    ax[0].axvline(x=mean_val, color = 'cyan', linestyle='dashed', linewidth = 2)
    ax[0].axvline(x=med_val, color = 'red', linestyle='dashed', linewidth = 2)
    ax[0].axvline(x=mod_val, color = 'yellow', linestyle='dashed', linewidth = 2)
    ax[0].axvline(x=max_val, color = 'gray', linestyle='dashed', linewidth = 2)

    # Plot the boxplot
    ax[1].boxplot(var_data, vert=False)
    ax[1].set_xlabel('Value')

    # Add a title to the Figure
    fig.suptitle(colname + ' Data Distribution')

    # Show the figure
    fig.show()

# Get the variable to examine

for col in ['Population growth (annual %)', 'Inflation, consumer prices (annual %)',
       'Rural population (% of total population)',
       'Exports of goods and services (% of GDP)', 'EXIM', 'FDI',
       'External debt stocks (% of GNI)',
       'Life expectancy at birth, total (years)',
       'Employment in agriculture (% of total employment) (modeled ILO estimate)']:
    show_distribution(dataIndia_df[col],col)

# Call the function
```

Output:
The population growth (annual %) feature has the following distribution properties: minimum, 1.02; mean, 1.54; median, 1.58; mode, 1.02; and maximum, 2.04. What it tells is that the population growth (annual %) is declining in India.

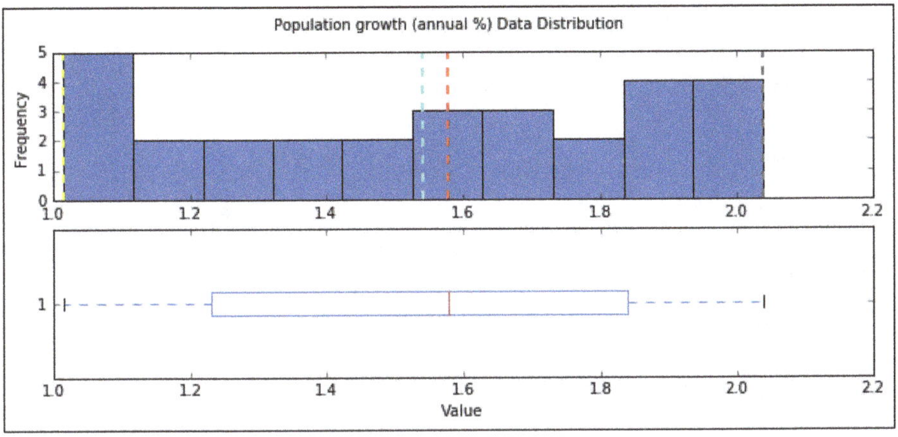

Employment in agriculture (% of total employment) (modeled ILO estimate): Employment in agriculture (% of total employment) is trending toward lower percentages. In 1990, 63% were employed in agriculture, and in 2019, that percentage is reduced to 43%. Here are the additional statistics: minimum, 42.60; mean, 54.45; median, 56.00; mode, 42.60; and maximum, 63.32.

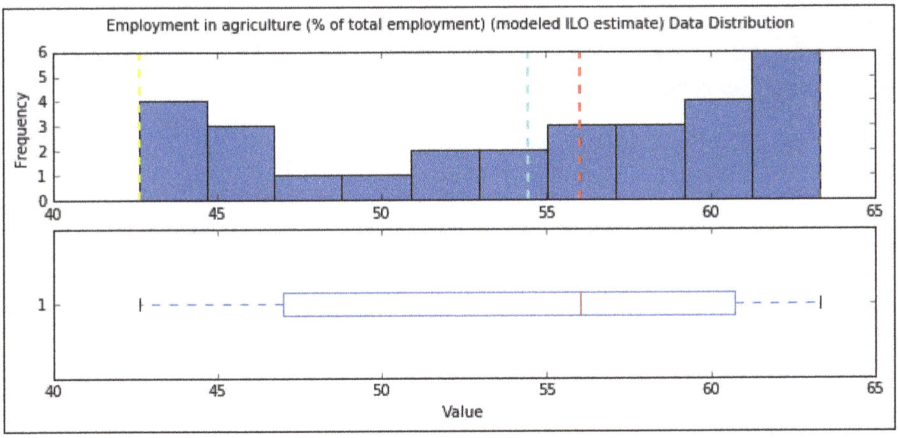

Step 5: Density map—the density map would provide skewness analysis:

```
def show_density(var_data,colname):
    from matplotlib import pyplot as plt

    fig = plt.figure(figsize=(10,4))

    # Plot density
    var_data.plot.density()

    # Add titles and labels
    plt.title(colname + ' Density')

    # Show the mean, median, and mode
    plt.axvline(x=var_data.mean(), color = 'cyan', linestyle='dashed', linewidth = 2)
    plt.axvline(x=var_data.median(), color = 'red', linestyle='dashed', linewidth = 2)
    plt.axvline(x=var_data.mode()[0], color = 'yellow', linestyle='dashed', linewidth = 2)

    # Show the figure
    plt.show()

# Get the density of Grade

for col in ['Population growth (annual %)', 'Inflation, consumer prices (annual %)',
       'Rural population (% of total population)',
       'Exports of goods and services (% of GDP)', 'EXIM', 'FDI',
       'External debt stocks (% of GNI)',
       'Life expectancy at birth, total (years)',
       'Employment in agriculture (% of total employment) (modeled ILO estimate)']:
    show_density(dataIndia_df[col],col)
```

Output:

Population growth (annual %): as you can see, the population growth has peaks during the early 1990s but has falling peaks thereafter.

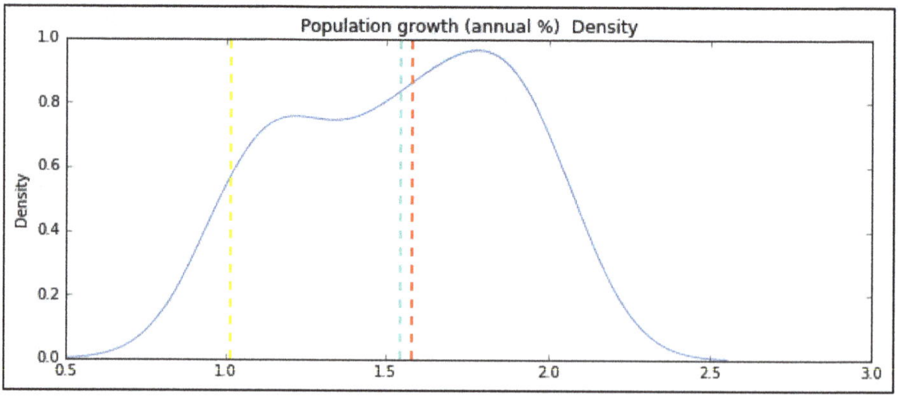

Step 6: Correlation analysis.

Develop correlation analysis for the Brazil feature engineer parameters:

```
plt.figure(figsize=(12, 9))

sns.heatmap(dataIndia_df.corr()
```

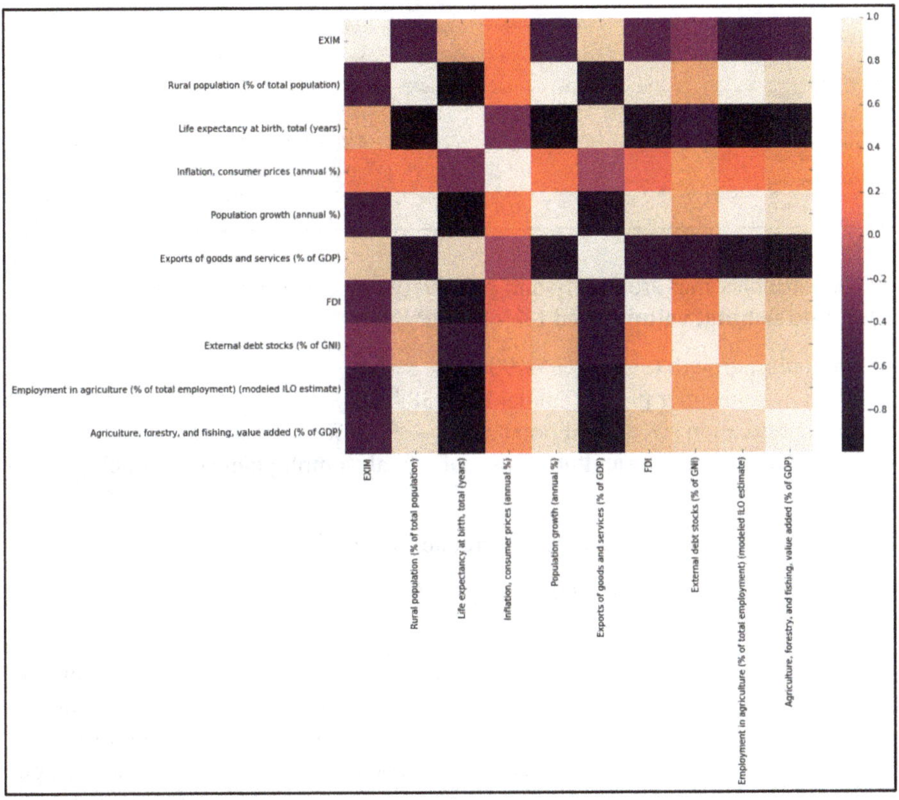

Output: The correlation values:

	EXIM	Rural population (% of total population)	Life expectancy at birth, total (years)	Inflation, consumer prices (annual %)	Population growth (annual %)	Exports of goods and services (% of GDP)	FDI	External debt stocks (% of GNI)	Employment in agriculture (% of total employment) (modeled ILO estimate)	Agriculture, forestry, and fishing, value added (% of GDP)
EXIM	1.000000	-0.532174	0.566667	0.250234	-0.571866	0.774753	-0.439615	-0.269279	-0.583494	-0.528515
Rural population (% of total population)	-0.532174	1.000000	-0.985783	0.264882	0.996230	-0.805345	0.925775	0.565761	0.993555	0.893548
Life expectancy at birth, total (years)	0.566667	-0.985783	1.000000	-0.306387	-0.992311	0.863025	-0.885684	-0.661219	-0.975435	-0.942278
Inflation, consumer prices (annual %)	0.250234	0.264882	-0.306387	1.000000	0.262580	-0.119088	0.138669	0.468869	0.221452	0.380179
Population growth (annual %)	-0.571866	0.996230	-0.992311	0.262580	1.000000	-0.835390	0.916997	0.572949	0.994608	0.905076
Exports of goods and services (% of GDP)	0.774753	-0.805345	0.863025	-0.119088	-0.835390	1.000000	-0.673676	-0.674164	-0.804817	-0.911185
FDI	-0.439615	0.925775	-0.885684	0.138669	0.916997	-0.673676	1.000000	0.345747	0.932313	0.737513
External debt stocks (% of GNI)	-0.269279	0.565761	-0.661219	0.468869	0.572949	-0.674164	0.345747	1.000000	0.499219	0.815124
Employment in agriculture (% of total employment) (modeled ILO estimate)	-0.583494	0.993555	-0.975435	0.221452	0.994608	-0.804817	0.932313	0.499219	1.000000	0.863539
Agriculture, forestry, and fishing, value added (% of GDP)	-0.528515	0.893548	-0.942278	0.380179	0.905076	-0.911185	0.737513	0.815124	0.863539	1.000000

The variables that are highly and positively correlated with "**Agriculture, forestry, and fishing, value added (% of GDP)**" are:

- Population growth (annual %)—90.05 %
- Inflation, consumer prices (annual %)—38.07%
- Rural population (% of total population)—89.35%
- Employment in agriculture (% of total employment) (modeled ILO estimate)—68.16%

The variables are highly negatively correlated with:

- Life expectancy at birth, total (years)—94.22%
- EXIM—52.85 %

The correlation makes sense as the value of agriculture added to the economy (%GDP) is depended upon the inflation rate—healthy inflation rate drives up the food production needs, and as a result, the agricultural economy would see high positive net effect. The rural population is the backbone of the economy as the contribution to the GDP depends upon the stronger base roots of a country and agriculture is such gross root infrastructure. The population growth and employment in agriculture both have a positive net effect on the overall economy. India witnessed financial crisis[49,50](1990–1991) and had a remarkable growth followed by Prime Minister P.V Narasihma Rao[51] and thereafter by the Chief Modern Indian Economic Prime Minister Manmohan Singh [14–16].

---

[49] What Caused the 1991 Currency Crisis in India? - https://www.imf.org/external/pubs/ft/staffp/2002/03/pdf/cerra.pdf

[50] Economic Crisis Forcing Once Self-Reliant India to Seek Aid - https://www.nytimes.com/1991/06/29/world/economic-crisis-forcing-once-self-reliant-india-to-seek-aid.html

[51] PV Narasimha Rao Remembered as Father of Indian Economic Reforms - https://www.voanews.com/archive/pv-narasimha-rao-remembered-father-indian-economic-reforms

Economic Crisis Forcing Once Self-Reliant India to Seek Aid

Facing a grave economic crisis, the newly elected Indian Government has begun urgent talks with the International Monetary Fund, seeking emergency aid of several billion dollars. Yet the conditions for such assistance are stirring anxiety in this nation, which historically prides itself on self-reliance. At issue is not only the economy but also India's sensitivity to Western involvement. What makes the nation's plight especially delicate is the realization that, without bowing to Western conditions for economic change, India could slide into collapse

The immediate issue is India's struggle to avoid defaulting on loans. It has never defaulted since independence 44 years ago. India's foreign debt has climbed to about $72 billion, making it the world's third-largest debtor after Brazil and Mexico. In 1980, its foreign debt was $20.5 billion. At the moment, Western officials say India has only $1.1 billion in its hard currency reserves, enough for 2 weeks of imports. Policy Changes Required

PV Narasimha Rao Remembered as Father of Indian Economic Reforms[52,53]

Prime Minister PV Narasimha Rao is credited with launching free market reforms that opened up India's stagnant, socialist economy. The shift turned the bankrupt nation into a regional economic powerhouse. According to government figures, India's economy has posted an average growth rate of 6% since 1990, and poverty rates have fallen by 10%

Manmohan Singh's 1991 Budget: The Day that Changed India Forever[54]

June 21 to July 24, 1991, witnessed an intellectual revolution, but it was an evolutionary one—both the crisis and the response were some years in the making. To borrow an analogy from the great historian of ideas, Isaiah Berlin, if Dr. Singh was the hedgehog who knew only one big thing, that is, economic reforms, Rao was the crafty and cunning fox who knew many things. It was this unusual jugalbandi that rescued India at its darkest moment when India could well have mirrored Greece in 2015. There are many lessons to be drawn from what the duo did and more importantly how they did it, lessons that have great contemporary relevance as well

Dr. Manmohan Singh entering Parliament in New Delhi with the budget documents in 1991. Photo: Shanker Chakravarthy/The Hindu Archives[55] [16, 17]

[52] PV Narasimha Rao Remembered as Father of Indian Economic Reforms - https://www.voanews.com/archive/pv-narasimha-rao-remembered-father-indian-economic-reforms

[53] 4 reforms that pulled India back after it ran out of money in 1991 - https://economictimes.indiatimes.com/news/economy/policy/4-reforms-that-pulled-india-back-after-it-ran-out-of-money-in-1991/articleshow/53308703.cms?from=mdr

[54] Manmohan Singh's 1991 Budget: the day that changed India forever - https://www.thehindu.com/business/Economy/Manmohan-Singhs-1991-Budget-the-day-that-changed-India-forever/article14505003.ece

[55] Manmohan Singh's 1991 Budget: the day that changed India forever - https://www.thehindu.com/business/Economy/Manmohan-Singhs-1991-Budget-the-day-that-changed-India-forever/article14505003.ece

Step 7: Regression.

Let's train and test split the dataset:

```
X, y = dataIndia_df [['Population growth (annual %)', 'Inflation, consumer prices (annual %)',
     'Rural population (% of total population)',
     'Exports of goods and services (% of GDP)', 'EXIM', 'FDI',
     'External debt stocks (% of GNI)',
     'Life expectancy at birth, total (years)',
     'Employment in agriculture (% of total employment) (modeled ILO estimate)']].values,
dataBrazil_df['Agriculture, forestry, and fishing, value added (% of GDP)'].values

X_train,X_test, y_train , y_test = train_test_split(X ,y, test_size = 0.2,random_state = 0)
```

Train size is 80% and test size is 20% of the dataset:

```
regressor = LinearRegression()
regressor.fit(X_train,y_train)
```

Output:

```
array([26.10518793, 16.16449943, 15.88528711, 18.34104865, 20.5531691 ,
     16.56496164])
```

Plot the regression line:

```
import matplotlib.pyplot as plt

%matplotlib inline

plt.scatter(y_test, y_pred)
plt.xlabel('Actual Labels')
plt.ylabel('Predicted Labels')
plt.title('Agriculture, forestry, and fishing, value added (annual % growth)')
# overlay the regression line
z = np.polyfit(y_test, y_pred, 1)
p = np.poly1d(z)
plt.plot(y_test,p(y_test), color='magenta')
plt.show()
```

Output:

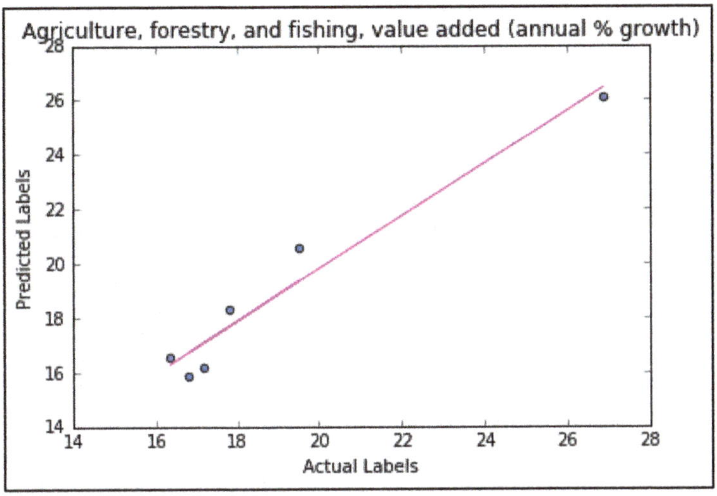

Check the actual and predicted values:

```
pred_y_df= pd.DataFrame({'Actual Value':y_test, 'Predicted value':y_pred,'Difference': y_test-
y_pred})
pred_y_df
```

Output:

	Actual Value	Predicted value	Difference
0	26.881801	26.105188	0.776613
1	17.191970	16.164499	1.027471
2	16.791935	15.885287	0.906647
3	17.814691	18.341049	-0.526358
4	19.536700	20.553169	-1.016469
5	16.363800	16.564962	-0.201161

The regression coefficients play an extreme role in analyzing the impact:

```
regressor.coef_
```

Output:

```
array([1.62300616e-01, 8.88532693e-02, 1.00226502e+00, -4.37032208e-01,
    1.02968594e+00, -2.89507841e-05, 7.77900794e-02, -3.03692608e-01,
    -2.43846219e-01])
```

Feature engineer variable	Coefficient	Macro-influence
Population growth (annual %)	0.162300616e	Population growth is a positive coefficient—implies India can transfer valuable agricultural labor to the modern sectors without affecting the agricultural growth. By using technological innovations in the agricultural sector, such as artificial intelligence augmented analytics and climate defense models enabling the agricultural productivity, India will witness a huge growth in agriculture
Inflation, consumer prices (annual %)	0.888532	Good news is inflation is not negative (cost-push[56] is not an exogenous influencer!), and demand-pull seems to be playing an important role [9]
Rural population (% of total population)	0.00122	India has been able to maintain a high percentage of its population in the rural sector, and it could be expected that its agricultural sector would constitute a larger percentage of its gross output—AI would have a very positive effect if it reached to the farmers at the gross root levels and will have a net positive effect on the overall economy
Exports of goods and services (% of GDP)	0.12940318	India has negative coefficient of exports of goods and services. What it implies is the domestic demand is more influential when it comes to agricultural sector. Additionally, technological innovations and positive competition from international market prepare agricultural market to be more efficient and competitive.[57] Artificial intelligence will enable agricultural sector to be more productive [10]
EXIM	0.45223404	Positive coefficient implies the ratio of India exports to imports is positive—in simple terms, more agricultural exports as compared to imports
FDIs	0.01348183	More competitive business markets and hence a possible thriving economy
External debt	0.07779	A positive external debt coefficient would force India to produce more of its own food, resulting in increased agricultural production and an increased contribution of agriculture to GDP. Therefore, the higher the external debt of the country, the greater the contribution of agriculture is expected to be
Life expectancy at birth, total (years)	-0.0303	Correlation analysis showed a negative life expectancy at birth— negative regression coefficient implies capital accumulation and tends to be higher, making market more competitive. Artificial intelligence can reinforce the economic growth

[56] Economics by Lipsey, Richard, Chrystal, Alec 13th Revised edition (2015) Paperback Paper back – January 1, 1601 – Oxford University Press; 13th Revised edition (January 1, 1601).

[57] Robert Carbaugh, International Economics 15th Edition, Cengage Learning; 15th edition (September 8, 2014), ISBN-13: 978-1285854359.

Feature engineer variable	Coefficient	Macro-influence
Employment in agriculture (% of total employment) (modeled ILO estimate)	−0.024	Negative employment in agriculture (% of total employment) coefficient implies the contribution of agriculture to the GDP is highly dependent on productivity enhancements—artificial intelligence plays an important role in increasing the productivity

Regression metrics—the following table summarizes regression metrics:

```
from sklearn.metrics import mean_squared_error, r2_score

mse = mean_squared_error(y_test, y_pred)
print("MSE:", mse)

rmse = np.sqrt(mse)
print("RMSE:", rmse)

r2 = r2_score(y_test, y_pred)
print("R2:", r2)
```

Output:

```
MSE: 0.6385937333934221
RMSE: 0.7991206000306975
R2: 0.951409176505843
```

Step 8: Durbin-Watson Test.

One of the assumptions of linear regression is that there is no correlation between the residuals. In other words, the residuals are assumed to be independent. One way to determine if this assumption is met is to perform a Durbin-Watson test, which is used to detect the presence of autocorrelation in the residuals of a regression:

```
from statsmodels.stats.stattools import durbin_watson

residual = (y_test - y_pred)
#perform Durbin-Watson test
durbin_watson(residual)
```

Output:

0.7923580342011368

- A test statistic of 2 indicates no serial correlation.
- The closer the test statistics is to 0, the more evidence of positive serial correlation.
- The closer the test statistics is to 4, the more evidence of negative serial correlation.

As a rule of thumb, test statistic values between the range of 1.5 and 2.5 are considered normal. However, values outside of this range could indicate that autocorrelation is a problem.

Explainability of Model

Set up explainer model:

```
# Explainers:
# 1. SHAP Tabular Explainer
from interpret.ext.blackbox import TabularExplainer

# OR

# 2. Mimic Explainer
#from interpret.ext.blackbox import MimicExplainer
# You can use one of the following four interpretable models as a global surrogate to the black box
model
from interpret.ext.glassbox import LGBMExplainableModel
from interpret.ext.glassbox import LinearExplainableModel
from interpret.ext.glassbox import SGDExplainableModel
from interpret.ext.glassbox import DecisionTreeExplainableModel

# OR

# 3. PFI Explainer
#from interpret.ext.blackbox import PFIExplainer
```

Create table explainer:

```
feature_names= ['Population growth (annual %)',
        'Inflation, consumer prices (annual %)','Rural population (% of total population)',
        'Life expectancy at birth, total (years)','Exports of goods and services (% of GDP)','EXIM','FDI'
        ,'External debt stocks (% of GNI)','Employment in agriculture (% of total employment)
(modeled ILO estimate)']
# 1. Using SHAP TabularExplainer
explainer = TabularExplainer(regressor,
                    X_train,
                    features = feature_names)
```

```
# Passing in test dataset for evaluation examples - note it must be a representative sample of the
original data
# x_train can be passed as well, but with more examples explanations will take longer although they
may be more accurate
global_explanation = explainer.explain_global(X_test)
```

Generate local values:

```
# Sorted SHAP values
print('ranked global importance values: {}'.format(global_explanation.get_ranked_global_values()))
# Corresponding feature names
print('ranked global importance names: {}'.format(global_explanation.get_ranked_global_names()))
# Feature ranks (based on original order of features)
print('global importance rank: {}'.format(global_explanation.global_importance_rank))

# Print out a dictionary that holds the sorted feature importance names and values
print('global importance rank: {}'.format(global_explanation.get_feature_importance_dict()))
```

Output:

```
ranked global importance values: [2.2557644825073115, 2.0103026219710345, 1.5577708468990104,
0.8890601096301248, 0.3957135764580931, 0.37685234819879954, 0.295717151281755l,
0.21569966034881025, 0.04737623109433651]
ranked global importance names: ['Rural population (% of total population)', 'Life expectancy at birth,
total (years)', 'Employment in agriculture (% of total employment) (modeled ILO estimate)', 'External
debt stocks (% of GNI)', 'Exports of goods and services (% of GDP)', 'EXIM', 'FDI', 'Inflation,
consumer prices (annual %)', 'Population growth (annual %)']
global importance rank: [2, 3, 8, 7, 4, 5, 6, 1, 0]
global importance rank: {'Rural population (% of total population)': 2.2557644825073115, 'Life
expectancy at birth, total (years)': 2.0103026219710345, 'Employment in agriculture (% of total
employment) (modeled ILO estimate)': 1.5577708468990104, 'External debt stocks (% of GNI)':
0.8890601096301248, 'Exports of goods and services (% of GDP)': 0.3957135764580931, 'EXIM':
0.37685234819879954, 'FDI': 0.295717151281755l, 'Inflation, consumer prices (annual %)':
0.21569966034881025, 'Population growth (annual %)': 0.04737623109433651}
```

Ranked global importance generated feature names and values.
Generate local importance values:

```
# Note: Do not run this cell if using PFIExplainer, it does not support local explanations
# feature shap values for all features and all data points in the training data
print('local importance values: {}'.format(global_explanation.local_importance_values))
```

Output values:

```
local importance values: [[0.0660123838564172, -0.1351711368320018, 3.2272498031502765,
2.875522606445263, -0.24933866365636306, -0.32701186606881283, 0.8699851832769616,
1.4894802014900157, -1.906241302641489], [-0.04468620893564764, 0.08975806156829318, -
1.8692678480122058, -3.551554959929999, 0.778869238902984, 0.34871798090382095, -
0.26978397615254945, -0.8735519805875255, 1.361298401526145], [-0.0678873404151942, -
0.1328341650585381, -2.9797774955024656, -2.8643472131196535, 0.40035801762579065,
0.31975280160354763, 0.05688252256878487, -1.2246206353210625, 2.1830598977561717],
[0.008342597645159868, -0.36260495197777, 0.5091070558636203, -0.631620439676396, -
0.37024372125306726, -0.2389386973662667, -0.34342372219781125, 0.048155084391954074, -
0.47242527554935104], [0.020394244676251977, -0.3155197304166316, 1.169599707213886,
0.9393971743196592, -0.497752368941228, -0.2286406739973081, -0.10116058832925796,
0.2883759372416321, -0.916225318059331], [-0.0769346110373482, -0.2583099162396268, -
3.7795849853014163, -1.1993733383352343, -0.07771944836912566, 0.7980520692530408, -
0.13306691516516556, -1.4101768187485595, 2.5073748858615748]]
```

```
# E.g., Explain the first data point in the test set
local_explanation = explainer.explain_local(X_test[0,:])
# Sorted local feature importance information; reflects the original feature order
sorted_local_importance_names = local_explanation.get_ranked_local_names()
sorted_local_importance_values = local_explanation.get_ranked_local_values()

print('sorted local importance names: {}'.format(sorted_local_importance_names))
print('sorted local importance values: {}'.format(sorted_local_importance_values))
```

Output:

```
sorted local importance names: ['Rural population (% of total population)', 'Life expectancy at birth, total (years)', 'External debt stocks (% of GNI)', 'FDI', 'Population growth (annual %)', 'Inflation, consumer prices (annual %)', 'Exports of goods and services (% of GDP)', 'EXIM', 'Employment in agriculture (% of total employment) (modeled ILO estimate)']
sorted local importance values: [3.2272498031502765, 2.875522606445263, 1.4894802014900157, 0.8699851832769616, 0.0660123838564172, -0.1351711368320018, -0.24933866365636306, -0.32701186606881283, -1.906241302641489]
```

Aggregate feature importance: rural population (% of total population), life expectancy at birth, employment in agriculture (% of total employment) (modeled ILO estimate), and external debt stocks (% of GNP, population growth) are major explainers of the model.

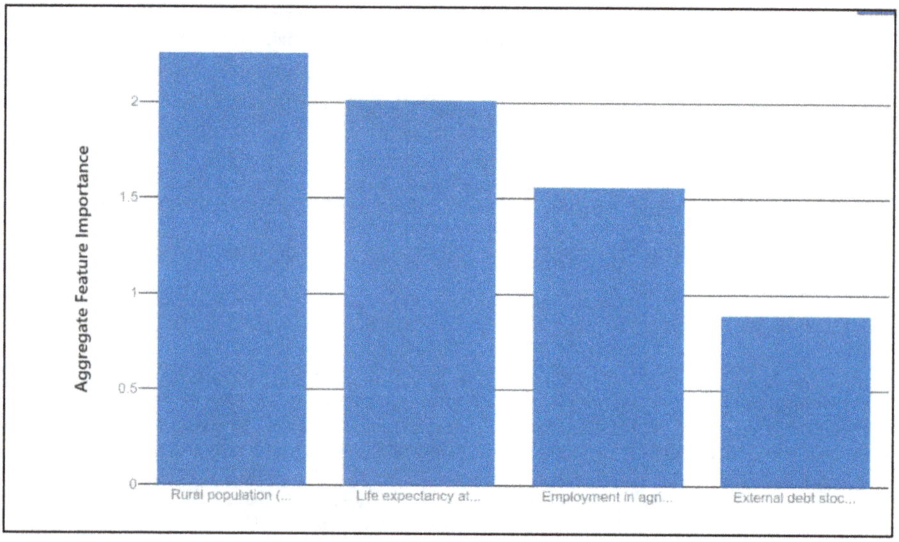

In summary, countries such as India and Brazil have huge positive movement and growth opportunities with the help of technological innovations in the agriculture. Given that the model has highly correlated with employment in agriculture and population at large, any improvement in agricultural worker productivity is a game changer and will address the emerging needs as well as the challenges faced by the agriculture. The artificial intelligence, digital transformation, empowering farmers, and development of innovative technological advancements such as Internet of things-enabled sensors (Cow Necklace[58]) will enable to operate at higher efficiencies and could gain economic scale as well as improvement in services and adjust development in industries.

After reading this chapter, you should be able to answer artificial intelligence, machine learning, and deep learning. You should be able to answer types of learning and should be able to evaluate performance metrics of classification and regression techniques. Finally, able to establish a relationship between a set of macrolevel indicators f (rural population, life expectancy, foreign direct investment, the rate of inflation, level of exports of all goods and services) and {the *ratio of agricultural exports to agricultural imports*} significantly determined the percentage contribution of agriculture to GDP.

References

1. Arthur Lewis, W.: Economic Development with Unlimited Supplies of Labour, First published: May 1954. https://doi.org/10.1111/j.1467-9957.1954.tb00021.x. Access Date: 17 Sept 2019
2. De Sormeaux, A., Pemberton, C.: Factors Influencing Agriculture's Contribution to GDP: Latin America and the Caribbean. https://ageconsearch.umn.edu/record/187331/files/Factors%20Influencing%20Agriculture_s%20Contribution%20to%20GDP%20Latin%20America%20and%20the%20Caribbean.pdf. Access Date: 22 Nov 2019
3. Bughin, J., Seong, J., Manyika, J., Chui, M., Joshi, R.: McKinsey Global Institute (MGI) – Notes from the AI frontier modeling the impact of AI on the world economy (September 2018). https://www.mckinsey.com/~/media/McKinsey/Featured%20Insights/Artificial%20Intelligence/Notes%20from%20the%20frontier%20Modeling%20the%20impact%20of%20AI%20on%20the%20world%20economy/MGI-Notes-from-the-AI-frontier-Modeling-the-impact-of-AI-on-the-world-economy-September-2018.ashx. Access Date: 8 Nov 2020
4. Boucekkine, R., Diene, B., Azomahou, T.: A closer look at the relationship between life expectancy and economic growth, August 14, 2007, May 2009. Int. J. Econ. Theory **5**(2), 201–244. https://doi.org/10.1111/j.1742-7363.2009.00105.x. Access Date: 8 Nov 2019
5. Epaphra, M., Mwakalasya, A.: Analysis of foreign direct investment, agricultural sector and economic growth in Tanzania. Mod. Econ. **8**, 111–140 (2017). https://doi.org/10.4236/me.2017.81008
6. Diego Arias (World Bank), Pedro Abel Vieira (Embrapa), Elisio Contini (Embrapa), Barbara Farinelli (World Bank) and Michael Morris (World Bank): Agriculture Productivity Growth in Brazil (September 24, 2017). http://documents1.worldbank.org/curated/en/268351520343354377/pdf/123948-WP-6-3-2018-8-39-22-AriasetalAgriculturalgrowthinBrazil.pdf. Access Date: 18 Sept 2020

[58] Class 10 Diagnostic Veterinarian Sensors—Hanumayamma Innovations and Technologies, Inc., http://www.hanuinnotech.com

7. Mari, A.: Brazil creates national AI innovation network (November 2, 2020). https://www.zdnet.com/article/brazil-creates-national-ai-innovation-network/. Access Date: 8 Jan 2021
8. Valdes, C., Hjort, K., Seeley, R.: Brazil's changing macroeconomic conditions: Impacts on agriculture (2016). https://ageconsearch.umn.edu/record/236207/files/Brazil%20Macro%20and%20Agriculture.pdf. Access Date: 8 Jan 2021
9. Lipsey, R., Chrystal, A.: Economics, 13th Rev. edn. Oxford University Press, Oxford (2015). Paperback Paper back – January 1, 1601
10. Carbaugh, R.: International Economics, 15th edn. Cengage Learning, Boston (September 8, 2014) ISBN-13: 978-1285854359
11. Perry, T.S.: Want a Really Hard Machine Learning Problem? Try Agriculture, Says John Deere Labs (October 4, 2019). https://spectrum.ieee.org/view-from-the-valley/robotics/artificial-intelligence/want-a-really-hard-machine-learning-problem-try-agriculture-say-john-deere-labs-leaders. Access Date: 2 Jan 2021
12. Sunder, S.: India economic survey 2018: Farmers gain as agriculture mechanisation speeds up, but more R&D needed. Updated: January 29, 2018. https://www.financialexpress.com/budget/india-economic-survey-2018-for-farmers-agriculture-gdp-msp/1034266/. Access Date: 2 Jan 2021
13. Gowan, J.: "Free" fertilizer is saving rural farmers (October 2, 2017). https://www.permaculturenews.org/2017/10/02/free-fertilizer-saving-rural-farmers/. Access Date: 25 Mar 2021
14. Cerra, V., Saxena, S.C.: What caused the 1991 currency crisis in India? IMF Staff Pap. **49**(3) (2002) International Monetary Fund, https://www.imf.org/external/pubs/ft/staffp/2002/03/pdf/cerra.pdf., Access Date: March 25 2021
15. Weinraub, B.: Economic crisis forcing once self-reliant India to seek aid (June 29, 1991). https://www.nytimes.com/1991/06/29/world/economic-crisis-forcing-once-self-reliant-india-to-seek-aid.html. Access Date: 8 Jan 2021
16. Pasricha, A.: PV Narasimha Rao remembered as father of Indian economic reforms (October 26, 2009). https://www.voanews.com/archive/pv-narasimha-rao-remembered-father-indian-economic-reforms. Access Date: 25 Mar 2021
17. Pandey, V.: ET Bureau, 4 reforms that pulled India back after it ran out of money in 1991 (1991). https://economictimes.indiatimes.com/news/economy/policy/4-reforms-that-pulled-india-back-after-it-ran-out-of-money-in-1991/articleshow/53308703.cms?utm_source=contentofinterest&utm_medium=text&utm_campaign=cppst. Access Date: 25 Mar 2021

Chapter 7
The Role of the Government and the AI Readiness

This chapter covers:

- The role of the government in agriculture credit and employment
- Factors that influence dissemination of technologies, ICT
- AI readiness
- Countries with the highest agricultural employment and the lowest AI readiness
- Use case: Brazilian economy and Indian economy
- Regression
- Regression and model explainability

The chapter introduces the role of the government credit to agriculture and the influence and impacts of it. Next, the chapter examines the influence of AI readiness framework on countries with the highest agricultural employment and recommends the impact of AI. The countries studied include Pakistan, Brazil, Uruguay, Indonesia, and Peru. Next, the chapter walks through machine learning and artificial intelligence use cases for milk production in India and the impact of government credit in achieving milk sufficiency in India.

Agricultural infrastructure is the backbone of a nation. Generations before us have relied on agriculture, and our future generations will be dependent on agriculture. It is one of the noble essential works we as a social society perform. Historically, agricultural infrastructure includes water supplies, irrigation channel development, construction of hydroelectric and electrical dams and power plants, fertilizer manufacturing facilities, agriculture equipment manufactures, and availability of natural and human resources. Now, in addition to traditional agricultural infrastructure, the data-enabled Information and Communication Technologies (ICT), digital communication, satellite, GPS, rural digital infrastructure, intelligent machines, data storages, availability of digital compute and storage, democratization of machine learning, and availability of insights to farmers are part of the core infrastructure of

C. Vuppalapati, *Machine Learning and Artificial Intelligence for Agricultural Economics*, International Series in Operations Research & Management Science 314,
https://doi.org/10.1007/978-3-030-77485-1_7

the agriculture. Data-intensive architecture, until now, is confined to corporate data centers, private incorporation, and national research laboratories. Nonetheless, the future data sources will no longer form corporate data centers but form multidisciplinary business and agricultural use cases. The democratization of compute to enable farmers to better plan and manage their farms is the need of the day. Data systems, sensors, climate sensors, weather events notification, timely crop planning updates, crop commodity forecasts, and advanced analytics are digital infrastructure that is needed in the twenty-first century.

> According to Gartner, by 2022, as a result of digital business projects, 75% of enterprise-generated data will be created and processed outside the traditional,[1,2] enterprise data center or cloud, up from less than 10% in 2018 [1–3]

Additionally, government and policy makers should embrace creation of digital infrastructure that enables symbiotic growth of agricultural yields with long-term research and development plus inclusion of productivity enhancement technologies such as artificial intelligence and ICT-based Internet of things sensors that collect data.

Finally, in addition to agricultural infrastructures, one of the most important aspects that enables farmers to increase their agriculture footprint and expand their farms is the availability of capital. Capital in the form of cheaper loans from central banks, credit availability through cooperative schemes, governmental subsidies, and farm insurances are the most important aspects of modern agriculture. Farmers all over the world receive government subsidies[3] [4]. In 2019, China spent an estimated $185.9 billion on farm subsidies, followed by the European Union ($101.3 billion), the United States ($48.9 billion), and India ($11 billion). But different countries give these subsidies in different ways.

On the other side, the growth and deepening of agriculture finance markets is constrained by a variety of factors, which include:

- Inadequate or ineffective policies
- High transaction costs to reach remote rural populations
- Covariance of production, market, and price risks
- Absence of adequate instruments to manage risks
- Low levels of demand due to fragmentation and incipient development of value chains
- Lack of expertise of financial institutions in managing agricultural loan portfolios

[1] Hewlett Packard Enterprise Commits $4 Billion to Accelerate the Intelligent Edge - https://www.hpe.com/us/en/newsroom/press-release/2018/06/hewlett-packard-enterprise-commits-4-billion-to-accelerate-the-intelligent-edge.html

[2] What Edge Computing Means for Infrastructure and Operations Leaders - https://www.gartner.com/smarterwithgartner/what-edge-computing-means-for-infrastructure-and-operations-leaders/

[3] IndiaFarmLawsandGlobalProblem-https://www.brookings.edu/opinions/indias-farm-laws-are-a-global-problem/

Agricultural Data

The demand for good-quality statistical data continues to increase. Timely and reliable statistics are key inputs to the broad development strategy. Improvements in the quality and quantity of data on all aspects of development are essential if we are to achieve the goal of a world without poverty.[4] Good data are needed to set baselines, identify effective public and private actions, set goals and targets, monitor progress, and evaluate impacts. They are also an essential tool of good government, providing means for people to assess what governments do and helping them to participate directly in the development process.

AI Technology Policy and Enablement at the Gross Root Levels

Big data surrounds us. Every minute, our smartphones collect billions of data points on our locations, search histories, lifestyle, and habits. The collected data is mined to provide exclusive recommendations that have business value for the companies and marketers. Such collected data has become one of the world's most valuable commodities for companies, a data revolution indeed. Artificial intelligence is the center of it and it is fueling the growth.

Monsanto's aggressive move into the data era was perhaps punctuated in October 2013 with their announced acquisition of the Climate Corporation for $930 million. One may ask why a firm, with its roots in fertilizers and pesticides, would spend nearly $1 billion on an information technology (IT) company. This aggressive acquisition demonstrates the evolution of the industry [1]. The real valuation is not due to tangible assets but based on data-infused insights—"the Climate Corporation is focused on unlocking new value for the farm through data science", commented Hugh Grant, the Chairman and Chief Executive Officer for Monsanto [1]. Data has surpassed oil as the world's most valuable resource. Data is giving rise to a new economy and has become a competitive and strategic asset for companies and nations alike.

However, data revolution has not reached every economic sector and every country yet. Agriculture sector in developing countries has been largely passed over. Not only is this a huge, missed opportunity for big data companies, it's a significant obstacle in the path toward sustainable food [5]. Richer countries are benefitting more from the new Information and Communication Technologies (ICT) possibilities than poorer countries that lack the resources for investment, training, and experimentation [5]. According to McKinsey consulting group, African countries spend about 1.1% of Gross Domestic Product (GDP) on investment in and use of Internet services, less than a third of what, on average, is spent by richer

[4] The World Bank - https://data.worldbank.org/about

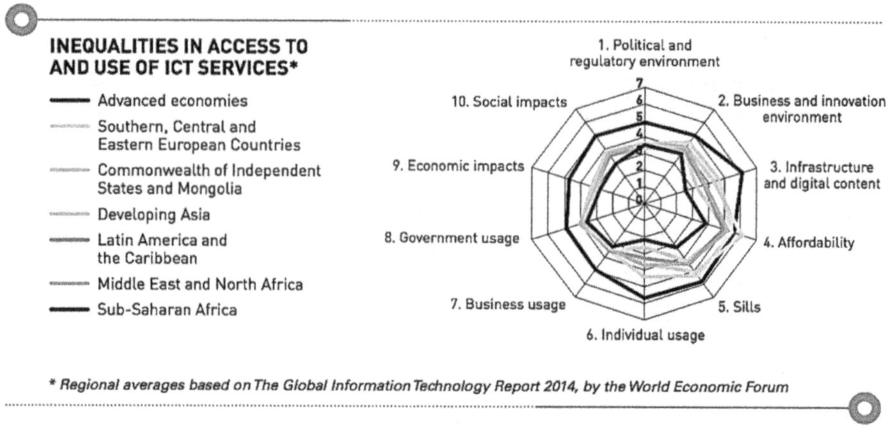

Fig. 7.1 Inequalities to access ICT

countries—meaning that the gap in the Internet availability and use is growing every year, as some regions accelerate ahead [5]. The graph (please see Fig. 7.1) below shows how advanced economies are ahead of the rest of the world on every indicator of access to, use of, and impact of the use of digital technologies, Information and Communication Technologies (ICT) [6].

Huge disparities in ICT spending are a reality between developed economies and poor nations; ironically, the ICT[5] plays significant role in overcoming societal issues such poverty and education (please see Fig. 7.1). There is direct correlation between ICT[6] penetration and decrease in global poverty [7], i.e., a clear correlation between falling global absolute poverty and rising ICT penetration (please see Fig. 7.2) [7]. The report finds that technological progress, measured as the share of ICT capital stock, has a statistically significant impact on inequality, and the effect of technological change was greater than that of financial globalization [7]. In other words, the improvement of life is more profound for the poor communities in the world by penetration of ICT than disbursement of loans via International Monetary Fund [8]. The deeper the ICT penetration, the greater access to compute resources and, thus, address the perennial issue in democratization of AI[7]—the data capture. Having deeper ICT technological penetration is a win-win for global communities [9].

[5] The Paradox of ICT's Impact on Growth and Inequality - https://blogs.cisco.com/gov/gitr2015

[6] ICTs, Income In equality, and Ensuring Inclusive Growth - https://alln-extcloud-storage.cisco.com/ciscoblogs/GITR-2015-Cisco-Chapter.pdf

[7] Chandrasekar Vuppalapati, Democratization of Artificial Intelligence for the Future of Humanity 1st Edition, CRC Press; 1st edition (January 18, 2021), ISBN-13: 978-0367524098.

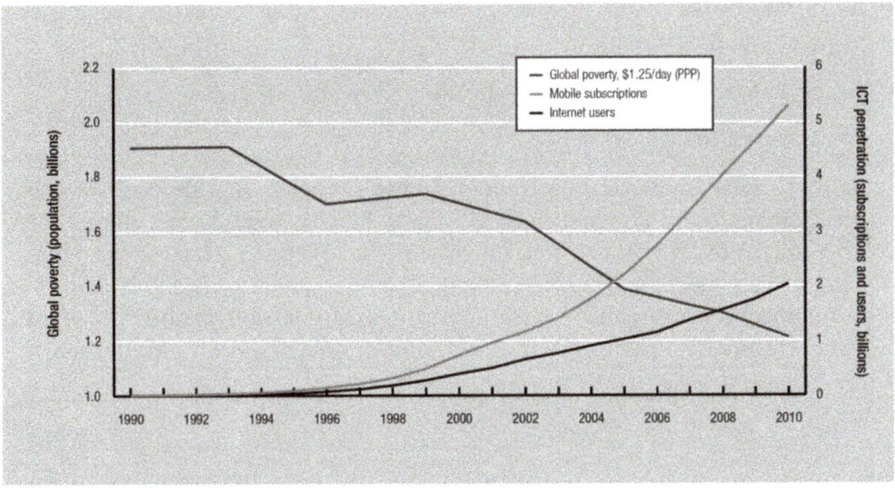

Sources: World Bank *PovCal* database (1990, 1993, 1996, 1999, 2002, 2005, 2008, 2010); authors' calculations and interpolation, ITU *World Telecommunication/ICT Indicators* database June 2013.

Fig. 7.2 ICT impact

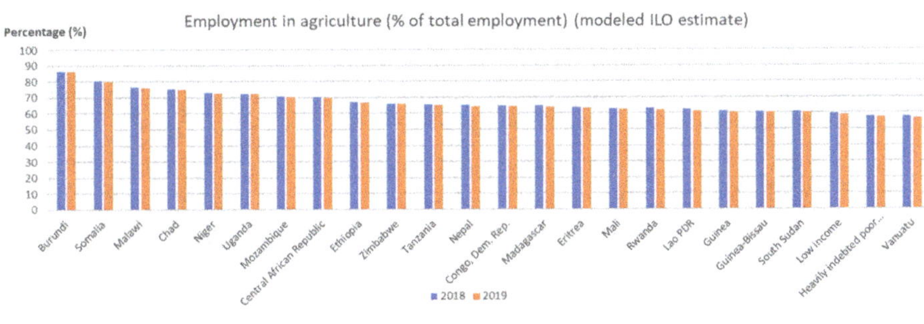

Fig. 7.3 Countries with highest employment in agriculture

AI Readiness and Lower Productivity

Countries with the highest percentage of employment in agriculture have some of the lowest AI readiness[8] (please see Fig. 7.3). The digital gap is causing lower productivity, rural poverty, social and political unrest, economic recessions, and increased burden to local and national governments (please see Fig. 7.1).

[8] McKinsey Global Institute (MGI) – Notes from the AI frontier Modeling the impact of AI on the world economy September-2018 - https://www.mckinsey.com/~/media/McKinsey/Featured%20Insights/Artificial%20Intelligence/Notes%20from%20the%20frontier%20Modeling%20the%20impact%20of%20AI%20on%20the%20world%20economy/MGI-Notes-from-the-AI-frontier-Modeling-the-impact-of-AI-on-the-world-economy-September-2018.ashx

As per McKinsey report, in terms of readiness for AI, the countries in the world are divided into four groups:[9]

- Group 1: Active Global Leaders—Leading the AI efforts [10].
  - China and the United States
- Group 2: Economies with strong comparative strength—A wide range of countries belongs to this group, including Canada, France, South Korea, and Sweden. They are relatively well positioned to capture the benefits of AI, given their generally robust foundation of enablers. Many of these economies are highly motivated to embrace AI because they have been experiencing slowing productivity growth.

Group 2: Economies with strong comparative strength		
Australia	Belgium	Canada
Estonia	Finland	Germany
Iceland	Israel	Japan
Netherlands	Norway	Singapore
South Korea	Sweden	United Kingdom

- Group 3: Economies with moderate foundations—This group, which includes India, Italy, and Malaysia, has a moderate ability to capture economic benefits from AI. While the potential for economic gains is broadly positive, these countries are in a weaker starting position than those in the first two groups, but they exhibit comparative strengths in specific areas on which they may be able to build. India, for instance, has somewhat underdeveloped digital infrastructure and currently has a relatively low automation potential, but it has more STEM students than the total number of STEM graduates produced by all G7 countries.

Chile	Costa Rica	Czech Republic
India	Italy	Lithuania
Malaysia	South Africa	Spain
Thailand	Turkey	

- Group 4: Economies that need to strengthen foundations—These countries are relatively challenged in their ability to capture the economic benefits of AI. They have somewhat limited automation potential because wages tend to be rather low, and therefore the incentive to substitute labor to boost productivity is weak. They also have relatively underdeveloped digital infrastructure, innovation and investment capacity, and digital skills and are comparatively isolated from global trade and data flows. These economies tend to prioritize stimulating economic growth, reducing poverty, and developing away from agriculture into basic, and then more advanced, manufacturing and services.

[9] McKinsey Global Institute (MGI) – Notes from the AI frontier Modeling the impact of AI on the world economy September-2018 - https://www.mckinsey.com/~/media/McKinsey/Featured%20Insights/Artificial%20Intelligence/Notes%20from%20the%20frontier%20Modeling%20the%20

Brazil	Bulgaria	Cambodia
Colombia	Greece	**Indonesia**
Pakistan	Peru	Tunisia
Uruguay	Zambia	

Both Group 3 and Group 4 countries exhibit the highest agriculture employment paradoxically unable to tap into AI due to infrastructure and policy issues. Let's look at these countries' performance over the last two decades (2000–2019) and impact on agriculture due to AI readiness: in general, the following are the trends all over the world in agriculture growth and employment:

World: Agriculture, Forestry, and Fishing, Value Added (Annual % Growth)[10]

As you can see, agriculture growth was 2.71% in 1971 and 2.606% in 2019 with peak growth achieved in 1984 of 5.341%, 5.235% in 1988, and 5.001% in 1996. The lows were 0.409% in 1972 and 0.339% in 1993 (please see Fig. 7.4).

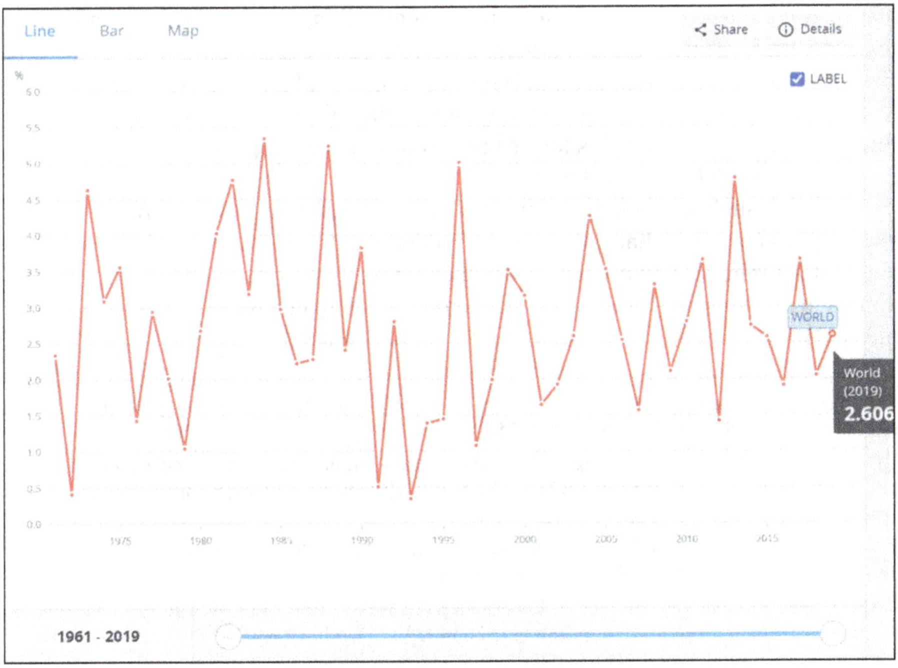

Fig. 7.4 World agricultural growth

impact%20of%20AI%20on%20the%20world%20economy/MGI-Notes-from-the-AI-frontier-Modeling-the-impact-of-AI-on-the-world-economy-September-2018.ashx

[10] World Agriculture, forestry, and fishing, value added (annual % growth) - https://data.worldbank.org/indicator/NV.AGR.TOTL.KD.ZG

Employment in Agriculture (% of Total Employment) (Modeled ILO Estimate)[11]

As you can see, employment in agriculture (% of total employment) was 43.69% in 1991 and 26.756% in 2019 (please see Fig. 7.5).

Brazil

AI Readiness: Group 4—economies that need to strengthen foundations

Brazil had 16.49% (2000) employment[12] in agriculture (% of total employment). That is, out of 100 total employment, 16 people belong to agriculture in 2000. Two decades later, that number is 9.07% (only 9 out of 100 people employed in agriculture) (2019). Agriculture productivity growth[13] in recent decades in Brazil has been mainly driven by investments in agriculture innovation, facilitation of sector financing, and trade liberalization. Nonetheless, low productivity growth is impacting Brazil's economic growth.[14] Agriculture productivity (please see Fig. 7.6) has room to grow further, improving productivity of lagging midsize farmers and regions and reforming agriculture policies toward agriculture financing [11].

Data Source: The World Bank[15]

Artificial intelligence can foster such productivity and growth. Given its value and importance, Brazilian government recently initiated AI policy[16] and created national AI innovation network[17]—focused on artificial intelligence (AI) with the aim of increasing the production capacity and competitiveness of local companies [12].

---

[11] Employment in agriculture (% of total employment) (modeled ILO estimate) - https://data.worldbank.org/indicator/SL.AGR.EMPL.ZS

[12] Data Source—Employment in agriculture (% of total employment) (modeled ILO estimate) - https://data.worldbank.org/indicator/SL.AGR.EMPL.ZS?end=2017&locations=1W&start=1981&view=chart

[13] API_NV.AGR.TOTL.KD.ZG_DS2_en_excel_v2_2063913.XLS - https://api.worldbank.org/v2/en/indicator/NV.AGR.TOTL.KD.ZG?downloadformat=excel

[14] Agriculture Productivity Growth in Brazil—Recent trends and future prospects - http://documents1.worldbank.org/curated/en/268351520343354377/pdf/123948-WP-6-3-2018-8-39-22-AriasetalAgriculturalgrowthinBrazil.pdf

[15] API_SL.AGR.EMPL.ZS_DS2_en_excel_v2_2059645.XLS - https://api.worldbank.org/v2/en/indicator/SL.AGR.EMPL.ZS?downloadformat=excel

[16] AI In Brazil - https://www.oecd.ai/dashboards/countries/Brazil

[17] Brazil creates national AI innovation network - https://www.zdnet.com/article/brazil-creates-national-ai-innovation-network/

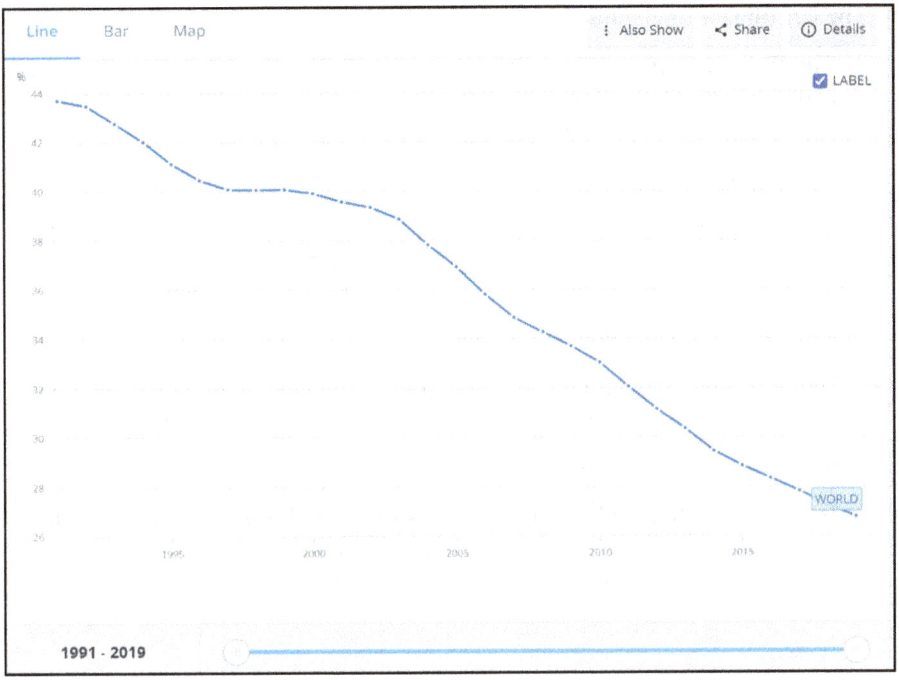

Fig. 7.5 Employment in agriculture

Fig. 7.6 Brazil annual agriculture growth and employment

Brazil AI Policies[18]

- Applied research centers in AI
  - Brazil has announced its aims to establish eight applied research centers hosted by established academic and research institutions.

[18] Brazil AI Policies - https://www.oecd.ai/dashboards/policy-initiatives?conceptUris=http:%2F%2Fkim.oecd.org%2FTaxonomy%2FGeographicalAreas%23Brazil

- The Brazilian artificial intelligence strategy
  - The Brazilian artificial intelligence strategy will cover relevant topics on AI and its impact on economy, ethics, development, education, and jobs, as well as coordinate specific public policies addressing such issues.
- Brazil's digital transformation strategy
  - Brazil's digital transformation strategy (E-Digital), published on March 2018, harmonizes and coordinates different governmental initiatives on digital issues to advance the Sustainable Development Goals in Brazil.
- General Personal Data Protection Law
  - The Brazilian data protection law includes information on the protection of personal data as well as a section on database ownership and database rights.
- National Internet of Things Plan: Degree 9854
  - Institutes the National Internet of Things Plan and provides for the management and monitoring chamber for the development of communication systems machine to machine and Internet of things.

Pakistan

AI Readiness: Group 4—economies that need to strengthen foundations	

Pakistan had 43.00% (2000) employment in agriculture (% of total employment). That is, out of 100 total employment, 43 people belong to agriculture in 2000. Two decades later, that number is 36.91% (ceil+ to 37%) (only 37 out of 100 people employed in agriculture) (2019). Agriculture productivity remains a concern in recent decades and AI will play an important role. It is more of a need than a luxury in Pakistan, as many small farmers rely on their livelihood on agriculture (please see Fig. 7.7).

To set wide adoption, Pakistan has launched Presidential Initiative for Artificial Intelligence & Computing[19] (PIAIC). The mission of PIAIC is to reshape Pakistan by revolutionizing education, research, and business by adopting latest, cutting-edge technologies. Experts are calling this the fourth industrial revolution. We want Pakistan to become a global hub for AI, data science, cloud native computing, edge computing, blockchain, augmented reality, and Internet of things.

[19] Presidential Initiative for Artificial Intelligence & Computing (PIAIC) - https://www.piaic.org/

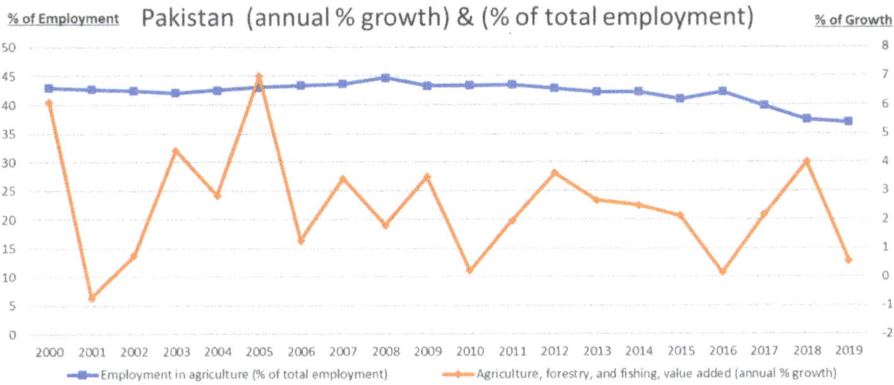

Fig. 7.7 Pakistan employment growth to agriculture

Indonesia

AI Readiness: Group 4—economies that need to strengthen foundations	

Indonesia is a leading producer of palm oil and a major global producer of rubber, copra, cocoa, and coffee. It is the second marine fisheries producer in the world, after China.[20] Indonesia had 45.28% (2000) employment in agriculture (% of total employment). That is, out of 100 total employment, 45 people belong to agriculture in 2000. Two decades later, that number is 28.5% (ceil+ to 29%) (only 29 out of 100 people employed in agriculture) (2019). Agriculture productivity remains a concern in recent decades and AI will play an important role. Indonesia has been known for centuries as an agricultural hub with an abundant crop potential, but it must improve its agricultural research to boost productivity and food security for its large and growing population.[21] According to a study conducted by the Center for Indonesian Policy Studies (CIPS), rice consumption in Indonesia will increase to 99.55 kg per capita by 2045 from 97.6 kg per capita in 2017. The figure assumes that the population will grow to about 300 million in 2045. However, challenges persist as Agriculture Ministry data shows that rice productivity was only 51.92%, and Indonesia only ranked 65 out of 113 in the overall Global Food Index (please see Fig. 7.8).

[20] Socio-economic context and role of agriculture - http://www.fao.org/3/i7696e/i7696e.pdf

[21] Indonesia needs more research to boost agricultural productivity - https://www.thejakartapost.com/news/2019/11/11/indonesia-needs-more-research-to-boost-agricultural-productivity.html

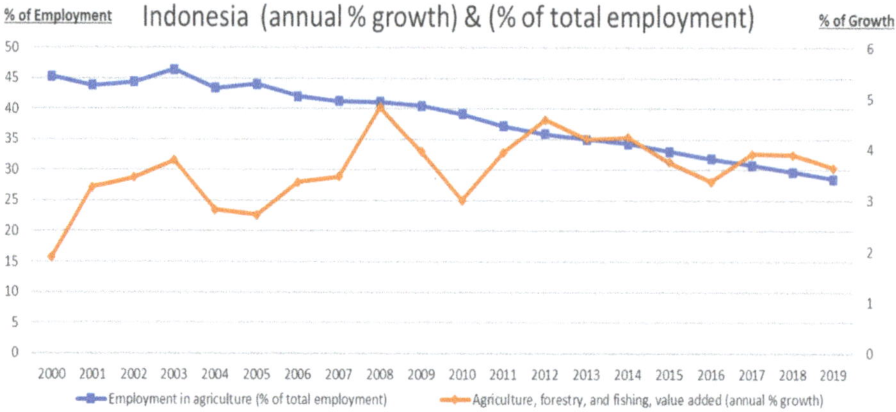

Fig. 7.8 Indonesia employment growth

To set wide AI adoption, Indonesia has national AI strategy that aims to bring productivity across the industry verticals.[22] Indonesia joins the ranks of countries worldwide with national AI strategies like Singapore, South Korea, and Canada. While economic development is often the key goal among many countries' AI strategies, Indonesia's plan stands out for leveraging AI to address specific problem areas. For instance, Indonesia aims to use AI to help tackle its struggles with poor nutrition among infants and to implement government services digitally. Malaysia and Thailand currently lack national AI strategies but have already implemented a variety of programs applying AI, including e-government services. AI will be a key area for Canada to engage with Southeast Asia's largest economy, especially as both tackle COVID-19's economic impact in an increasingly digital world.[23]

Uruguay

AI Readiness: Group 4—economies that need to strengthen foundations	

Uruguay, a small country, an agricultural generator with a temperate climate that is sandwiched between South American giants Argentina and Brazil, leads the world

[22] Indonesia new national strategy - https://www.thejakartapost.com/news/2020/08/13/indonesia-sets-sights-on-artificial-intelligence-in-new-national-strategy.html

[23] New National AI Policy—Indonesia Leans on Artificial Intelligence for Development - https://www.asiapacific.ca/asia-watch/indonesia-leans-artificial-intelligence-development

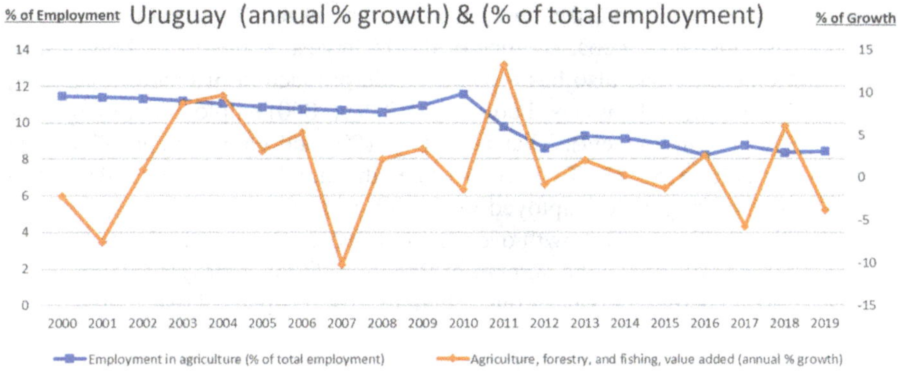

Fig. 7.9 Uruguay employment and annual growth

in arable land per person, with 15 million hectares[24] [13]. It is also likely the only country with four cows per person, each of them equipped with an electronic ear chip to trace every cut of beef in this nation of steak lovers. Crop production in Uruguay has never been as important as livestock raising. Only about 8% of the land area was dedicated to growing crops in the mid-1980s, compared with 75% dedicated to livestock.[25]

Uruguay had 11.47% (2000) employment in agriculture (% of total employment). That is, out of 100 total employment, 12 people belong to agriculture in 2000. Two decades later, that number is 8.5% (ceil+ to 9%) (only 9 out of 100 people employed in agriculture) (2019). Agriculture productivity is not as high as it is in Uruguay, where advanced productivity techniques are applied (please see Fig. 7.9).

Peru

AI Readiness: Group 4—economies that need to strengthen foundations	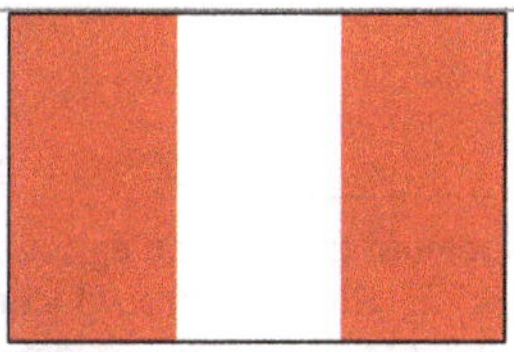

Peru's climate and different geographical zones make it an important agricultural nation. Of the 120 domesticated plants Peru has provided the world, potato is the most important. There are more than 3,000 varieties of potatoes found in Peru, making it the world's genetic center for the crop.[26] Peru is one of the 5 largest producers

[24] Little Uruguay has big plans for smart agriculture - https://phys.org/news/2014-12-uruguay-big-smart-agriculture.html

[25] Agriculture Uruguay - http://countrystudies.us/uruguay/55.htm

[26] Peru - https://www.nationsencyclopedia.com/economies/Americas/Peru-AGRICULTURE.html#ixzz6pjtaWvhN

of avocado, blueberry, artichoke, and asparagus, one of the 10 largest producers in the world of coffee and cocoa, and one of the 15 largest producers in the world of potato and pineapple and also has a considerable production of grape, sugarcane, rice, banana, maize, and cassava. Peru had 45.27% (2000) employment in agriculture (% of total employment). That is, out of 100 total employment, 45 people belong to agriculture in 2000. Two decades later, that number is 28.5% (ceil+ to 9%) (only 29 out of 100 people employed in agriculture) (2019). Peruvian agriculture has experienced impressive growth over the past two decades,[27] which has contributed to the steady decline in the number of Peruvians living in poverty, yet millions of the country's smallholders have missed out on that prosperity[28] [14, 15].

Government as Enabler of Digital Infrastructure

A mix of short- and long-term factors is contributing to higher food prices.[29] On long-term trends level, one reason increasing food prices is reduced agricultural research and development by government and international institutions.[30] Although private sector funding of research has grown, private sector research has generally focused on innovations that private companies could sell to producers. These have often been cost reducing rather than yield-enhancing technological developments. Focusing on yield-enhancing agricultural techniques with a mix of Information and Communication Technologies (ICT) could lead a sustained and long-term stable agricultural development that the world is desperately looking and gets embraced with. Technologies such as Internet of things (IoT), low-cost sensor development, smart agricultural technologies, and machine learning and artificial technologies are the enablers of future proof and data-centric technologies that not only address food yield requirements but also help us to fight climate change and global warming (please see Fig. 7.10).

---

[27] Gaining momentum in Peruvian agriculture: opportunities to increase productivity and enhance competitiveness (English) - https://documents.worldbank.org/en/publication/documents-reports/documentdetail/107451498513689693/gaining-momentum-in-peruvian-agriculture-opportunities-to-increase-productivity-and-enhance-competitiveness

[28] As Peru's agricultural production grows, smallholders long for better markets - https://blogs.worldbank.org/latinamerica/peru-s-agricultural-production-grows-smallholders-long-better-markets

[29] Global Agricultural Supply and Demand: Factors Contributing to the Recent Increase in Food Commodity Prices - https://www.hsdl.org/?view&did=485652

[30] Global Agricultural Supply and Demand: Factors Contributing to the Recent Increase in Food Commodity Prices - https://www.hsdl.org/?view&did=485652

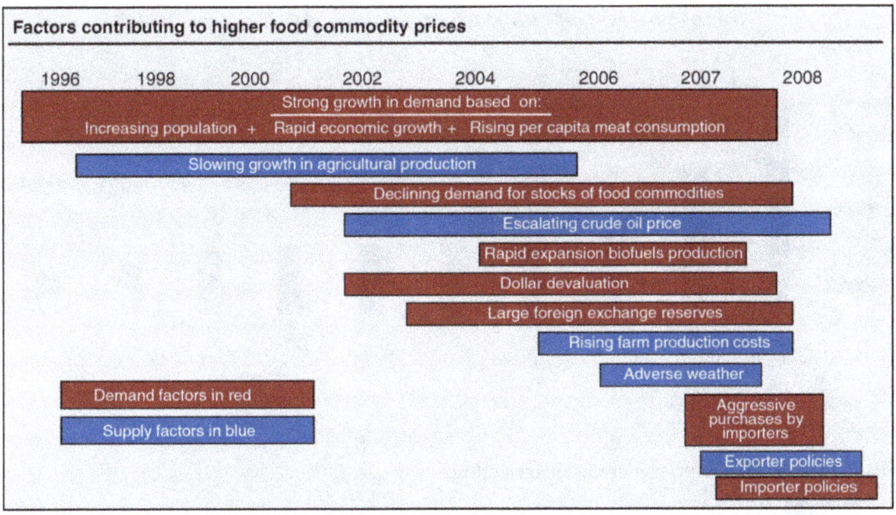

Fig. 7.10 Factors contributing to higher food prices [16]

Credit to Agriculture

There is an ever-increasing need to invest in agriculture due to a drastic rise in global population and changing dietary preferences of the growing middle class in emerging markets toward higher-value agricultural products. In addition, climate risks increase the need for investments to make agriculture more resilient to such risks. Estimates suggest that demand for food will increase by 70% by 2050, and at least $80 billion annual investments will be needed to meet this demand, most of which needs to come from the private sector. Financial sector institutions in developing countries lend a disproportionately lower share of their loan portfolios to agriculture compared to the agriculture sector's share of GDP. Agriculture finance empowers poor farmers to increase their wealth and food production to be able to feed nine billion people by 2050.

In addition to supply of digital, farm equipment, capital availability to farmers is very critical to have a sustained and healthy agriculture. Countries across the world have invested in digital banking and announced credit to spur infrastructure improvement and investment. Here are the top 10 2019 countries that have extended credit to agriculture:[31] China, India, the United States, France, Germany, Australia, Italy, the Netherlands, New Zealand, and the Republic of Korea (please see Fig. 7.11).

[31] Top 10 Countries—2019 - http://www.fao.org/faostat/en/#data/IC/visualize

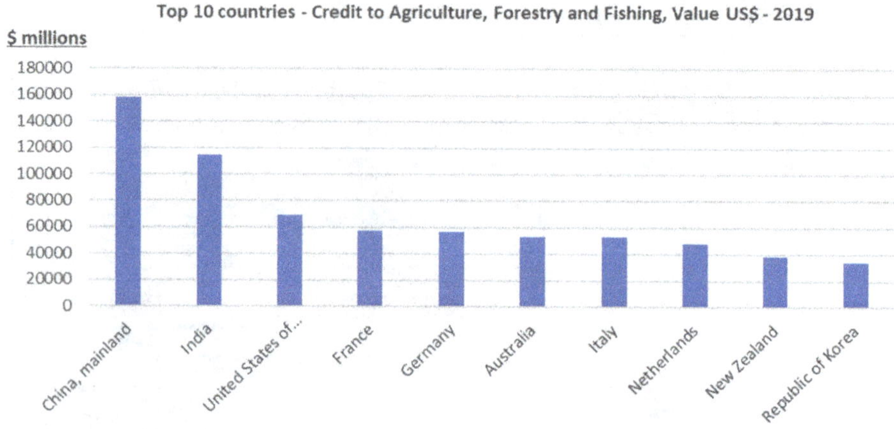

Fig. 7.11 Top 10 countries—credit to agriculture

The United States

The US Federal Government offers farm loan program to help farmers and ranchers get the financing they need to start, expand, or maintain a family farm.[32] The examples of farm loans include Operation Loans, Farm Ownership Loans, Microloans, Youth loans, Native American Tribal Loans, Emergency Loans, and Targeted Loan Funding. The ISDA announces the farmers' rate on a quarterly.[33]

China

China is perhaps the most prominent example of a developing country that has transitioned from taxing to supporting agriculture.[34] In recent year, 2013, Chinese price supports, and subsidies have risen at an accelerating pace after they were linked to rising production costs [17].

Please find timeline of Chinese agricultural support programs:

Year	Policy measure	Commodities
2000	Pilot reforms of rural taxes and fees	
	Soybean seed subsidy and pilot grain subsidy programs in several regions	Soybeans, rice, wheat, corn

[32] Farm Loan Program - https://www.fsa.usda.gov/programs-and-services/farm-loan-programs/

[33] USDA Announces March 2021 Lending Rates for Agricultural Producers - https://www.fsa.usda.gov/news-room/news-releases/2021/usda-announces-march-2021-lending-rates-for-agricultural-producers

[34] Growth and Evolution in China's agricultural support policies - https://www.ers.usda.gov/webdocs/publications/45115/39368_err153.pdf?v=41491

Year	Policy measure	Commodities
2004–2006	Direct payment to grain producers General input subsidy Improved seed subsidy Machinery subsidy Transfer payments to grain counties Reform of grain marketing system Eliminated agricultural tax, specialty crop, and animal slaughter Taxes Rice and wheat price supports	Rice, wheat, corn, soybeans
2007	Package of pork industry subsidies introduced and expanded Seed subsidy for cotton and rapeseed Transfer payments to oilseed and pork counties	Pork, cotton, rapeseed
2008	General input subsidy linked to input prices Support prices for corn, soybeans, rapeseed Strategy of raising price supports annually adopted	Soybeans, rice, wheat, corn, rapeseed
2009	Hog price intervention program	Pork
2011	Cotton price support Grassland protection program	Cotton Cattle and sheep

India

India is one of the fastest growing G20 economies, largely reflecting an ambitious reform agenda under implementation since 2014. Against this background, agriculture is a key sector in terms of its contribution to both employment and GDP[35] [18]. Sustained by improved access to inputs such as fertilizers and seeds, as well as better irrigation and credit coverage, production has been increasing on average at about 3.6% annually since 2011. The sector has also been diversifying from grains toward pulses, fruit, vegetables, and livestock products, largely driven by evolving demographics, urbanization, and changing demand patterns. India has achieved a significant fall in the proportion of the population that is undernourished, from around 24% in 1990–1992 to 15% in 2014–2016. Moreover, it has also emerged as a major agricultural exporter of several key commodities, currently being the largest exporter of rice globally and the second largest of cotton.

India sets agricultural loan target at $226 billion in 2020–2021 to boost farm credit and infrastructure to spur output[36] [19]. India also proposed to impose extra taxes on imports of some commodities, including edible oils, pulses, gold, silver, and some alcoholic beverages to finance more infrastructure facilities in rural areas.

[35] Agricultural Policies in India - https://www.oecd-ilibrary.org/docserver/9789264302334-en.pdf?expires=1615502002&id=id&accname=guest&checksum=0ECD4B14C9B8B233FE8A2AA18F8CED19

[36] India to Boost Farm Credit and Infrastructure to Spur Output - https://www.bloomberg.com/news/articles/2021-02-01/india-plans-to-enhance-farm-credit-to-boost-agricultural-output

The impact of agricultural credit is applicable to the entire economy. The multiplier effect can be seen in the investment of agriculture.[37] The following macroeconomic indicators have a positive effect on:

- Real GDP growth
- Unemployment rate (% of labor force)
- Overall budgetary surplus/deficit (% of GDP)
- Exports/Imports

As farmers go, so goes the nation.[38]
How did agriculture help the United States recover from the Great Depression in the spring of 1933 [20]?

France

The French agriculture sector is a dominant force within the EU-28, accounting for 18% of EU-28 agricultural output in 2018. With 456,500 farms, the French agriculture sector also accounted for 3.9% of the total employment in the EU-28 in 2018. The value of French agricultural production reached EUR 76.6 billion in 2018. The major agricultural produce that place France among the top producers in the EU are cereals (21.7% of the EU-28), wine (46%), industrial crops (19.7%), cattle (23%), and dairy (16%). Nearly 60% of the agricultural output in France is generated through crop production, with wine representing 17.4% and cereals 14%. The livestock sector generates 34.5% of agricultural output, 9 with dairy accounting for 12.5% of this subsector and cattle 9.9%. With a value of EUR 63.5 billion in 2018, agriculture exports accounted for nearly 13% of France's total exports (please see Fig. 7.12). France has provided $60 million agriculture credit to farmers. The major purposes of bank loans include working capital, investment in new machinery and equipment, purchase of land, investment on land (orchards, vineyards), and renting additional land.[39]

Agriculture finance and agricultural insurance are strategically important for eradicating extreme poverty and boosting shared prosperity. Globally, there are an estimated 500 million smallholder farming households—representing 2.5 billion people—relying, to varying degrees, on agricultural production for their

[37] As farmers go, so goes the nation - https://www.aeaweb.org/research/farmer-recovery-great-depression

[38] As farmers go, so goes the nation - https://www.aeaweb.org/research/farmer-recovery-great-depression

[39] Financial needs in the agriculture and agri-food sectors in France - https://www.fi-compass.eu/sites/default/files/publications/financial_needs_agriculture_agrifood_sectors_France_0.pdf

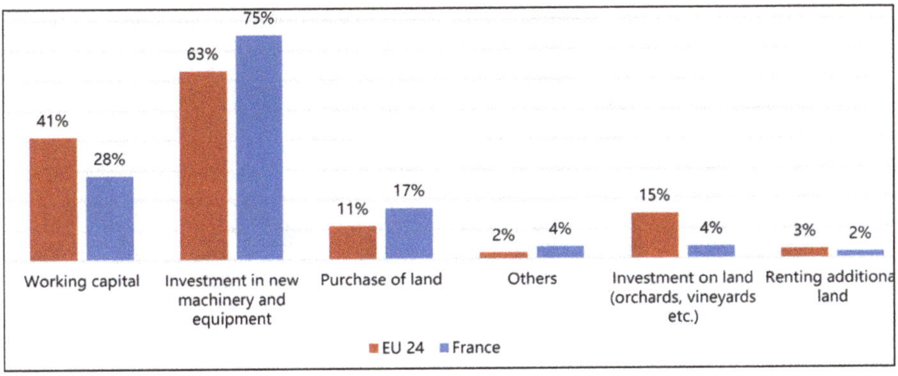

Fig. 7.12 France credit to agriculture

livelihoods. The benefits of our work include the following: growing income of farmers and agricultural SMEs through commercialization and access to better technologies, increasing resilience through climate smart production, risk diversification and access to financial tools, and smoothing the transition of noncommercial farmers out of agriculture and facilitating the consolidation of farms, assets, and production (financing structural change).

History of Credit to Agriculture

2010

During 2010, the top countries that have budgets and offered credit to agriculture include China, India, the United States of America, the Netherlands, Italy, Australia, Germany, the Republic of Korea, New Zealand, and Spain (please see Fig. 7.13).

2000

During 2000, the top countries that have budgets and offered credit to agriculture include the United States, Germany, Australia, the United Kingdom, Spain, Canada, India, Argentina, New Zealand, and Iran (please see Fig. 7.14).

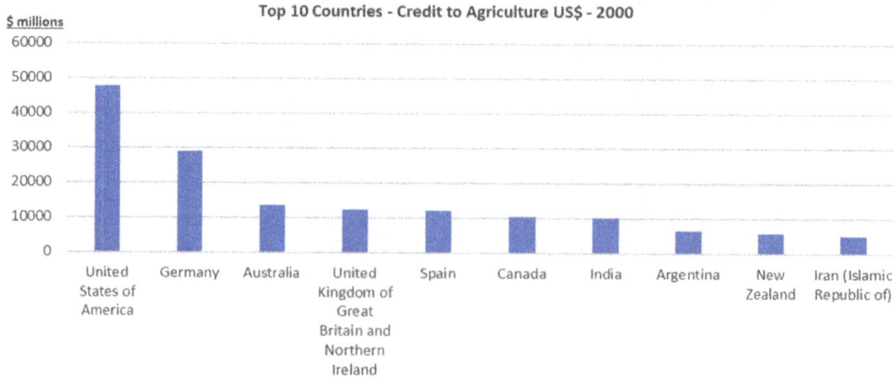

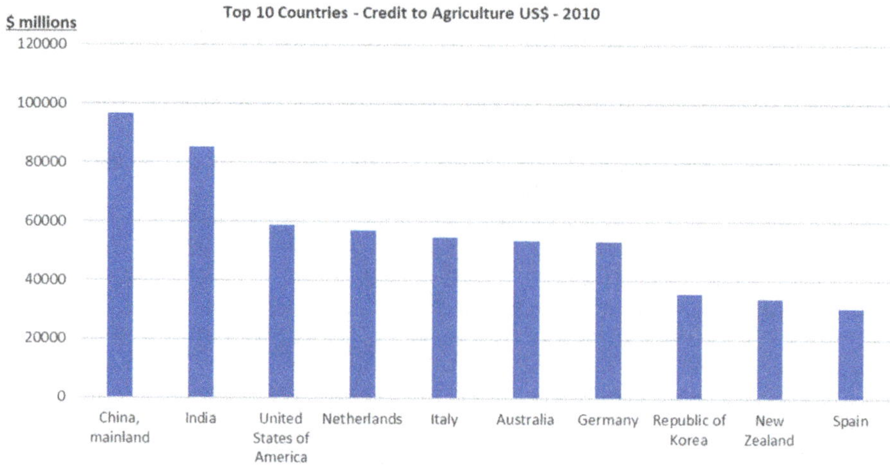

Fig. 7.13 Credit to agriculture—2010

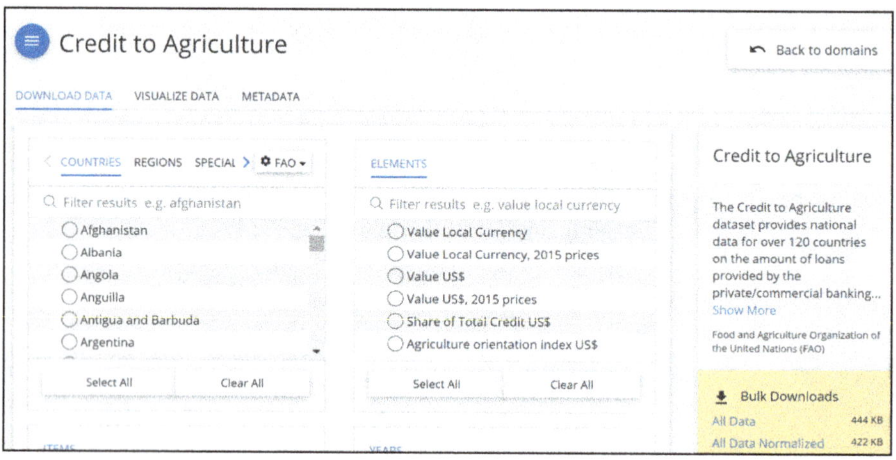

Fig. 7.14 Credit to agriculture

1991

During 1991, the top countries that have budgets and offered credit to agriculture include Germany, the United States, Canada, Iran, Thailand, Israel, New Zealand, Egypt, Tunisia, and Nigeria (please see Fig. 7.15).

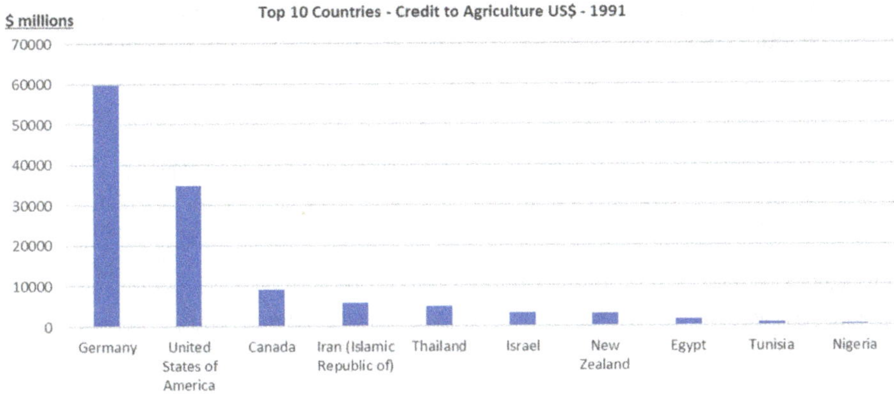

Please note: Iran wasn't there in the list in 2010.

The World Bank: Accessing Finance

The accessing finance indicator measures laws and regulations on the use of warehouse receipts as collateral to obtain credit, as well as key aspects related to financial inclusion. Farmers need access to credit to purchase inputs such as seed and fertilizer, as well as to finance harvesting, processing, and transporting operations. For domestic farmers who may not have traditional immovable collateral, warehouse receipt financing can be an effective tool to access credit.

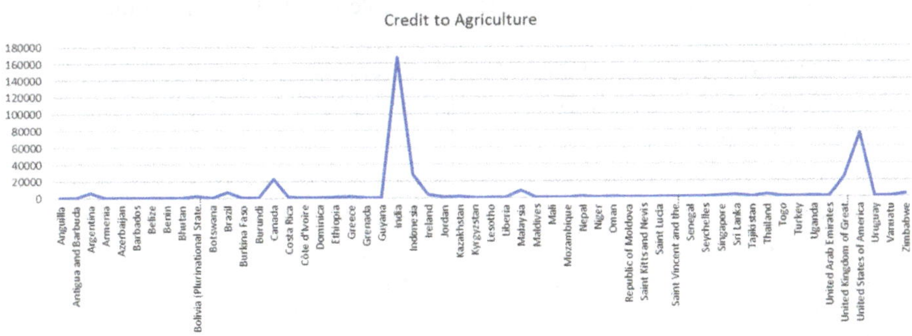

Fig. 7.15 World—credit to agriculture

The Credit to Agriculture dataset provides national data for over 120 countries on the amount of loans provided by the private/commercial banking sector to producers in agriculture, forestry, and fisheries, including household producers, cooperatives, and agrobusinesses. For some countries, the three subsectors of agriculture, forestry, and fishing are completely specified. In other cases, complete disaggregation is not available. The dataset also provides statistics on the total credit to all industries, indicators on the share of credit to agricultural producers, and an agriculture orientation index (the agriculture share of credit, over the agriculture share of GDP).

Data Sources:[40]

Countries: select countries that you would like to know contributed to credit to agriculture (please see Fig. 7.14) Selected option: select all countries	Elements Select value US $	Items: Credit to agriculture
Years Select all years	Out Type: Table	File Type: CSV
Thousand separator in "Show Data" None	Output formatting Select flags, codes, and units	

Please find FAO Credit to Agriculture[41] (please see Figs. 7.15 and 7.16).

Credit to Agriculture, Forestry, and Fishing Value US$ by country[42]

Government: A Major Risk Bearer

The government is a major risk bearer in agricultural prices. The establishment of minimum support price or minimum guarantee price ensures the farmers are not affected by the fluctuations in the price movements of commodities.[43]

Government Credit to Farmers has a huge advantage for both small farmers and farm operators:[44]

- Large federal payments to farmers mitigate the negative effect of lower prices on farm financial conditions and will play a key role in stabilizing farm income and balance sheets.
- Government credit to farmers will provide diverse market opportunities to small farmers and in turn enable small farmers to repay debt through off-farm economic conditions.

[40] Credit to Agriculture - http://www.fao.org/faostat/en/#data/IC

[41] Credit to Agriculture - http://www.fao.org/faostat/en/#data/IC

[42] Credit to Agriculture - http://www.fao.org/faostat/en/#data/IC/visualize

[43] The Futures Market for Farm Commodities – What it means to a Farmer - https://files.stlouisfed.org/files/htdocs/publications/review/74/08/Futures_Aug1974.pdf

[44] Agriculture Credit - https://www.govinfo.gov/content/pkg/CHRG-107shrg81040/html/CHRG-107shrg81040.htm

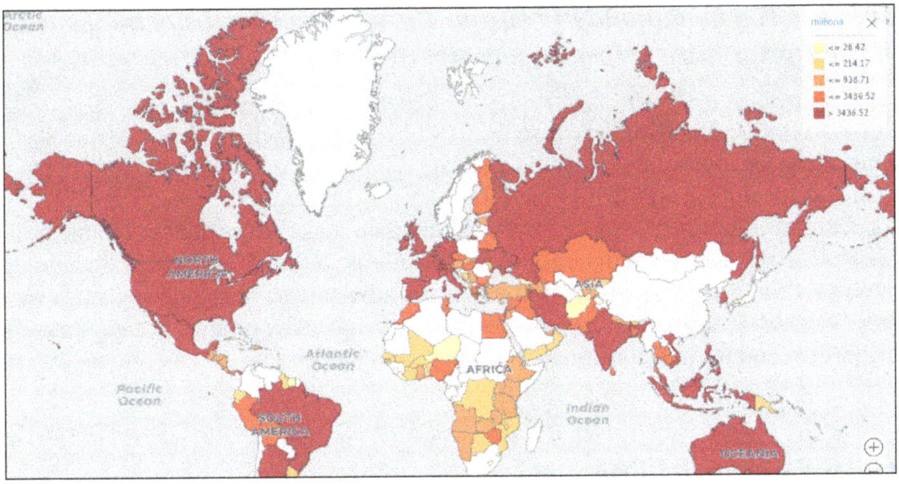

Fig. 7.16 Heatmap—credit to agriculture

- Third, government payments not only contribute to farm income but also impact farm debt and assets. Additionally, government payments also help farmers to meet debt repayment obligations.
- Finally, farmers and their lenders continue to benefit from strong balance sheets bolstered by high levels of government support. Thanks largely to this support, there remains no widespread evidence of financial distress even as the sector experiences the effects of COVID-19 and low prices for many farm commodities.

Technology Policies

Season-Average Price Forecasts[45]

Farmers and policy makers are interested in season-average forecasts of prices received by farmers for corn, soybeans, and wheat. These forecasts are also needed to compute Price Loss Coverage (PLC) payment rates and Agricultural Risk Coverage (ARC) payments that began in 2014/2015 under the 2014 Farm Act and continued under the 2018 Farm Act. For more information on the PLC and ARC programs, see Agricultural Act of 2014: Highlights and Implications and Agricultural Improvement Act of 2018: Highlights and Implications and USDA's Farm Service Agency (FSA).

[45] Season-Average Price Forecasts - https://www.ers.usda.gov/data-products/season-average-price-forecasts/

Title I: Crop Commodity Program Provisions After Enactment of the Agriculture Improvement Act of 2018

The Agriculture Improvement Act of 2018 (2018 Farm Bill) was not a watershed legislation for crop commodity programs.[46] Following a major overhaul of programs in the 2014 Farm Bill (which took 4 years to enact), the House and Senate Agriculture committees chose to tweak rather than make major changes to the programs in 2018 and to continue some modifications to the 2014 Farm Bill made in the Bipartisan Budget Act of 2018. Despite the adjustments being modest, however, they do address issues with commodity programs raised by producers and others in the farm economy, and some may increase benefits.

Major Commodity Programs

The Price Loss Coverage (PLC) program provides income support payments to farmers when the effective price falls below a certain level under the 2018 Farm Bill. Price Loss Coverage (PLC) and Agriculture Risk Coverage (ARC) were first authorized in the Agricultural Act of 2014. The Marketing Assistance Loan Program is the descendant of the commodity loan program established under the Agricultural Adjustment Act of 1938. Participants in these programs must be actively engaged in farming, must meet the Adjusted Gross Income (AGI) eligibility limits, and, where payment imitations were continued by the 2018 Farm Bill, may not receive payments above the established limit.

PLC program payments[47] are issued when the effective price of a covered commodity is less than the effective reference price for that commodity. The effective price equals the higher of the market year average price (MYA) or the national average loan rate for the covered commodity. PLC payments are issued to owners of historical base acres and are not tied to current production of covered commodities. Covered commodities include wheat, corn, sorghum, barley, oats, seed cotton, long- and medium-grain rice, certain pulses, soybeans/other oilseeds, and peanuts (please see Table 7.1).

[46] Title I: Crop Commodity Program Provisions After Enactment of the Agriculture Improvement Act of 2016 - https://www.ers.usda.gov/topics/farm-economy/farm-commodity-policy/title-i-crop-commodity-program-provisions-after-enactment-of-the-agriculture-improvement-act-of-2018/

[47] Title I: Crop Commodity Program Provisions After Enactment of the Agriculture Improvement Act of 2018 - https://www.ers.usda.gov/topics/farm-economy/farm-commodity-policy/title-i-crop-commodity-program-provisions-after-enactment-of-the-agriculture-improvement-act-of-2018/

Table 7.1 Commodity prices

Commodity	Unit	2018 Farm Bill Reference Prices: Crop Years 2019–2023**	2018 Farm Bill National Average Loan Rates: Crop Years 2019–2023
Wheat	Bushel	5.50	3.38
Rice	Hundredweight	14.00	7.00
Corn	Bushel	3.70	2.20
Oats	Bushel	2.40	2.00
Barley***	Bushel	4.95	2.50
Sorghum	Bushel	3.95	2.20
Seed cotton	lb.	.367	****
Peanuts	Ton	535	355
Soybeans	Bushel	8.40	6.20
Other oilseeds*****	Hundredweight	20.15	10.09
Dry peas	Hundredweight	11.00	6.15
Lentils	Hundredweight	19.97	13.00
Chickpeas, small	Hundredweight	19.04	10.00
Chickpeas, large	Hundredweight	21.54	14.00
Cotton, upland******	lb.	–	0.45–0.52
Cotton, extra-long staple	lb.	–	0.95
Wool, graded	lb.	–	1.15
Wool, non-graded	lb.	–	0.40
Mohair	lb.	–	4.20
Honey	lb.	–	0.69

** Eeffective reference prices may be higher if average market year prices are higher than statutory reference prices.
*** Barley price is based on all-barley price.
**** Seed cotton is unginned cotton - a seed cotton loan is a recourse loan that must be repaid at full principal plus interest.
*****Other oilseeds are sunflower seed, rapeseed, canola, safflower, flaxseed, mustard seed, crambe, sesame seed, and other oilseeds designated by the U.S. Department of Agriculture. Not all other oilseeds have the same reference price.
******Loan rate for upland cotton is set by statutory formulas and cannot be less than $0.45 per pound or more than $0.52 per pound.

Agriculture Credit Policy: India

Agriculture is a key sector of Indian economy in view of its contribution to employment and GDP. Agricultural credit plays a vital role in farm sector development and facilitates adoption of new technologies.[48] However, any amount of credit even at the most reasonable rates cannot guarantee higher productivity or adequate income

[48] Report of the Internal Working Group to Review Agricultural Credit - https://www.rbi.org.in/Scripts/PublicationReportDetails.aspx?UrlPage=&ID=942#:~:text=From%20%E2%82%B937.71%20billion%20in,%E2%82%B986254.25%20billion%20and%20approximately

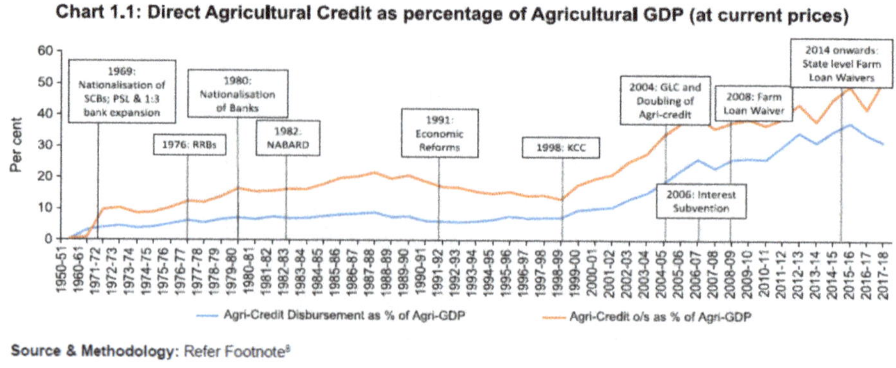

Fig. 7.17 Direct credit to agriculture [22]

among farmers, as the success depends upon many other supporting factors including the availability of agricultural inputs, services, and remunerative markets for the products [21].

The agricultural credit policies designed and implemented in India are mainly supply driven through targeted ground-level credit, interest subvention scheme, and directed lending by way of regulatory prescription under Priority Sector Lending guidelines. These policies along with other policy interventions at the level of Government and Reserve Bank of India (RBI) have yielded commendable results in the field of agricultural credit. However, agricultural sector still faces challenges such as lack of capital formation, regional disparity, dependence of farmers especially small and marginal farmers, tenant farmers, landless laborers and sharecroppers on non-institutional sources of credit at significantly higher rates, non-realization of the fair price for agricultural produce causing farmers' distress, and farm loan waivers impacting credit culture and weakening state finances (please see Fig. 7.17).

The agriculture sector is an important component of the Indian economy as it provides livelihood to a large section of the population. According to Census 2011, out of the total workers of 481.7 million, there are 118.7 million cultivators and 144.3 million agricultural laborers, which means approximately 55% of the total workers were employed in agriculture and allied sector. However, the percentage share of workers engaged in agriculture sector has been declining. As per Labour Bureau Report 2015–2016, 46.1% of the working population was employed in agriculture and allied sector.

1. Further, as per an ILO estimate, employment in agriculture sector as percentage of the total employment was approximately 44% in the year 2018[49] [22].
2. As per Agriculture Census 2015–2016, the total number of operational holdings in the country was 146 million, and the total operated area was 157.14 million

[49] Report of the Internal Working Group to Review Agricultural Credit - https://www.rbi.org.in/Scripts/PublicationReportDetails.aspx?UrlPage=&ID=942#:~:text=From%20%E2%82%B937.71%20billion%20in,%E2%82%B986254.25%20billion%20and%20approximately

hectares in 2015–2016. The small and marginal holdings taken together (0.00–2.00 ha) constituted 86.21%, while their share in the operated area stood at 47.34% in 2015–2016. The average size of land holding in 2015–2016 was 1.08 ha.

3. Agriculture plays a significant role in the development of the Indian economy. However, the contribution of agriculture to GDP has gone down from 52% in the 1950s to 30% in the 1990s and further below 20% from 2010 onward. In 2018–2019, the share of Agriculture and Allied GVA in the overall GVA at 2011–2012 prices was 14.4% and at current prices was 16.14%.

Special Thrust Programs

The government plays an important role in promoting agriculture and to the economic development of the country's most important industry, the agriculture. To create development and technology transfer program for rural poor agriculture farms, the governments of countries create special thrust programs to promote the ownership and technology enablement of rice cultivation. The outcome of special thrust programs[50] is the creation of central agricultural universities as well as fertilizer and equipment supply to farmers [22].

Poverty Headcount Ratio at $1.90 a Day (2011 PPP) (% of Population): World

Poverty headcount ratio at $1.90 a day is the percentage of the population living on less than $1.90 a day at 2011 international prices. As a result of revisions in PPP exchange rates, poverty rates for individual countries cannot be compared with poverty rates reported in earlier editions.

[50] Special Thrust Programs - http://www.fao.org/3/x6905e/x6905e09.htm

[51] Poverty Index - https://data.worldbank.org/indicator/SI.POV.DDAY?contextual=max&end=2017&locations=1W&start=1988&view=chart

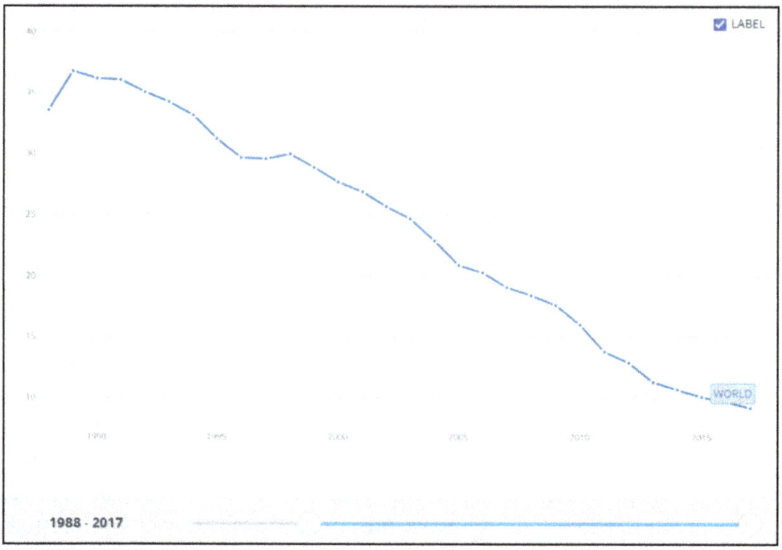

Source: The World Bank[51]

Development Relevance

The World Bank Group is committed to reducing extreme poverty to 3% or less, globally, by 2030. Monitoring poverty is important on the global development agenda as well as on the national development agenda of many countries. The World Bank produced its first global poverty estimates for developing countries for the World Development Report 1990: Poverty (World Bank 1990) using household survey data for 22 countries (Ravallion, Datt, and van de Walle 1991). Since then, there has been considerable expansion in the number of countries conducting household income and expenditure surveys. The World Bank's Development Research Group maintains a database that is updated annually as new survey data become available (and thus may contain more recent data or revisions) and conducts a major reassessment of progress against poverty every year. PovcalNet is an interactive computational tool that allows users to replicate these internationally comparable \$1.90, \$3.20, and \$5.50 a day global, regional, and country-level poverty estimates and to compute poverty measures for custom country groupings and for different poverty lines. The Poverty and Equity Data Portal provides access to the database and user-friendly dashboards with graphs and interactive maps that visualize trends in key poverty and inequality indicators for different regions and countries. The country dashboards display trends in poverty measures based on the national poverty lines alongside the internationally comparable estimates, produced from and consistent with PovcalNet.

General Comments: The World Bank's internationally comparable poverty monitoring database now draws on income or detailed consumption data from more than 1600 household surveys across 164 countries in 6 regions and 25 other high-income countries (industrialized economies). While income distribution data are published for all countries with data available, poverty data are published for low- and middle-income countries and countries eligible to receive loans from the World Bank (such as Chile) and recently graduated countries (such as Estonia) only. The aggregated numbers for low- and middle-income countries correspond to a total of six regions in PovcalNet, which include low- and middle-income countries and countries eligible to receive loans from the World Bank (such as Chile) and recently graduated countries (such as Estonia). See PovcalNet (iresearch.worldbank.org/PovcalNet/WhatIsNew.aspx) for definitions of geographical regions and industrialized countries.

Limitations and Exceptions: Despite progress in the last decade, the challenges of measuring poverty remain. The timeliness, frequency, quality, and comparability of household surveys need to increase substantially, particularly in the poorest countries. The availability and quality of poverty monitoring data remains low in small states, countries with fragile situations, and low-income countries and even some middle-income countries. The low frequency and lack of comparability of the data available in some countries create uncertainty over the magnitude of poverty reduction. Besides the frequency and timeliness of survey data, other data quality issues arise in measuring household living standards. The surveys ask detailed questions on sources of income and how it was spent, which must be carefully recorded by trained personnel. Income is generally more difficult to measure accurately, and consumption comes closer to the notion of living standards. And income can vary over time even if living standards do not. But consumption data are not always available: the latest estimates reported here use consumption data for about two-thirds of countries. However, even similar surveys may not be strictly comparable because of differences in timing or in the quality and training of enumerators. Comparisons of countries at different levels of development also pose a potential problem because of differences in the relative importance of the consumption of nonmarket goods. The local market value of all consumption in kind (including own production, particularly important in underdeveloped rural economies) should be included in the total consumption expenditure but may not be. Most survey data now include valuations for consumption or income from own production, but valuation methods vary.

Long Definition: Poverty headcount ratio at $1.90 a day is the percentage of the population living on less than $1.90 a day at 2011 international prices. As a result of revisions in PPP exchange rates, poverty rates for individual countries cannot be compared with poverty rates reported in earlier editions.

Periodicity: Annual

Statistical Concept and Methodology: International comparisons of poverty estimates entail both conceptual and practical problems. Countries have different definitions of poverty, and consistent comparisons across countries can be difficult. Local poverty lines tend to have higher purchasing power in rich countries, where more generous standards are used, than in poor countries. Since the World

Development Report 1990, the World Bank has aimed to apply a common standard in measuring extreme poverty, anchored to what poverty means in the world's poorest countries. The welfare of people living in different countries can be measured on a common scale by adjusting for differences in the purchasing power of currencies. The commonly used $1 a day standard, measured in 1985 international prices and adjusted to local currency using purchasing power parities (PPPs), was chosen for the World Development Report 1990 because it was typical of the poverty lines in low-income countries at the time. As differences in the cost of living across the world evolve, the international poverty line has to be periodically updated using new PPP price data to reflect these changes. The last change was in October 2015, when we adopted $1.90 as the international poverty line using the 2011 PPP. Prior to that, the 2008 update set the international poverty line at $1.25 using the 2005 PPP. Poverty measures based on international poverty lines attempt to hold the real value of the poverty line constant across countries, as is done when making comparisons over time. The $3.20 poverty line is derived from typical national poverty lines in countries classified as lower middle income. The $5.50 poverty line is derived from typical national poverty lines in countries classified as upper middle income. Early editions of World Development Indicators used PPPs from the Penn World Tables to convert values in local currency to equivalent purchasing power measured in US dollars. Later editions used 1993, 2005, and 2011 consumption PPP estimates produced by the World Bank. The current extreme poverty line is set at $1.90 a day in 2011 PPP terms, which represents the mean of the poverty lines found in 15 of the poorest countries ranked by per capita consumption. The new poverty line maintains the same standard for extreme poverty—the poverty line typical of the poorest countries in the world—but updates it using the latest information on the cost of living in developing countries. As a result of revisions in PPP exchange rates, poverty rates for individual countries cannot be compared with poverty rates reported in earlier editions. The statistics reported here are based on consumption data or, when unavailable, on income surveys. Analysis of some 20 countries for which income and consumption expenditure data were both available from the same surveys found income to yield a higher mean than consumption but also higher inequality. When poverty measures based on consumption and income were compared, the two effects roughly cancelled each other out: there was no significant statistical difference.

Topic: Poverty—Poverty Rates

Import Duties

Changes in import duties and import taxes are some policy tools available to influence local or national agriculture. For example, India would like to improve local agricultural industry by raising import taxes on commodities. India is considering raising import taxes on edible oils as the country seeks to become self-reliant by boosting local oilseed production with the help of tax revenues, two government

[52] India plans to hike edible oil import taxes to boost local supply, sources say - https://economictimes.indiatimes.com/news/economy/foreign-trade/india-plans-to-hike-edible-oil-import-taxes-to-boost-local-supply-sources-say/articleshow/76266851.cms?utm_source=contentofinterest&utm_medium=text&utm_campaign=cppst

[53] Milk production and per capita availability of milk in India - https://www.nddb.coop/information/stats/milkprodindia

and two industry officials told Reuters. The tax hike could curb the world's biggest vegetable oil importer's purchases and weigh on Malaysian palm oil, along with soy and sunflower oil prices while propping up local prices of oilseeds such as rapeseed, soybean, and ground nut.[52]

Use Case: The Role of Government Spending in Milk Production

Let's look at the role of government spending in ensuring milk availability per capita and achieving minimum availability of milk to all.[53] To analyze the role of government spending in improving milk productivity, the following datasets are used:

Data Sources

The following major sources are used to yield the importance of agriculture credit to improve the outcome of dairy farmers:

Data source	Description	
Milk Production—National Dairy Development Board (NDDB)[54]	The National Dairy Development Board (NDDB) of India was created on the conviction that socioeconomic progress lies largely on the development of rural India The Dairy Board was created to promote, finance, and support producer-owned and controlled organizations. NDDB's programs and activities seek to strengthen farmer-owned institutions and support national policies that are favorable to the growth of such institutions. Fundamental to NDDB's efforts are cooperative strategies and principles	
Food and Agriculture Organization of the United States[55]	The Food and Agriculture Organization (FAO) is a specialized agency of the United Nations that leads international efforts to defeat hunger. Our goal is to achieve food security for all and make sure that people have regular access to enough high-quality food to lead active, healthy lives. With over 194 member states, the FAO works in over 130 countries worldwide. We believe that everyone can play a part in ending hunger[56]	

Please find the summary of data sources and list of key features in the following figure:

[54] National Dairy Development Board - https://www.nddb.coop/information/stats/milkprodindia

[55] FAO STAT - http://www.fao.org/faostat/en/#data

[56] About FAOSTAT - http://www.fao.org/about/en

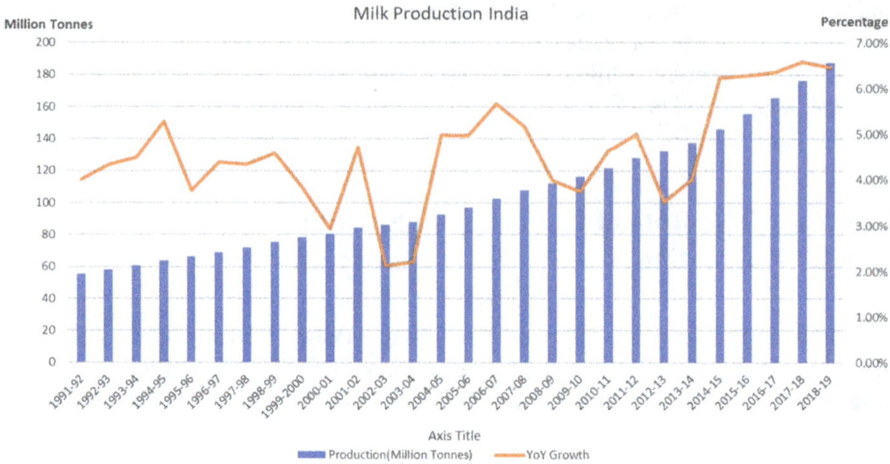

Fig. 7.18 Milk production in India

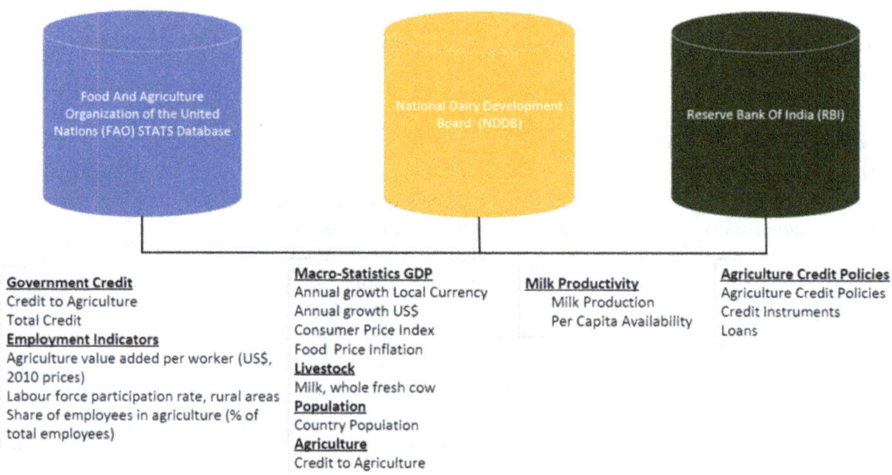

Milk Production Data

The milk production data is retrieved from the NDDB website. The data contains milk production from 1991 to 2019. Additionally, the milk production growth rate is increasing year after year—with an average growth of 4.59% (please see Fig. 7.18).

[57] FAO STAT Livestock Primary - http://www.fao.org/faostat/en/#data/QL

[58] FAO Macro indicators data - http://www.fao.org/faostat/en/#data

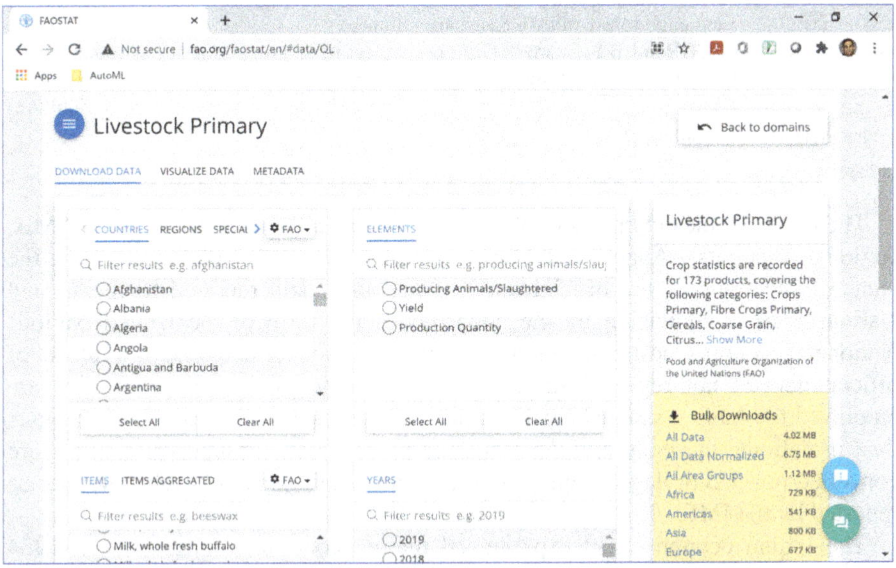

Fig. 7.19 Live Stock FAO STATS

Livestock Data

The livestock data is retrieved from the FAOSTAT database. Livestock Primary[57] are retrieved for India (please see Fig. 7.19).

Countries: select India	Elements: select production quantity	Items (FAO Coding) aggregated: Select item: milk, whole fresh cow.
Years: select all years	Output type: table File type: CSV	

Macro Indicators Data

Select Macro Indicators from Macro-Statistics.[58]

[59] UNSD AMA - https://unstats.un.org/unsd/snaama/

Countries: select India	Elements: select value US$, share of GDP in US $, Share of GDP in Local Currency, Annual Growth US$	Items : Select item: gross domestic product value in US $ (millions), value added (agriculture), and gross output (agriculture)
Years: select all years	Output type: table File type: CSV	

The Gross Domestic Product Value in US $ is from the data from UNSD AMA, National Accounts—Analysis of Main Aggregates (AMA).[59] Please note that the annual collections of the official national accounts data reported to the United Nations Statistics Division by the countries in the form of the United Nations National Accounts Questionnaires serve as the basis for the database. If a full set of official data is not reported for a specific country, estimation procedures are employed to obtain estimates for the entire time series. Once the national accounts data in national currency are complete, the current and constant price series are converted into US dollars by applying the corresponding market exchange rates as reported by the IMF.

The Indian economy had experienced major economic contractions in 1984 (−0.41%), 1991 (−11.53%), 1993 (−2.19%), and 2012 (−0.59%). The good story, on the other hand, was that the Indian economy had witnessed major double-digit expansions in 1973 (19.39%), 1974 (12.81%), 1977 (16.81%), 1978 (15.53%), 1979 (10.57%), 1980 (23.00%), 1987 (10.59%), 1988 (10.51%), 1994 (14.01%), 1995 (13.51%), 2003 (16.91%), 2004 (17.29%), 2005 (17.14%), 2006 (14.02%), 2007 (26.16%), 2010 (26.95%), 2011 (12.12%), and 2017 (14.82%).

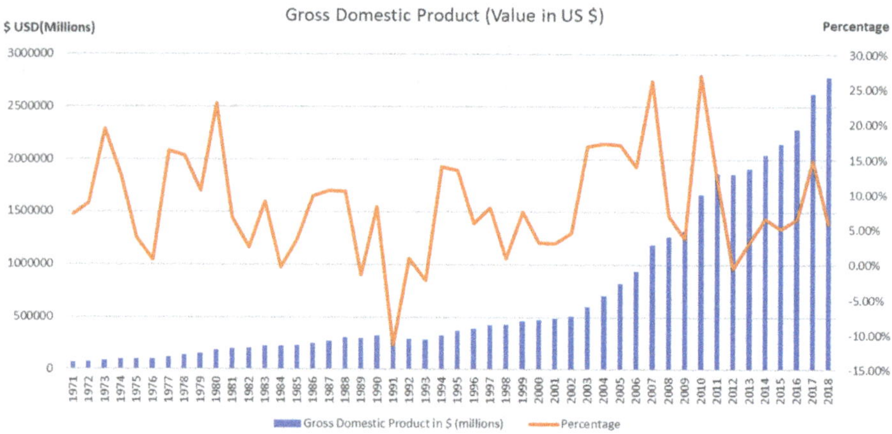

Massive expansions happen in the years 1980, 2007, and 2010.

Gross output of agriculture is seen in 2016 (only 2012–2016 data available).

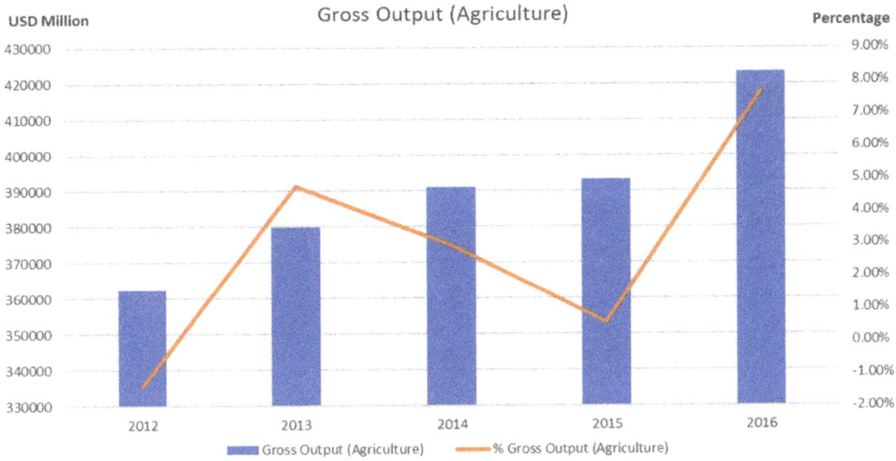

Consumer Price Indices Data

Select Consumer Price Index from Prices Data element.[60]

Countries: select India[61]	Moths: select all	Items : Select item: food price inflation (%)
Years: select all years	Output type: table File type: CSV	

Please find food price inflation[62] in India from 2001 to 2020. On an average, it can be observed that the food price inflation is around 7.5%.

[60] FAO Consumer Price Indices - http://www.fao.org/faostat/en/#data

[61] CPI - http://www.fao.org/faostat/en/#data/CP

[62] Food Inflation in India - https://www.elibrary.imf.org/view/IMF071/22825-9781513541259/22825-9781513541259/ch03.xml?language=en&redirect=true

[63] Total Population - http://www.fao.org/faostat/en/#data/OA

Total Population Data

The total population of India is collected from the FAO site.[63]

Countries: select India	Element: select all	Items : Select item: food price inflation (%)
Years: select all years	Output type: table File type: CSV	

The population of India is 1.3 billion in 2018 and 1.366 billion in 2019.

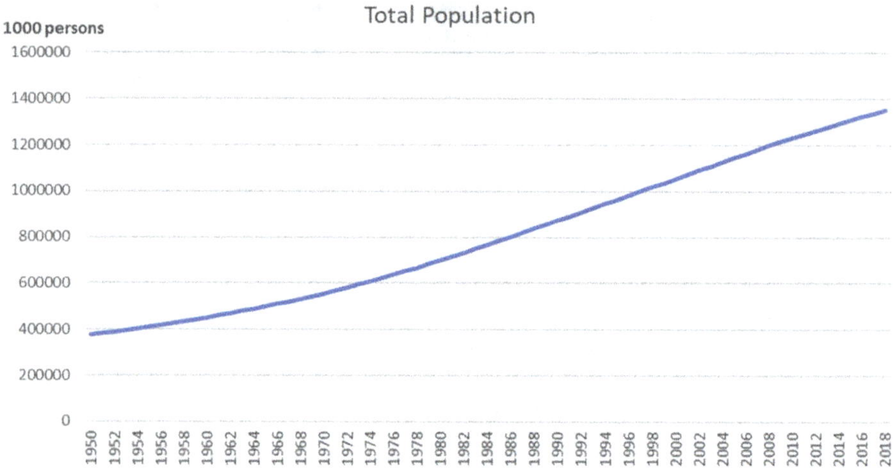

Credit to Agriculture

The credit to agriculture is collected from the FAO site (under investment).[64]

Countries: select India	Element: select value US $	Items : Select item: credit to agriculture and credit to agriculture and fishery
Years: select all years	Output type: Table File type: CSV	

The Indian Government is increasing credit to agriculture at an annual rate of 11.54% (average credit from 1997 to 2018) with a rapid increase in credit disbursed in 2005 (23.95%), 2006 (21.18%), and 2007 (20.27%).

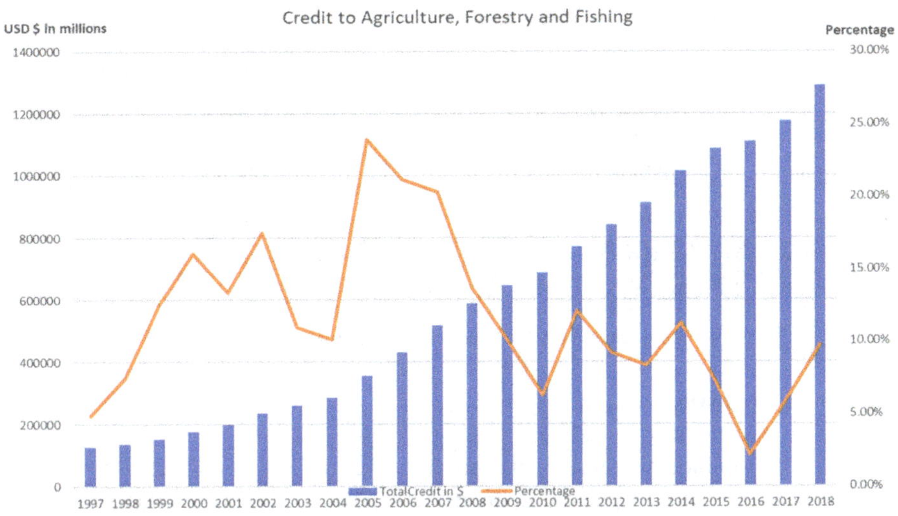

The ratio of outstanding Agri-Credit to Agri-GDP jumped from 0.6% in 1950–1951 to 9.81% in 1971–1972. Post-1972, the ratio shows an upward trend up to 1987–1988 increasing to 21.76%. The impressive achievement of agricultural credit against agricultural GDP during the 1950s–1980s is on account of nationalization of banks and introduction of RRBs, which expanded the reach of formal credit in the country. However, the reverse trend in the ratio started from 1990 to 1991 onward, and it fell to 13.34% in 1998–1999. Post-1999, the ratio increased steeply and reached up to 39.55% in 2006–2007, which indicates that introduction of KCC was a big booster for agricultural credit and brought about a sea change in improving the reach of credit to the farming community. Many of the other policy initiatives started in 2004–2005 also played an important role. In later years, despite a fluctuating trend, it rose to 49.63% in 2015–2016 and 51.56% in 2017–2018. The chart reveals that the trend of both the outstanding Agri-credit and disbursement as percentage of Agri-GDP are largely similar except in certain periods where there is divergence between the two. The reasons could be announcement of loan waivers which negatively impacted the repayment behavior of the borrowers and made the banks averse to fresh lending.[65]

Impact of Farm Loan Waivers on Agricultural Credit

The economic rationale for loan waivers comes from alleviating the debt overhang of beneficiaries, thus enabling them to undertake productive investment and boost real economic activity (investment, production, and consumption). Loan waiver of highly indebted farming households can potentially free up lines of credit, enabling them to make new investments, provided supply of bank credit to them is not affected by the changed risk profile of the household. The economic benefit of loan

[64] Total Population - http://www.fao.org/faostat/en/#data/OA

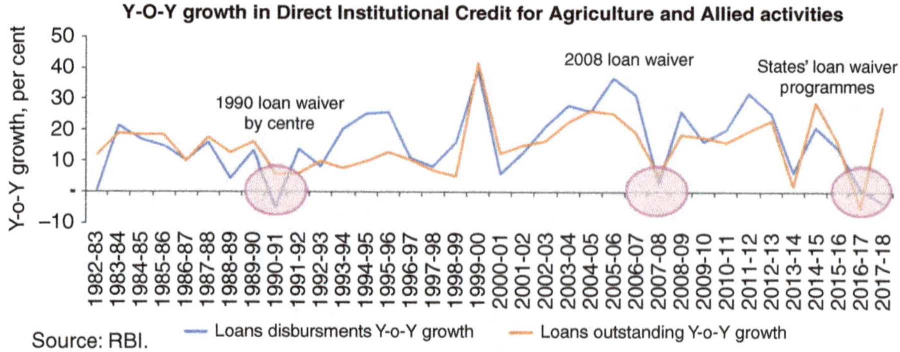

Fig. 7.20 Y-o-Y growth in direct investment [22]

waiver on households, therefore, can only be realized if credit allocation to beneficiary households is not impacted. Loan waivers, however, do impact the credit flow to agriculture due to moral hazard among both beneficiaries and non-beneficiaries of the bailout. This impact could be both in loan performance, as borrowers choose to default strategically in anticipation of future bailouts, and in credit allocation, as banks reallocate lending to lower-risk borrower segments[66] [22] (Fig. 7.20).

> The agriculture sector of China is characterized by smallholders' economy where the arable land in terms of hectares per person is 0.086 as against 0.118 in India, yet the agricultural productivity in China is very high compared to India. The primary reasons are[67] [22]:
>
> Structural initiatives
> Consolidation of farmlands
> Outsourcing of farming operations
> Institutional support to agriculture

Model Development

> The software code for this model can be found on GitHub Link: FAOSTAT Data Analysis on Milk Production data. Ipynb (Jupyter Notebook)

[65] Report of the Internal Working Group to Review Agricultural Credit - https://www.rbi.org.in/Scripts/PublicationReportDetails.aspx?UrlPage=&ID=942#CP12

[66] Report of the Internal Working Group to Review Agricultural Credit (Section 3.3) - https://www.rbi.org.in/Scripts/PublicationReportDetails.aspx?UrlPage=&ID=942#CHII34

[67] Reserve Bank of India (RBI) Report - https://www.rbi.org.in/Scripts/PublicationReportDetails.aspx?UrlPage=&ID=942#CHII34

Step 1: *Load Required Libraries*

The first step in predicting milk production based on historical GDP, populations, government expenditure, credit to agriculture, and temperature data is to explore the data. The goal of this exploration is to try to understand the relationships between its attributes; in particular, any apparent correlation between the features and the label your model will try to predict. This may require some work to detect and fix issues in the data (such as dealing with missing values, errors, or outlier values), deriving new feature columns by transforming or combining existing features (a process known as feature engineering), normalizing numeric features (values you can measure or count) so they're on a similar scale, and encoding categorical features (values that represent discrete categories) as numeric indicators. Let's start by loading the bicycle sharing data as a pandas DataFrame and viewing the first few rows.

```
import pandas as pd
import numpy as np
import plotly.graph_objects as go
import seaborn as sns; sns.set()
import matplotlib.pyplot as plt
from scipy import stats
from scipy.stats import pearsonr
import matplotlib.pyplot as plt
from sklearn.preprocessing import MinMaxScaler
from sklearn.model_selection import train_test_split
from sklearn.linear_model import LinearRegression
from sklearn.ensemble import RandomForestClassifier
from sklearn.metrics import r2_score
from sklearn import metrics

import eli5
import matplotlib.pyplot as plt
import matplotlib as mpl
import numpy as np
import scipy.stats as spstats
import seaborn as sns

%matplotlib inline
mpl.style.reload_library()
mpl.style.use('classic')
mpl.rcParams['figure.facecolor'] = (1, 1, 1, 0)
mpl.rcParams['figure.figsize'] = [6.0, 4.0]
mpl.rcParams['figure.dpi'] = 100
```

Step 2: *Import Data from the NDDB and FAO Data Sources*

Load data from the NDDB and FAO data sources:

```
# Importing data
milk_production = pd.read_excel("milk_india_nddp.xlsx")
MacroIndi_df = pd.read_csv("FAOSTAT_MacroIndicators.csv")
ConsumerPrices_df = pd.read_csv("FAOSTAT_ConsumerPrices.csv")
AnnualPopulation_df  = pd.read_csv("FAOSTAT_AnnualPopulation.csv")
CropLiveStock_df = pd.read_csv("FAOSTAT_Livestock.csv")
GovtExpenditure_df = pd.read_csv("FAOSTAT_GovtExcp.csv")
CreditAgriculture_df  = pd.read_csv("FAOSTAT_creditAgriculture.csv")
Temperature_df = pd.read_csv("FAOSTAT_Temperature.csv")
```

Run milk production data frame:

```
milk_production
```

	Year	Production (Million Tonnes)	Per Capita Availibilty (gms/day)
0	1991-92	55.6	178
1	1992-93	58.0	182
2	1993-94	60.6	186
3	1994-95	63.8	192
4	1995-96	66.2	195
5	1996-97	69.1	200
6	1997-98	72.1	205
7	1998-99	75.4	210
8	1999-2000	78.3	214
9	2000-01	80.6	217
10	2001-02	84.4	222

Validate the statistical properties of milk production data frame:

```
milk_production.describe()
```

Output:

	Production (Million Tonnes)	Per Capita Availibilty (gms/day)
count	28.000000	28.000000
mean	104.217857	254.785714
std	37.660236	60.813228
min	55.600000	178.000000
25%	74.575000	208.750000
50%	94.800000	237.000000
75%	129.025000	292.250000
max	187.700000	394.000000

The statistics reveal some information about the distribution of the data in each of the numeric fields, including the number of observations (there are 28 records), the mean and standard deviation, minimum and maximum values, and the quartile values (the threshold values for 25%, 50% which is also the median, and 75% of the data). From this, we can see that the mean number of milk production (millions of tons) is around 104; but there's a comparatively small standard deviation, indicating a not much variance in the yearly milk production!

We might get a clearer idea of the distribution of production (millions of tons) values by visualizing the data. Common plot types for visualizing numeric data distributions are histograms and box plots:

```
import pandas as pd
import matplotlib.pyplot as plt

# This ensures plots are displayed inline in the Jupyter notebook
%matplotlib inline

# Get the label column
label = milk_production['Production (Million Tonnes)']


# Create a figure for 2 subplots (2 rows, 1 column)
fig, ax = plt.subplots(2, 1, figsize = (9,12))

# Plot the histogram
ax[0].hist(label, bins=100)
ax[0].set_ylabel('Frequency')

# Add lines for the mean, median, and mode
ax[0].axvline(label.mean(), color='magenta', linestyle='dashed', linewidth=2)
ax[0].axvline(label.median(), color='cyan', linestyle='dashed', linewidth=2)

# Plot the boxplot
ax[1].boxplot(label, vert=False)
ax[1].set_xlabel('Production (Million Tonnes)')

# Add a title to the Figure
fig.suptitle('Production (Million Tonnes) Distribution')

# Show the figure
fig.show()
```

Output:

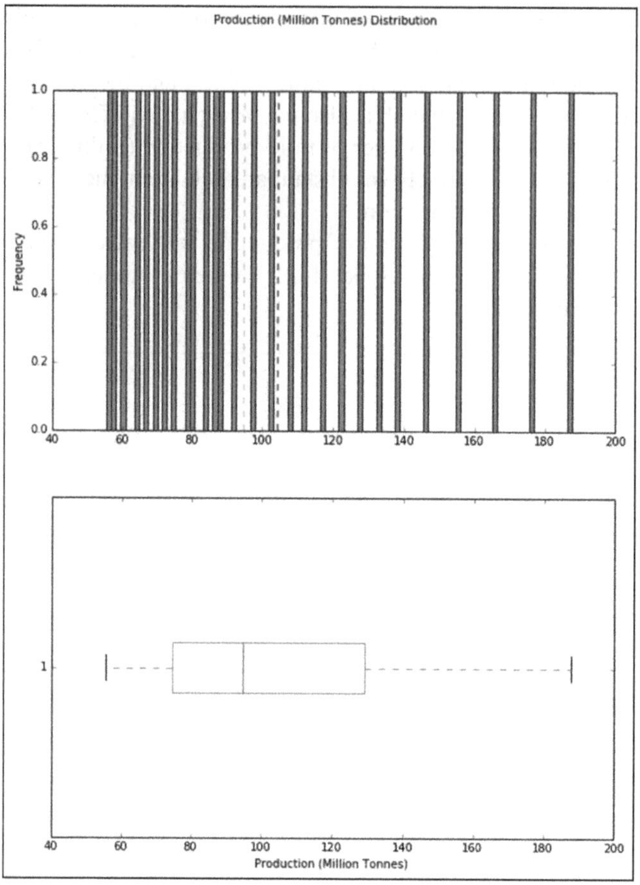

The plots show that the Production (Million Tonnes) ranges from 40 to just over 200 million tons. However, the mean (and median) number of Production (Million Tonnes) is closer to the mid of that range, with most of the data between 50 and around 100 million tons. We can do the same kind of visual exploration of the numeric features. Let's create a histogram for each of these:

```
# Plot a histogram for each numeric feature
for col in ['Per Capita Availibilty (gms/day)']:
    fig = plt.figure(figsize=(9, 6))
    ax = fig.gca()
    feature = milk_production[col]
    label = milk_production['Production (Million Tonnes)']
    correlation = feature.corr(label)
    plt.scatter(x=feature, y=label)
    plt.xlabel(col)
    plt.ylabel('Production (Million Tonnes)')
    ax.set_title('Production (Million Tonnes) vs ' + col + '- correlation: ' + str(correlation))
plt.show()
```

Output: It is very clear that the more Production, the more availability per capita, as can be seen in the below figure:

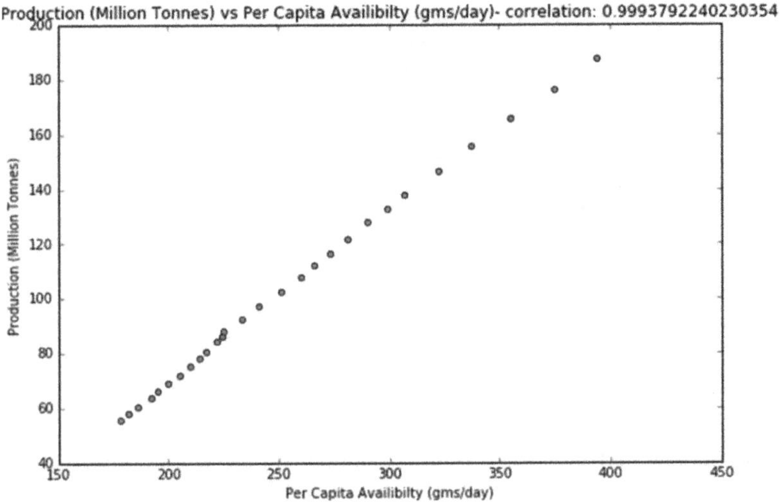

Step 3: *Combine All Data Frames (CPI, Food Inflation, Government Expenditure, Credit, Annual Population, and Temperature)*

Combine all the relevant data frames (CPI, food inflation, government expenditure, credit, annual population, and temperature) to observe the impact of agriculture production to credit to agriculture:

```
combined_df=
pd.concat([ConsumerPrices_df2,ConsumerPrices_df_infla2,GovtExpenditure_df2,CreditAgriculture_d
f3,GDP,AnnualPopulation_df3,Temperature_final1,milk_production2],axis = 1)
combined_df
```

Output:

Year	ConsumerPrice	Food price inflation	Total outlays (Central Government) in millions	TotalCredit in $	Gross Domestic Product in $	Total Population	January-Temperature	February-Temperature	March-Temperature	April-Temperature	May-Temperature
2001	37.948152	8.847675	76787.57	199158.0	490658.80	1075000.085	0.216	0.999	0.354	-0.253	0.224
2002	39.578843	8.833892	83739.83	233949.0	512774.62	1093317.189	0.475	0.562	0.849	0.868	0.846
2003	41.085163	8.833900	92081.53	259564.0	599470.43	1111523.144	-0.017	0.895	0.183	0.866	0.598
2004	42.632938	8.833900	108418.87	285889.0	703128.96	1129623.456	0.246	0.468	1.866	0.923	-0.096
2005	44.443283	8.833908	128893.50	354353.0	823611.58	1147609.927	0.453	0.599	0.747	-0.187	0.039
2006	47.019445	8.833942	146872.64	429393.0	939066.42	1165486.291	0.738	2.533	0.045	0.189	0.128
2007	50.015937	8.833933	186856.93	516422.0	1184724.49	1183209.472	0.824	0.510	-0.018	0.880	0.201
2008	54.191904	8.833933	225698.62	587078.0	1267470.12	1200669.765	0.276	-0.817	0.726	-0.058	-0.095
2009	60.089257	8.833967	223368.26	646199.0	1315230.23	1217726.215	1.584	1.657	1.062	0.850	0.388
2010	67.293593	8.878700	278811.89	686790.0	1669619.77	1234281.170	0.209	0.982	2.192	1.990	1.052
2011	73.254703	6.068508	275853.25	769552.0	1871918.01	1250287.943	-0.435	0.354	0.570	-0.491	0.253
2012	80.076508	8.834050	305060.70	840271.0	1860877.24	1265780.247	-0.252	0.023	0.269	0.203	0.650
2013	88.810968	13.117058	319371.81	910271.0	1917053.70	1280842.125	0.009	0.496	0.560	0.085	0.775
2014	94.453300	7.183483	315993.67	1012713.0	2042939.29	1295600.772	0.190	-0.459	-0.361	0.261	-0.224
2015	100.000012	5.284225	332983.20	1085751.0	2146758.61	1310152.403	-0.073	0.970	-0.186	-0.505	0.640
2016	104.941008	5.373317	362306.22	1109041.0	2286233.13	1324517.249	1.100	1.777	1.607	1.670	0.581
2017	107.554992	1.822208	398756.21	1173744.0	2625091.07	1338676.785	0.786	1.286	0.592	1.198	0.842

Correlation analysis of Production to all other features provides the following details:

	Production (million tonnes)
ConsumerPrice	**0.986393725**
Food price inflation	−0.607933523
Total outlays (Central Government) in millions	**0.972665589**
TotalCredit in $	**0.988803292**
Gross domestic product in $	**0.98282657**
Total population	**0.977391412**
January-temperature	0.06717972
February-temperature	0.058512766
March-temperature	−0.06428574
April-temperature	0.135252268
May-temperature	**0.304058012**
June-temperature	0.153432952
July-temperature	−0.080236483
August-temperature	**0.398449773**
September-temperature	0.264227315
October-temperature	**0.509863234**
November-temperature	0.039355132
December-temperature	0.019214868
Production (million tonnes)	1

It is clear that ConsumerPrice (98.6%), Total outlays (Central Government) in millions (97.2%), TotalCredit in $ (98.8%), Gross Domestic Product in $ (98.2%), and Total Population (97.7%) have a huge positive correlation with milk production. This makes sense as milk in some parts of India is a discretionary spending, and having a booming economy (higher GDP) will have a positive correlation. Additionally, the credit to agriculture plays an important role in India as the country went through economic distress and drought cycles (3 successive years of drought between 2000–2001 and 2002–2003 and 2014–2015 with poor rainfall performance[68]).

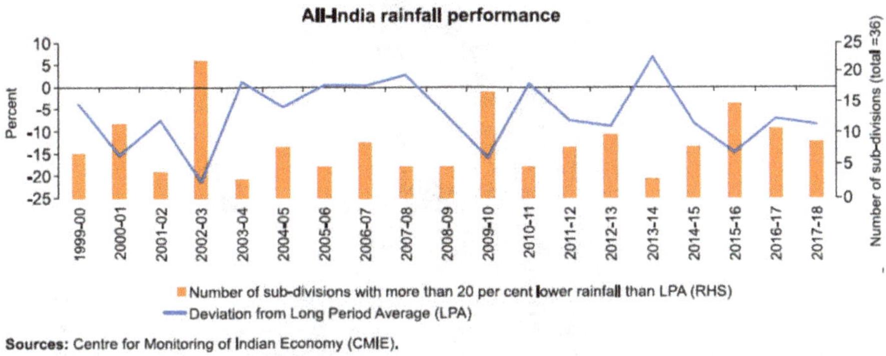

Next, milk production is negatively correlated with food price inflation (−60.7%) and highly correlated with **October temperature** and moderately correlated with May and August temperatures.

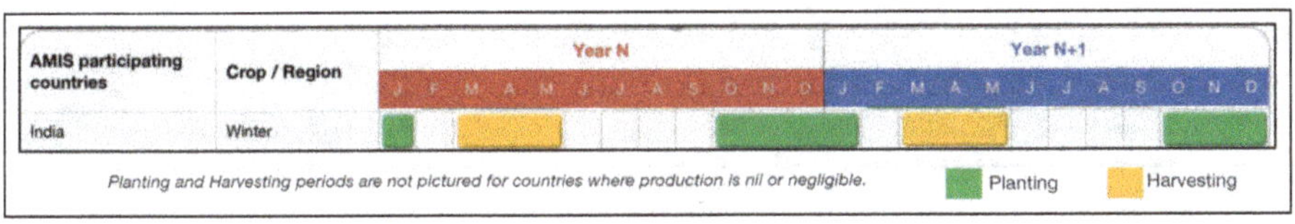

One potential reason for high correlation with October temperatures is the beginning of planting season as per Indian agriculture calendar (Appendix).[69] The same could be observed from DataFrame correlation map.

```
plt.figure(figsize=(12, 9))
sns.heatmap(combined_df1.corr())
```

[68] RBI Report—Section 3.2 - https://www.rbi.org.in/Scripts/PublicationReportDetails.aspx?UrlPage=&ID=942#CHI134

[69] AMIS Online Calendar - http://www.amis-outlook.org/fileadmin/user_upload/amis/docs/resources/AMIS-online-crop-calendar_REDUCED3.pdf

Output:

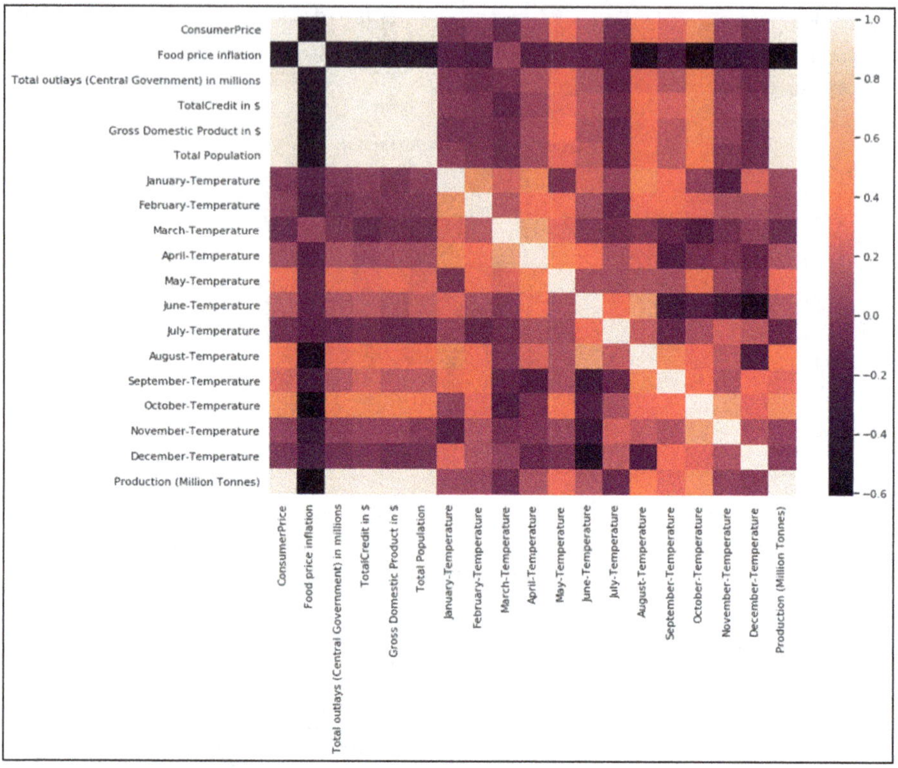
ConsumerPrice
Food price inflation
Total outlays (Central Government) in millions
TotalCredit in $
Gross Domestic Product in $
Total Population
January-Temperature
February-Temperature
March-Temperature
April-Temperature
May-Temperature
June-Temperature
July-Temperature
August-Temperature
September-Temperature
October-Temperature
November-Temperature
December-Temperature
Production (Million Tonnes)
1.0
0.8
0.6
0.4
0.2
0.0
−0.2
−0.4
−0.6

Step 4: *Feature Distribution*

By plotting the feature distribution, we can see the data trends:

```
# Create a function that we can re-use
def show_distribution(var_data,colname):
    from matplotlib import pyplot as plt

    # Get statistics
    min_val = var_data.min()
    max_val = var_data.max()
    mean_val = var_data.mean()
    med_val = var_data.median()
    mod_val = var_data.mode()[0]


print('Minimum:{:.2f}\nMean:{:.2f}\nMedian:{:.2f}\nMode:{:.2f}\nMaximum:{:.2f}\n'.format(min_v
al,
                                                                                        mean_val,
                                                                                        med_val,
                                                                                        mod_val,
                                                                                        max_val))

    # Create a figure for 2 subplots (2 rows, 1 column)
    fig, ax = plt.subplots(2, 1, figsize = (10,4))

    # Plot the histogram
    ax[0].hist(var_data)
    ax[0].set_ylabel('Frequency')

    # Add lines for the mean, median, and mode
    ax[0].axvline(x=min_val, color = 'gray', linestyle='dashed', linewidth = 2)
    ax[0].axvline(x=mean_val, color = 'cyan', linestyle='dashed', linewidth = 2)
    ax[0].axvline(x=med_val, color = 'red', linestyle='dashed', linewidth = 2)
    ax[0].axvline(x=mod_val, color = 'yellow', linestyle='dashed', linewidth = 2)
    ax[0].axvline(x=max_val, color = 'gray', linestyle='dashed', linewidth = 2)

    # Plot the boxplot
    ax[1].boxplot(var_data, vert=False)
    ax[1].set_xlabel('Value')

    # Add a title to the Figure
    fig.suptitle(colname + ' Data Distribution')

    # Show the figure
    fig.show()


# Get the variable to examine

for col in ['ConsumerPrice','Food price inflation','Total outlays (Central Government) in millions',
        'TotalCredit in $','Gross Domestic Product in $','Total Population','January-
Temperature','February-Temperature']:
    show_distribution(combined_df1[col],col)
```

Output:

Consumer Price Distribution:

The consumer price distribution is weighted toward the left side of the distribution with minimum of 37.95, mean of 66.67, median of 60.09, mode of 37.95, and maximum of 107.55.

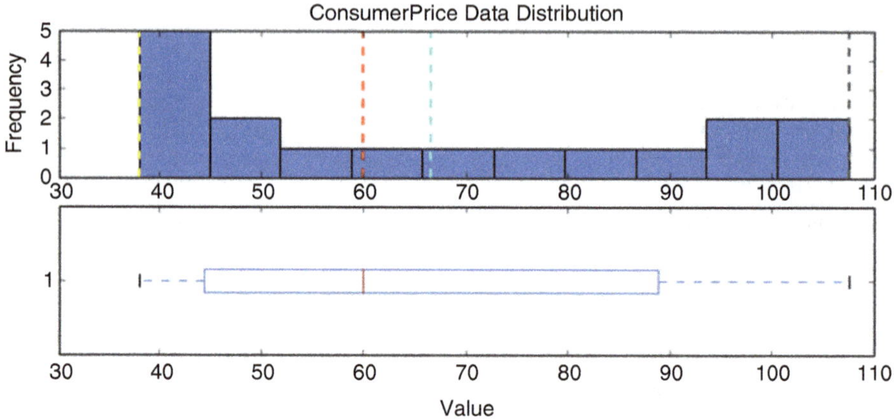

Food Price Inflation:

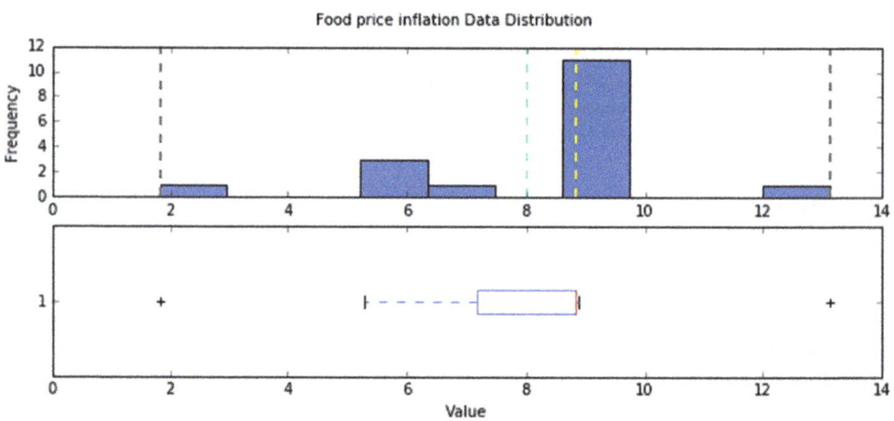

The food price inflation distribution is weighted toward the center of the distribution with minimum of 1.82, mean of 8.00, median of 8.83, mode of 8.83, and maximum of 13.12.

Total Outlays (Central Government) in Millions:

The total outlays (central government) distribution is weighted toward the center of the distribution with minimum of 76,787.57 million, mean of 227,167.92, median of 225,698.62, mode of 767,87.57, and maximum of 398,756.21.

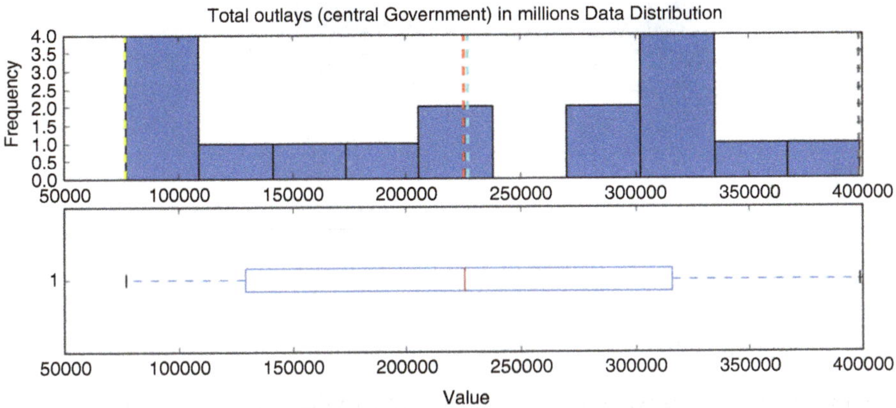

Total Credit in $:

The total credit in $ distribution is weighted toward the center of the distribution with minimum of 199,158.00 million, mean of 652,949.29, median of 646,199.00, mode of 199,158.00, and maximum of 1,173,744.00.

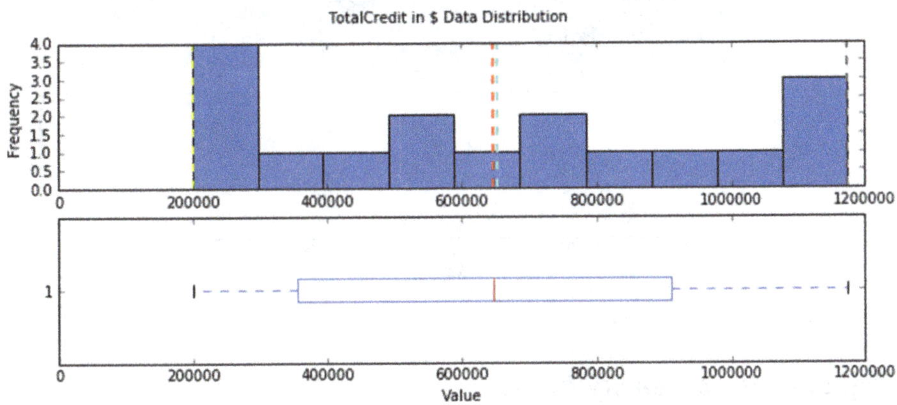

Gross Domestic Product in $:

The gross domestic product in $ distribution is weighted toward the center with minimum of 490,658.80 million, mean of 1,426,860.38, median of 1,315,230.23, mode of 490,658.80, and maximum of 2,625,091.07.

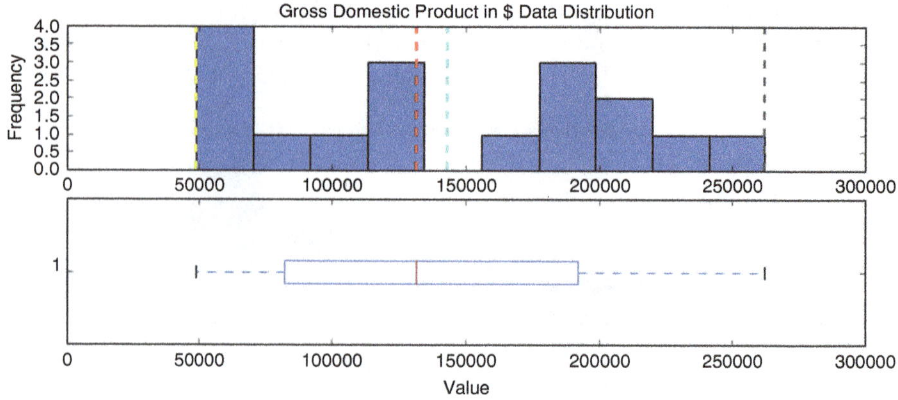

Total Population

The total population is weighted toward the center of the distribution with minimum of 1,075,000.08 billion people, mean of 1,213,194.37, median of 1,217,726.22, mode of 1,075,000.08, and maximum of 1,338,676.78.

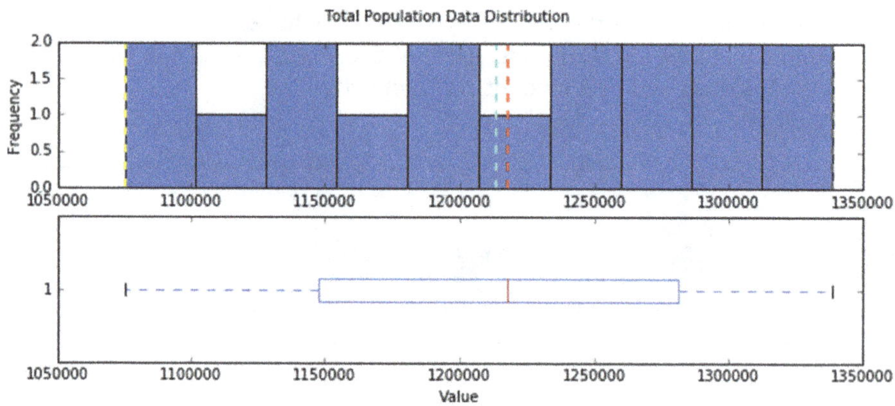

Step 5: *Feature Density Function*

By plotting the feature distribution, we can see the data skewness:

```
def show_density(var_data,colname):
  from matplotlib import pyplot as plt

  fig = plt.figure(figsize=(10,4))

  # Plot density
  var_data.plot.density()

  # Add titles and labels
  plt.title(colname + '  Density')

  # Show the mean, median, and mode
  plt.axvline(x=var_data.mean(), color = 'cyan', linestyle='dashed', linewidth = 2)
  plt.axvline(x=var_data.median(), color = 'red', linestyle='dashed', linewidth = 2)
  plt.axvline(x=var_data.mode()[0], color = 'yellow', linestyle='dashed', linewidth = 2)

  # Show the figure
  plt.show()

# Get the density of Grade

for col in ['ConsumerPrice','Food price inflation','Total outlays (Central Government) in millions',
        'TotalCredit in $','Gross Domestic Product in $','Total Population','January-
Temperature','February-Temperature'
        ,'March-Temperature','April-Temperature','May-Temperature','June-Temperature'
        ,'July-Temperature','August-Temperature','September-Temperature','October-Temperature'
        ,'November-Temperature','December-Temperature']:
  show_density(combined_df1[col],col)
```

Output:

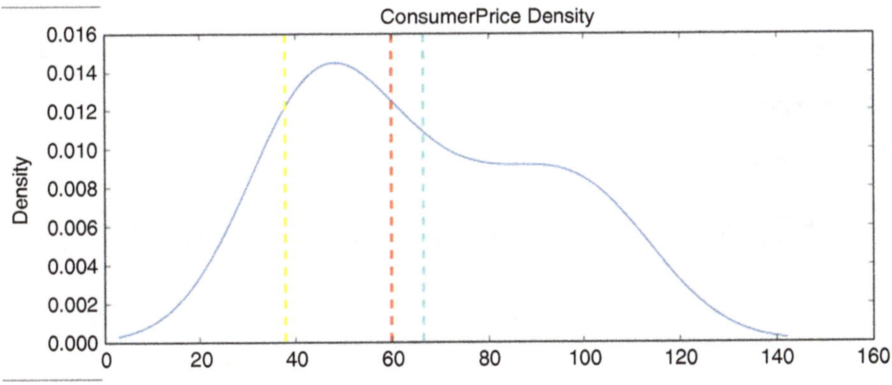

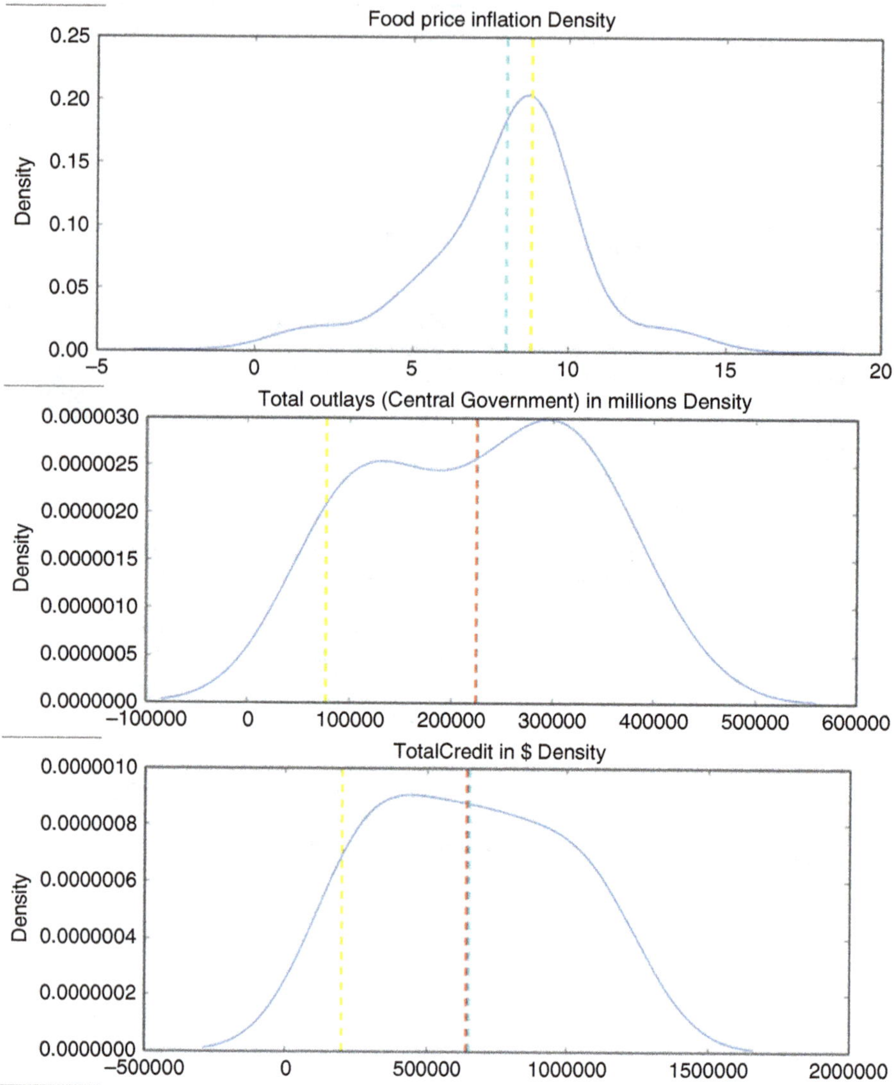
Food price inflation Density
Density
0.25
0.20
0.15
0.10
0.05
0.00
−5
0
5
10
15
20
Total outlays (Central Government) in millions Density
Density
0.0000030
0.0000025
0.0000020
0.0000015
0.0000010
0.0000005
0.0000000
−100000
0
100000
200000
300000
400000
500000
600000
TotalCredit in $ Density
Density
0.0000010
0.0000008
0.0000006
0.0000004
0.0000002
0.0000000
−500000
0
500000
1000000
1500000
2000000

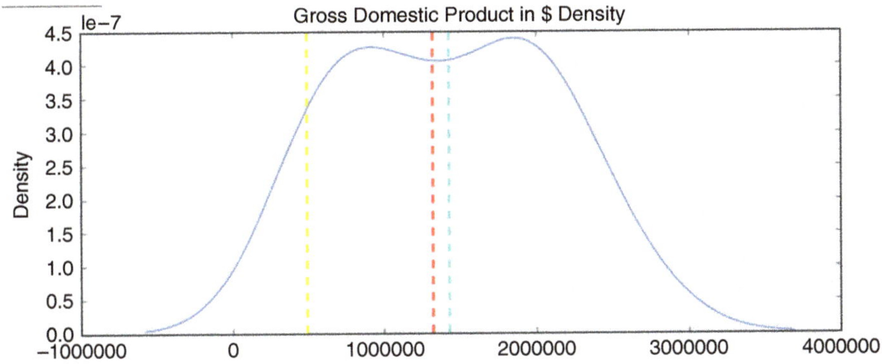

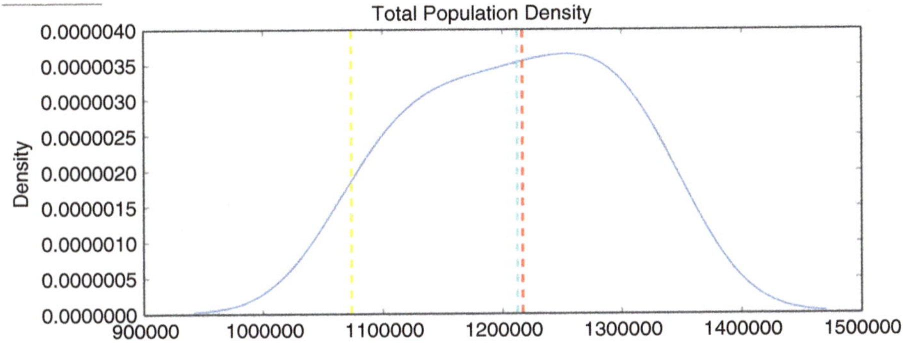

Step 6: *Train and Test Data*

Start training the model by splitting combined data into train and test datasets:

```
X_train,X_test, y_train , y_test = train_test_split(X ,y, test_size = 0.2,random_state = 0)
```

Step 7: *Regression model*

Prepare regression model:

```
regressor = LinearRegression()
regressor.fit(X_train,y_train)

# predicting the test
y_pred = regressor.predict(X_test)
y_pred
```

Output:

array([84.96847413, 107.83811424, 121.68480236, 124.93528283])

Plot the regression actual to predicted values:

```
import matplotlib.pyplot as plt

%matplotlib inline

plt.scatter(y_test, y_pred)
plt.xlabel('Actual Labels')
plt.ylabel('Predicted Labels')
plt.title('Milk Production (in Million Tonnes)')
# overlay the regression line
z = np.polyfit(y_test, y_pred, 1)
p = np.poly1d(z)
plt.plot(y_test,p(y_test), color='magenta')
plt.show()
```

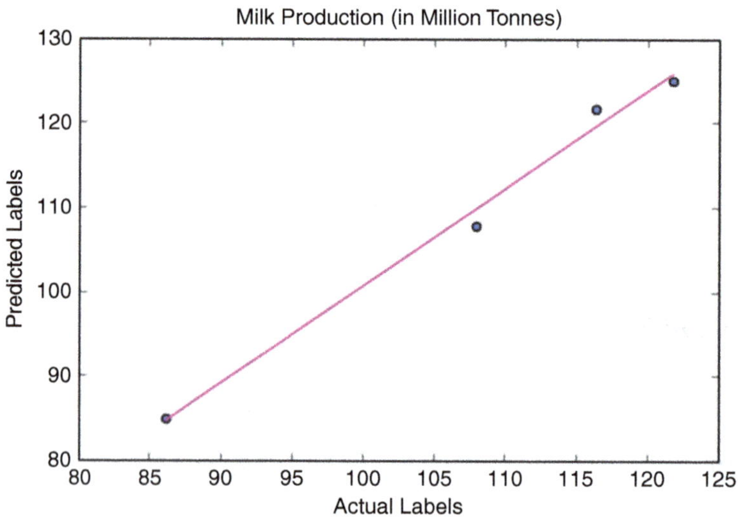


```
pred_y_df = pd.DataFrame({'Actual Value':y_test, 'Predicted value':y_pred,'Difference': y_test-
y_pred})
pred_y_df
```

Output:

	Actual Value	Predicted value	Difference
0	86.2	84.968474	1.231526
1	107.9	107.838114	0.061886
2	116.4	121.684802	-5.284802
3	121.8	124.935283	-3.135283

Regression accuracy:

```
from sklearn.metrics import mean_squared_error, r2_score

mse = mean_squared_error(y_test, y_pred)
print("MSE:", mse)

rmse = np.sqrt(mse)
print("RMSE:", rmse)

r2 = r2_score(y_test, y_pred)
print("R2:", r2)
```

Output:

MSE: 9.819905047720587
RMSE: 3.133672772916883
R^2: 0.9466474422771733

```
y_test
np.round(y_pred,2)
```

array([86.2, 107.9, 116.4, 121.8])
array([84.97, 107.84, 121.68, 124.94])

Step 8: Explainability of the *Model*

Permutation is important with Eli5 library. Eli5 is a Python library which allows to visualize and debug various machine learning models using unified API. It has a built-in support for several ML frameworks and provides a way to explain black-box models:[70]

```
eli5.show_weights(regressor)
```

[70] Machine Learning Explainability - https://maelfabien.github.io/machinelearning/Explainability/#example

Output:

y top features

Weight?	Feature
+18.816	<BIAS>
+2.995	x13
+2.947	x5
+1.725	x8
+0.966	x9
+0.705	x14
+0.520	x12
+0.208	x6
+0.173	x0
+0.000	x4
+0.000	x2
+0.000	x3
-0.199	x7
-0.587	x16
-1.113	x11
-1.208	x10
-1.370	x1
-3.224	x15

```
eli5.show_prediction(regressor,X1.iloc[1],feature_names = list(X1.columns))
```

y (score **86.443**) top features

Contribution?	Feature
+60.649	Total Population
+18.816	<BIAS>
+11.483	TotalCredit in $
+6.838	ConsumerPrice
+1.498	April-Temperature
+1.400	January-Temperature
+1.348	September-Temperature
+0.817	May-Temperature
+0.629	Gross Domestic Product in $
+0.572	October-Temperature
+0.192	August-Temperature
+0.117	February-Temperature
-0.169	March-Temperature
-0.289	June-Temperature
-0.821	December-Temperature
-1.751	July-Temperature
-2.785	November-Temperature
-12.100	Food price inflation

In our model, the most important feature was Total Population, Total Credit to Agriculture in $, Consumer Price, April-Temperature, January-Temperature, September-Temperature, May-Temperature, GDP, and October-Temperature. A negative value for importance occurs when the feature is not important at all.

Step 9: Explainability of the *Model—Azure*

Set up Azure interpretable explainer:

```
# Explainers:
# 1. SHAP Tabular Explainer
from interpret.ext.blackbox import TabularExplainer

# OR

# 2. Mimic Explainer
#from interpret.ext.blackbox import MimicExplainer
# You can use one of the following four interpretable models as a global surrogate to the black box model
from interpret.ext.glassbox import LGBMExplainableModel
from interpret.ext.glassbox import LinearExplainableModel
from interpret.ext.glassbox import SGDExplainableModel
from interpret.ext.glassbox import DecisionTreeExplainabl
```

Set feature names:

```
feature_names= ['ConsumerPrice','Food price inflation',
        'TotalCredit in $','Gross Domestic Product in $','Total Population','January-Temperature','February-Temperature'
        ,'March-Temperature','April-Temperature','May-Temperature','June-Temperature'
        ,'July-Temperature','August-Temperature','September-Temperature','October-Temperature'
        ,'November-Temperature','December-Temperature']
# 1. Using SHAP TabularExplainer
explainer = TabularExplainer(regressor,
                    X_train,
                    features = feature_names)
```

Passing in test dataset for evaluation examples—note it must be a representative sample of the original data—x_train can be passed as well, but with more examples, explanations will take longer although they may be more accurate:

```
global_explanation = explainer.explain_global(X_test)
```

Sorted SHAP values:

```
print('ranked global importance values: {}'.format(global_explanation.get_ranked_global_values()))
# Corresponding feature names
print('ranked global importance names: {}'.format(global_explanation.get_ranked_global_names()))
# Feature ranks (based on original order of features)
print('global importance rank: {}'.format(global_explanation.global_importance_rank))
# Print out a dictionary that holds the sorted feature importance names and values
print('global importance rank: {}'.format(global_explanation.get_feature_importance_dict()))
```

Output:

```
ranked global importance values: [8.479717619046122, 2.8073439969009337, 2.5729516268249584,
1.6032007339482686, 1.5052450373847186, 1.461186913764645, 0.7851565637135702,
0.7313041141054973, 0.5283709466677231, 0.5119516765294958, 0.45076837146397597,
0.34325369562905905, 0.26801176888571676, 0.2047192445147778, 0.15150414368647436,
0.12075086104463989, 0.08134933972852693]
ranked global importance names: ['TotalCredit in $', 'ConsumerPrice', 'Total Population', 'January-
Temperature', 'Food price inflation', 'April-Temperature', 'September-Temperature', 'November-
Temperature', 'June-Temperature', 'Gross Domestic Product in $', 'July-Temperature', 'May-
Temperature', 'December-Temperature', 'October-Temperature', 'March-Temperature', 'August-
Temperature', 'February-Temperature']
global importance rank: [2, 0, 4, 5, 1, 8, 13, 15, 10, 3, 11, 9, 16, 14, 7, 12, 6]
global importance rank: {'TotalCredit in $': 8.479717619046122, 'ConsumerPrice':
2.8073439969009337, 'Total Population': 2.5729516268249584, 'January-Temperature':
1.6032007339482686, 'Food price inflation': 1.5052450373847186, 'April-Temperature':
1.461186913764645, 'September-Temperature': 0.7851565637135702, 'November-Temperature':
0.7313041141054973, 'June-Temperature': 0.5283709466677231, 'Gross Domestic Product in $':
0.5119516765294958, 'July-Temperature': 0.45076837146397597, 'May-Temperature':
0.34325369562905905, 'December-Temperature': 0.26801176888571676, 'October-Temperature':
0.2047192445147778, 'March-Temperature': 0.15150414368647436, 'August-Temperature':
0.12075086104463989, 'February-Temperature': 0.08134933972852693}
```

Feature shap values for all features and all data points in the training data:

```
# feature shap values for all features and all data points in the training data
print('local importance values: {}'.format(global_explanation.local_importance_values))
```

Output:

```
local importance values: [[-5.341140563521542, -1.489861609429397, -22.56155657627088, -
1.2188009337399222, -7.179989094808798, 0.6660355990299791, -0.029039929237928797, -
0.06216299619829101, 0.9798313337261013, 0.4966634799218407, -0.08040124353374618, -
1.2759896375770816, -0.06847512206222571, -0.5069242671078077, 0.14283998657392813, -
0.19243539930344733, -0.3276916919636196], [-3.5379045152347497, -1.4899186802064521, -
8.69657198015546, -0.3940039321631461, -2.1934801759197224, 1.6945595993019382, -
0.03983805474114106, 0.11019595340833574, 1.0005347995342107, -0.12648829011326834,
0.03559962574962408, -0.0722290579554028, 0.02090629821281026, -0.3691330345030491, -
0.2990153191429933, -0.10216930993945875, 0.12756117573266645], [-1.7975185409834513, -
1.489964336828098, -2.3265610200634472, -0.233812305010129, -0.2787657969279013,
3.9343253305531953, 0.19834367510856002, -0.1045072364130678, 0.9487761350139371,
0.054177416858150806, -1.2271181700953964, -0.3091985620768607, 0.3207497836703436,
1.3113209109593331, -0.1601867781123178, 0.4200844928093296, -0.12855527628743774], [-
0.5528123678639904, -1.5512355230749266, -0.33418089969470016, 0.20118953520478608,
0.6395714396434121, -0.11788240690796092, 0.05817569982647788, -0.3291503887262029,
2.915605386784331, 0.6956855956229764, -0.7703647472921259, -0.1456562282465587,
0.07287224023317998, -0.9532480422840911, 0.2168348942298719, -2.2105272543697536,
0.48823893155914333]]
```

```
# Note: Do not run this cell if using PFIExplainer, it does not support local explanations
# You can pass a specific data point or a group of data points to the explain_local function

# E.g., Explain the first data point in the test set
local_explanation = explainer.explain_local(X_test[0,:])

# E.g., Explain the first five data points in the test set
# local_explanation_group = explainer.explain_local(x_test[0:4,:])
```

Sorted local feature importance information; reflects the original feature order:

```
sorted_local_importance_names = local_explanation.get_ranked_local_names()
sorted_local_importance_values = local_explanation.get_ranked_local_values()

print('sorted local importance names: {}'.format(sorted_local_importance_names))
print('sorted local importance values: {}'.format(sorted_local_importance_values))
```

Output:

```
sorted local importance names: ['April-Temperature', 'January-Temperature', 'May-Temperature',
'October-Temperature', 'February-Temperature', 'March-Temperature', 'August-Temperature', 'June-
Temperature', 'November-Temperature', 'December-Temperature', 'September-Temperature', 'Gross
Domestic Product in $', 'July-Temperature', 'Food price inflation', 'ConsumerPrice', 'Total Population',
'TotalCredit in $']
sorted local importance values: [0.9798313337261013, 0.6660355990299791, 0.4966634799218407,
0.14283998657392813, -0.029039929237928797, -0.06216299619829101, -0.06847512206222571, -
0.08040124353374618, -0.19243539930344733, -0.3276916919636196, -0.5069242671078077, -
1.2188009337399222, -1.2759896375770816, -1.489861609429397, -5.341140563521542, -
7.179989094808798, -22.56155657627088]
```

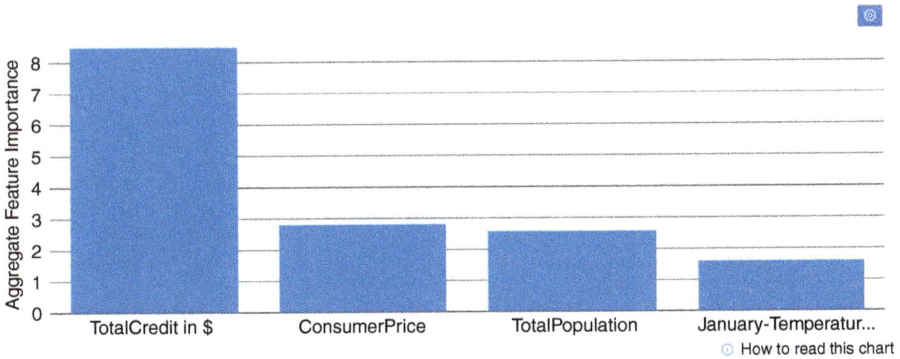

Step 10: *Model Deployment*

First save the model using Joblib library:

```
import joblib
# To dump
f = 'regressor_model.joblib'
joblib.dump(regressor, f)
```

The output of the above code is the generation model file.
Load the saved model in Python code:

```
"""
Created on Mon Mar 15 17:40:35 2021

@author: CHVUPPAL
"""

# Import libraries
import numpy as np

import joblib
# To dump
f = 'regressor_model.joblib'
# To load
regressorfromefile = joblib.load(f)


WIF_X_test = [[ 3.95788425e+01, 8.83389167e+00, 11.48508e+05,
       5.12774620e+05, 1.09331719e+06, 4.75000000e-01,
       5.62000000e-01, 8.49000000e-01, 8.68000000e-01,
       8.46000000e-01, 2.39000000e-01, 1.57400000e+00,
       3.70000000e-01, 4.50000000e-01, 8.11000000e-01,
       8.64000000e-01, 1.39800000e+00],
      [ 5.00159375e+01, 8.83393333e+00, 11.23815000e+05,
       1.18472449e+06, 1.18320947e+06, 8.24000000e-01,
       5.10000000e-01, -1.80000000e-02, 8.80000000e-01,
       2.01000000e-01, 1.43000000e-01, 4.92000000e-01,
       5.42000000e-01, 4.96000000e-01, 1.84000000e-01,
       8.36000000e-01, 6.23000000e-01],
      [ 6.00892575e+01, 8.83396667e+00, 10.99653000e+05,
       1.31523023e+06, 1.21772622e+06, 1.58400000e+00,
       1.65700000e+00, 1.06200000e+00, 8.50000000e-01,
       3.88000000e-01, 1.18800000e+00, 7.05000000e-01,
       1.11900000e+00, 1.05700000e+00, 3.81000000e-01,
       6.74000000e-01, 1.05900000e+00],
      [ 6.72935925e+01, 8.87870000e+00, 10.760100000e+05,
       1.66961977e+06, 1.23428117e+06, 2.09000000e-01,
       9.82000000e-01, 2.19200000e+00, 1.99000000e+00,
       1.05200000e+00, 8.10000000e-01, 5.58000000e-01,
       6.42000000e-01, 3.01000000e-01, 9.16000000e-01,
       1.49000000e+00, 9.00000000e-03]]

print(WIF_X_test)


y_wif_pred = regressorfromefile.predict(WIF_X_test)
print(y_wif_pred)
```

Please find Spyder interface:

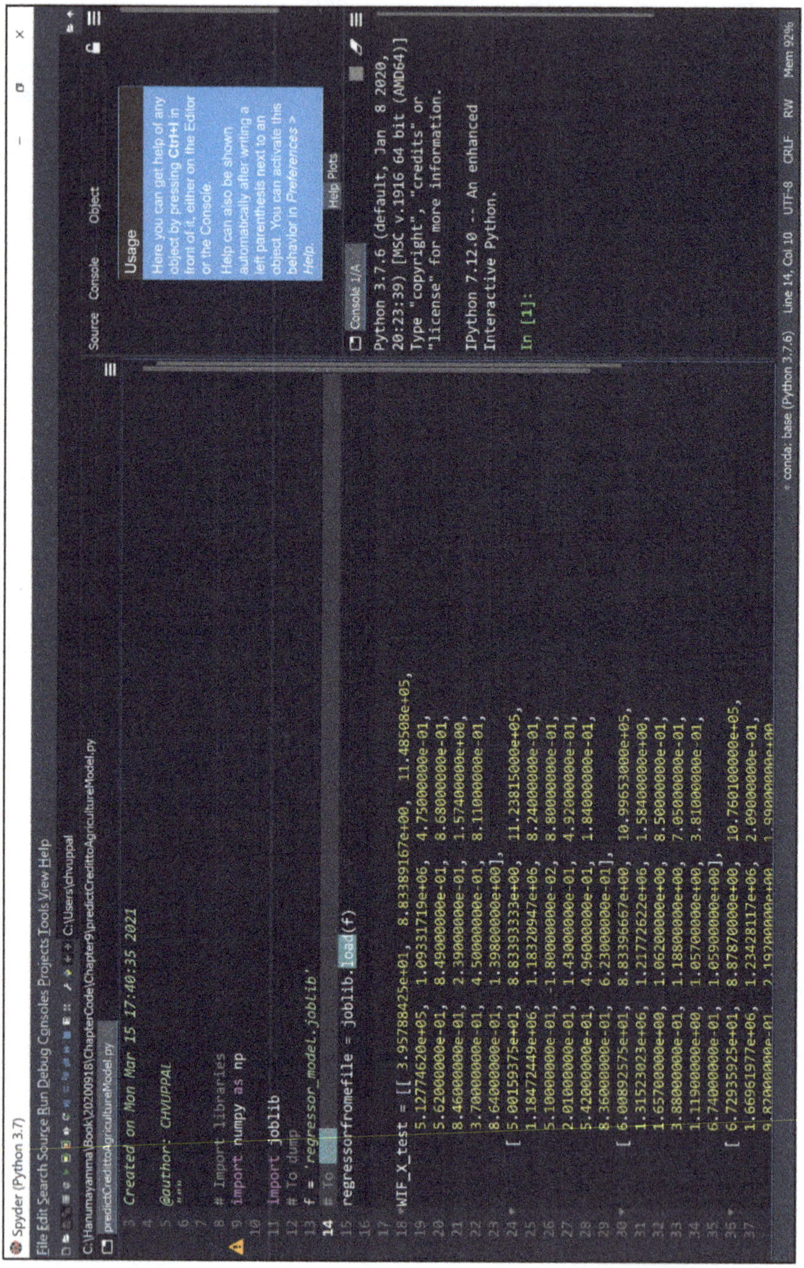

Finally, execute the code:

```
python predictCredittoAgricultureModel.py
```

Output:

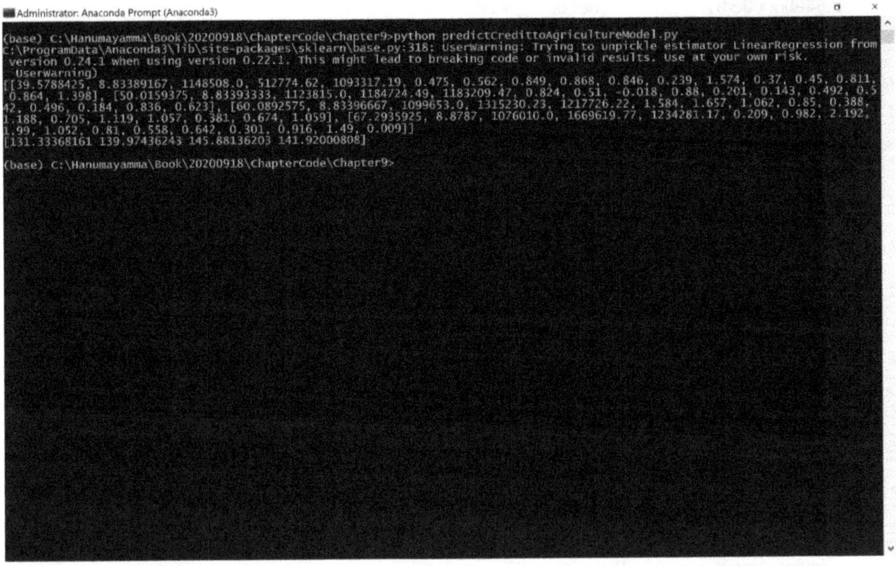

What If Analysis

The advantage of developing predictive models is to analyze the impact of various macroeconomic indicators on the class variable or label—"Milk Production Million Tonnes".

Consumer Price Increase

What if the CPI increases 15% year after year for the next 4 years (2021, 2022, 2023, and 2024)? The impact on the production would be production needs to increase comapreletvely than the case when percentage does not change:

```
WIF_X_test = [[ 12.3668e+01, 8.83389167e+00, 2.33949000e+05,
    5.12774620e+05, 1.09331719e+06, 4.75000000e-01,
    5.62000000e-01, 8.49000000e-01, 8.68000000e-01,
    8.46000000e-01, 2.39000000e-01, 1.57400000e+00,
    3.70000000e-01, 4.50000000e-01, 8.11000000e-01,
    8.64000000e-01, 1.39800000e+00],
   [ 14.2449375e+01, 8.83393333e+00, 5.16422000e+05,
    1.18472449e+06, 1.18320947e+06, 8.24000000e-01,
    5.10000000e-01, -1.80000000e-02, 8.80000000e-01,
    2.01000000e-01, 1.43000000e-01, 4.92000000e-01,
    5.42000000e-01, 4.96000000e-01, 1.84000000e-01,
    8.36000000e-01, 6.23000000e-01],
   [ 16.357792575e+01, 8.83396667e+00, 6.46199000e+05,
    1.31523023e+06, 1.21772622e+06, 1.58400000e+00,
    1.65700000e+00, 1.06200000e+00, 8.50000000e-01,
    3.88000000e-01, 1.18800000e+00, 7.05000000e-01,
    1.11900000e+00, 1.05700000e+00, 3.81000000e-01,
    6.74000000e-01, 1.05900000e+00],
   [ 18.81145925e+01, 8.87870000e+00, 6.86790000e+05,
    1.66961977e+06, 1.23428117e+06, 2.09000000e-01,
    9.82000000e-01, 2.19200000e+00, 1.99000000e+00,
    1.05200000e+00, 8.10000000e-01, 5.58000000e-01,
    6.42000000e-01, 3.01000000e-01, 9.16000000e-01,
    1.49000000e+00, 9.00000000e-03]]
```

What if production changes:

```
y_wif_pred = regressor.predict(WIF_X_test)
y_wif_pred
```

Output:

array([100.97144707, 126.13080724, 141.50382473, 143.68988935])

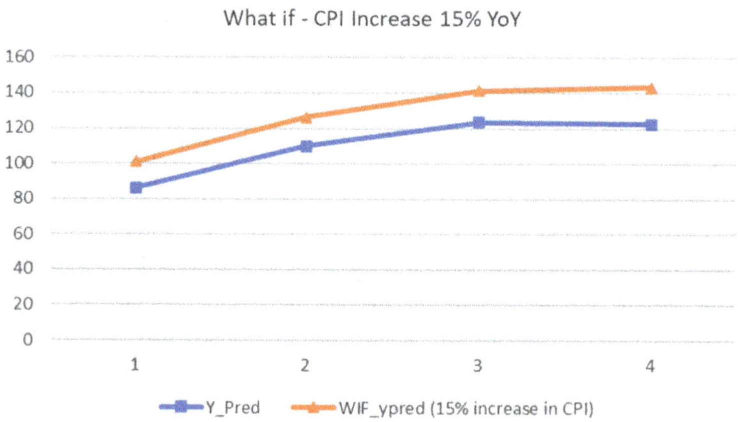

Consumer Price Decrease

What if the CPI decreases 10% year after year for the next 4 years (2021, 2022, 2023, and 2024)? The impact on the production would be less than the case when CPI maintains or does not change:

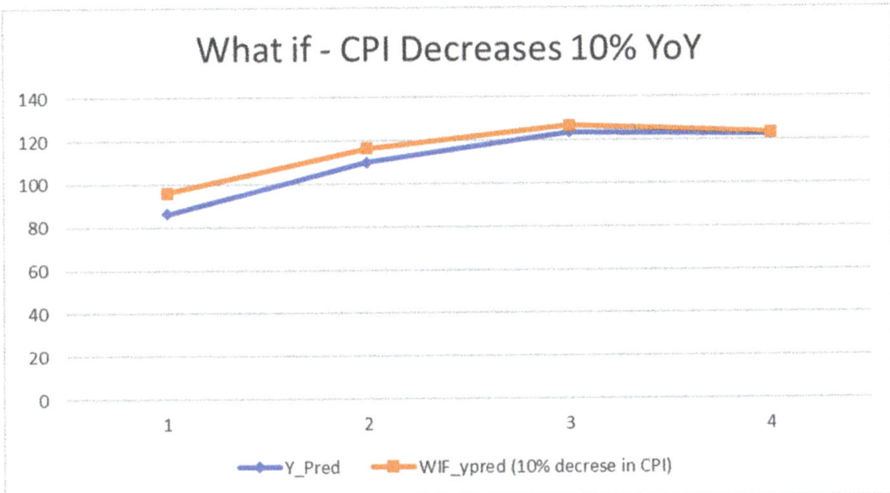

Credit to Agriculture Increases

What if credit to agriculture increases like in the years of 2003 at the rate of 23% per year for the next 4 years? One reason can be attributed to overcoming the impact of COVID-19 on agriculture!

```
## ['ConsumerPrice','Food price inflation',
##        'TotalCredit in $','Gross Domestic Product in $','Total Population','January-
Temperature','February-Temperature'
##        ,'March-Temperature','April-Temperature','May-Temperature','June-Temperature'
##        ,'July-Temperature','August-Temperature','September-Temperature','October-Temperature'
##        ,'November-Temperature','December-Temperature']

WIF_X_test = [[ 3.95788425e+01, 8.83389167e+00, 14.43705e+05,
      5.12774620e+05, 1.09331719e+06, 4.75000000e-01,
      5.62000000e-01, 8.49000000e-01, 8.68000000e-01,
      8.46000000e-01, 2.39000000e-01, 1.57400000e+00,
      3.70000000e-01, 4.50000000e-01, 8.11000000e-01,
      8.64000000e-01, 1.39800000e+00],
    [ 5.00159375e+01, 8.83393333e+00, 17.75757000e+05,
      1.18472449e+06, 1.18320947e+06, 8.24000000e-01,
      5.10000000e-01, -1.80000000e-02, 8.80000000e-01,
      2.01000000e-01, 1.43000000e-01, 4.92000000e-01,
      5.42000000e-01, 4.96000000e-01, 1.84000000e-01,
      8.36000000e-01, 6.23000000e-01],
    [ 6.00892575e+01, 8.83396667e+00, 21.84181000e+05,
      1.31523023e+06, 1.21772622e+06, 1.58400000e+00,
      1.65700000e+00, 1.06200000e+00, 8.50000000e-01,
      3.88000000e-01, 1.18800000e+00, 7.05000000e-01,
      1.11900000e+00, 1.05700000e+00, 3.81000000e-01,
      6.74000000e-01, 1.05900000e+00],
    [ 6.72935925e+01, 8.87870000e+00, 26.865430000e+05,
      1.66961977e+06, 1.23428117e+06, 2.09000000e-01,
      9.82000000e-01, 2.19200000e+00, 1.99000000e+00,
      1.05200000e+00, 8.10000000e-01, 5.58000000e-01,
      6.42000000e-01, 3.01000000e-01, 9.16000000e-01,
      1.49000000e+00, 9.00000000e-03]]
```

Predict with what if analysis:

```
y_wif_pred = regressor.predict(WIF_X_test)
y_wif_pred
```

Output:

```
array([145.82321462, 171.97446788, 199.11464101, 220.9718652 ])
```

If the government pursues an increase in credit to agriculture at the rate of 23% YoY, it would result in **huge growth** and substantial benefits to agriculture and the overall economy in general.

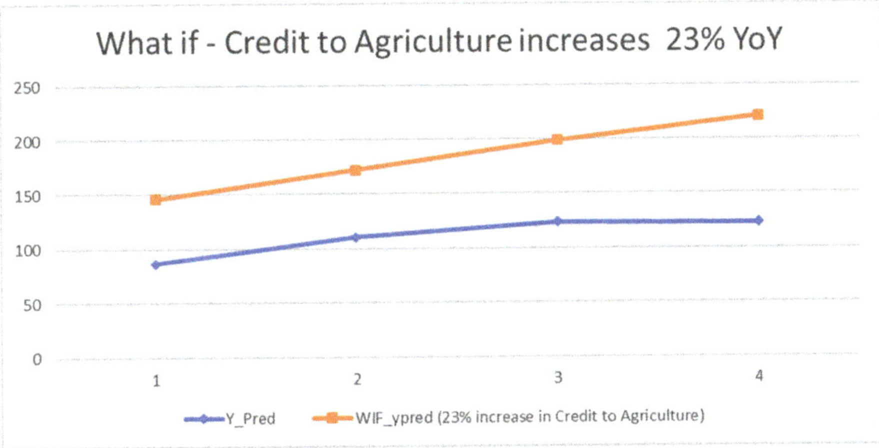

What if the credit to agriculture is maintained at the levels of 2016 (2.15%):

```
WIF_X_test = [[ 3.95788425e+01, 8.83389167e+00, 11.98979e+05,
    5.12774620e+05, 1.09331719e+06, 4.75000000e-01,
    5.62000000e-01, 8.49000000e-01, 8.68000000e-01,
    8.46000000e-01, 2.39000000e-01, 1.57400000e+00,
    3.70000000e-01, 4.50000000e-01, 8.11000000e-01,
    8.64000000e-01, 1.39800000e+00],
   [ 5.00159375e+01, 8.83393333e+00, 12.24757000e+05,
    1.18472449e+06, 1.18320947e+06, 8.24000000e-01,
    5.10000000e-01, -1.80000000e-02, 8.80000000e-01,
    2.01000000e-01, 1.43000000e-01, 4.92000000e-01,
    5.42000000e-01, 4.96000000e-01, 1.84000000e-01,
    8.36000000e-01, 6.23000000e-01],
   [ 6.00892575e+01, 8.83396667e+00, 12.51089000e+05,
    1.31523023e+06, 1.21772622e+06, 1.58400000e+00,
    1.65700000e+00, 1.06200000e+00, 8.50000000e-01,
    3.88000000e-01, 1.18800000e+00, 7.05000000e-01,
    1.11900000e+00, 1.05700000e+00, 3.81000000e-01,
    6.74000000e-01, 1.05900000e+00],
   [ 6.72935925e+01, 8.87870000e+00, 12.779880000e+05,
    1.66961977e+06, 1.23428117e+06, 2.09000000e-01,
    9.82000000e-01, 2.19200000e+00, 1.99000000e+00,
    1.05200000e+00, 8.10000000e-01, 5.58000000e-01,
    6.42000000e-01, 3.01000000e-01, 9.16000000e-01,
    1.49000000e+00, 9.00000000e-03]]
```

Predict with what if analysis:

```
y_wif_pred = regressor.predict(WIF_X_test)
y_wif_pred
```

Output:

```
array([133.81101444, 144.9290281 , 153.31448947, 151.83395334])
```

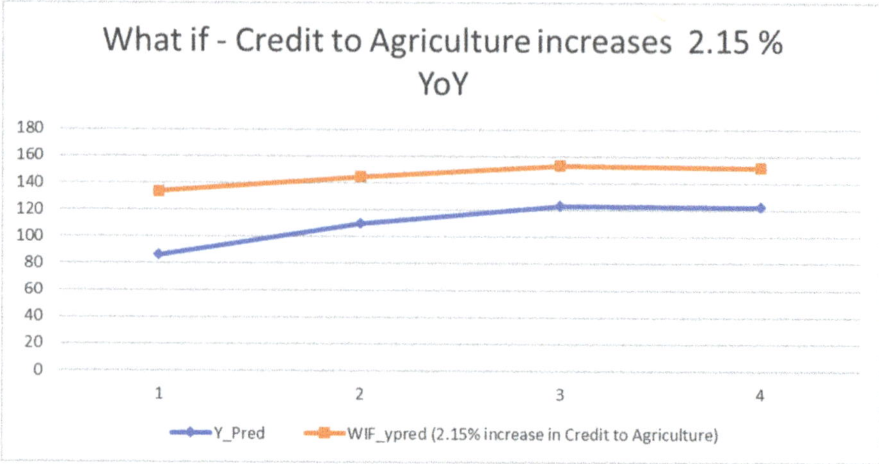

If the government pursues an increase in credit to agriculture at the rate of 2.15% YoY, it would result in maintaining the growth and benefits at the current rate to agriculture and the overall economy in general.

Credit to Agriculture Decreases

What if the government reduces the credit to agriculture by 2.15%? The production would go down and would result into lesser overall economic growth.

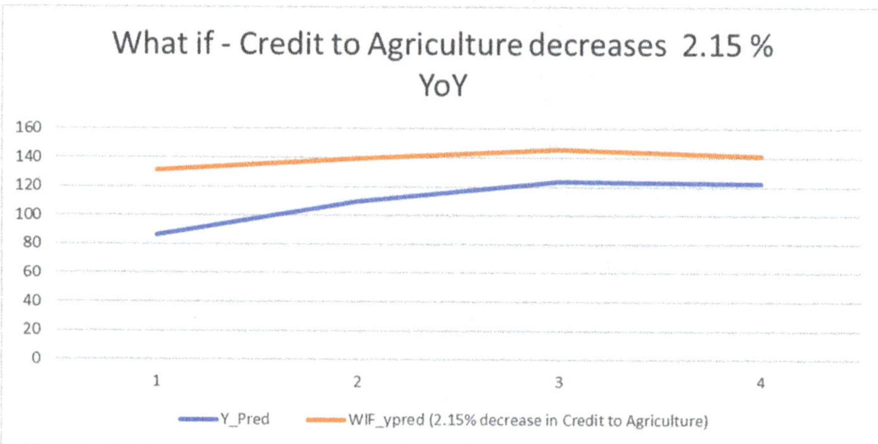

And it is clear that the credit to agriculture is crucial to achieve enough milk to capita and enable farming sustainability.

In conclusion, it is clear that the time has arrived to have a digital counterpart at policy governance for the federal and state governments. That is, in addition to have human agents (politicians, beaurocrats) who make policy, we need to have machine learning-enabled intelligent systems to be make the policies and thereafter make a judicious analysis of the pros and cons of the decision based purely on data to the future governance. It is clear that we, as human beings, need augmented intelligent agents to help us to make better data decisions, which have been our blind spots so far. Don't get me wrong; we, as human beings, are very well-deserved agents capable of making decisions based on emotions, a trite that makes us purely human beings.

After reading this chapter, you should be able to analyze the importance of government credit to agriculture and the role it plays in the development of agricultural economy, AI readiness and the impact of AI enablement in high agricultural employment countries, and the role of ICT and its influence on poverty. In addition, you should be able to develop data and machine learning models to include agriculture credit as part of ML model development!

References

1. Thomas, R., McSharry, P.: Big Data Revolution: What farmers, doctors and insurance agents teach us about discovering big data patterns, 1st edn. Wiley, Chichester (March 2, 2015) ISBN-10: 9781118943717
2. van der Meulen, R.: Edge computing promises near real-time insights and facilitates localized actions (October 3, 2018). https://www.gartner.com/smarterwithgartner/what-edge-computing-means-for-infrastructure-and-operations-leaders/. Access Date: 10 Mar 2021
3. Cearley, D., Burke, B., Walker, M., Gill, B., Bittman, T.: Top 10 Strategic Technology Trends for 2018: Cloud to the Edge (2018). Published: 8 March 2018, URL: https://www.gartner.com/en/documents/3865403/top-10-strategic-technology-trends-for-2018-cloud-to-the0. Access Date: 18 Sept 2019
4. Basu, K.: India's farm laws are a global problem, Friday, (February 19, 2021). https://www.brookings.edu/opinions/indias-farm-laws-are-a-global-problem/. Access Date: 10 Mar 2021
5. Ranganathan, J., Waite, R., Searchinger, T., Hanson, C.: How to Sustainably Feed 10 Billion People by 2050, in 21 Charts, (2018). Published: December 5, 2018. URL: https://www.wri.org/blog/2018/12/how-sustainably-feed-10-billion-people-2050-21-charts. Access Date: 18 Sept 2019

6. Hardjono, T., Shrier, D.L., Pentland, A.: Trust::Data: A New Framework for Identity and Data Sharing. Edited by Hardjono, T. CreateSpace Independent Publishing Platform, Scotts Valley. (October 21, 2016), ISBN-10: 9781539114215
7. Dutta, S., Geiger, T., Lanvin, B.: The Global Information Technology Report 2015. Published: 2015 URL: http://www3.weforum.org/docs/WEF_Global_IT_Report_2015.pdf. Access Date: 22 Nov 2019.
8. Melamed, C.: A World That Counts, 2020. Published: 11 November 2018. URL: https://www.undatarevolution.org/draft-report. Access Date: 22 Nov 2019.
9. Vuppalapati, C.: Democratization of Artificial Intelligence for the Future of Humanity, 1st edn. CRC Press, Boca Raton (January 18, 2021) ISBN-13: 978-0367524098
10. Chui, M., Manyika, J., Miremadi, M., Henke, N., Chung, R., Nel, P., Malhotra, S.: Notes from the AI frontier: Applications and value of deep learning (April 2018). https://www.mckinsey.com/featured-insights/artificial-intelligence/notes-from-the-ai-frontier-applications-and-value-of-deep-learning. Access Date: 18 Sept 2019
11. Diego Arias (World Bank), Pedro Abel Vieira (Embrapa), Elisio Contini (Embrapa), Barbara Farinelli (World Bank) and Michael Morris (World Bank): Agriculture productivity growth in Brazil recent trends and future prospects (September 24, 2017). http://documents1.worldbank.org/curated/en/268351520343354377/pdf/123948-WP-6-3-2018-8-39-22-AriasetalAgriculturalgrowthinBrazil.pdf. 18 Sept 2020
12. Mari, A.: Brazil creates national AI innovation network (November 2, 2020). https://www.zdnet.com/article/brazil-creates-national-ai-innovation-network/. Access Date: 8 Jan 2021
13. Cibils, A.I.: Little Uruguay has big plans for smart agriculture (December 24, 2014). https://phys.org/news/2014-12-uruguay-big-smart-agriculture.html. Access Date: 10 Dec 2020
14. Morris, M.: Gaining Momentum in Peruvian Agriculture: Opportunities to Increase Productivity and Enhance Competitiveness (English). World Bank Group, Washington, DC (2017) http://documents.worldbank.org/curated/en/107451498513689693/Gaining-momentum-in-Peruvian-agriculture-opportunities-to-increase-productivity-and-enhance-competitiveness
15. Dudenhoefer, D.: As Peru's agricultural production grows, smallholders long for better markets (March 15, 2018). https://blogs.worldbank.org/latinamerica/peru-s-agricultural-production-grows-smallholders-long-better-markets. Access Date: Feb 2021
16. Trostle, R.: Global agricultural supply and demand: Factors contributing to the recent increase in food commodity prices. https://www.hsdl.org/?view&did=485652. Access Date: 22 Nov 22 2020
17. Gale, F.: Growth and evolution in china's agricultural support policies (August 2013). https://www.ers.usda.gov/webdocs/publications/45115/39368_err153.pdf?v=41491. Access Date: 22 Nov 22 2020
18. OECD/ICRIER: Agricultural Policies in India OECD Food and Agricultural Reviews. OECD Publishing, Paris (2018). https://doi.org/10.1787/9789264302334-en
19. Parija, P.: India to boost farm credit and infrastructure to spur output (January 31, 2021). https://www.bloomberg.com/news/articles/2021-02-01/india-plans-to-enhance-farm-credit-to-boost-agricultural-output. Access Date: 22 Nov 22 2020
20. Smith, T.: As farmers go, so goes the nation how did agriculture help the US recover from the Great Depression in the spring of 1933?, (25 March 2019). https://www.aeaweb.org/research/farmer-recovery-great-depression. Access Date: Mar 2020
21. Lugar, R.G.: Committee on Agriculture, Agriculture Credit Nutrition, and Forestry (May 16, 2001). https://www.govinfo.gov/content/pkg/CHRG-107shrg81040/html/CHRG-107shrg81040.htm
22. Reserve Bank of India: Report of the Internal Working Group to Review Agricultural Credit (13 September 2019). https://www.rbi.org.in/Scripts/PublicationReportDetails.aspx?UrlPage=&ID=942#LET. Access Date: 18 Sept 2020

Section IV
Future

Chapter 8
Future

Agriculture is filled with uncertainties, risks, losses, and back-breaking work. Yet, the returns on agriculture, especially for small farmers, are miniscule or, in some cases, non-existent. No wonder, many small farms have disappeared in the United States, and the same trend can be witnessed worldwide. There are two kinds of risks that farmers face: internal and external farm risks. Internal farm risk is easily controlled by the farmers. The risks are related to soil, fertilizers, phenological stages of a crop, and personal/family issues. The risks related to external farm, for example, commodity price variations, macroeconomic conditions, real-time price models' unintended consequences, trade wars, sudden changes in people tastes and perception of a food commodity, and global climate change, are something beyond farmers' control.

The purpose of the book is to develop comprehensive machine learning and artificial intelligence models to factor in macrolevel economic influencers and able to develop predictive models to reduce risk to farmers and ensure stability and sustainability. In addition, the book aims to impart knowledge of the models and wealthiness of data estate that we as a humanity would be developing during our journey and to hand over the data-rich futuristic well-harnessed and thoroughly industrialized machine learning and artificial intelligence models to our future generations for them to prepare for any challenges and have a sustainable future. We wanted to make the world a better place for our future generations than what we inherited.

C. Vuppalapati, *Machine Learning and Artificial Intelligence for Agricultural Economics*, International Series in Operations Research & Management Science 314,
https://doi.org/10.1007/978-3-030-77485-1_8

Appendixes

Appendix A: Data and Agricultural Statistics Services – Mission Statements

Service to the Nation and the Humanity

Data is asset of the future. It serves, unlike any other perishable resource, and provides guidance into the future by capturing the trends and historical time series data. Of all resources that are available to the world, data is the most important!

Every data source should have following mission statements and should be in service of people of the world.

USDA NASS

The National Agricultural Statistics Service[1] provides timely, accurate, and useful statistics in service to US agriculture.

Core Values:

- Trust
  - Trust Among data providers: NASS will continue the relationship of mutual respect and our legal pledge of confidentiality with our respondents who provide their personal and agricultural data in support of US agriculture.
- Credibility

[1] NASS Mission Statement – https://www.nass.usda.gov/About_NASS/Mission_Statement/index.php

C. Vuppalapati, *Machine Learning and Artificial Intelligence for Agricultural Economics*, International Series in Operations Research & Management Science 314,
https://doi.org/10.1007/978-3-030-77485-1

Credibility among data users – NASS will maintain credibility with our data users by issuing complete and accurate statistical information about agriculture.

- Objectivity
  - Objectivity of statistics – NASS will produce and disseminate statistical information on an equitable and transparent basis so that the public understands that the data are objective and impartial.
- Relevance
  - Relevance to policy issues – We will remain knowledgeable about the issues and requirements of the agriculture industry, public policy, and Federal programs and will remain able to provide objective, accurate, and timely information relevant to those needs.
- Commitment
  - Commitment to service – Each USDA NASS employee and representative is committed to these core values and takes great pride in working together to serve American farmers and ranchers, rural communities, and all who rely on the information we produce.

Website Last Modified: 05/04/2018

The World Bank

The mission statement of the World Bank is working for a world free of poverty. And data plays an integral part of the achieving mission statement at the World Bank.[2]

Appendix B: Department of Commerce

Bureau of Economic Analysis – The United States Department of Commerce

- NIPA handbook: Concepts and Methods of the U.S. National Income and Product Accounts

  NIPA Handbook: Concepts and Methods of the U.S. National Income and Product Accounts | U.S. Bureau of Economic Analysis (BEA)

[2] The World Bank – https://data.worldbank.org/about

Appendix C: Family Agricultural Farms

The United States of America

Ninety seven percent of US farms are family owned.[3]

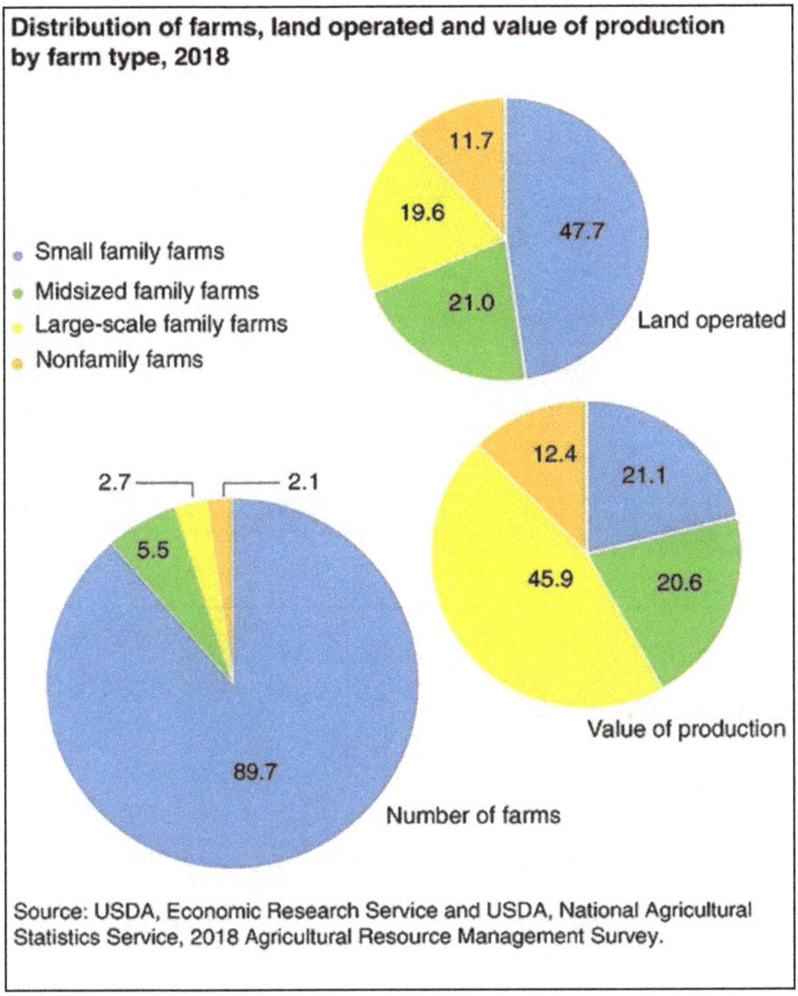

The more than 2 million farms in the United States vary greatly in size and characteristics. For example, annual gross revenue can range from as little as $1000 to more than $5 million.[4]

[3] U.S. Family-Owned Farms – https://texasfarmbureau.org/percent-of-us-farms-are-familyowned/

[4] A look at America's Family Farms – https://www.usda.gov/media/blog/2020/01/23/look-americas-family-farms

Agricultural Data Surveys

USDA Nass

The Unites States Department of Agriculture – National Analytics Statistics Services:

Online survey form.[5]

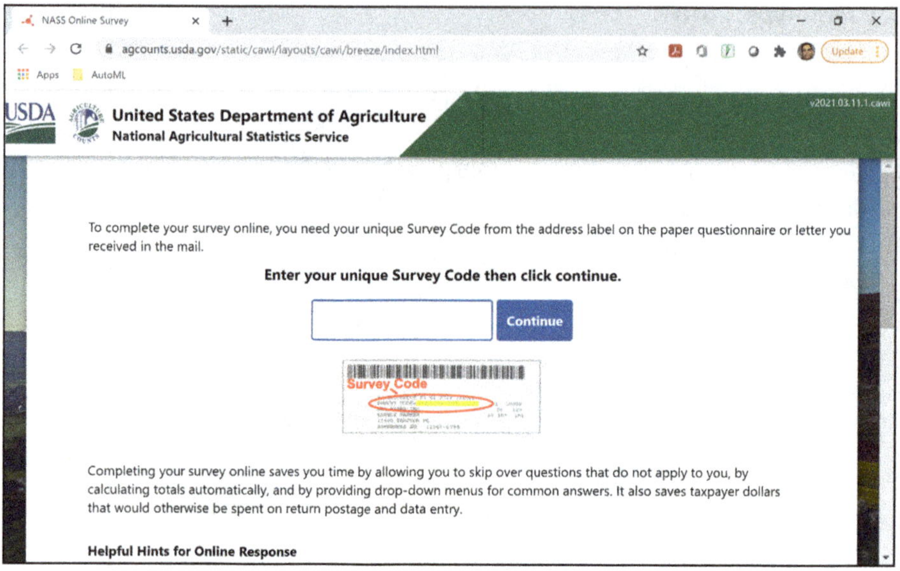

Appendix D: Data Sources

UN Data Marts

Source: http://data.un.org/Explorer.aspx?d=FAO

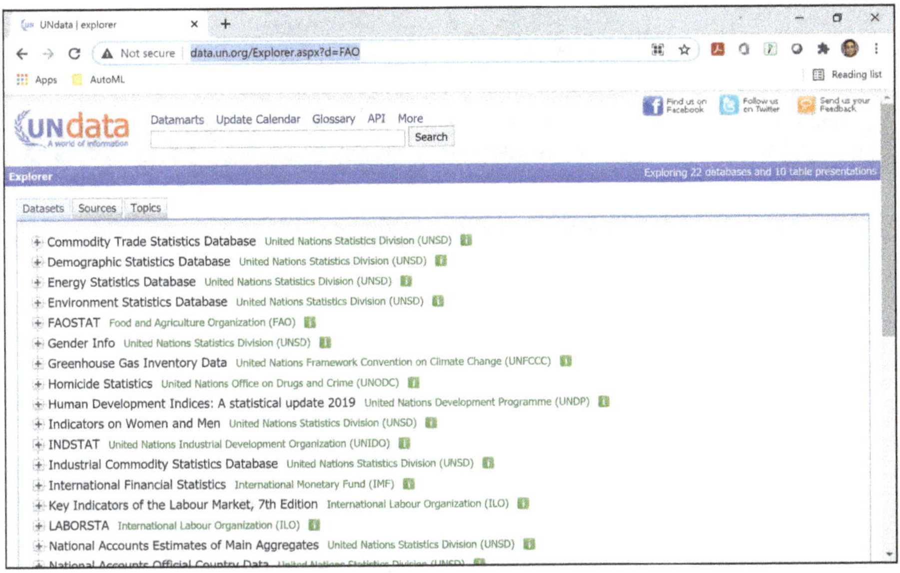

IHS Global Economy Data

IHS[6] provides global economic database covering 200+ countries. The data includes: balance of payments, cyclical indicators, finance and financial markets, government finance, housing and construction, output, capacity and capacity utilization, merchandise trade, national accounts, labor market, population, prices, and wholesale and retail trade.

US Commodities Futures Data

Investing.com future prices.[7]

[5] NASS Online Survey – https://www.agcounts.usda.gov/static/cawi/layouts/cawi/breeze/index.html

[6] IHS Global Economy Data – https://ihsmarkit.com/products/global-economic-data.html

[7] Investing.com – https://www.investing.com/commodities/us-soybeans-historical-data

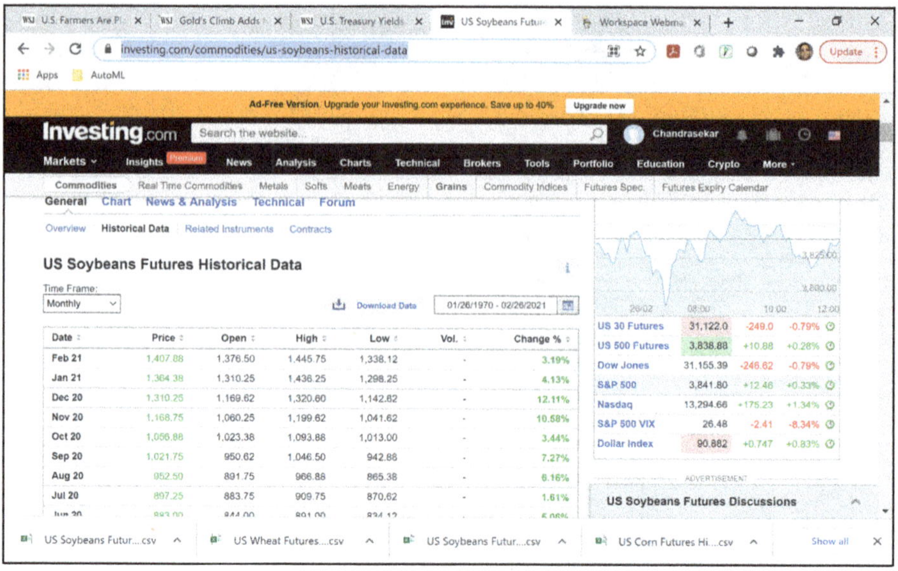

IMF Data Access to Macroeconomic and Financial Data

The International Monetary Fund (IMF[8]) covers following topics:

- External sector
- Fiscal sector
- Financial sector
- Real sector
- Cross domain

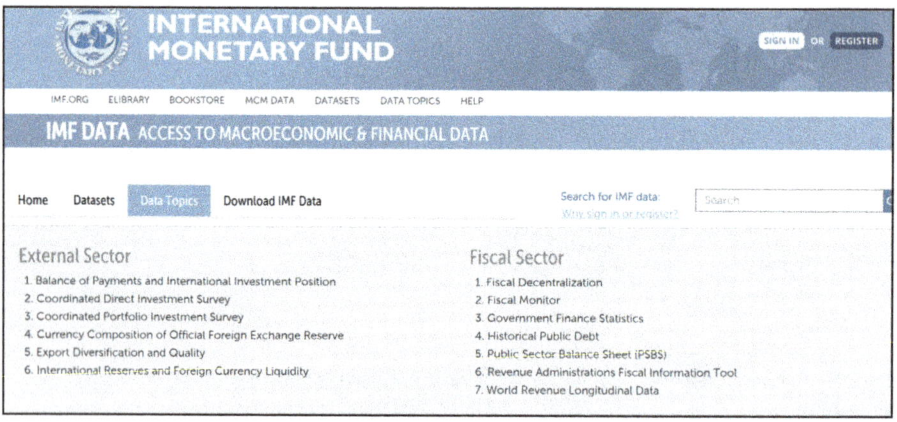

[8] IMF – https://data.imf.org/?sk=388DFA60-1D26-4ADE-B505-A05A558D9A42&sId=1479329328660

Some of the macro-level indicators: Real GDP, percentage change from previous year, CPI, percentage change from previous year, external trade, and real effective exchange rate.

https://data.imf.org/?sk=4C514D48-B6BA-49ED-8AB9-52B0C1A0179B&sId=1502896793454

IMF Country Index Weights

Country indexes and weights provide macro-level consumption and social behavior of people in term purchase preferences, food and nonalcoholic beverages, alcoholic beverages, clothing, and others.

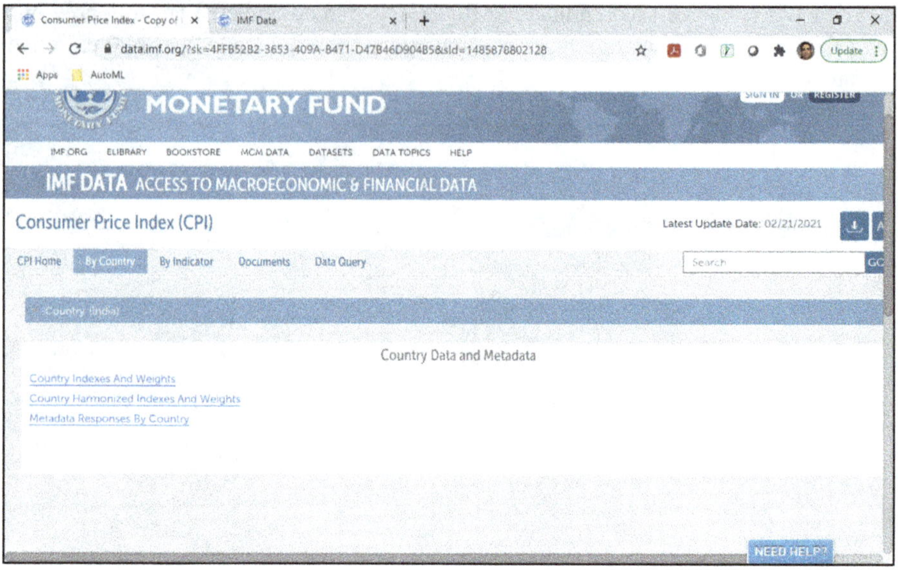

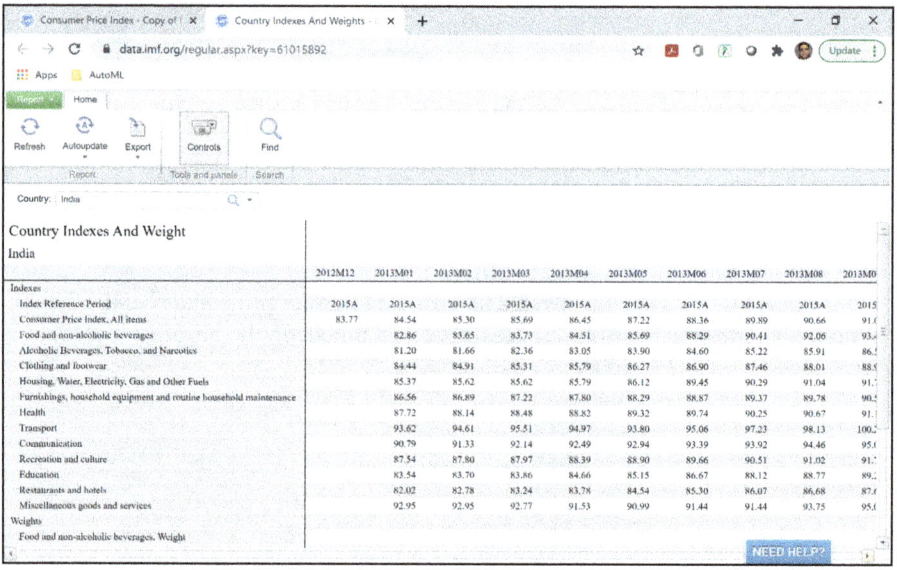

The World Bank Data

Macroeconomic world bank data.[9]

[9] The World Bank Data – https://www.worldbank.org/en/topic/macroeconomics

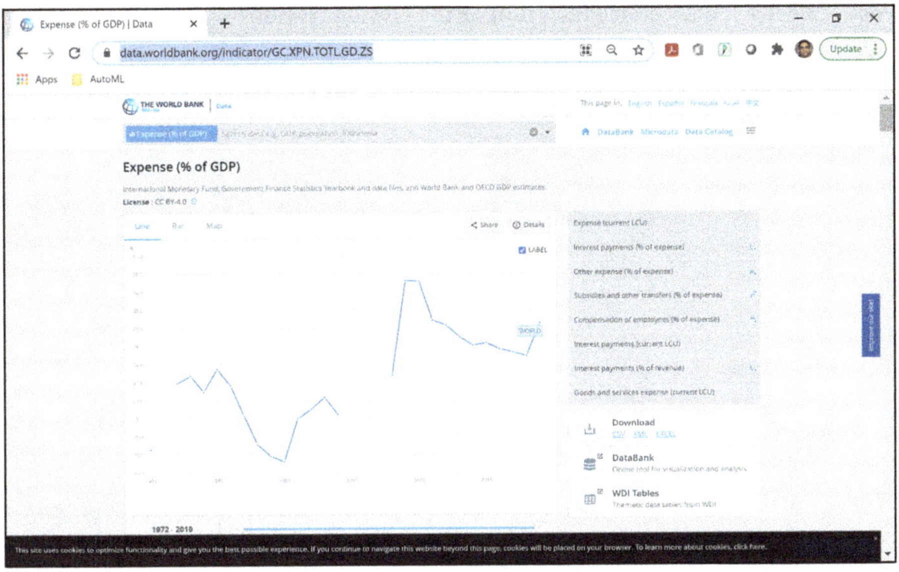

Source: https://data.worldbank.org/indicator/GC.XPN.TOTL.GD.ZS

The World Bank Development Indicators

World Development Indicator tool[10]:

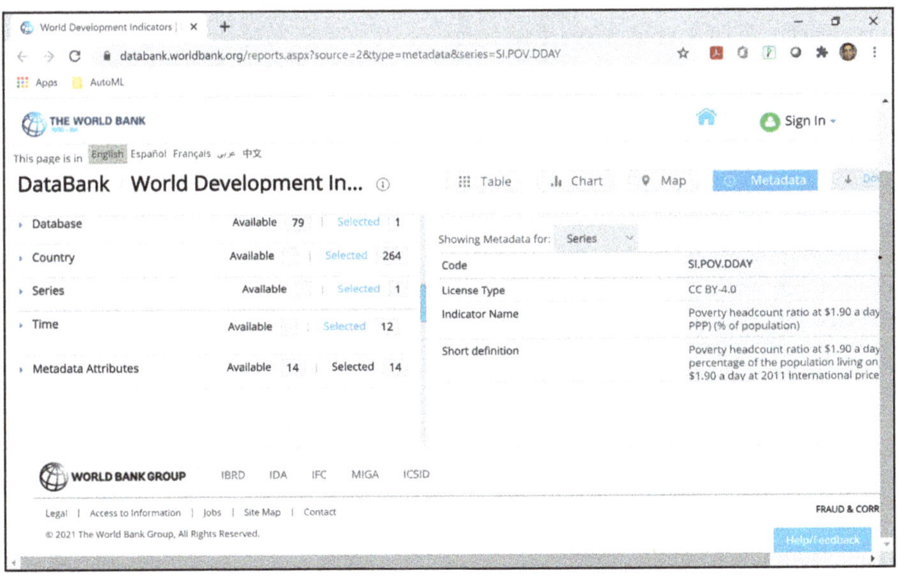

[10] The World Bank Development Indicator – https://databank.worldbank.org/reports.aspx?source=2&type=metadata&series=SI.POV.DDAY

Food and Agriculture Organization of the United Nations

FAOSTAT Data[11]: FAO provides following data domains: production, prices, emission-agriculture, inputs, trade, population, emissions-land use, investment, macroeconomic data, and others.

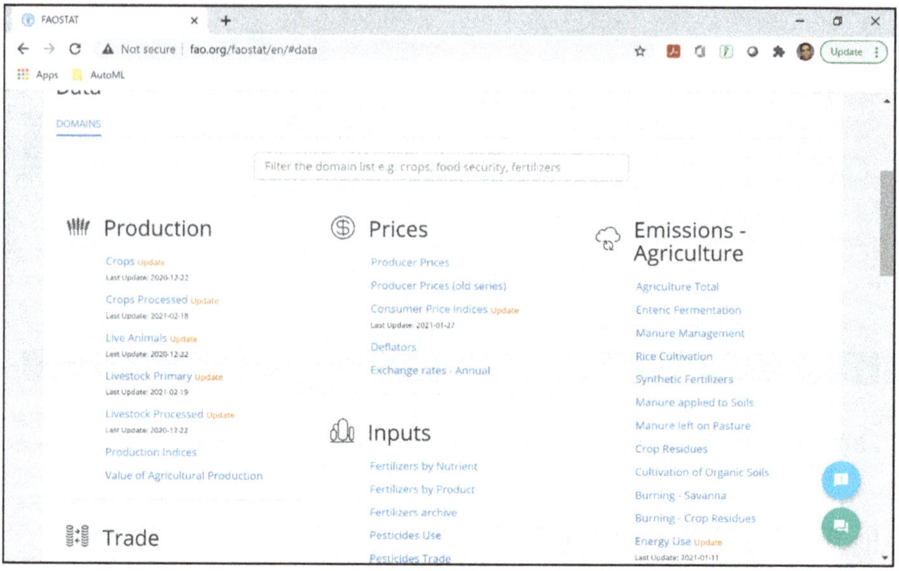

Appendix E: Conversion Factors

Here are a few useful conversion factors[12]:

- 1 bushel of wheat = 60 pounds or 77.2 kilograms/hectoliter
- 1 metric ton = 2204.622 pounds
- Bushels × 0.0272155 = metric tons
- Metric tons × 36.7437 = bushels
- Price per bushel x 36.7437 = price per metric ton
- Price per metric ton × 0.0272155 = price per bushel
- 1 acre = 0.4047 hectares
- 1 hectare = 2.4710 acres
- Bushels/acre × 0.06725 = metric tons/hectare
- Metric tons/hectare × 14.87 = bushels/acre

[11] FAPSTAT – http://www.fao.org/faostat/en/#data

[12] The USDA Conversion Factors – https://www.ers.usda.gov/data-products/wheat-data/documentation/

Appendix F: USDA Datasets

USDA Datasets[13]:

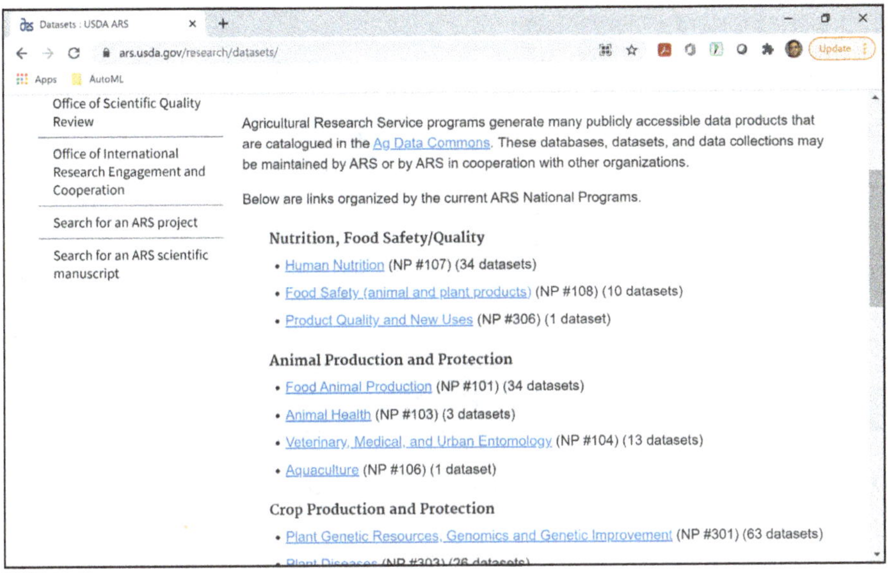

DATA.GOV

Explore data[14] that can help inform agriculture investment, innovation, and policy strategy. If you are interested in agricultural production, food security, rural development, nutrition, natural resources, and regional food systems, this page is for you.

---

13 USDA Datasets – https://www.ars.usda.gov/research/datasets/

14 Data.gov – https://www.data.gov/food/

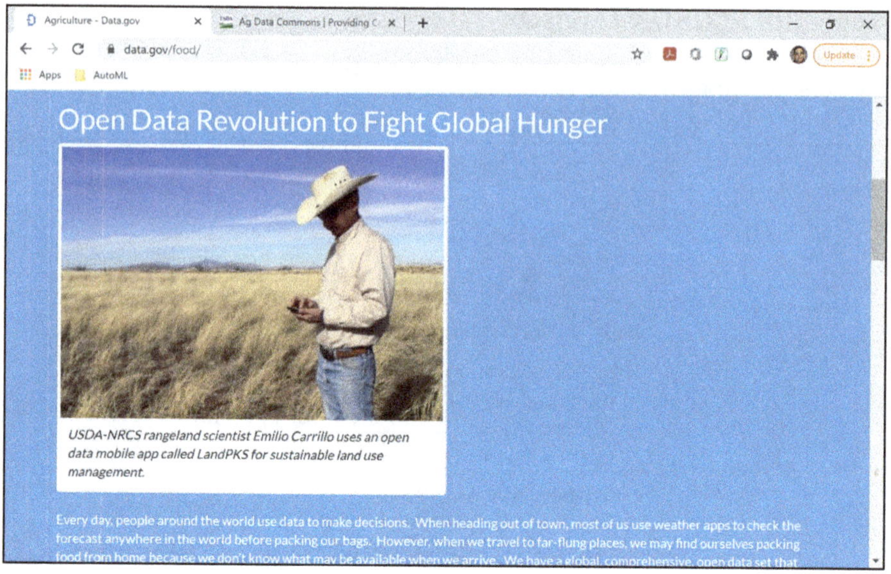

Wheat Data – https://www.ers.usda.gov/data-products/wheat-data.aspx

Wheat Data Disappearance and End Stocks

This data product contains statistics on wheat – including the five classes of wheat: hard red winter, hard red spring, soft red winter, white, and durum and rye. Includes data published in the monthly Wheat Outlook and previously annual Wheat Yearbook. Data are monthly, quarterly, and/or annual, depending upon the data series. Most data are on a marketing-year basis, but some are calendar year.[15]

[15] Wheat Data – https://www.ers.usda.gov/data-products/wheat-data/

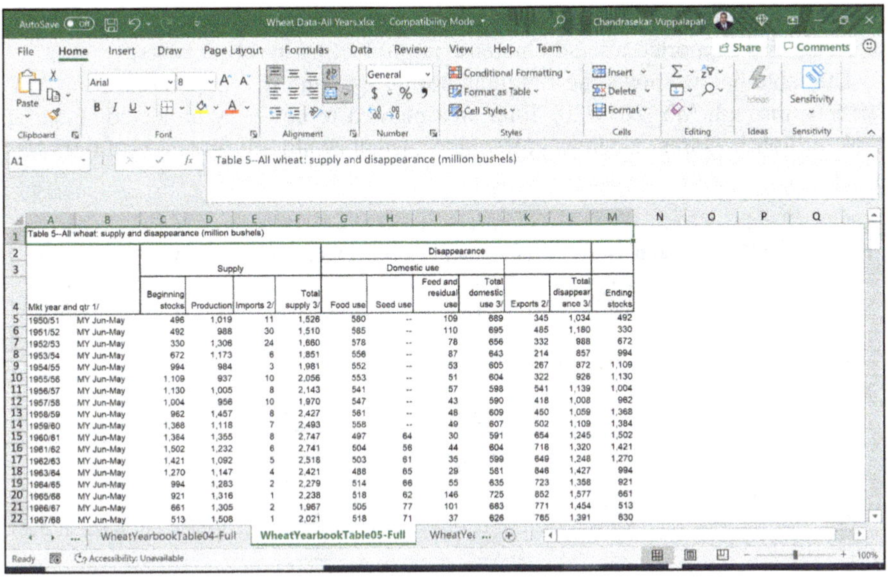

Mkt year and qtr 1/		Beginning stocks	Production	Imports 2/	Total supply 3/	Food use	Seed use	Feed and residual use	Total domestic use 3/	Exports 2/	Total disappearance 3/	Ending stocks
		Supply				Disappearance: Domestic use						
1950/51	MY Jun-May	496	1,019	11	1,526	580	--	109	689	345	1,034	492
1951/52	MY Jun-May	492	988	30	1,510	585	--	110	695	485	1,180	330
1952/53	MY Jun-May	330	1,306	24	1,660	578	--	78	656	332	988	672
1953/54	MY Jun-May	672	1,173	6	1,851	556	--	87	643	214	857	994
1954/55	MY Jun-May	994	984	3	1,981	552	--	53	605	267	872	1,109
1955/56	MY Jun-May	1,109	937	10	2,056	553	--	51	604	322	926	1,130
1956/57	MY Jun-May	1,130	1,005	8	2,143	541	--	57	598	541	1,139	1,004
1957/58	MY Jun-May	1,004	956	10	1,970	547	--	43	590	418	1,008	962
1958/59	MY Jun-May	962	1,457	8	2,427	561	--	48	609	450	1,059	1,368
1959/60	MY Jun-May	1,368	1,118	7	2,493	558	--	49	607	502	1,109	1,384
1960/61	MY Jun-May	1,384	1,355	8	2,747	497	64	30	591	654	1,245	1,502
1961/62	MY Jun-May	1,502	1,232	6	2,741	504	56	44	604	718	1,320	1,421
1962/63	MY Jun-May	1,421	1,092	5	2,518	503	61	35	599	649	1,248	1,270
1963/64	MY Jun-May	1,270	1,147	4	2,421	488	65	29	581	846	1,427	994
1964/65	MY Jun-May	994	1,283	2	2,279	514	66	55	635	723	1,358	921
1965/66	MY Jun-May	921	1,316	1	2,238	518	62	146	725	852	1,577	661
1966/67	MY Jun-May	661	1,305	2	1,967	505	77	101	683	771	1,454	513
1967/68	MY Jun-May	513	1,508	1	2,021	518	71	37	626	765	1,391	630

Hard Red Wheat Contracts

https://www.barchart.com/futures/quotes/KEN20/interactive-chart

Brazil Coffee[16] Annual

ATO/Sao Paulo projects Brazilian coffee production[17] for marketing year (MY) 2019/20 (July–June) at 59.30 million 60-kg bags, a 5.5 million bags drop compared to record historical production in the former season (64.8 million bags). In spite of good weather conditions in the majority of growing regions, Arabica coffee trees are mostly in the off year of the biennial production cycle, thus producing less. Coffee exports for MY 2018/19 are estimated at 39.72 million bags, again reaching historical export levels, due to large production levels and steady competitiveness of the Brazilian product.

Arabica production is forecast at 41 million bags, a 15% reduction compared to the previous crop. Good weather conditions have generally prevailed in most coffee growing regions, supporting fruit setting, development, and filling.

Robusta/Conillon production is expected to reach 18.3 million bags, an increase of 1.7 million bags compared to the current MY. Espirito Santo. The major Robusta/Conillon producing state has been favored by abundant rainfall. Meanwhile, the

[16] Brazil Coffee Annual – https://www.fas.usda.gov/data/brazil-coffee-annual-4

[17] Brazil Coffee – https://apps.fas.usda.gov/newgainapi/api/report/downloadreportbyfilename?filename=Coffee%20Annual_Sao%20Paulo%20ATO_Brazil_5-16-2019.pdf

state of Rondonia's increased use of good crop management practices and clonal seedlings has supported a steady increase in production in recent years.

The table below shows the Brazilian coffee area and tree population from MY 2015/16 through MY 2019/20. Total area planted to coffee is projected stable at 2.39 million hectares, whereas coffee tree inventory slightly increase to 6.93 billion trees.

Brazilian Coffee Area and Tree Population (million trees, thousand hectares, trees/hectare)					
	MY 15/16	MY 16/17	MY 17/18	MY 18/19	MY 19/20
Total Trees	6,860	6,860	6,940	6,890	6,930
Non-Bearing	1,125	1,125	1,300	1,150	1,230
Bearing	5,735	5,735	5,640	5,740	5,700
Total Area	2,410	2,410	2,400	2,395	2,390
Non-Bearing	340	340	380	335	350
Harvested	2,070	2,070	2,020	2,060	2,040
Trees/ha	2,846	2,846	2,892	2,877	2,900
Non-Bearing	3,309	3,309	3,421	3,433	3,514
Bearing	2,771	2,771	2,792	2,786	2,794
Source: ATO/Sao Paulo					

Brazil's coffee yield for MY 2019/20 is forecast at 29.07 bags/hectare, a decrease of 8% compared to MY 2018/19 (31.46 bags/hectare). The drop in yields due to the off-year of the biennial production cycle for Arabica trees was partially offset by the recovery of Robusta/Conillon production. The graph bellows illustrates the evolution of Brazilian coffee yields since the 2010/11 crop.[18]

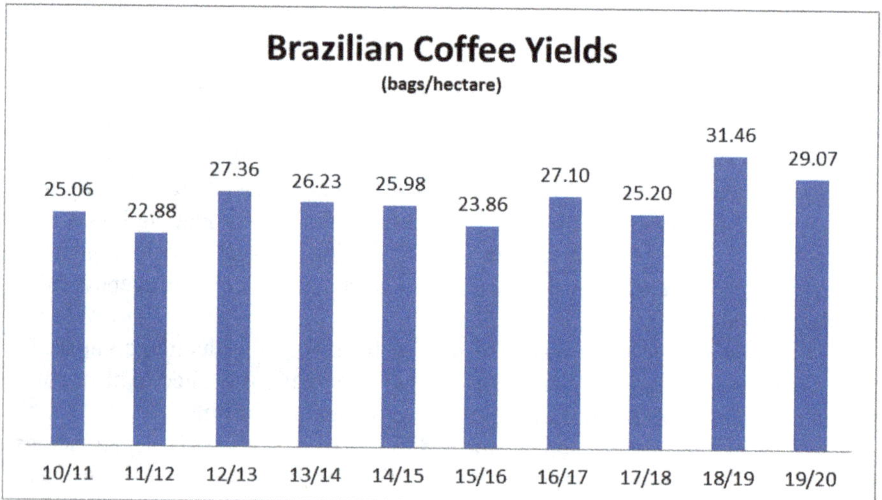

Source: ATO/Sao Paulo

[18] Brazil Coffee – https://www.fas.usda.gov/data/brazil-coffee-annual-4

Dollars/Bushel: Dollars/Ton Converter

Dollars/Bushel Converter[19]:

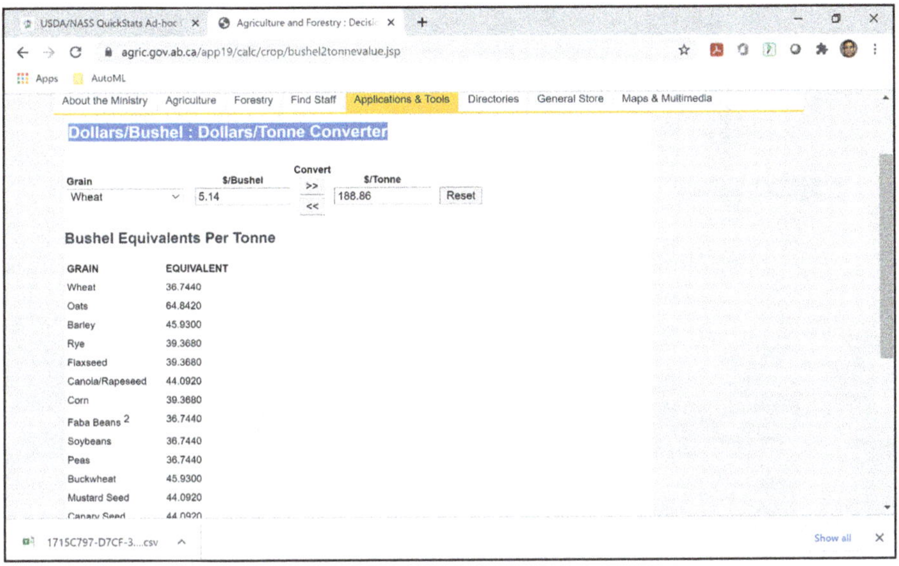

Appendix G: NOAA – Storm Events Database

Storm Events Database

The Storm Events Database contains the records used to create the official NOAA Storm Data publication, documenting[20]:

- The occurrence of storms and other significant weather phenomena having sufficient intensity to cause loss of life, injuries, significant property damage, and/or disruption to commerce
- Rare, unusual, weather phenomena that generate media attention, such as snow flurries in South Florida or the San Diego coastal area
- Other significant meteorological events, such as record maximum or minimum temperatures or precipitation, that occur in connection with another event

The database currently contains data from January 1950 to November 2020, as entered by NOAA's National Weather Service (NWS). Due to changes in the data collection and processing procedures over time, there are unique periods of record available depending on the event type. NCEI has performed data reformatting and standardization of event types but has not changed any data values for locations,

[19] Dollar Bushel Converter – https://www.agric.gov.ab.ca/app19/calc/crop/bushel2tonnevalue.jsp

[20] Storm Events Database – https://www.ncdc.noaa.gov/stormevents/

fatalities, injuries, damage, narratives, and any other event specific information. Please refer to the Database Details page for more information.

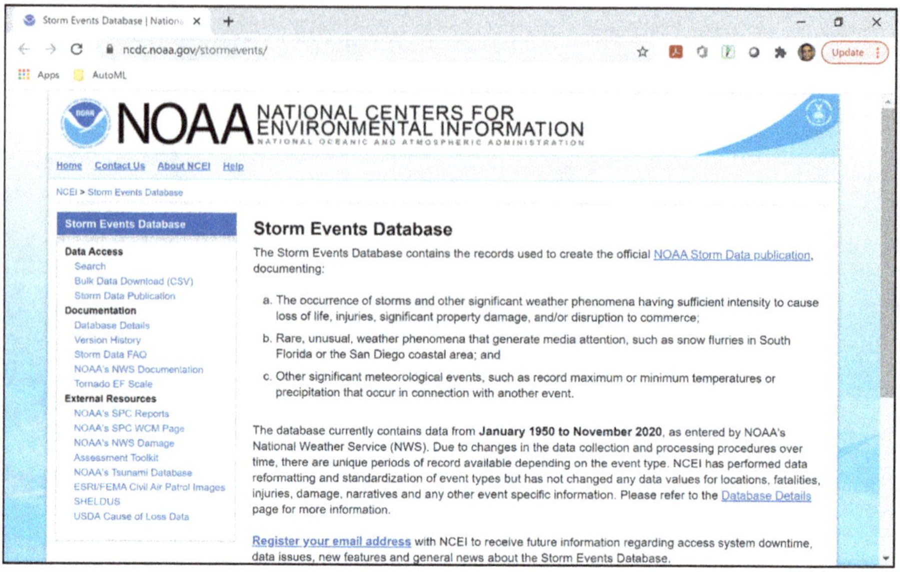

Storm Data Exports Format: describe the storm database meta data.[21]

Storm Data Event Table

The only events permitted in Storm Data are listed in the table below. The chosen event name should be the one that most accurately describes the meteorological event leading to fatalities, injuries, damage, etc. However, significant events, such as tornadoes, having no impact or causing no damage, should also be included in Storm Data.[22]

[21] Storm Data Export Format – https://www1.ncdc.noaa.gov/pub/data/swdi/stormevents/csvfiles/Storm-Data-Export-Format.pdf

[22] NATIONAL WEATHER SERVICE INSTRUCTION 10-1605 STORM DATA PREPARATION https://www.nws.noaa.gov/directives/sym/pd01016005curr.pdf

Astronomical Low Tide	Z	Lightning	C
Avalanche	Z	Marine Dense Fog	M
Blizzard	Z	Marine Hail	M
Coastal Flood	Z	Marine Heavy Freezing Spray	M
Cold/Wind Chill	Z	Marine High Wind	M
Debris Flow	C	Marine Hurricane/Typhoon	M
Dense Fog	Z	Marine Lightning	M
Dense Smoke	Z	Marine Strong Wind	M
Drought	Z	Marine Thunderstorm Wind	M
Dust Devil	C	Marine Tropical Depression	M
Dust Storm	Z	Marine Tropical Storm	M
Excessive Heat	Z	Rip Current	Z
Extreme Cold/Wind Chill	Z	Seiche	Z
Flash Flood	C	Sleet	Z
Flood	C	Sneaker Wave	Z
Frost/Freeze	Z	Storm Surge/Tide	Z
Funnel Cloud	C	Strong Wind	Z
Freezing Fog	Z	Thunderstorm Wind	C
Hail	C	Tornado	C
Heat	Z	Tropical Depression	Z
Heavy Rain	C	Tropical Storm	Z
Heavy Snow	Z	Tsunami	Z
High Surf	Z	Volcanic Ash	Z
High Wind	Z	Waterspout	M
Hurricane (Typhoon)	Z	Wildfire	Z
Ice Storm	Z	Winter Storm	Z
Lake-Effect Snow	Z	Winter Weather	Z
Lakeshore Flood	Z		

Legend: There are three designators: C - County/Parish; Z - Zone; and M – Marine Zone.

Appendix H: National Dairy Development Board (NDDB) India

The National Dairy Development Board's (NDDB)[23] creation is rooted in the conviction that our nation's socioeconomic progress lies largely on the development of rural India.

The Dairy Board was created to promote, finance, and support producer-owned and controlled organizations. NDDB's programs and activities seek to strengthen farmer-owned institutions and support national policies that are favorable to the growth of such institutions. Fundamental to NDDB's efforts are cooperative strategies and principles.

NDDB's efforts transformed India's rural economy by making dairying a viable and profitable economic activity for millions of milk producers while addressing the country's need for self-sufficiency in milk production.

NDDB has been reaching out to dairy farmers by implementing other income generating innovative activities and offering them sustainable livelihood.

[23] About NDDB – https://www.nddb.coop/

Milk production[24] and per capita availability of milk in India[25]

Year	Production (million tons)	Per capita availability (g/day)
1991–92	55.6	178
1992–93	58.0	182
1993–94	60.6	186
1994–95	63.8	192
1995–96	66.2	195
1996–97	69.1	200
1997–98	72.1	205
1998–99	75.4	210
1999–2000	78.3	214
2000–01	80.6	217
2001–02	84.4	222
2002–03	86.2	224
2003–04	88.1	225
2004–05	92.5	233
2005–06	97.1	241
2006–07	102.6	251
2007–08	107.9	260
2008–09	112.2	266
2009–10	116.4	273
2010–11	121.8	281
2011–12	127.9	290
2012–13	132.4	299
2013–14	137.7	307
2014–15	146.3	322
2015–16	155.5	337
2016–17	165.4	355
2017–18	176.3	375
2018–19	187.7	394

Appendix I: Worldwide – Artificial Intelligence (AI) Readiness

IN TERMS OF READINESS FOR AI, COUNTRIES APPEAR TO FALL INTO FOUR GROUPS[26] – Varying conditions among countries imply different degrees of AI adoption and absorption, and therefore economic impact.

---

[24] Milk Production in India – https://www.nddb.coop/information/stats/milkprodindia

[25] NDDB Daily Milk Data – https://www.nddb.coop/sites/default/files/statistics/Mp%20India-ENG-2019.pdf

[26] AI Readiness – https://www.mckinsey.com/~/media/McKinsey/Featured%20Insights/Artificial%20Intelligence/Notes%20from%20the%20frontier%20Modeling%20the%20impact%20of%20AI%20on%20the%20world%20economy/MGI-Notes-from-the-AI-frontier-Modeling-the-impact-of-AI-on-the-world-economy-September-2018.ashx

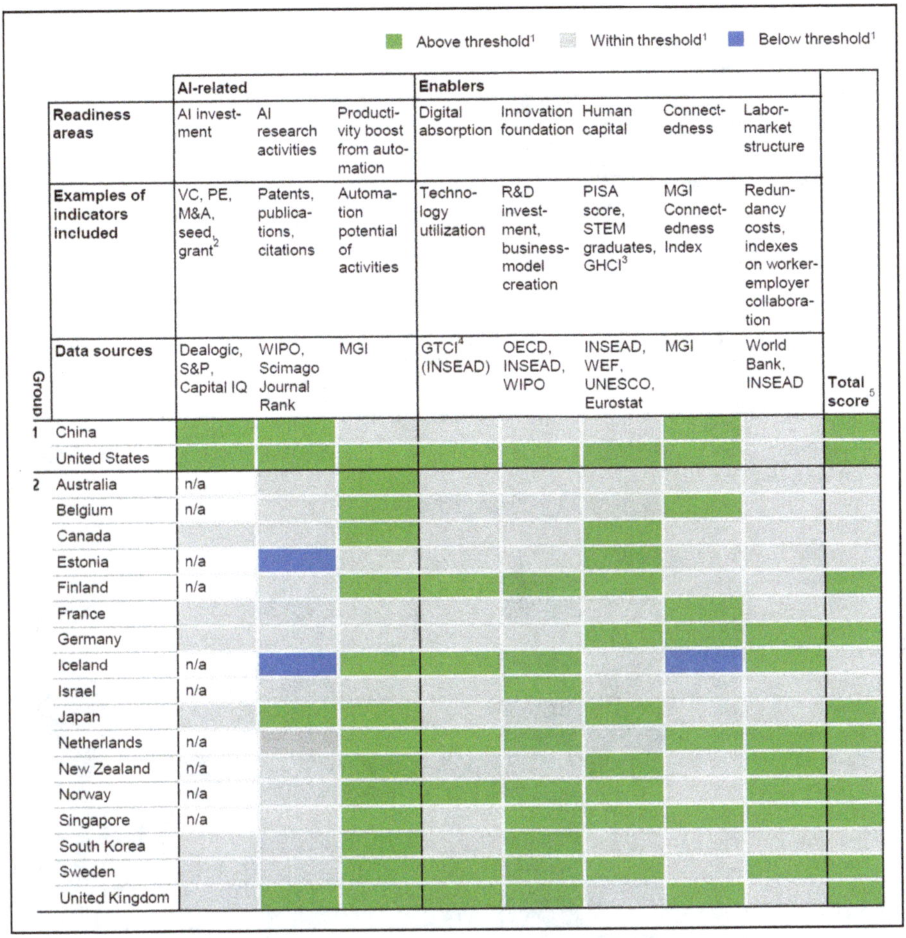

IN TERMS OF READINESS FOR AI, COUNTRIES APPEAR TO FALL INTO FOUR GROUPS.[27]

[27] AI Readiness – https://www.mckinsey.com/~/media/McKinsey/Featured%20Insights/Artificial%20Intelligence/Notes%20from%20the%20frontier%20Modeling%20the%20impact%20of%20AI%20on%20the%20world%20economy/MGI-Notes-from-the-AI-frontier-Modeling-the-impact-of-AI-on-the-world-economy-September-2018.ashx

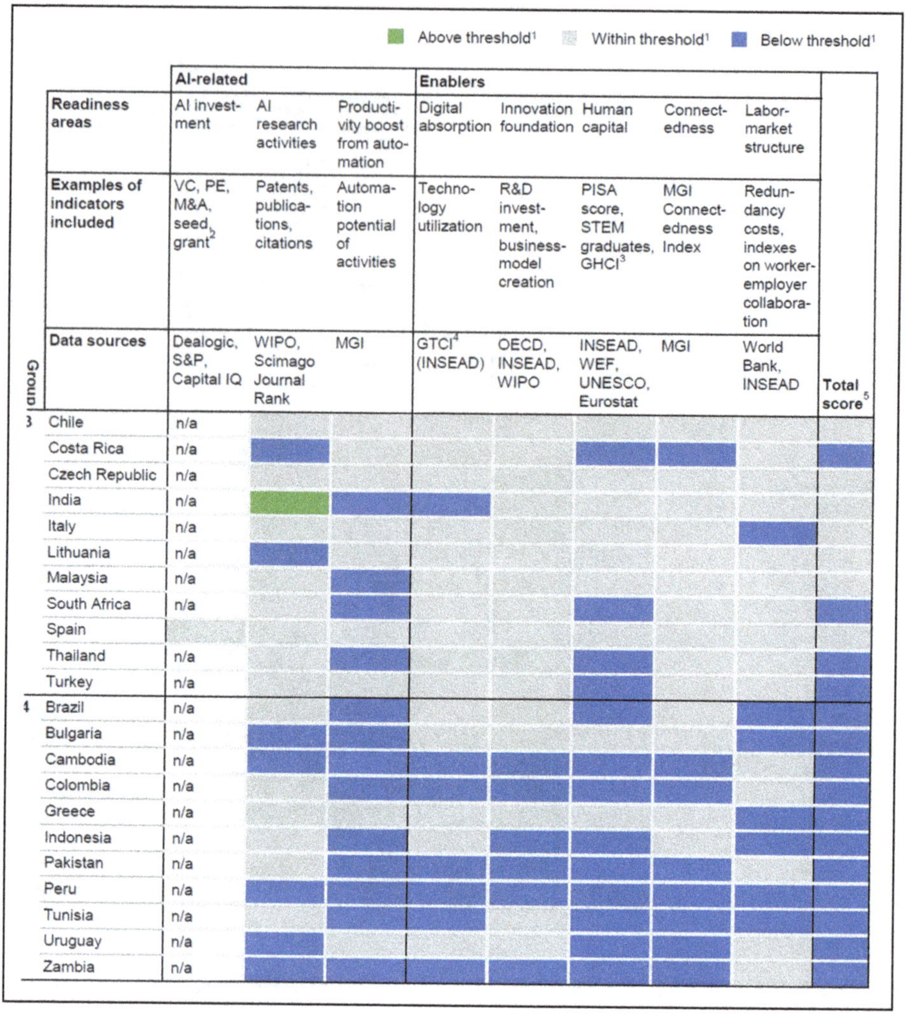

Appendix J: Food Aids

PL-480 or Food for Peace

The Government of India has received generous PL 480 from the United States during 1956–1968. USAID and PL–480, 1961–1969[28]. The administrations of John F. Kennedy and Lyndon B. Johnson marked a revitalization of the US foreign assistance program, signified a growing awareness of the importance of humanitarian aid

[28] USAID – PL480 – https://history.state.gov/milestones/1961-1968/pl-480

as a form of diplomacy, and reinforced the belief that American security was linked to the economic progress and stability of other nations.

Johnson with Gandhi, March 28, 1966. (White House Photo Office).

The U.S. Bureau of Labor Statistics

BLS Data Finder.[29]

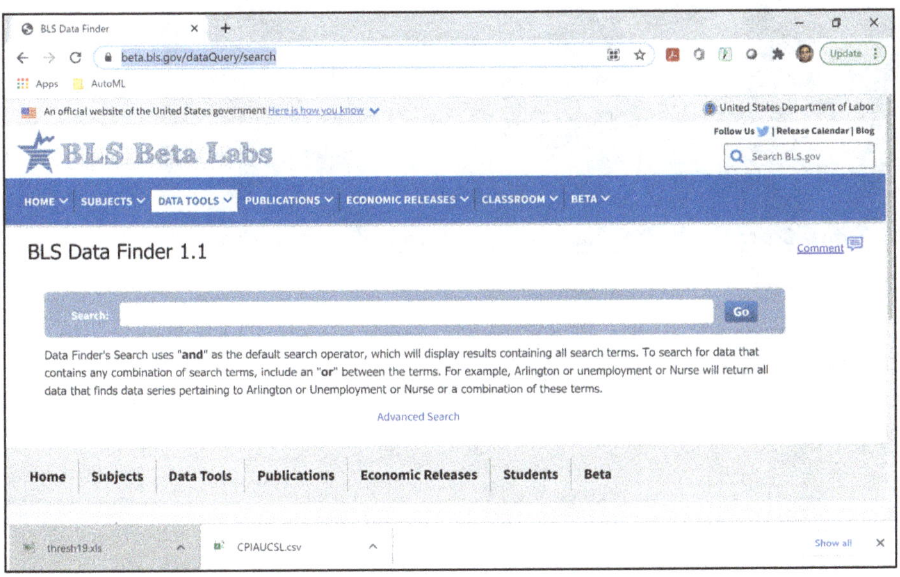

[29] BLS Data Finder – https://beta.bls.gov/dataQuery/search

Appendix K: The United Nations – 17 Sustainable Development Goals (SDGs)

The 17 Sustainable Development Goals (SDGs), which are an urgent call for action by all countries – developed and developing – in a global partnership. They recognize that ending poverty and other deprivations must go together with strategies that improve health and education, reduce inequality, and spur economic growth – all while tackling climate change and working to preserve our oceans and forests.[30]

[30] The 17 Goals – https://sdgs.un.org/goals

Appendix L: The Statistical Distributions of Commodity Prices in Both Real and Nominal Terms

Statistical properties of commodities[31]:

Decade	Variable	Mean	Std dev	Coeff. of variation	Skewness	Kurtosis
1970/80	banana -n	216.8	64.9	29.9	0.6	-0.6
	banana -r	543.5	89.0	16.4	0.5	-0.0
	cocoa -n	1777.9	1202.5	67.6	0.7	-1.0
	cocoa -r	3970.1	1765.4	44.5	0.9	-0.2
	coffee -n	2219.8	1412.4	63.6	1.2	0.8
	coffee -r	5159.0	2172.5	42.1	1.9	4.1
	cotton -n	1243.3	438.2	35.2	-0.0	-1.4
	cotton -r	3059.3	666.7	21.8	1.7	2.9
	jute -n	322.7	57.9	17.9	1.0	-0.2
	jute -r	845.0	201.4	23.8	0.3	-1.4
	maize -n	98.6	26.7	27.1	-0.4	-0.8
	maize -r	226.2	51.8	22.9	0.4	-0.7
	palmoil -n	466.0	164.8	35.4	-0.1	-1.2
	palmoil -r	1028.6	247.7	24.1	1.3	1.8
	rapeoil -n	496.7	170.7	34.4	0.0	-0.5
	rapeoil -r	1108.4	299.0	27.0	1.3	1.7
	rapeseed -n	263.6	81.3	30.8	-0.3	-0.6
	rapeseed -r	592.2	148.5	25.1	1.2	0.9
	rice -n	285.4	127.2	44.6	0.6	-0.0
	rice -r	682.2	254.6	37.3	2.0	3.3
	rubber -n	676.0	271.5	40.2	0.5	-0.5
	rubber -r	1641.9	344.7	21.0	1.1	2.1
	sisal – n	476.6	271.9	57.0	0.9	0.0
	sisal -r	1141.2	553.3	48.5	1.6	1.5
	soybeans -n	237.9	69.5	29.2	0.0	-0.1
	soybeans -r	540.2	149.1	27.6	2.4	9.0
	soymeal -n	196.3	68.9	35.1	1.5	5.4
	soymeal – r	451.0	181.5	40.3	3.4	14.5
	sugar -n	243.4	198.8	81.7	2.5	7.4
	sugar -r	595.4	424.3	71.3	2.5	7.1
	sunflmeal -n	153.3	44.8	29.2	0.8	2.4
	sunflmeal -r	353.6	118.6	33.5	2.6	8.4
	tea -n	1532.3	607.3	39.6	1.7	3.7
	tea -r	3802.9	892.6	23.5	2.4	8.4
	wheat -n	123.5	43.8	35.5	-0.1	-1.0
	wheat -r	284.0	87.5	30.8	1.7	2.5
1980/90	banana -n	406.3	83.4	20.5	1.2	2.2
	banana -r	557.4	104.4	18.7	0.7	-0.1
	cocoa -n	2060.1	428.7	20.8	0.3	0.6

[31] The statistical distributions of commodity prices in both real and nominal terms—http://www.fao.org/3/y4344e/y4344e0i.htm

Decade	Variable	Mean	Std dev	Coeff. of variation	Skewness	Kurtosis
	cocoa -r	2882.9	769.7	26.7	-0.1	-0.5
	coffee -n	2892.9	569.4	19.7	0.9	1.2
	coffee -r	4031.8	990.4	24.6	0.2	-0.3
	cotton -n	1625.3	319.7	19.7	-0.4	-0.3
	cotton -r	2254.3	524.6	23.3	-0.3	-1.1
	jute -n	361.1	138.7	38.4	2.5	5.8
	jute -r	507.7	247.0	48.6	2.7	6.4
	maize -n	112.8	22.6	20.0	-0.3	-0.5
	maize -r	154.5	42.2	27.3	0.1	-1.1
	palmoil -n	448.4	157.0	35.0	0.9	0.5
	palmoil -r	622.5	273.7	44.0	0.9	0.2
	rapeoil -n	454.1	124.7	27.5	0.9	0.6
	rapeoil -r	626.1	232.4	37.1	1.0	0.1
	rapeseed -n	267.5	58.7	21.9	0.1	-0.0
	rapeseed -r	369.1	115.3	31.2	0.1	-0.7
	rice -n	294.4	92.7	31.5	1.0	0.2
	rice -r	404.1	132.7	32.8	1.0	0.4
	rubber -n	997.3	203.3	20.4	1.2	0.8
	rubber -r	1379.5	313.5	22.7	0.8	-0.5
1980/90	sisal – n	594.5	85.8	14.4	1.3	1.1
	sisal -r	822.7	145.6	17.7	0.4	-0.1
	soybeans -n	259.6	43.1	16.6	0.6	-0.6
	soybeans -r	351.0	71.6	20.4	0.5	-0.5
	soymeal -n	219.3	41.8	19.1	0.4	-0.4
	soymeal – r	294.4	56.1	19.0	0.5	-0.3
	sugar -n	236.6	168.3	71.1	1.9	3.7
	sugar -r	320.9	224.0	69.8	2.0	3.6
	sunflmeal -n	143.2	38.5	26.9	0.3	-0.5
	sunflmeal -r	193.7	58.6	30.3	0.6	-0.9
	tea -n	2131.0	564.5	26.5	1.8	2.8
	tea -r	2991.0	1035.2	34.6	1.6	2.0
	wheat -n	149.2	23.1	15.5	-0.5	-0.7
	wheat -r	203.2	39.8	19.6	-0.4	-1.2
1990/00	banana -n	480.5	104.3	21.7	0.7	0.5
	banana -r	514.0	115.3	22.4	0.7	0.3
	cocoa -n	1340.0	226.5	16.9	0.0	-0.9
	cocoa -r	1434.4266.9	18.6	0.5	-0.5	
	coffee -n	2095.6	805.6	38.4	0.9	0.1
	coffee -r	2238.4	860.0	38.4	0.9	0.2
	cotton -n	1632.6	307.1	18.8	0.3	0.0
	cotton -r	1741.3	306.7	17.6	-0.1	-0.5
	jute -n	329.7	80.7	24.5	0.6	-0.6
	jute -r	351.1	80.2	22.8	0.5	-0.8
	maize -n	111.2	23.9	21.5	2.0	4.9

Decade	Variable	Mean	Std dev	Coeff. of variation	Skewness	Kurtosis
	maize -r	118.1	24.1	20.4	1.6	3.7
	palmoil -n	477.6	125.7	26.3	0.3	-1.3
	palmoil -r	509.1	139.1	27.3	0.4	-0.9
	rapeoil -n	507.2	100.2	19.8	0.1	-1.3
	rapeoil -r	541.5	117.1	21.6	0.2	-1.2
	rapeseed -n	248.9	47.4	19.0	-0.0	-1.5
	rapeseed -r	265.7	55.1	20.7	0.0	-1.3
	rice -n	285.5	41.6	14.6	0.1	-0.3
	rice -r	304.1	44.4	14.6	0.2	-0.5
	rubber -n	993.4	66.9	6.7	0.2	-1.0
	rubber -r	1061.2	71.4	6.7	-0.2	-0.5
	sisal – n	700.4	109.6	15.6	-0.2	-0.6
	sisal -r	749.0	121.5	16.2	-0.3	-0.3
	soybeans -n	250.6	33.7	13.5	0.5	-0.2
	soybeans -r	267.1	39.4	14.8	0.2	-0.5
	soymeal -n	205.5	40.0	19.5	0.6	-0.2
	soymeal – r	218.8	43.8	20.0	0.6	-0.2
	sugar -n	234.7	51.3	21.9	-0.1	-0.4
	sugar -r	251.0	55.6	22.2	-0.1	-0.1
	sunflmeal -n	116.8	24.5	21.0	-0.2	-0.7
	sunflmeal -r	124.3	26.3	21.2	-0.3	-0.7
	tea -n	2002.6	334.3	16.7	1.1	1.6
	tea -r	2147.0	409.0	19.0	1.2	2.1
	wheat -n	146.4	32.2	22.0	1.1	1.3
	wheat -r	155.5	32.3	20.8	0.7	0.8

r-real, n-nominal

The kurtosis for a normal distribution is 3.

Appendix M: Poverty Thresholds for 2019 by Size of Family and Number of Related Children Under 18 Years

Size of family unit	Weighted average thresholds	Related children under 18 years								
		None	One	Two	Three	Four	Five	Six	Seven	Eight or more
One person (unrelated individual):	13,011									
Under age 65..................................	13,300	13,300								
Aged 65 and older............................	12,261	12,261								
Two people:	16,521									
Householder under age 65...............	17,196	17,120	17,622							
Householder aged 65 and older........	15,468	15,453	17,555							
Three people......................................	20,335	19,998	20,578	20,598						
Four people..	26,172	26,370	26,801	25,926	26,017					

Size of family unit	Weighted average thresholds	Related children under 18 years								
		None	One	Two	Three	Four	Five	Six	Seven	Eight or more
Five people	31,021	31,800	32,263	31,275	30,510	30,044				
Six people	35,129	36,576	36,721	35,965	35,239	34,161	33,522			
Seven people	40,016	42,085	42,348	41,442	40,811	39,635	38,262	36,757		
Eight people	44,461	47,069	47,485	46,630	45,881	44,818	43,470	42,066	41,709	
Nine people or more	52,875	56,621	56,895	56,139	55,503	54,460	53,025	51,727	51,406	49,426

Source: U.S. Census Bureau

Appendix N: Crop Calendar

The Crop Calendar[32] is a tool that provides timely information about seeds to promote local crop production. It contains information on planting, sowing, and harvesting periods of locally adapted crops in specific agroecological zones. It also provides information on the sowing rates of seed and planting material and the main agricultural practices.

This tool supports farmers and agriculture extensionists across the world in taking appropriate decisions on crops and their sowing period, respecting the agroecological dimension. It also provides a solid base for emergency planning of the rehabilitation of farming systems after disasters. The Crop Calendar provides information for more than 130 crops, located in 283 agroecological zones of 44 countries.

US CROP Calendar

The United States Crop Calendar[33] includes: barley (Fall), barley (Spring), corn, cotton, ground nut, oats (Fall), oats (Spring), rapeseed, rice, rye, sorghum, soybeans, sunflower, wheat (Spring), and wheat (Winter).

[32] Crop Calendar – http://www.fao.org/agriculture/seed/cropcalendar/welcome.do

[33] US Crop Calendar – https://ipad.fas.usda.gov/rssiws/al/crop_calendar/us.aspx

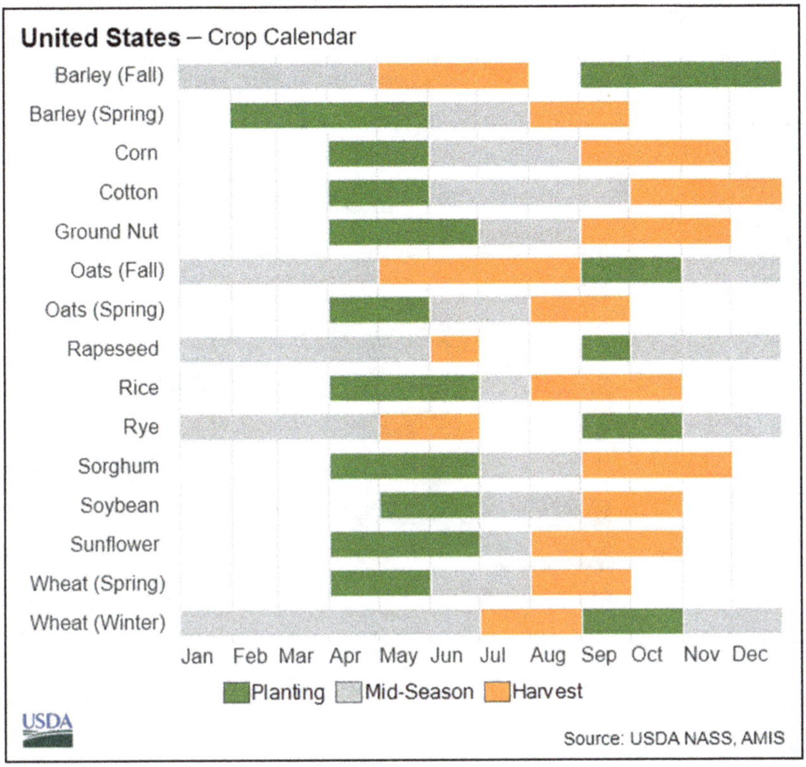

Canada Crop Calendar

The Canada Crop Calendar[34] includes: barley, canola, corn, oats, soybean, wheat (Spring), and wheat (Winter).

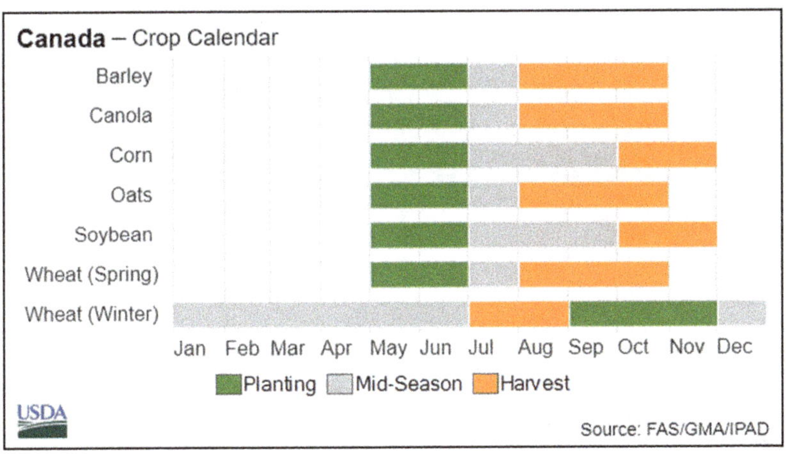

[34] Canada Crop Calendar – https://ipad.fas.usda.gov/rssiws/al/crop_calendar/can.aspx

Brazil Crop Calendar

The Brazil Crop Calendar[35] includes: barley, corn, cotton, rice, sorghum, and wheat.

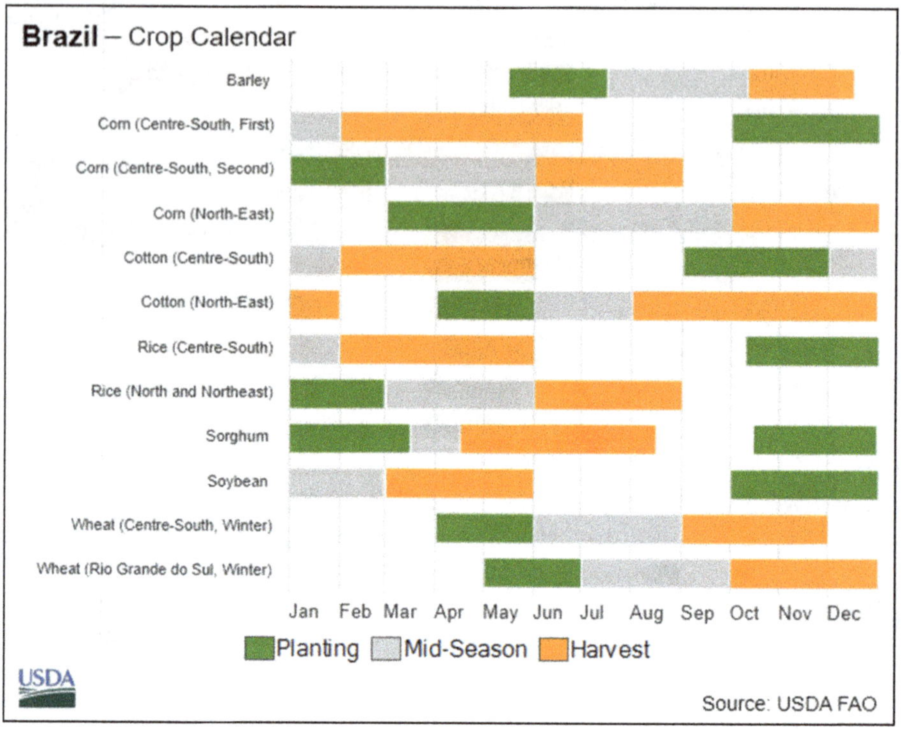

Global Coffee Harvest Calendar

Global Coffee Harvest Calendar.[36]

[35] Brazil Calendar – https://ipad.fas.usda.gov/rssiws/al/crop_calendar/br.aspx

[36] Global Coffee Harvest Calendar – https://coffee-mind.com/wp-content/uploads/2013/10/Coffee-harvest-calendar.pdf

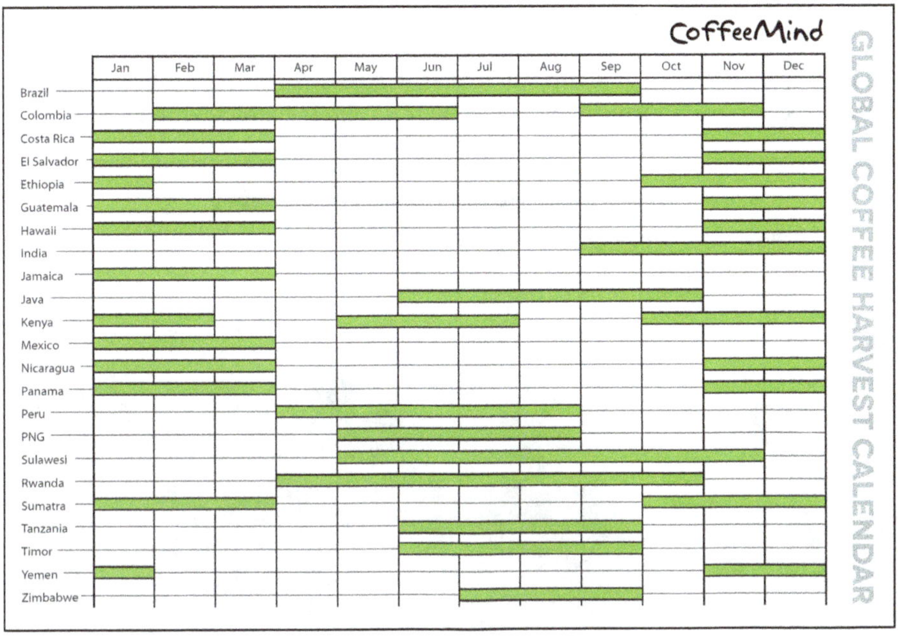

World Specialty Coffee Harvest Chart

World Specialty Coffee Harvest Chart[37]:

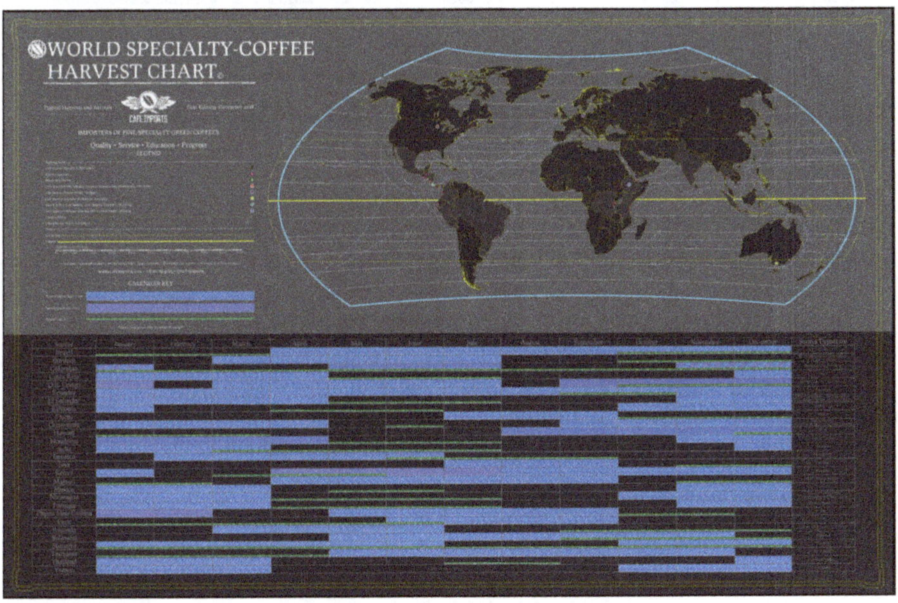

[37] Specialty Chart – https://images.cafeimports.com/Cafe_Imports_Harvest_Chart_Web.png

The India Meteorological – Crop Calendar

Crop Weather Calendar,[38] Crop Advisories, Climate Smart Agriculture, Weather Based ... Watershed Organization Trust (WOTR), the India Meteorological.

Appendix O: G20 – The Agricultural Market Information System (AMIS)

The Agricultural Market Information System (AMIS)[39] is a G20 initiative to enhance food market transparency and encourage coordination of policy action in response to market uncertainty.

The initial focus of AMIS is on four crops that are particularly important in international food markets, namely wheat, maize, rice, and soybeans.

AMIS seeks to strengthen collaboration and dialogue among main producing, exporting, and importing countries.

---

[38] WOTR – https://wotr-website-publications.s3.ap-south-1.amazonaws.com/102_Complete-Crop-Calendar.pdf

[39] AMIS Online Calendar – http://www.amis-outlook.org/fileadmin/user_upload/amis/docs/resources/AMIS-online-crop-calendar_REDUCED3.pdf

Wheat Crops

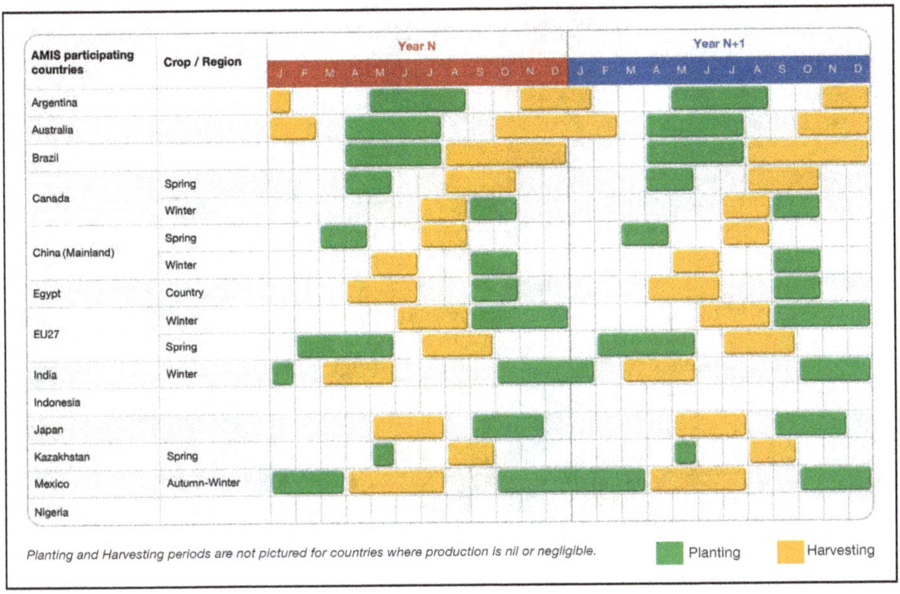

Rice

G20 participating countries rice calendar.[40]

[40]AMIS Online Calendar – http://www.amis-outlook.org/fileadmin/user_upload/amis/docs/resources/AMIS-online-crop-calendar_REDUCED3.pdf

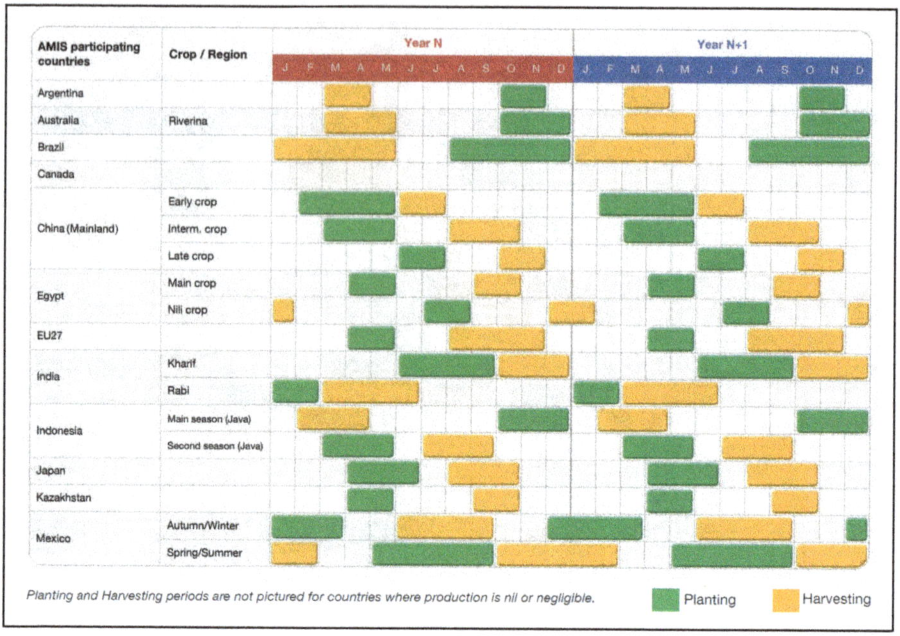

Appendix P: National Feed Security Mission (NFSM) – Ministry of Agriculture and Farmers' Welfare

Crop calendar[41]:

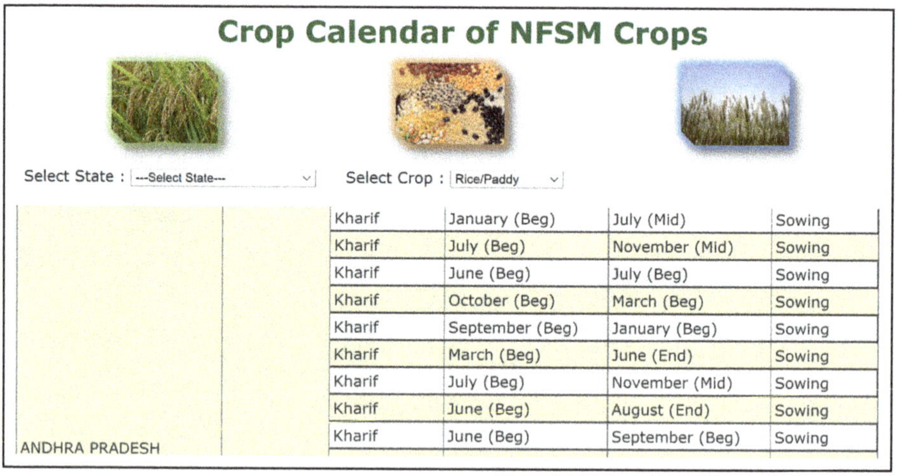
Crop Calendar of NFSM Crops

Select State : ---Select State---   Select Crop : Rice/Paddy

		Kharif	January (Beg)	July (Mid)	Sowing
		Kharif	July (Beg)	November (Mid)	Sowing
		Kharif	June (Beg)	July (Beg)	Sowing
		Kharif	October (Beg)	March (Beg)	Sowing
		Kharif	September (Beg)	January (Beg)	Sowing
		Kharif	March (Beg)	June (End)	Sowing
		Kharif	July (Beg)	November (Mid)	Sowing
		Kharif	June (Beg)	August (End)	Sowing
ANDHRA PRADESH		Kharif	June (Beg)	September (Beg)	Sowing

[41] NFSM Crops – https://nfsm.gov.in/nfmis/rpt/calenderreport.aspx

Indian Crop Calendar 1956 – Third Edition

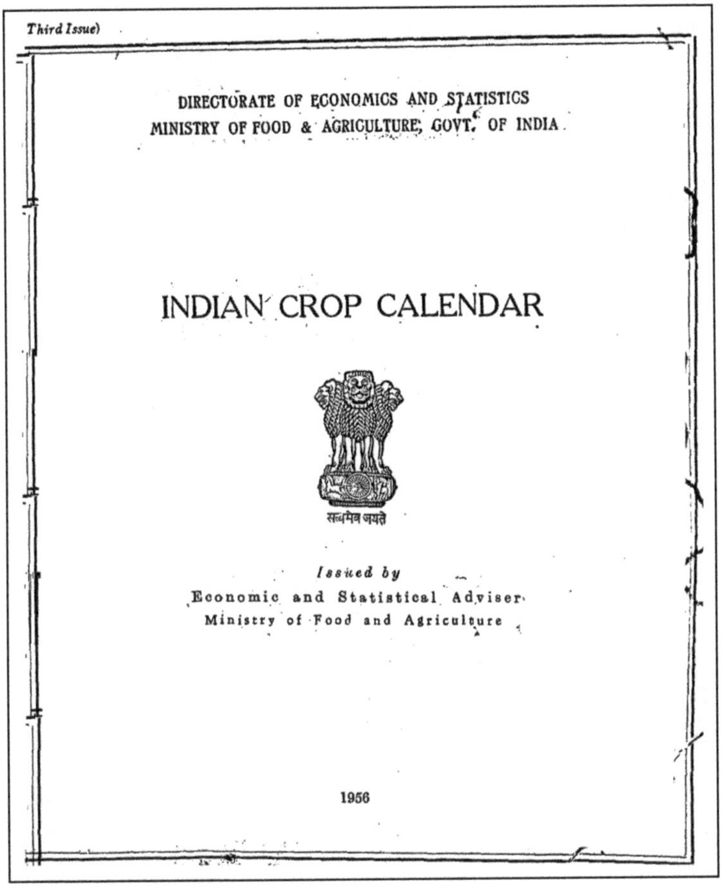
Third Issue)

DIRECTORATE OF ECONOMICS AND STATISTICS
MINISTRY OF FOOD & AGRICULTURE, GOVT. OF INDIA.

INDIAN CROP CALENDAR

सत्यमेव जयते

Issued by
Economic and Statistical Adviser
Ministry of Food and Agriculture

1956

1956 Edition.[42]

[42] Indian Calendar 1956 – https://dspace.gipe.ac.in/xmlui/bitstream/handle/10973/37203/GIPE-050467-Contents.pdf?sequence=2&isAllowed=y

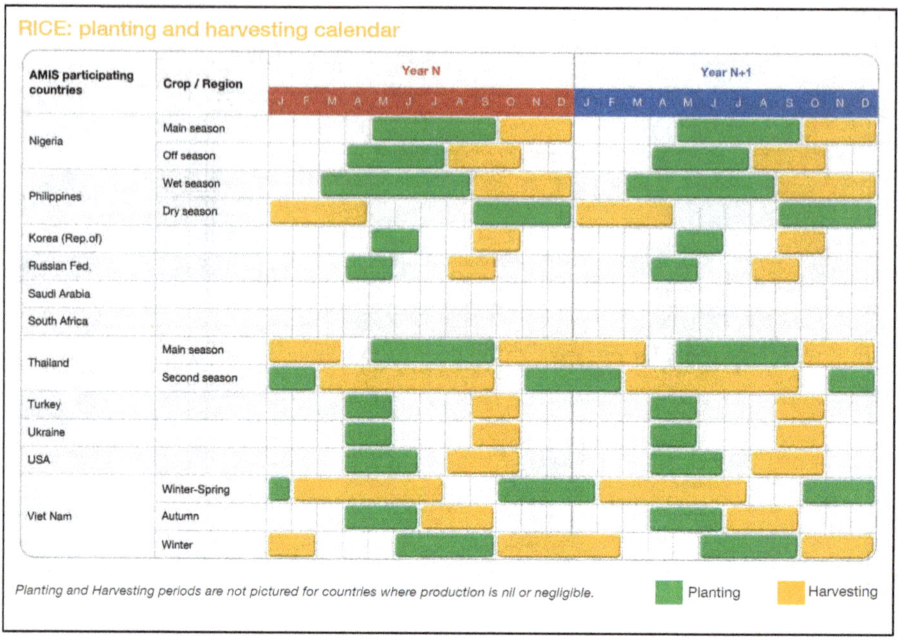
RICE: planting and harvesting calendar
AMIS participating countries
Crop / Region
Year N
Year N+1
J F M A M J J A S O N D J F M A M J J A S O N D
Nigeria
Main season
Off season
Philippines
Wet season
Dry season
Korea (Rep.of)
Russian Fed.
Saudi Arabia
South Africa
Thailand
Main season
Second season
Turkey
Ukraine
USA
Viet Nam
Winter-Spring
Autumn
Winter
Planting and Harvesting periods are not pictured for countries where production is nil or negligible.
Planting
Harvesting

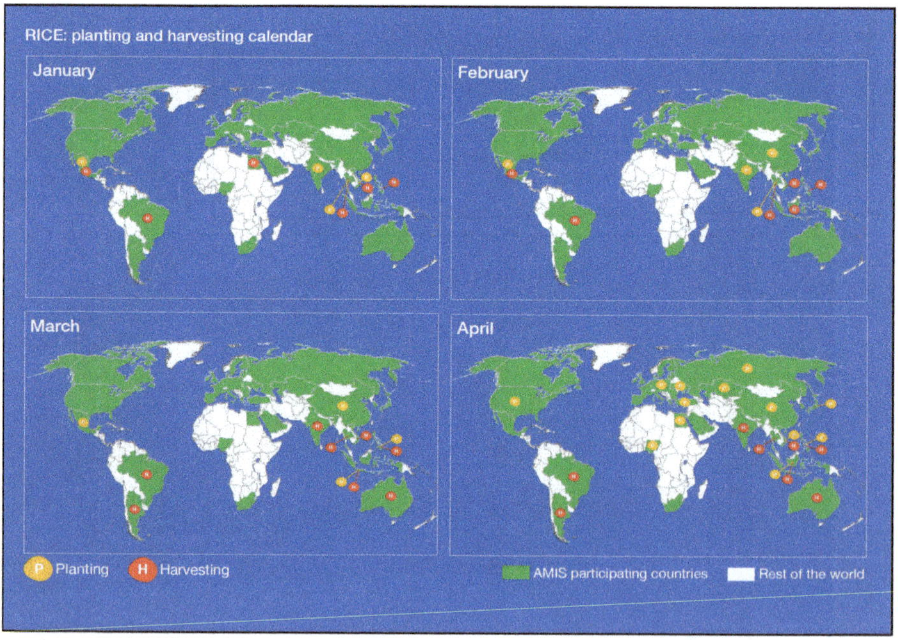
RICE: planting and harvesting calendar
January
February
March
April
P Planting
H Harvesting
AMIS participating countries
Rest of the world

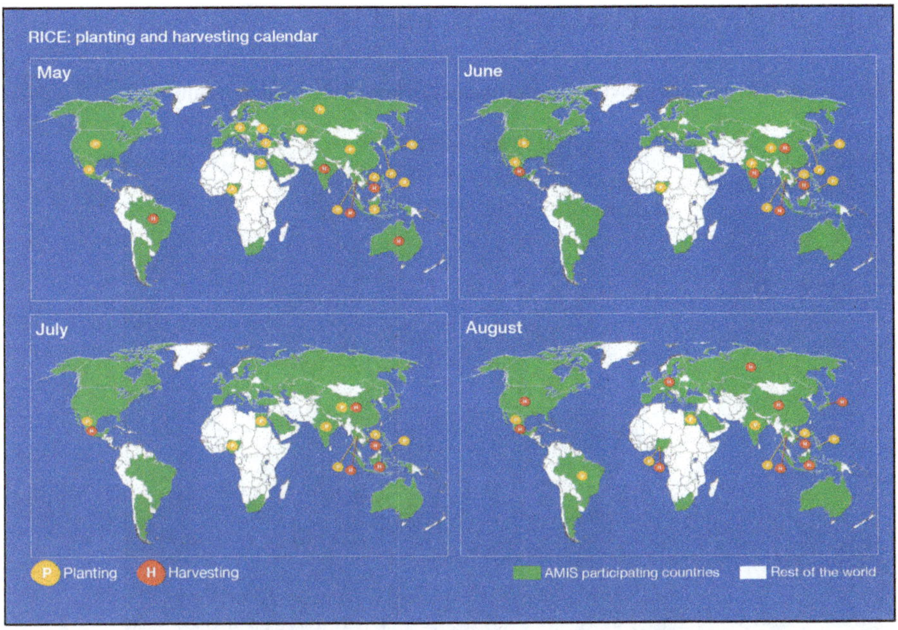
RICE: planting and harvesting calendar
May
June
July
August
P Planting
H Harvesting
AMIS participating countries
Rest of the world

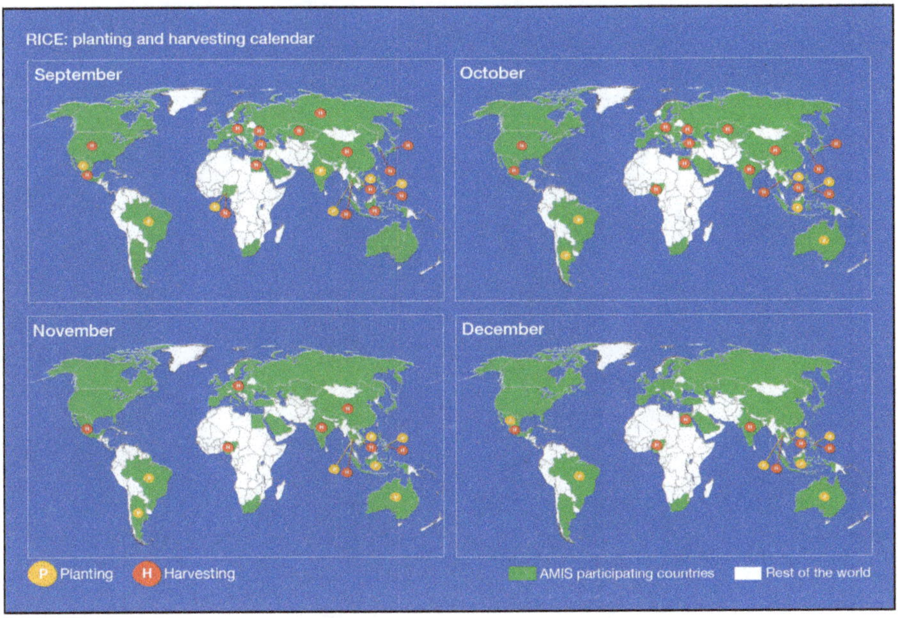
RICE: planting and harvesting calendar
September
October
November
December
P Planting
H Harvesting
AMIS participating countries
Rest of the world

Appendix Q: Rice Production Manual

Rice production manual provides comprehensive view of rice and agriculture.[43]

[43] Rice Production Manual – http://rice.ucanr.edu/files/288581.pdf

Index

C. Vuppalapati, *Machine Learning and Artificial Intelligence for Agricultural Economics*, International Series in Operations Research & Management Science 314,
https://doi.org/10.1007/978-3-030-77485-1

T

U

V

www.ingramcontent.com/pod-product-compliance
Lightning Source LLC
Chambersburg PA
CBHW051557250825
31638CB00003B/69
9783030774844